LONDON MATHEMATICAL SOCIETY LECTURE NOTE SERIES

Managing Editor: Professor N.J. Hitchin, Mathematics Institute,
University of Oxford, 24–29 St Giles, Oxford OX1 3TG, United Kingdom

The titles below are available from booksellers, or, in case of difficulty, from Cambridge University Press.

46 *p*-adic Analysis: a short course on recent work, N. KOBLITZ
59 Applicable differential geometry, M. CRAMPIN & F.A.E. PIRANI
66 Several complex variables and complex manifolds II, M.J. FIELD
86 Topological topics, I.M. JAMES (ed)
87 Surveys in set theory, A.R.D. MATHIAS (ed)
88 FPF ring theory, C. FAITH & S. PAGE
89 An F-space sampler, N.J. KALTON, N.T. PECK & J.W. ROBERTS
90 Polytopes and symmetry, S.A. ROBERTSON
92 Representation of rings over skew fields, A.H. SCHOFIELD
93 Aspects of topology, I.M. JAMES & E.H. KRONHEIMER (eds)
96 Diophantine equations over function fields, R.C. MASON
97 Varieties of constructive mathematics, D.S. BRIDGES & F. RICHMAN
98 Localization in Noetherian rings, A.V. JATEGAONKAR
99 Methods of differential geometry in algebraic topology, M. KAROUBI & C. LERUSTE
100 Stopping time techniques for analysts and probabilists, L. EGGHE
104 Elliptic structures on 3-manifolds, C.B. THOMAS
105 A local spectral theory for closed operators, I. ERDELYI & WANG SHENGWANG
107 Compactification of Siegel moduli schemes, C.-L. CHAI
109 Diophantine analysis, J. LOXTON & A. VAN DER POORTEN (eds)
113 Lectures on the asymptotic theory of ideals, D. REES
114 Lectures on Bochner-Riesz means, K.M. DAVIS & Y.-C. CHANG
116 Representations of algebras, P.J. WEBB (ed)
119 Triangulated categories in the representation theory of finite-dimensional algebras, D. HAPPEL
121 Proceedings of *Groups - St Andrews 1985*, E. ROBERTSON & C. CAMPBELL (eds)
128 Descriptive set theory and the structure of sets of uniqueness, A.S. KECHRIS & A. LOUVEAU
130 Model theory and modules, M. PREST
131 Algebraic, extremal & metric combinatorics, M.-M. DEZA, P. FRANKL & I.G. ROSENBERG (eds)
132 Whitehead groups of finite groups, ROBERT OLIVER
133 Linear algebraic monoids, MOHAN S. PUTCHA
134 Number theory and dynamical systems, M. DODSON & J. VICKERS (eds)
137 Analysis at Urbana, I, E. BERKSON, T. PECK, & J. UHL (eds)
138 Analysis at Urbana, II, E. BERKSON, T. PECK, & J. UHL (eds)
139 Advances in homotopy theory, S. SALAMON, B. STEER & W. SUTHERLAND (eds)
140 Geometric aspects of Banach spaces, E.M. PEINADOR & A. RODES (eds)
141 Surveys in combinatorics 1989, J. SIEMONS (ed)
144 Introduction to uniform spaces, I.M. JAMES
146 Cohen-Macaulay modules over Cohen-Macaulay rings, Y. YOSHINO
148 Helices and vector bundles, A.N. RUDAKOV *et al*
149 Solitons, nonlinear evolution equations and inverse scattering, M. ABLOWITZ & P. CLARKSON
150 Geometry of low-dimensional manifolds 1, S. DONALDSON & C.B. THOMAS (eds)
151 Geometry of low-dimensional manifolds 2, S. DONALDSON & C.B. THOMAS (eds)
152 Oligomorphic permutation groups, P. CAMERON
153 L-functions and arithmetic, J. COATES & M.J. TAYLOR (eds)
155 Classification theories of polarized varieties, TAKAO FUJITA
156 Twistors in mathematics and physics, T.N. BAILEY & R.J. BASTON (eds)
158 Geometry of Banach spaces, P.F.X. MÜLLER & W. SCHACHERMAYER (eds)
159 Groups St Andrews 1989 volume 1, C.M. CAMPBELL & E.F. ROBERTSON (eds)
160 Groups St Andrews 1989 volume 2, C.M. CAMPBELL & E.F. ROBERTSON (eds)
161 Lectures on block theory, BURKHARD KÜLSHAMMER
162 Harmonic analysis and representation theory, A. FIGA-TALAMANCA & C. NEBBIA
163 Topics in varieties of group representations, S.M. VOVSI
164 Quasi-symmetric designs, M.S. SHRIKANDE & S.S. SANE
166 Surveys in combinatorics, 1991, A.D. KEEDWELL (ed)
168 Representations of algebras, H. TACHIKAWA & S. BRENNER (eds)
169 Boolean function complexity, M.S. PATERSON (ed)
170 Manifolds with singularities and the Adams-Novikov spectral sequence, B. BOTVINNIK
171 Squares, A.R. RAJWADE
172 Algebraic varieties, GEORGE R. KEMPF
173 Discrete groups and geometry, W.J. HARVEY & C. MACLACHLAN (eds)
174 Lectures on mechanics, J.E. MARSDEN
175 Adams memorial symposium on algebraic topology 1, N. RAY & G. WALKER (eds)
176 Adams memorial symposium on algebraic topology 2, N. RAY & G. WALKER (eds)
177 Applications of categories in computer science, M. FOURMAN, P. JOHNSTONE & A. PITTS (eds)
178 Lower K- and L-theory, A. RANICKI
179 Complex projective geometry, G. ELLINGSRUD *et al*
180 Lectures on ergodic theory and Pesin theory on compact manifolds, M. POLLICOTT
181 Geometric group theory I, G.A. NIBLO & M.A. ROLLER (eds)

182 Geometric group theory II, G.A. NIBLO & M.A. ROLLER (eds)
183 Shintani zeta functions, A. YUKIE
184 Arithmetical functions, W. SCHWARZ & J. SPILKER
185 Representations of solvable groups, O. MANZ & T.R. WOLF
186 Complexity: knots, colourings and counting, D.J.A. WELSH
187 Surveys in combinatorics, 1993, K. WALKER (ed)
188 Local analysis for the odd order theorem, H. BENDER & G. GLAUBERMAN
189 Locally presentable and accessible categories, J. ADAMEK & J. ROSICKY
190 Polynomial invariants of finite groups, D.J. BENSON
191 Finite geometry and combinatorics, F. DE CLERCK *et al*
192 Symplectic geometry, D. SALAMON (ed)
194 Independent random variables and rearrangement invariant spaces, M. BRAVERMAN
195 Arithmetic of blowup algebras, WOLMER VASCONCELOS
196 Microlocal analysis for differential operators, A. GRIGIS & J. SJÖSTRAND
197 Two-dimensional homotopy and combinatorial group theory, C. HOG-ANGELONI, W. METZLER & A.J. SIERADSKI (eds)
198 The algebraic characterization of geometric 4-manifolds, J.A. HILLMAN
199 Invariant potential theory in the unit ball of $\mathbf{C}^n$, MANFRED STOLL
200 The Grothendieck theory of dessins d'enfant, L. SCHNEPS (ed)
201 Singularities, JEAN-PAUL BRASSELET (ed)
202 The technique of pseudodifferential operators, H.O. CORDES
203 Hochschild cohomology of von Neumann algebras, A. SINCLAIR & R. SMITH
204 Combinatorial and geometric group theory, A.J. DUNCAN, N.D. GILBERT & J. HOWIE (eds)
205 Ergodic theory and its connections with harmonic analysis, K. PETERSEN & I. SALAMA (eds)
206 An introduction to noncommutative differential geometry and its physical applications, J. MADORE
207 Groups of Lie type and their geometries, W.M. KANTOR & L. DI MARTINO (eds)
208 Vector bundles in algebraic geometry, N.J. HITCHIN, P. NEWSTEAD & W.M. OXBURY (eds)
209 Arithmetic of diagonal hypersurfaces over finite fields, F.Q. GOUVÊA & N. YUI
210 Hilbert C*-modules, E.C. LANCE
211 Groups 93 Galway / St Andrews I, C.M. CAMPBELL *et al* (eds)
212 Groups 93 Galway / St Andrews II, C.M. CAMPBELL *et al* (eds)
214 Generalised Euler-Jacobi inversion formula and asymptotics beyond all orders, V. KOWALENKO, N.E. FRANKEL, M.L. GLASSER & T. TAUCHER
215 Number theory 1992–93, S. DAVID (ed)
216 Stochastic partial differential equations, A. ETHERIDGE (ed)
217 Quadratic forms with applications to algebraic geometry and topology, A. PFISTER
218 Surveys in combinatorics, 1995, PETER ROWLINSON (ed)
220 Algebraic set theory, A. JOYAL & I. MOERDIJK
221 Harmonic approximation, S.J. GARDINER
222 Advances in linear logic, J.-Y. GIRARD, Y. LAFONT & L. REGNIER (eds)
223 Analytic semigroups and semilinear initial boundary value problems, KAZUAKI TAIRA
224 Computability, enumerability, unsolvability, S.B. COOPER, T.A. SLAMAN & S.S. WAINER (eds)
225 A mathematical introduction to string theory, S. ALBEVERIO, J. JOST, S. PAYCHA, S. SCARLATTI
226 Novikov conjectures, index theorems and rigidity I, S. FERRY, A. RANICKI & J. ROSENBERG (eds)
227 Novikov conjectures, index theorems and rigidity II, S. FERRY, A. RANICKI & J. ROSENBERG (eds)
228 Ergodic theory of $\mathbf{Z}^d$ actions, M. POLLICOTT & K. SCHMIDT (eds)
229 Ergodicity for infinite dimensional systems, G. DA PRATO & J. ZABCZYK
230 Prolegomena to a middlebrow arithmetic of curves of genus 2, J.W.S. CASSELS & E.V. FLYNN
231 Semigroup theory and its applications, K.H. HOFMANN & M.W. MISLOVE (eds)
232 The descriptive set theory of Polish group actions, H. BECKER & A.S. KECHRIS
233 Finite fields and applications, S. COHEN & H. NIEDERREITER (eds)
234 Introduction to subfactors, V. JONES & V.S. SUNDER
235 Number theory 1993–94, S. DAVID (ed)
236 The James forest, H. FETTER & B. GAMBOA DE BUEN
237 Sieve methods, exponential sums, and their applications in number theory, G.R.H. GREAVES, G. HARMAN & M.N. HUXLEY (eds)
238 Representation theory and algebraic geometry, A. MARTSINKOVSKY & G. TODOROV (eds)
239 Clifford algebras and spinors, P. LOUNESTO
240 Stable groups, FRANK O. WAGNER
241 Surveys in combinatorics, 1997, R.A. BAILEY (ed)
242 Geometric Galois actions I, L. SCHNEPS & P. LOCHAK (eds)
243 Geometric Galois actions II, L. SCHNEPS & P. LOCHAK (eds)
244 Model theory of groups and automorphism groups, D. EVANS (ed)
245 Geometry, combinatorial designs and related structures, J.W.P. HIRSCHFELD, S.S. MAGLIVERAS & M.J. DE RESMINI (eds)
246 p-Automorphisms of finite p-groups, E.I. KHUKHRO
247 Analytic number theory, Y. MOTOHASHI (ed)
248 Tame topology and o-minimal structures, LOU VAN DEN DRIES
249 The atlas of finite groups: ten years on, ROBERT CURTIS & ROBERT WILSON (eds)
250 Characters and blocks of finite groups, G. NAVARRO
251 Gröbner bases and applications, B. BUCHBERGER & F. WINKLER (eds)
252 Geometry and cohomology in group theory, P. KROPHOLLER, G. NIBLO, R. STÖHR (eds)
253 The q-Schur algebra, S. DONKIN
254 Galois representations in arithmetic algebraic geometry, A.J. SCHOLL & R.L. TAYLOR (eds)
255 Symmetries and integrability of difference equations, P.A. CLARKSON & F.W. NIJHOFF (eds)
260 Groups St Andrews 1997 in Bath, I, C.M. CAMPBELL *et al* (eds)
261 Groups St Andrews 1997 in Bath, II, C.M. CAMPBELL *et al* (eds)

London Mathematical Society Lecture Note Series. 260

Groups St Andrews 1997 in Bath, I

Edited by

C. M. Campbell
University of St Andrews

E. F. Robertson
University of St Andrews

N. Ruskuc
University of St Andrews

G. C. Smith
University of Bath

CAMBRIDGE
UNIVERSITY PRESS

CAMBRIDGE UNIVERSITY PRESS
Cambridge, New York, Melbourne, Madrid, Cape Town,
Singapore, São Paulo, Delhi, Mexico City

Cambridge University Press
The Edinburgh Building, Cambridge CB2 8RU, UK

Published in the United States of America by Cambridge University Press, New York

www.cambridge.org
Information on this title: www.cambridge.org/9780521655880

First published 1999

A catalogue record for this publication is available from the British Library

ISBN 978-0-521-65588-0 Paperback

CONTENTS

Volume I

Contents of Volume II

INTRODUCTION

An international conference Groups St Andrews 1997 in Bath was held on the campus of the University of Bath during the period 26 July – 9 August 1997. Some 299 mathematicians from 41 countries were involved in the meeting, as well as 82 family members and partners. This was the fifth meeting of the four-yearly Groups St Andrews Conferences, and the series continues to flourish. The shape of the conference was similar to the previous conferences in that the first week was dominated by five series of talks, each surveying an area of rapid contemporary development in group theory. The main speakers were Laszlo Babai (Chicago), Martin Bridson (Oxford), Chris Brookes (Cambridge), Cheryl Praeger (Western Australia) and Aner Shalev (Jerusalem). The second week featured two special days, a Burnside Day and a Lyndon Day. Our thanks are due to Efim Zelmanov (Yale) and Chuck Miller III (Melbourne), respectively, for helping organise the programmes for these days. In addition the week contained a wide variety of research talks. In the evenings throughout the conference, and during the rest periods, there was an extensive social programme, only some of which was disrupted by rain. There was also much extemporised music-making in the Senior Common Room in the evenings.

These Proceedings contain the written evidence of the academic achievements of the conference. The five main speakers have all provided substantial survey articles, giving a wide perspective on their fields. In the case of Laszlo Babai, the article is written jointly with Bob Beals (Arizona), one of the invited speakers of the second week. Sixteen other papers in these Proceedings are written by authors who gave one hour invited lectures. A rigorous journal-style refereeing process was applied to all the research articles and survey articles submitted for publication in the Proceedings.

Less easy to quantify, but much more important, were the exchanges of ideas and joint work that was done, both at the conference and as a result of meetings at the conference. As the largest regular meeting on group theory in the world, this series has provided a continuing stimulus to research in group theory.

There are many who helped to make the conference a memorable occasion; in particular we thank Mrs Nada Harvey for secretarial assistance, and Mr John McDermott, Mr Aaron Wilson and many other students from the Department of Mathematical Sciences of the University of Bath whose help in the day to day running of the conference was invaluable. Many of Geoff Smith's colleagues from the academic staff of the Department of Mathematical Sciences gave valuable help and we particularly thank Professor John Toland for guidance on conference organization, and Dr Fran Burstall for TEX and LATEX assistance to the the conference and its newspaper, The Daily Group Theorist.

We record our thanks also to the then Head of School, Professor Alastair Spence, and to the Pro-Vice Chancellor, Professor I M Jamieson, for smoothing the path of the conference organization. For her constructive attitude and industrious work with the conference accommodation we thank Marian Short from the conference office of the University of Bath.

It is a pleasure to acknowledge the excellent level of support we received from various funding bodies. The Edinburgh Mathematical Society and the London Mathematical Society provided financial help which defrayed the main speakers' travel, registration and accommodation expenses. The London Mathematical Society additionally sponsored three visitors from Moscow State University, Professor A. Yu. Ol'shanskii, and two research students Ivan Arzhantsev and Gulnara Arzhantseva. This funding provided support for travel, accommodation, conference fees and subsistence. The Royal Society of London treated the conference as two meetings, because of its exceptional size, and because it is held only once in four years. They therefore generously supported four visitors from the former Soviet Union.

We also gratefully acknowledge the financial support from the University of Bath Initiative Fund, the Department of Mathematics of the University of Bath and the School of Mathematical and Computational Sciences of the University of St Andrews. We thank the Bath and North-East Somerset Council for hosting a civic reception at the Roman Baths. Finally, it is a pleasure to thank Olga Tabachnikova both for her assistance in countless aspects of conference organization, and for the loan of Geoff Smith.

Colin Campbell
Ed Robertson
Nik Ruškuc
Geoff Smith

RADICAL RINGS AND PRODUCTS OF GROUPS

BERNHARD AMBERG* and YAROSLAV SYSAK†[1]

*Fachbereich Mathematik der Universität Mainz, D-55099 Mainz, Germany
†Institute of Mathematics, Ukrainian National Academy of Sciences, 252601 Kiev, Ukraine

1 Introduction

Groups which can be written as a product $G = AB$ of two of its subgroups A and B have been studied by many authors; see for instance the monograph [8]. In these investigations groups of the form $G = AB = AM = BM$ where M is a normal subgroup of G play a particular role. If M is abelian and the intersections $A \cap M$ and $B \cap M$ are trivial, there is an interesting connection with radical rings (in sense of Jacobson) and their generalisations to so-called radical modules. In the following we will discuss in detail some aspects of this connection and, in addition, consider certain structural questions about radical rings.

The notation is standard and can be found in [36] and [37] for the group-theoretical terminology and in [27] and [40] for the ring-theoretical terms.

2 Fundamentals

2.1 Triply factorized groups

Consider a group $G = AB$ which is the product of two subgroups A and B. To study such groups it is desirable to find subgroups of G which are likewise the product of a subgroup of A with a subgroup of B. We recall some elementary facts, which can be found in [8], Section 1.1.

A subgroup S of the group $G = AB$ is called *factorized* if $S = (A \cap S)(B \cap S)$ and $A \cap B \subseteq S$. It is easy to see that a subgroup S of $G = AB$ is factorized if and only if, whenever $ab \in S$ with $a \in A$ and $b \in B$, then $a \in S$. In particular, the group $G = AB$ itself is factorized.

Obviously the intersection of an arbitrary set of factorized subgroups of $G = AB$ is a factorized subgroup of G. Therefore, if U is a subgroup of $G = AB$, then there exists a smallest factorized subgroup $X(U)$ of G containing U; this subgroup $X(U)$ is called the *factorizer* of U in G. Clearly, a subgroup U of $G = AB$ is factorized if and only if $U = X(U)$.

If N is a normal subgroup of $G = AB$, then $X(N) = AN \cap BN = N(A \cap BN) = N(B \cap AN) = (A \cap BN)(B \cap AN)$ (see [8], Lemma 1.1.4). This means that in many investigations of factorized groups one has to consider *triply factorized* groups of the form $G = AB = AM = BM$ where M is a normal subgroup of G. If M is

[1]The second author likes to thank the Deutsche Forschungsgemeinschaft for financial support and the Department of Mathematics of the University of Mainz for its excellent hospitality during the preparation of this paper.

abelian, then $C = (A \cap M)(B \cap M)$ is normal in G. Factoring out this normal subgroup we arrive at a triply factorized group

$$G = AB = AM = BM \text{ with the "intersection property" } A \cap M = B \cap M = 1.$$

In this case A and B are complements of M in the factorized group $G = AB$, and M is a $\mathbb{Z}A$-module with special properties. Note that A and B are not conjugate unless $A = B = G$.

2.2 Radical rings

Let R be an associative ring, not necessarily with an identity element. The set of all elements of R forms a semigroup with identity element $0 \in R$ under the operation $a \circ b = a + b + ab$ for all a and b of R. The group of all invertible elements of this semigroup is called the *adjoint group* of R and denote by R°. Following Jacobson ([27], p. 4) a ring R is called *radical* if $R = R^\circ$, which means that R coincides with its Jacobson radical. Obviously such a ring does not have an identity element. If the radical ring R is embedded in the usual way into a ring R_1 with identity 1, then R° is isomorphic with the subgroup $1 + R$ of the group of units of R_1. It is easy to see that this property even characterizes radical rings. It should be also noted that subrings of a radical ring need not be radical, but right and left ideals, homomorphic images and cartesian products of radical rings are always radical. The next theorem gives an other important property of radical rings (see [27], p. 11).

Theorem 2.1 *If R is a radical ring, then the ring $M_n(R)$ of all $n \times n$ matrices over R is likewise radical.*

A ring R is a *nil ring* if every element a of R is *nilpotent*, i.e. there exists a positive integer $n = n(a)$ such that $a^n = 0$. It is clear that every nil ring is radical, and it is not difficult to see that a radical ring is nil if and only if each of its subrings is radical.

Example 2.2 For every prime p let W_p be the set of all rational numbers whose numerator is divisible by p, but not its denominator. Then W_p is a radical ring, as $1 + W_p$ is a group. But W_p has no non-trivial nilpotent elements, and so is not nil. Moreover, every finitely generated subring of W_p is not radical.

For each positive integer n let R^n be a subring of R which is generated by all products of n elements of R. Clearly R^n is an ideal of R. A ring R is *nilpotent* if $R^{n+1} = 0$ for some non-negative integer n; the smallest such n is called the *class* of R. For instance, for every prime p and every natural number n the ring $p\mathbb{Z}/p^{n+1}\mathbb{Z}$ is nilpotent of class n. Trivial examples of nilpotent rings are the *null rings*, i.e. those of class 1. A proper subclass of the class of nil rings which contains all nilpotent rings is the class of *locally nilpotent* rings, in which every finitely generated subring is nilpotent. The first example of a nil ring which is not locally nilpotent was given by Golod (see for instance [40], Theorem 6.2.9). Properly contained between the nilpotent rings and the locally nilpotent rings is the class of *T-nilpotent* rings, in which every non-trivial homomorphic image has a non-trivial annihilator. The direct sum of the nilpotent rings $p\mathbb{Z}/p^{n+1}\mathbb{Z}$ for every non-negative integer n is a T-nilpotent ring which is not nilpotent. Observe that a ring R is nilpotent, T-nilpotent or locally nilpotent if and only if for each positive integer n the matrix ring $M_n(R)$ has the same property, respectively. However it is unknown at present *whether for a nil ring R the ring $M_2(R)$ is also nil.* This question is equivalent to the famous problem of Koethe that every nil one-sided ideal of a ring is contained in a nil ideal of this ring (see [41]).

The relation among all above-mentioned classes of rings can be seen from the following list, where each class is a proper subclass of the preceding one.

- radical rings
- nil rings
- locally nilpotent rings
- T-nilpotent rings
- nilpotent rings
- null rings

Observe also that if a ring R is nilpotent of class n, T-nilpotent or locally nilpotent, then its adjoint group R° is nilpotent of class at most n, hypercentral or locally nilpotent, respectively, and if R is a null ring, then R° is isomorphic with its additive group R^+. As Example 2.2 shows, the converse of this statement is false.

2.3 Construction of triply factorized groups

Radical rings may be used to construct examples of triply factorized groups in the following way (see [45] and [8], Section 6.1).

Let P be a right ideal of the radical ring R and let $M = R/P$ as a right R-module. The adjoint group $A = R^\circ$ operates on M via the rule $m^a = m + ma$ for a in A and m in M. The *associated group* $G(M) = A \ltimes M$ is the semidirect product of M by A. If $B = \{am \mid m = a + P,\ a \in A\}$ is the diagonal of $G(M)$, then B is a subgroup of $G(M)$ and

$$G = G(M) = A \ltimes M = B \ltimes M = AB.$$

Moreover, B is isomorphic to R° and the intersection $A \cap B$ is isomorphic to P°. If in particular $P = 0$, then the normal subgroup M of $G(M)$ is isomorphic to the additive group R^+ of R and $A \cap B = 1$. In this case, the group $G(R)$ can also be represented as a matrix group over R.

Indeed, consider the ring of 2×2-matrices of the form $\begin{pmatrix} 0 & R \\ 0 & R \end{pmatrix}$ over a radical ring R. This is a left ideal of the matrix ring $M_2(R)$ and so is a radical ring by Theorem 2.1. If this ring is regarded as a subring of the matrix ring $M_2(R_1)$, then the adjoint group of $M_2(R)$ is isomorphic with the multiplicative subgroup

$$\Gamma(R) = \begin{pmatrix} 1 & R \\ 0 & 1+R \end{pmatrix}$$

of the group of units of $M_2(R_1)$. It is easy to see that the group $\Gamma(R)$ is isomorphic to $G(R)$, its subgroups $A_R = \begin{pmatrix} 1 & 0 \\ 0 & 1+R \end{pmatrix}$ and $M_R = \begin{pmatrix} 1 & R \\ 0 & 1 \end{pmatrix}$ are isomorphic to A and M, respectively, and the subgroup B_R of $\Gamma(R)$ consisting of all 2×2-matrices of the form $\begin{pmatrix} 1 & r \\ 0 & 1+r \end{pmatrix}$ where $r \in R$ is isomorphic to B.

We mention some elementary relations between certain properties of the ring R and those of the group $\Gamma(R)$. Observe first that if S is a radical subring of R, then $\Gamma(S)$ is a subgroup of $\Gamma(R)$. Moreover, a subgroup G of $\Gamma(R)$ has the decomposition $G = AB = AM = BM$ with $A = A_R \cap G$, $B = B_R \cap G$ and $M = M_R \cap G$, if and only if $G = \Gamma(S)$ for some radical subring S of R.

Lemma 2.3 *Let S be a radical subring of a radical ring R. Put $A_S = A_R \cap \Gamma(S)$, $B_S = B_R \cap \Gamma(S)$ and $M_S = M_R \cap \Gamma(S)$. Then the following statements are valid.*

1) $\Gamma(S) = A_S B_S = A_S M_S = B_S M_S$.
2) *The subring S is a right, left or two-sided ideal of R if and only if M_S is normal in $\Gamma(R)$, $\Gamma(S)$ is normal in $A_S M_R$ or $\Gamma(S)$ is normal in $\Gamma(R)$, respectively.*
3) *If ${}^{\perp}R$ and $R^{\perp}$ are the left and right annihilators of R respectively, then $C_{M_R}(A_R) = \begin{pmatrix} 1 & {}^{\perp}R \\ 0 & 1 \end{pmatrix}$ and $C_{A_R}(M_R) = \begin{pmatrix} 1 & 0 \\ 0 & 1+R^{\perp} \end{pmatrix}$. In particular, if R has no divisors of zero, then $C_{A_R}(M_R) = C_{M_R}(A_R) = 1$.*
4) *The ring R is nilpotent of class n or locally nilpotent if and only if the group $\Gamma(R)$ is nilpotent of class n or locally nilpotent, respectively.*
5) *The ring R is a Fitting ring, i.e. R is a sum of its nilpotent ideal, if and only if the group $\Gamma(R)$ is a product of nilpotent normal subgroups of the form $G = AB = AM = BM$ where $A = A_R \cap G$, $B = B_R \cap G$ and $M = M_R \cap G$.*

Conversely, if $G = AB = AM = BM$ is a triply factorized group with three abelian subgroups A, B and M where M is normal in G and $A \cap B = A \cap M = B \cap M = 1$, then there exist a commutative radical ring R and an isomorphism α from G onto $\Gamma(R)$ such that $\alpha(A) = A_R$, $\alpha(M) = M_R$ and $\alpha(B) = B_R$ (for details see [8], Proposition 6.1.4). This result cannot immediately be extended to triply factorized groups with nilpotent subgroups A, B and M, as the following example shows.

Example 2.4 Let M be an elementary abelian group of order 8 generated by elements m_1, m_2, m_3 of order 2, and let A be a dihedral group of order 8 generated by elements a and b such that $a^4 = b^2 = (ab)^2 = 1$. The operation of A on M be as follows:

$$m_1{}^a = m_1{}^b = m_1, m_2{}^a = m_1 m_2, m_2{}^b = m_2, m_3{}^a m_2 m_3, m_3{}^b = m_1 m_2 m_3.$$

Let B be the subgroup of the semidirect product $G = AM$ of M by A generated by the elements am_3 and bm_2. Then $G = AB = AM = BM$ is a nilpotent group with $A \cap B = A \cap M = B \cap M = 1$ and there exists no radical ring R such that G is isomorphic to $\Gamma(R)$.

Indeed, assume that there exists such a radical ring R. It is well-known that the Jacobson radical of a ring with minimal condition on left or right ideals is nilpotent (see [27], p. 38). This implies in particular that the finite radical ring R is nilpotent and so $R^{\perp} \neq 0$. Therefore $C_A(M) \neq 1$ by Lemma 2.3.3. On the other hand, direct computation shows that $C_A(M) = 1$, a contradiction.

2.4 Examples

Some examples of triply factorized groups with additional properties can be constructed for various radical rings and can be found in [45] and [8]. Their properties follow from those of the radical rings under consideration and from Lemma 2.3.

First, using the Jacobson radical of the algebra $F[[x]]$ of formal power series in an indeterminate x over the field F of p elements, which coincides with the ideal $xF[[x]]$ (see [27], p. 21), as an example of a radical F-algebra, we obtain the following.

Example 2.5 There exists a metabelian group G with the following properties.

(i) $G = AB = AM = BM$ where A and B are torsion-free abelian subgroups and M is an infinite elementary abelian normal p-subgroup for some prime p.

(ii) $A \cap B = A \cap M = B \cap M = 1$.

(iii) 1 is the only normal subgroup of G contained in A or B.

Next, let R be the augmentation ideal of the group algebra FA of a Prüfer p-group A over the field F with p elements. Then R is a nil F-algebra (see for instance [33], Lemma 8.1.17) whose additive group R^+ is an infinite elementary abelian p-group and the adjoint group R° is a divisible abelian p-group of infinite rank. For every r of R there is an s in R such that $1 + r = (1 + s)^p = 1 + s^p$ and so $r = s^p = ss^{p-1} \in R^2$. Thus $R = R^2$. It easy to see that the ideal rFA of R is nilpotent for every r of R and so R is generated by its nilpotent ideals. This leads to the following.

Example 2.6 There exists a metabelian p-group G with the following properties.

(i) $G = AB = AM = BM$ where A and B are divisible abelian subgroups and $M = G'$ is an elementary abelian normal subgroup of G.

(ii) $A \cap B = A \cap M = B \cap M = 1$.

(iii) G has trivial centre and is generated by its nilpotent normal subgroups.

(iv) 1 is the only normal subgroup of G contained in A or B.

(v) There is no proper normal subgroup of G containing A or B.

Finally, for every prime p there exists a commutative nil ring R of characteristic p, whose adjoint and additive groups are both elementary abelian p-groups, and such that $R^2 = R$. This leads to the following example (see [8], Theorem 6.1.3).

Example 2.7 (Holt and Howlett [26]) For every prime p there exists a countable locally finite group G of exponent p^2 with the following properties.

(i) $G = AB = AM = BM$, where A, B, and M are elementary abelian p-subgroups and M is normal in G.

(ii) $A \cap B = A \cap M = B \cap M = 1$.

(iii) G has trivial centre and is generated by its nilpotent normal subgroups.

(iv) 1 is the only normal subgroup of G contained in A or B.

(v) There is no proper normal subgroup of G containing A or B.

2.5 Adjoint groups factorized by three pairwise permutable subgroups

Let R be a radical ring and R_1 a ring obtained from R by adjoining an identity 1. Denote by $\Gamma_2(R)$ the group of all invertible 2×2-matrices over R_1 that are congruent to the identity matrix modulo R, and by $t_{ij}(r)$ the transvection with the element r in the position (i, j). In fact,

$$\Gamma_2(R) = \begin{pmatrix} 1+R & R \\ R & 1+R \end{pmatrix}.$$

It is easy to see that this group is isomorphic to the adjoint group of the matrix ring $M_2(R)$ which is radical by Theorem 2.1. The following assertion can be verified directly.

Proposition 2.8 (Sysak [46]) *Let A be the subgroup of $\Gamma_2(R)$ consisting of all its diagonal matrices, $B = t_{12}(-1)At_{12}(1)$ and $C = t_{21}(-1)At_{21}(1)$. Then $\Gamma_2(R) = ABC$ and the subgroups A, B, C commute pairwise, i.e. $AB = BA$, $AC = CA$ and $BC = CB$.*

It is clear that if the ring R is commutative, then the subgroups A, B and C of $\Gamma_2(R)$ are abelian. Hence, as a consequence of Proposition 2.8 we obtain the existence of insoluble linear groups over a field that are the product of three pairwise permutable abelian subgroups (see [46]). Note that every group which is a product of two abelian subgroups must be metabelian by a well-known theorem of Ito (see [8], Theorem 2.1.1).

Corollary 2.9 *Let F be a field whose multiplicative group is non-periodic. Then in F there exists a radical subring $R \neq 0$, and the group $\Gamma_2(R)$ contains a non-abelian free subgroup. In particular, $\Gamma_2(R)$ is an insoluble linear group over F which is a product of three pairwise permutable abelian subgroups.*

The next example follows from Proposition 2.8 and shows that for every prime p and each natural number d there exists a finite p-group that is decomposable into a product of three pairwise permutable cyclic subgroups and whose derived length is greater than d.

Example 2.10 Let p be a prime and $R = p\mathbb{Z}_{p^m}$ for the odd p or $R = 4\mathbb{Z}_{2^m}$ for $p = 2$ where $m \geq 3$. Then the factor group $\Gamma_2(R)/Z(\Gamma_2(R))$ is a product of three pairwise permutable cyclic subgroups of orders p^{m-1} or 2^{m-2}, respectively, and its derived length is at least $log_2(m)$.

2.6 Some general questions

The above construction raises the following questions:

(1) What can be said about the structure of the adjoint group of a radical ring?

(2) Which relations exist among the groups R°, R^+, and $G(R)$ for a radical ring R?

(3) How can these results be used in the study of factorized groups?

(4) Which triply factorized groups with special properties can be constructed using various radical rings?

(5) Which results about factorized groups can be used to study radical rings?

3 The structure of radical rings

3.1 Finiteness conditions

The following finiteness conditions play an important role in the theory of infinite groups with generalized solubility conditions (see [36]). It is easy to find abelian groups to show that all these classes are distinct.

- A group G has *min* (*max*) if it satisfies the minimum (maximum) condition on subgroups.
- G is a *minimax group* if it has a finite series whose factors satisfy the minimum or maximum condition on subgroups.
- G has *finite Prüfer rank* $r = r(G)$ if every finitely generated subgroup of G can be generated by r elements and r is the least such number.
- G has *finite abelian section rank* if every elementary abelian p-section of G is finite for every prime p.
- G has *finite torsion-free rank* if it has a series of finite length whose factors are periodic or infinite cyclic. The number of infinite cyclic factors in any such series is an invariant of G called its *torsion-free rank* $r_0(G)$.

In particular a group G is *periodic* if and only if it has torsion-free rank $r_0(G) = 0$. If N is a normal subgroup of the group G with finite torsion-f ree rank, then it is easy to see that $r_0(G) = r_0(N) + r_0(G/N)$. Thus the torsion-free rank is an excellent tool for induction arguments.

3.2 The adjoint group of a radical ring

Since the Jacobson radical of a ring with minimal condition on left or right ideals is nilpotent, the adjoint group of every finite radical ring is nilpotent. The following theorem gives some information on the structure of the adjoint group of an arbitrary radical ring. Recall that a group G is an *SN-group* if it has a (general) series with abelian factors (see [37], p. 365). If $\mathfrak{X}$ is a group-theoretical property, then a group G is a *locally $\mathfrak{X}$-group,* if every finitely generated subgroup of G has the property $\mathfrak{X}$.

Theorem 3.1 (Amberg, Dickenschied and Sysak [4]) *The adjoint group R° of every radical ring R is an SN-group in which every (locally) finite subgroup is (locally) nilpotent.*

Using similar arguments as in [4], it can even be proved that in every periodic subgroup G of the adjoint group R° of a radical ring R each two elements of coprime orders are permutable. However, we do not know at present whether such a periodic subgroup is the direct product of its primary components. This is in fact the case if the annihilator of R is trivial or R° has no central Prüfer subgroups. The matter in question can be reduced to the following more general problem which is of independent interest. *Does every central extension G of a Prüfer p-group by a periodic p'-group split if G is an SN-group?* It should be noted that if G is not an SN-group, then an example of Adjan ([1], p. 276, VII.1.9) shows that there exists such a non-split extension.

Since SN-groups which satisfy *min* are locally finite (see [36], vol. 1, p. 71, Corollary), Theorem 3.1 also implies that every subgroup with *min* of the adjoint group of a radical ring is also locally nilpotent. However, it follows from a result of Neroslavskii that subgroups of the adjoint group of a radical ring which have *max* need not be locally nilpotent. In fact, in [32] a radical algebra over the finite field $GF(p)$ for some prime $p \neq 2$ is given in which the adjoint group R° contains a subgroup $G \simeq \langle a\rangle \ltimes \langle b\rangle$ with $a^{-1}ba = b^2$ and where a has infinite order and b has order p. In particular G is polycyclic, but not nilpotent. This example also shows that there exist radical rings whose adjoint group does not have a central series.

It should be noted that if a polycyclic subgroup G of the adjoint group of a radical ring R generates R as a ring, then R must be a nilpotent ring and so G is a nilpotent group. This is a consequence of a deep result of Roseblade (see [39], Theorem B). Moreover, if G is locally polycyclic with finite torsion-free rank, then R is a locally nilpotent ring and so G is a locally nilpotent group, as follows from an observation by Brown (see the remarks preceding Corollary 4.6 in [17]). It would be interesting to know *whether the last statement can be extended to the case when the group G is (locally) minimax.* A number of locally nilpotent subgroups of the adjoint group of certain radical rings are also pointed out in [4]. In particular, the adjoint group of a nil ring contains many locally nilpotent subgroups, some of which can be found by the following result.

Let $\mathfrak{X}$ be a group-theoretical property such that every $\mathfrak{X}$-group has an ascending series with locally finite or locally nilpotent factors.

Theorem 3.2 (Amberg, Dickenschied and Sysak [4]) *Let G be a subgroup of the adjoint group of a nil ring R. Then G is locally nilpotent if one of the following conditions holds:*

(i) *G is a locally $\mathfrak{X}$-group,*

(ii) *Every finitely generated subgroup of G has finite Prüfer rank.*

Moreover, if G is locally nilpotent, then the subring of R generated by G is a locally nilpotent ring.

In particular, every maximal locally nilpotent subgroup of the adjoint group of a nil ring is in fact a subring of this ring.

Observe also that examples of simple radical rings with $R^2 = R$ exist (see [42]), whereas *examples of simple nil rings are unknown.*

3.3 Finiteness condition on the adjoint group

It is interesting to investigate the relation between the adjoint group and the additive group of a radical ring R. One of the first results in this direction was obtained by Watters [51], who showed that the adjoint group of a radical ring R has *max* if and only if its additive group is finitely generated, and in this case R is nilpotent. The following theorem extends this result to the class of minimax groups.

Theorem 3.3 (Amberg and Dickenschied [3]) *Let $\mathfrak{X}$ be a class of minimax groups which is closed under the forming of subgroups, epimorphic images and extensions. The following conditions of the radical ring R are equivalent:*

(1) *The additive group R^+ is an $\mathfrak{X}$-group,*

(2) *The associated group $G(R)$ is an $\mathfrak{X}$-group,*

(3) *The adjoint group R° is an $\mathfrak{X}$-group.*

In this case R is a nilpotent ring.

Consider again the radical ring W_2 of all rational numbers with odd denominator and even numerator from Example 2.2. Clearly the additive group of W_2 has Prüfer rank 1, but its adjoint group has infinite torsion-free rank, since $1 + W_2$ is generated by -1 and all odd primes. Therefore there is no complete analogue of Theorem 3.3 for the other finiteness conditions mentioned above. However, the following theorem shows that these finiteness conditions are inherited from the adjoint group of a radical ring R to its additive group and that they imply some nilpotency conditions of the ring R.

Theorem 3.4 (Amberg and Dickenschied [3]) *Let R be a radical ring. Then the following holds:*

(a) *If R° has finite torsion-free rank n, then also $r_0(R^+) = n$, and R is a nil ring,*

(b) *If R° has finite abelian section rank, then so does R^+, and R is a T-nilpotent ring of class $cl(R) \leq \omega + r_0(R^+)$,*

(c) *If R° has finite Prüfer rank, then so does R^+, and $r(R^+)$ is bounded by a function which only depends on $r(R^\circ)$.*

Here a ring R is called *T-nilpotent of class $cl(R) = \alpha$* if $B_\alpha(R) = R$ and α is the least ordinal with this property, where the *transfinite annihilator series* of a ring R is defined by

$$B_0(R) = 0, \qquad B_{\alpha+1}(R) = \{a \in R \mid aR + Ra \subseteq B_\alpha(R)\}$$

for each ordinal α and for each limit ordinal λ

$$B_\lambda(R) = \bigcup_{\beta<\lambda} B_\alpha(R).$$

It is reasonable to ask *which is the best bound for the Prüfer rank in Theorem 3.4(c)*. Note that this is unknown even if R is commutative. *Is perhaps $r(R^+) \leq 2r(R^\circ)$ in this latter case?*

The following result shows that the converse of Theorem 3.4 holds for nil rings. It is easy to see that the adjoint group of a nil ring is a p-group (torsion-free) if and only if, its additive group is a p-group (torsion-free).

Theorem 3.5 (Dickenschied [20]) *If R is a nil-ring, then the following holds:*

(a) *If R^+ has finite torsion-free rank n, then also $r_0(R^\circ) = n$,*

(b) *If R^+ has finite abelian section rank, then so does R°,*

(c) *If R^+ has finite Prüfer rank, then so does R°, and $r(R^\circ) \leq 3r(R^+)$. If R^+ contains no element of order 2 then even $r(R^\circ) \leq 2r(R^+)$.*

Statements (b) and (c) also hold for radical rings whose additive group is periodic (see [20]). On the other hand, if F is the field with p elements and $F[[x]]$ is the ring of formal power series in an indeterminate x over F, then the Jacobson radical $xF[[x]]$ of $F[[x]]$ gives an example of a radical ring R, whose additive group is an elementary abelian p-group, but whose adjoint group is torsion-free and has infinite torsion-free rank.

Which groups can occur as the adjoint group of a radical ring? This problem was in part discussed in [30]. Obviously, every abelian group A is the adjoint group of the null ring on A. For finite groups it suffices to consider only p-groups. Finite abelian groups which occur as the adjoint group of some commutative nilpotent p-algebra were discussed by Eggert in [23]. He proved that a finite abelian p-group A is the adjoint group of such an algebra if $r(A^{p^n}) \geq p \cdot r(A^{p^{n+1}})$ for every n with $A^{p^n} \neq 1$ and conjectured that the converse of this statement is also true. An affirmative answer to this conjecture would imply that the above-mentioned inequality $r(R^+) \leq 2r(R^\circ)$ holds for every commutative radical ring R.

Kaloujnine has shown in [28] that every p-group of class 2 for every odd prime p is isomorphic to the adjoint group of some nilpotent ring. In fact, all groups of order p, p^2 and p^3 occur as the adjoint of some nilpotent ring, but a group of order

p^4 if and only if it is nilpotent of class ≤ 2. Ault and Watters conjectured in [16] that any nilpotent group G of class 2 is the adjoint group of a nilpotent ring of class 2. This conjecture was disproved in [25], where it was also shown, that this holds if the factor group G/G' is a direct product of cyclic groups, or is radicable, or is torsion, or is torsion-free and completely decomposable. However, it seems to be unknown at present *whether every nilpotent group of class 2 can occur as the adjoint group of some nilpotent ring of class ≥ 2 or at least of some radical ring.* In [24] Gorlov gives a characterisation of all finite nilpotent algebras with metacyclic adjoint group and in [34] Popovich and Sysak classify all radical algebras in which every subgroup of the adjoint group is a subalgebra.

3.4 Engel and Lie conditions

If $a_1, a_2, \ldots$ are elements of the ring R, the Lie-commutators $[a_1, \ldots, a_{n+1}]$ are defined by $[a_1, a_2] = a_1a_2 - a_2a_1$ and $[a_1, \ldots, a_{n+1}] = [[a_1, \ldots, a_n], a_{n+1}]$ for all $n \geq 2$. The ring R is called *Lie-nilpotent of class* at most n, if $[a_1, \ldots, a_{n+1}] = 0$ for all $a_1, \ldots a_{n+1}$ of R. Moreover, the ring R is *Engel* if $[a, b, \ldots, b] = 0$ for each pair of elements a and b in R, and *n-Engel* if b appears exactly n-times. Note that Engel and n-Engel groups are defined in a corresponding way where the usual group commutator replaces the Lie-commutator.

Du has shown in [22] that a radical ring R is Lie-nilpotent of class n if and only if its adjoint group R° is nilpotent of class n, where n is any positive integer. The question arises *whether a radical ring R is n-Engel if and only if its adjoint group R° is n-Engel.*

For $n = 1$ this is trivially true. The cases $n = 2$ and $n = 3$ are treated in the dissertation of O. Dickenschied [21]. In particular it is proved there that the ring R is 2-Engel, if and only if, its adjoint group R° is a 2-Engel group; in this case, R is Lie-nilpotent of class ≤ 3 and thus R° is nilpotent of class ≤ 3. It is also shown that if R is 3-Engel without elements of additive order 2, then R is a 3-Engel ring if and only if R° is a 3-Engel group. Notice that subgroups of the adjoint group of a radical ring and in particular the adjoint group itself can be n-Engel for a positive integer n only if they are locally nilpotent. This is in fact a consequence of Theorem 3.1 and Zelmanov's solution of the Restricted Burnside Problem. Furthermore, it follows from a result of Shalev [44] that, if a radical ring is an n-Engel algebra over a field of prime characteristic, then the adjoint group R° of R is m-Engel for some m depending on n, and so R° is also locally nilpotent. It is easy to check that every nil ring is an Engel ring. Thus the question arises *what can be said about the adjoint group R° of a nil ring R. Is R° perhaps always an Engel group?* On the other hand, if for a radical ring R the adjoint group of the matrix ring $M_2(R)$ or even the associated group $G(R)$ is Engel, then R is a nil ring (see [4]).

There are also some results on a radical ring R whose associated Lie ring, i. e. the ring R with the operation $[a, b] = ab - ba$, is soluble. By a theorem of Smirnov and Zalesskii [43], such a ring R has a nilpotent ideal I whose factor ring R/I is centre-by-metabelian as a Lie ring. However, the adjoint group R° of R need not be soluble in this case. Indeed, if F is a field of characteristic 2, whose multiplicative group is non-periodic, and R is a non-zero radical subring of F, then the matrix ring $M_2(R)$ is radical and its associated Lie ring is centre-by-metabelian, but the adjoint group of this ring is isomorphic to the group $\Gamma_2(R)$ and so is not soluble by Corollary 2.9. Moreover, it is proved in [35] that for each field F of characteristic $p \geq 0$ there exists a finite-dimensional nilpotent F-algebra which satisfies the identity $x^{2^p} = 0$ whenever $p > 2$ and $x^8 = 0$ when $p = 2$ and which is Lie centre-by-metabelian but fails to have a centre-by-metabelian adjoint group.

On the other hand, the adjoint group of every radical ring with metabelian associated Lie ring is likewise metabelian by a result of Krasil'nikov in [29]. Here it is also asked *whether a radical ring with metabelian adjoint group must have a metabelian associated Lie ring.* For nil rings this question was answered affirmatively. We remark that, using this, it can be proved that every radical ring which is finitely generated as a ring and whose adjoint group is metabelian is nilpotent and so is metabelian as a Lie ring. In particular, a radical ring whose adjoint group is finitely generated and metabelian is nilpotent. In this connection the question arises *whether every radical ring with finitely generated soluble adjoint group must be nilpotent.* Note that any radical algebra R over an infinite field cannot be a counterexample to this question. Indeed, in the opposite case R is finitely generated as an algebra and satisfies a non-trivial rational, namely group commutator identity. Therefore R must satisfy a non-trivial polynomial identity by a theorem of Valitskas [50] and so R is nilpotent by theorems of Amitsur-Procesi and Braun (see [40], Theorems 6.3.3 and 6.3.39), contrary to the assumption.

4 Radical modules

4.1 Reduction to group rings

Let A be a group and J a commutative ring with 1, and let M be a right JA-module. A map d from A to M is called a *derivation* from A to M if $d(gh) = d(g)h + d(h)$ for every two elements g and h in A. The module M is *radical* if there exists a surjective derivation from A onto M.

For example, let R be a radical ring. Then the additive group R^+ of R becomes a $\mathbb{Z}R^\circ$-module via the rule $a^b = a + ab$ for every two elements a and b in R. The identity map from R to R is a bijective derivation from R° onto R^+. Thus the $\mathbb{Z}R^\circ$-module R^+ is radical.

The following theorem describes radical modules from a group-theoretical and a ring-theoretical point of view.

Theorem 4.1 (Sysak [47]) *Let A be a group and M a right JA-module for some commutative ring J with* 1. *Then the following conditions are equivalent:*

(1) *The module M is radical,*

(2) *In the augmentation ideal $\Delta_J(A)$ there exists a right ideal $\mathfrak{p}$ of JA such that $\Delta_J(A) = A - 1 + \mathfrak{p}$, and the JA-module M is isomorphic with the JA-module $\Delta_J(A)/\mathfrak{p}$,*

(3) *In the semidirect product* $A \ltimes M$ *there exists a subgroup* B *such that* $A \ltimes M = B \ltimes M = AB$.
Moreover, if d *is a surjective derivation from* A *onto* M*, then the right ideal* $\mathfrak{p}$ *and the subgroup* B *can be chosen such that* $\operatorname{Ker} d = A \cap B = A \cap (1 + \mathfrak{p})$.

Of course, if the group A is abelian, then $\mathfrak{p}$ is an ideal of JA and the factor ring $\Delta_J(A)/\mathfrak{p}$ is radical. Moreover, if $A \cap (1 + \mathfrak{p}) = 1$, then A is isomorphic to the adjoint group of $\Delta_J(A)/\mathfrak{p}$. Hence *a group* A *can occur as the adjoint group of a radical* J*-algebra* R *if and only if in the group ring* JA *there exists an ideal* $\mathfrak{a}$ *such that* $\Delta_J(A) = A - 1 + \mathfrak{a}$ *and* $A \cap (1 + \mathfrak{a}) = 1$. Observe that in Theorem 4.1 the right ideal $\mathfrak{p}$ need not be an ideal of JA. For example, if F is the field with two elements and $A = \langle a, b \mid a^4 = b^2 = (ab)^2 = 1 \rangle$ is the dihedral group of order 8, then the right ideal $\mathfrak{p} = (1 + a + a^2 + ab)FA$ is not any ideal in FA and it is easy to verify that $\Delta_F(A) = A - 1 + \mathfrak{p}$ and $A \cap (1 + \mathfrak{a}) = 1$.

4.2 Hyperabelian groups

The relation between the structure of the group A and that of a radical JA-module has been studied successfully only in the case of a hyperabelian group A. Recall that a group A is *hyperabelian* if every non-trivial epimorphic image of A has a non-trivial abelian normal subgroup.

Theorem 4.2 (see Sysak [45, 47, 48, 49], Wilson [52, 53]) *Let* A *be a hyperabelian group and* M *a radical right* JA*-module. If* A *is*

(i) *a* π*-group for some set of primes* π*,*

(ii) *a minimax group,*

(iii) *a group with finite abelian section rank,*

(iv) *a group with finite Prüfer rank,*

(v) *a group with finite torsion-free rank,*

then the additive group M^+ *of* M *has the same property.*

If A is a hyperabelian group with finite torsion-free rank $r_0(A)$ it can even be shown that the (finite) torsion-free rank of M^+ in Theorem 4.2 satisfies $r_0(M^+) \leq r_0(A)$ and if A has finite Prüfer rank $r(A)$, then $r(M^+)$ is bounded by a polynomial function of $r(A)$ (see also Theorem 3.4).

As follows from Theorem 3.4(a), each radical ring whose adjoint group is periodic must be nil and so its additive group is also periodic. Therefore the question arises *what can be said about the additive group of a radical module over a periodic group. Is it always a periodic group ?* If every 2-generated subgroup of G is finite, then the answer to this question is positive.

Theorem 4.3 (Sysak[49]) *Let* G *be a group in which every two-generated subgroup is finite and let* M *be a radical* $\mathbb{Z}G$*-module. Then the additive group of* M *is periodic.*

Theorem 4.2 leads to the following general result about factorized hyperabelian groups, the first statement of which was first proved by Chernikov [19] and Sysak [45], the next three by Wilson [52, 53] and Sysak [47, 49], the last by Sysak [48].

Corollary 4.4 *Let the hyperabelian group $G = AB$ be the product of two subgroups A and B. If A and B have one of the properties (i) - (v) of Theorem 4.2, then G has likewise the same property*

It is unknown at present *whether this result for at least one of the properties (ii) - (v) can be extended to the case when the group G is radical.* Here, as usual, a group G is *radical* if every non-trivial homomorphic image of G has a non-trivial locally nilpotent normal subgroup. The next theorem follows from Theorem 4.3 and implies that a radical group which is the product of two periodic subgroups is likewise periodic and therefore locally finite. Recall that the *Hirsch-Plotkin radical* of a group G is its maximal locally nilpotent normal subgroup (see [37]).

Theorem 4.5 (Sysak [49]) *If the group $G = AB$ is the product of two locally finite subgroups A and B, then the Hirsch-Plotkin radical of G is locally finite.*

4.3 Groups with a locally nilpotent triple factorization

In a series of papers Amberg, Franciosi and de Giovanni have shown that under certain finiteness conditions a triply factorized group $G = AB = AM = BM$ of two subgroups A and B and a normal subgroup M satisfies some nilpotency condition if the three subgroups A, B and M satisfy this same nilpotency condition (see [5], [6], [7]). For instance, a group $G = AB = AM = BM$ with finite abelian section rank is locally nilpotent (and hence hypercentral) if A, B and M are locally nilpotent (see [8], Theorem 6.3.8). The proof of the following more general result uses radical modules and in particular Theorem 4.1.

Theorem 4.6 (Amberg and Sysak [12]) *Let the hyperabelian group $G = AB = AM = BM$ be the product of three locally nilpotent subgroups A, B and M, where M is normal in G. If G has finite torsion-free rank, then G is locally nilpotent.*

The *Hirsch-Plotkin series* of the group G is defined by

$$R_0 = 1,\ \ R_{\alpha+1}/R_\alpha \text{ is the Hirsch-Plotkin radical of } G/R_\alpha$$

for every ordinal α, and for limit ordinals λ

$$R_\lambda = \bigcup_{\beta<\lambda} R_\beta.$$

Theorem 4.6 has the following consequence which is proved as in the case of a hyperabelian group with finite abelian section rank (see [8], Corollary 6.3.9).

Corollary 4.7 *Let the hyperabelian group $G = AB$ with finite torsion-free rank be the product of two locally nilpotent subgroups A and B. Then each term of the Hirsch-Plotkin series of G is factorized. In particular the Hirsch-Plotkin radical $R = R(G)$ of G is factorized, i.e. satisfies $R = (A \cap R)(B \cap R)$ and $A \cap B \subseteq R$.*

Another result on groups with a locally nilpotent triple factorization is the following generalization of results in [5], [7], [9], and [13] (see also [8], Theorems 6.3.7, 6.3.8, 6.5.13 and 6.5.14).

Theorem 4.8 (Amberg and Sysak [15]) *Let $\mathfrak{X}$ be a class of soluble-by-finite minimax groups which is closed under the forming of subgroups, epimorphic images and extensions, and let the group $G = AB = AM = BM$ be the product of three (locally nilpotent)-by-$\mathfrak{X}$-subgroups A, B and M, where M is normal in G. If M has an ascending G-invariant series with minimax factors, then G is (locally nilpotent)-by-$\mathfrak{X}$.*

The following example derives from Example 2.2 and shows that if the hyperabelian group G in Theorem 4.6 does not have finite torsion-free rank, then G need not even be locally polycyclic and its Hirsch-Plotkin radical need not be factorized, although A, B and M are abelian (see also [8], Theorem 6.1.2).

Example 4.9 There exists a countable torsion-free metabelian group G with the following properties:

(i) $G = AB = AM = BM$, where A and B are torsion-free abelian subgroups of infinite rank and M is an abelian normal subgroup of G with Prüfer rank 1,

(ii) $A \cap B = A \cap M = B \cap M = 1$,

(iii) G is not locally polycyclic and M is the Hirsch-Plotkin radical of G,

(iv) G is not (locally nilpotent)-by-minimax.

This example also shows that Theorem 4.8 cannot be extended to the case when M has an ascending G-invariant series whose factors have finite Prüfer rank.

4.4 Locally soluble groups

In connection with Theorem 4.2 the question arises which of statements (i) - (v) of this theorem can be extended to the case when the group A is locally soluble. The following theorem shows that the statements (i) and (iii) do not hold in general in this case.

Theorem 4.10 (Sysak [49]) *There exists a locally finite-soluble group A and a radical JA-module M such that*

(i) *M^+ is an infinite elementary-abelian q-group for some prime q,*

(ii) *A has no elements of order q.*

Moreover, it can be arranged that either

(iii) *A is a p-group where p is any odd prime if* $q = 2$, *and* $p = 2$ *if* $q = 3$, *or*

(iv) $q = 2$ *and the maximal p-subgroups of A are finite abelian for all odd primes p.*

As a consequence, we immediately obtain the next two results about triply factorized groups.

Corollary 4.11 *For each prime p there exists a locally finite-soluble group G of the form* $G = AB = AM = BM$ *with* $A \cap M = B \cap M = 1$, *where A and B are (locally nilpotent) p-subgroups of G and M is a normal elementary-abelian q-subgroup with* $q \neq p$. *In particular G is not a p-group (and not locally nilpotent).*

Corollary 4.12 *There exists a locally soluble group G of the form* $G = AB = AM = BM$ *with* $A \cap M = B \cap M = 1$, *where the subgroups A and B have finite abelian section rank and M is an infinite elementary-abelian normal 2-subgroup of G. In particular G does not have finite abelian section rank.*

The preceding two corollaries of Theorem 4.10 show that Corollary 4.4 does not hold in general for locally soluble groups. It is an open question whether statement (v) of Theorem 4.2 holds for these groups. In particular, *is every locally soluble product of two periodic groups likewise periodic?* On the other hand, the following shows that statements (ii) and (iv) of Theorem 4.2 are valid for locally soluble groups.

Theorem 4.13 (Amberg and Sysak [11] and [14]) *If the locally soluble group* $G = AB$ *is the product of two subgroups A and B, then the following holds:*

(a) *If A and B are minimax groups, then G is a soluble minimax group,*

(b) *If A and B have finite Prüfer rank, then G has finite Prüfer rank,*

(c) *If A and B are hyperabelian with finite abelian section rank, then every periodic subgroup of G has finite abelian section rank.*

It remains open *whether every locally soluble product of two hyperabelian subgroups with finite abelian section rank must have finite abelian section rank.* The following theorem shows that this at least holds for radical groups.

Theorem 4.14 (Amberg and Sysak [14]) *Let the radical and locally soluble group* $G = AB$ *be the product of two subgroups A and B with finite abelian section rank. Then G is hyperabelian with finite abelian section rank.*

By Corollary 4.12 there exists a locally finite-soluble group $G = AB = AM = BM$ where M is a minimal normal subgroup of G and so an elementary abelian q-subgroup for some prime q and A and B are locally finite p-subgroups for some prime $p \neq q$, but G is not locally nilpotent. Here A and B are locally conjugate, but of course not conjugate in G. Since G is a periodic radical group, this example shows that Theorem 4.6 also becomes false in general when the factorized group

G is no longer hyperabelian. It also follows that the Hirsch-Plotkin radical of a radical and locally finite-soluble product of two locally nilpotent subgroups need not be factorized.

Remark 4.15 Let the locally soluble group $G = AB$ with finite Prüfer rank be the product of two subgroups A and B. Zaitsev and Robinson have shown that the Prüfer rank $r(G)$ of G is bounded by a polynomial function f of the Prüfer ranks $r(A)$ and $r(B)$ of A and B (see [55] and [38] or [8], Theorem 4.3.5). Is f even a linear function? This problem can be reduced to the case when G is a finite p-group. Some results concerning this problem can be found in [10].

If A and B are abelian, the commutator subgroup G' of G is abelian by Ito's theorem. The factorizer $X(G')$ of G' is a triply factorized group of the form $G = AB = AM = BM$ where A, B and M are abelian subgroups of the finite p-group G and M is normal in G. We may also suppose that $A \cap M = B \cap M = 1$ and M is an elementary abelian p-group. It was pointed out in section 2.3 that there is a radical ring R (which is a nilpotent algebra over the field with p elements in this case) such that the associated group $G(R) \simeq G$. This means that the above question is actually a problem about commutative radical rings.

References

[1] Adjan, S.I., The Burnside problem and identities in groups, Springer-Verlag, Berlin (1978).

[2] Amberg, B., Triply factorized groups. In: Groups - St. Andrews 1989, London Mathematical Society Lecture Notes Series 159 (1991), 1-13, Cambridge University Press.

[3] Amberg, B., Dickenschied, O., On the adjoint group of a radical ring, Bull. Canad. Math. Soc. 38 (1995), 262-270.

[4] Amberg, B., Dickenschied, O., Sysak Ya.P., Subgroups of the adjoint group of a radical ring, Canad. J. Math. 50 (1998), to appear.

[5] Amberg, B., Franciosi, S., de Giovanni, F., Groups with a nilpotent triple factorization, Bull. Austral. Math. Soc. 37 (1988), 69-79.

[6] Amberg, B., Franciosi, S., de Giovanni, F., Groups with an FC-nilpotent triple factorization, Ricerche Mat. 36 (1987), 103-114.

[7] Amberg, B., Franciosi, S., de Giovanni, F., Triply factorized groups, Comm. Algebra 18 (1990), 789-809.

[8] Amberg, B., Franciosi, S., de Giovanni, F., Products of Groups, Clarendon Press, Oxford (1992).

[9] Amberg, B., Franciosi, S., de Giovanni, F., FC-nilpotent products of hypercentral groups, Forum Math. 7 (1995), 307-316.

[10] Amberg, B., Kazarin, L.S., On the rank of a product of two finite p-groups and nilpotent p-algebras, to appear.

[11] Amberg, B., Sysak, Ya.P., Locally soluble products of minimax groups, Proceedings of "Groups - Korea 1994", de Gruyter Verlag, Berlin, 9-14.

[12] Amberg, B., Sysak, Ya.P., Groups with finite torsion-free rank which have a locally nilpotent triple factorization, J. Algebra 178 (1995), 136-148.

[13] Amberg, B., Sysak, Ya.P., On groups with a locally nilpotent triple factorization, Publ. Math. Debrecen 50 (1996), 1-7.

[14] Amberg, B., Sysak, Ya.P., Locally soluble products of two subgroups with finite rank, Comm. Algebra 26 (1996), 2421-2445.
[15] Amberg, B., Sysak, Ya.P., On groups with a locally nilpotent triple factorization II, Preprint no. 14 (1997), Fachbereich Mathematik, University of Mainz.
[16] Ault, J.C., Watters, J.F., Circle groups of nilpotent rings, Amer. Math. Monthly 80 (1973), 48-52.
[17] Brown, K.A., The Nullstellensatz for certain group rings, J. London Math. Soc. 26 (1982), 425-434.
[18] Catino, F., On the centers of a radical ring, Arch. Math. 60 (1993), 330-333.
[19] Chernikov, N.S., Products of groups of finite free rank (in Russian), in: Groups and Systems of their Subgroups, Kiev (1983), 42-56.
[20] Dickenschied, O., On the adjoint group of some radical rings, Glasgow Math. J. 39 (1997), 35-41.
[21] Dickenschied, O., On the structure of radical rings, Dissertation, Universität Mainz (1997).
[22] Du, X., The centers of a radical ring, Canad. Math. Bull. 35 (1992), 174-179.
[23] Eggert, N.H., Quasi-regular groups of finite commutative nilpotent algebras, Pacific J. Math. 36 (1971), 631-634.
[24] Gorlov, B.O., Finite nilpotent algebras with metacyclic adjoint group, Ukrain. Mat. Zh. 47 (1995), 1426-1431.
[25] Hales, A.W., Passi, I.B.S., The second augmentation quotient of an integral group ring, Arch. Math. 31 (1978), 259-265.
[26] Holt, D.F., Howlett, R.B., On groups which are the product of two abelian groups, J. London Math. Soc. (2) 31 (1984), 265-271.
[27] Jacobson, N., Structure of Rings. Amer. Math. Soc. Colloq. Publ. 37, Providence R.I. (1964).
[28] Kaloujnine, L., Zum Problem der Klassifikation der endlichen metabelschen p-Gruppen, Wiss. Z. Humboldt-Univ. Berlin Math.-Nat. Reihe 4 (1954/55), 1-7.
[29] Krasil'nikov, A.N., On the group of units of a ring whose associated Lie ring is metabelian, Russian Math. Surweys 47 (1992), 214 - 215.
[30] Kruse, R.L., Price, D.T., Nilpotent rings, Gordon and Breach, New York (1969).
[31] Laue, H., On the associated Lie ring and the adjoint group of a radical ring, Can. Math. Bull. 27 (1994), 215-222.
[32] Neroslavskii, O.M., Structures that are connected with radical rings (in Russian), Vesci Akad. Nauk BSSR Ser. Fiz.-Mat. Nauk 134 (1973), 5-10.
[33] Passman, D.S., The algebraic structure of group rings, Wiley Interscience, New York (1977).
[34] Popovich, S.V., Sysak, Y.P., Radical algebras in which every subgroup of the adjoint group is a subalgebra, Ukrain. Mat. Zh. 49 (1997), no. 12.
[35] Riley, D.M., Tasić, V., The transfer of a commutator law from a nil-ring to its adjoint group, Canad. Math. Bull. 40 (1997), 103-107.
[36] Robinson, D.J.S., Finiteness Conditions and Generalized Soluble Groups, Vol. 1 and 2, Springer Verlag, Berlin (1972).
[37] Robinson, D.J.S., A course in the theory of groups, Springer, Berlin (1982).
[38] Robinson, D.J.S., Soluble products of nilpotent groups, J. Algebra 98 (1986), 183-196.
[39] Roseblade, J.E., Group rings of polycyclic groups, J. Pure Appl. Algebra 3 (1973), 307-328.
[40] Rowen, L.H., Ring Theory, Vol. 1 and 2, Academic Press, San Diego - London (1988).
[41] Rowen, L.H., Koethe's conjecture. In: Ring Theory 1989, Israel Math. Conf. Proc. 1 (1989), 193-202.
[42] Sasiada, E., Cohn, P.M., An example of a simple radical ring, J. Algebra 5 (1967),

373-377.

[43] Smirnov, M.B., Zalesskii, E., Associative rings satisfying the identity of Lie solubility, Vestsi Akad. Nauk. BSSR, Ser. Fiz.-Mat. Nauk 2 (1982), 15-20.

[44] Shalev, A., On associative algebras satisfying the Engel condition, Israel J. Math. 67 (1989), 287-290.

[45] Sysak, Ya.P., Products of infinite groups (in Russian), Akad. Nauk Ukrain. Inst. Mat. Kiev, Preprint 82.53 (1982).

[46] Sysak, Ya.P., Linear groups that are decomposable into products of locally nilpotent subgroups, Uspekhi Mat. Nauk 45 (1990), 165-166.

[47] Sysak, Ya.P., Radical modules over groups of finite rank (in Russian), Akad. Nauk Ukrain. Inst. Mat. Kiev, Preprint 89.18 (1989).

[48] Sysak, Ya.P., Radical modules over hyperabelian groups of finite torsion-free rank, J. Group Theory 1 (1998), 189–202.

[49] Sysak, Ya.P., Some examples of factorized groups and their relation to ring theory. In: Infinite Groups 1994, Proc. Internat. Conference, 257-269, Walter de Gruyter, Berlin–New York 1995.

[50] Valitskas, A.I., Embedding rings in radical rings and rational identities of radical algebras, Transl. Ser. 2, Amer. Math. Soc. 156 (1993), 125-195.

[51] Watters, J.F., On the adjoint group of a radical ring, J. London Math. Soc. 43 (1968), 725-729.

[52] Wilson, J.S., Soluble products of minimax groups and nearly surjective derivations, J. Pure Appl. Algebra 53 (1988), 297-318.

[53] Wilson, J.S., Soluble groups which are products of groups of finite rank, J. London Math. Soc. (2) 40 (1989), 405-419.

[54] Tasic, V., On unit groups of Lie centre-by-metabelian algebras, J. Pure Appl. Algebra 78 (1992), 195-201.

[55] Zaitsev, D.I., Factorizations of polycyclic groups, Mat. Zametki 29 (1981), 481-490.

HOMOGENEOUS INTEGRAL TABLE ALGEBRAS OF DEGREES TWO, THREE AND FOUR WITH A FAITHFUL ELEMENT

ZVI ARAD

Department of Mathematics, Bar-Ilan University, Ramat Gan, Israel

Introduction

The algebras mentioned in the title were introduced and recently classified in a series of papers [1], [6], [3] [8], [9], [14] and [15]. The purpose of this article is to give a survey of the main results of this topic.

The main theorems will be stated in this survey after we review the necessary definitions and describe the algebras which arise.

Homogeneous Integral Table Algebras (HITA) of degree 2 are classified in Theorem 1. HITA of degree 3 are classified in Theorems 2, 3 and 4 and of degree 4 in Theorem 5.

Most of the basic definitions which are reviewed here may be found in [1], [2], [7] and [11]. Throughout, $\mathbb{C}$ denotes the complex numbers, $\mathbb{R}$ the reals, $\mathbb{R}^+$ the positive reals and $\mathbb{Z}^+$ the positive integers.

Definition 1.1 ([1], [2], [8], [11]) Let $\mathbf{B} = \{b_1, b_2, \ldots, b_k\}$ be a basis of a finite dimensional, associative and commutative algebra A over the complex field $\mathbb{C}$, with identity element $1 = 1_A = b_1$. Then $(A, \mathbf{B})$ is a *table algebra* and $\mathbf{B}$ is a *table basis* if and only if the following hold:

(i) For all i, j, m, $b_i b_j = \sum_{m=1}^{k} \beta_{ijm} b_m$, with β_{ijm} a nonnegative real number.

(ii) A has an algebra automorphism (denoted by $\overline{}$) of order dividing 2, such that $b_i \in \mathbf{B}$ implies that $\overline{b_i} \in \mathbf{B}$. (Then $\bar{i}$ is defined by $b_{\bar{i}} = \overline{b_i}$, and $b_i \in \mathbf{B}$ is called *real* if $i = \bar{i}$.)

(iii) For all i, j, $\beta_{ij1} \neq 0$ if and only if $j = \bar{i}$.

By [1, Lemma 2.9], there is an algebra homomorphism $f : A \to \mathbb{C}$ such that $f(b_i) = f(\overline{b_i}) \in \mathbb{R}^+$ for all i. Such a map f is uniquely determined by the orthogonality relations which hold for $(A, \mathbf{B})$. The positive real numbers $f(b_i)$, $1 \leq i \leq k$, are called the *degrees* of $(A, \mathbf{B})$.

Definition 1.2 ([11]) A table algebra $(A, \mathbf{B})$ is called *integral* iff all the structure constants β_{ijm} and all the degrees $f(b_i)$ are rational integers.

We abbreviate "integral table algebra" as ITA. Any finite group G yields two examples of ITA's: $(Z(\mathbb{C}G), \mathrm{Cla}(G))$, the center of the group algebra, with table basis the set of sums $\hat{C}$ of G-conjugacy classes C, with automorphism $\overline{}$ extended linearly from inversion in G, and with degrees $f(\hat{C}) = |C|$ for all $\hat{C} \in \mathrm{Cla}(G)$; and $(Ch(G), \mathrm{Irr}(G))$, the ring of complex valued class functions on G, with table basis the set of irreducible characters of G, with automorphism $\overline{}$ extended linearly from complex conjugation of characters, and with degrees $f(\chi) = \chi(1)$ for all $\chi \in \mathrm{Irr}(G)$. Another example is the Bose-Mesner algebra of a commutative association scheme, with table basis the set of adjacency matrices [7, Section II.2].

Definition 1.3 ([14]) [14] A table algebra $(A, \mathbf{B})$ is called *homogeneous* (of degree λ) iff $|\mathbf{B}| > 1$ and, for some fixed $\lambda \in \mathbb{R}^+$, degree $f(b) = \lambda$ for all $b \in \mathbf{B} \setminus \{1\}$.

Any table algebra may be *rescaled* (replacing each table basis element by a positive scalar multiple) to one which is homogeneous, and any ITA can be rescaled to a homogeneous ITA [14, Theorem 1]. Therefore a classification theorem for all homogeneous integral table algebras (HITA) is an impossible mission. But as this survey shows one can classify such HITA for small degrees.

Definition For $a \in A$ we define

$$\mathrm{Supp}_{\mathbf{B}}(a) = \{b_i \mid b_i \in \mathbf{B} \text{ such that } \lambda_i \neq 0 \text{ where } a = \sum_{i=1}^{K} \lambda_i b_i\}.$$

Let $(A, \mathbf{B})$ be a table algebra. A nonempty subset $\mathbf{C} \subseteq \mathbf{B}$ is called a *table subset* (or a *C-subset*) of $\mathbf{B}$ iff $\mathrm{Supp}_{\mathbf{B}}(b_i b_j) \subseteq \mathbf{C}$ for all $b_i, b_j \in \mathbf{C}$. Any table subset is stable under $\overline{}$ and contains 1_A [1, Proposition 2.7], [11, Proposition 2.19]. For any $c \in \mathbf{B}$, the set $\mathbf{B}_c$ defined by

$$\mathbf{B}_c := \bigcup_{n=1}^{\infty} \mathrm{Supp}_{\mathbf{B}}(c^n)$$

is easily seen to be a table subset of $\mathbf{B}$, called the *table subset generated by c*.

An element c of $\mathbf{B}$ is called *faithful* iff $\mathbf{B}_c = \mathbf{B}$. For any finite group G, $\hat{C} \in \mathrm{Cla}(G)$ is faithful iff $\langle C \rangle = G$, and $\chi \in \mathrm{Irr}(G)$ is faithful iff χ is faithful in the usual character-theoretic sense. Also, $c \in \mathbf{B}$ is called *linear* iff $\mathrm{Supp}_{\mathbf{B}}(c^n) = \{1\}$ for some $n > 0$. This is equivalent to $\mathrm{Supp}_{\mathbf{B}}(c\bar{c}) = \{1\}$ [1, Proposition 3.2]. (Note $\hat{C} \in \mathrm{Cla}(G)$ is linear iff $C \subseteq Z(G)$ iff $|C| = 1$, and $\chi \in \mathrm{Irr}(G)$ is linear iff $\chi(1) = 1$.) A table subset of $\mathbf{B}$ is called *abelian* iff each of its elements is linear. The set of all linear elements of $\mathbf{B}$, denoted $\mathbf{L}(\mathbf{B})$, is a table subset [1, Proposition 3.2].

Two table algebras $(A, \mathbf{B})$ and $(A', \mathbf{B}')$ are called *isomorphic* (denoted $\mathbf{B} \simeq \mathbf{B}'$) when there exists an algebra isomorphism $\psi : A \to A'$ such that $\psi(\mathbf{B})$ is a rescaling of $\mathbf{B}'$; and the algebras are called *exactly isomorphic* (denoted $\mathbf{B} \simeq_x \mathbf{B}'$) when $\psi(\mathbf{B}) = \mathbf{B}'$ [11, Section 1]. So $\mathbf{B} \simeq_x \mathbf{B}'$ means that $\mathbf{B}$ and $\mathbf{B}'$ yield the same structure constants.

If H is any finite abelian group then $(\mathbb{C}H, H)$ is an abelian table algebra, and [1, Theorem A] shows that any abelian table algebra is a rescaling of one of this form. Note that $(\mathbb{C}H, H)$ is a homogeneous ITA of degree 1. On the other hand, if $(A, \mathbf{B})$ is a homogeneous ITA of degree 1, then by [11, Proposition 5.9], $\mathbf{B}$ is abelian. Thus by [1, Theorem A], $(A, \mathbf{B}) \simeq_x (\mathbb{C}H, H)$ for some finite abelian group H.

Homogeneous integral table algebras of degree 2 with a faithful real element

In [9] one can also find some examples of homogeneous table algebras which are needed for the classification theorem of homogeneous integral table algebras of degree 2.

Example 1.4 ([14]) Let H be a finite abelian group which admits a fixed-point-free action by the cyclic group $\mathbb{Z}_n$, for some $n > 0$. Let $G = \mathbb{Z}_n \ltimes H$, and

$$\mathbf{O} = \mathbf{O}(n, H) := \{\hat{C} \in \mathrm{Cla}(G) \mid C \subseteq H\},$$

the set of sums over the orbits of $\mathbb{Z}_n$ on H. Then $\mathbf{O}$ is a table subset of $\mathrm{Cla}(G)$. For $A = \langle \mathbf{O} \rangle$, $(A, \mathbf{O})$ is an ITA which is homogeneous of degree n. In the special case $\mathbf{O}(2, \mathbb{Z}_{2n+1})$, G is the dihedral group of order $2(2n+1)$.

Example 1.5 ([14]) Let G be the dihedral group of order $4n$ for some $n \in \mathbb{Z}^+$, and let $\mathbb{Z}_{2n} \simeq H \triangleleft G$. Let $\mathbf{B} = \{\hat{C} \in \mathrm{Cla}(G) \mid C \subseteq H\}$. Thus for $H = \langle x \rangle$, $\mathbf{B} = \{1, c_1, c_2, \ldots, c_n\}$, where $c_i = x^i + x^{-i}$ for $1 \leq i < n$ and $c_n = x^n$. Define $\mathbf{D}_{2n} := \{1, c_1, c_2, \ldots, c_{n-1}, 2c_n\}$, a rescaling of $\mathbf{B}$, and let $A = \langle \mathbf{D}_{2n} \rangle$. Then $(A, \mathbf{D}_{2n})$ is a homogeneous ITA of degree 2.

Example 1.6 ([14]) Let $(A, \mathbf{B})$ be a table algebra which is homogeneous of degree $\lambda \in \mathbb{R}^+$. Let H be an abelian group and $C := \mathbb{C}H$. Then $A \otimes_{\mathbb{C}} C$, with basis $\mathbf{B} \otimes H := \{b \otimes h \mid b \in \mathbf{B},\ h \in H\}$ is a table algebra, where $\overline{b \otimes h} = \bar{b} \otimes h^{-1}$ and $f(b \otimes h) = f(b)$ for all $b \in \mathbf{B}$, $h \in H$ [11, Example 1.5]. Define

$$\mathbf{B} \otimes H' := \{b \otimes h \mid b \in \mathbf{B} \setminus \{1\},\ h \in H\} \cup \{1 \otimes \lambda h \mid h \in H\},$$

a rescaling of $\mathbf{B} \otimes H$. Then $(A \otimes C, \mathbf{B} \otimes H')$ is homogeneous of degree λ; and if $(A, \mathbf{B})$ is an ITA then so is $(A \otimes C, \mathbf{B} \otimes H')$. We may regard $\{1\} \otimes H'$ as simply $H' := \{1\} \cup \{\lambda h \mid h \in H \setminus \{1\}\}$.

It was proved in [9, Proposition 3.3] that $\mathbf{O}(2, \mathbb{Z}_{2n+1}) \otimes \mathbb{Z}'_m$ has a faithful element for all $n, m \in \mathbb{Z}^+$, but that $\mathbf{D}_{2n} \otimes \mathbb{Z}'_m$ has a faithful element iff m is odd. Furthermore, $\mathbf{O}(2, \mathbb{Z}_{2n+1}) \otimes \mathbb{Z}'_m \simeq_x \mathbf{D}_{2s} \otimes \mathbb{Z}'_l$ for some $n, m, s, l \in \mathbb{Z}^+$ iff $l = m/2$ and is odd and $s = 2n+1$.

Example 1.7 ([9]) There are families $\mathbf{H}(2n, m), \mathbf{J}(2n, m), \mathbf{M}(n, m)$ and $\mathbf{N}(n, m)$, parametrized by $n, m \in \mathbb{Z}^+$ (m even in the case of $\mathbf{J}(2n, m)$ and $\mathbf{N}(n, m)$, and $n \geq 2$ for $\mathbf{M}(n, m)$ and $\mathbf{N}(n, m)$), which are table bases for classes of homogeneous ITA's of degree 2. Their explicit definitions are given in [9, Section 3] and we note here only that $|\mathbf{H}(2n, m)| = (2n+1)m, |\mathbf{J}(2n, m)| = (2n+1)m/2, |\mathbf{M}(n, m)| = (n+1)m, |\mathbf{N}(n, m)| = (n+1)m/2$; and that $\mathbf{H}(2n, m)$ and $\mathbf{J}(2n, m)$ each have $\mathbf{D}_{2n}$ as a table subset, while $\mathbf{M}(n, m)$ and $\mathbf{N}(n, m)$ are nilpotent [11, Definition 1.16]. Proposition 3.3 of [9] lists all the isomorphic pairs which occur among the members of these families. In particular, $\mathbf{H}(2n, m) \simeq_x \mathbf{D}_{4n} \otimes \mathbb{Z}'_m$ when m is odd, but when m is even neither $\mathbf{H}(2n, m)$ nor $\mathbf{J}(2n, m)$ is exactly isomorphic to any $\mathbf{O}(2, \mathbb{Z}_{2s+1}) \otimes \mathbb{Z}'_t$ or $\mathbf{D}_{2s} \otimes \mathbb{Z}'_t$.

Now we state the main result of [9].

Theorem 1 *Let* $(A, \mathbf{B})$ *be an integral table algebra which is homogeneous of degree* 2, *and such that* $\mathbf{B}$ *contains a faithful element. Then* $\mathbf{B}$ *is exactly isomorphic to one of the following, for some* $n, m \in \mathbb{Z}^+ : \mathbb{Z}'_m, \mathbf{O}(2, \mathbb{Z}_{2n+1}) \otimes \mathbb{Z}'_m$, *or for* $n \geq 2$, *one of* $\mathbf{H}(2n, m)$, $\mathbf{J}(2n, m)$, *or* $\mathbf{N}(n, m)$.

The proof is given in Section 4 of [9].

Homogeneous integral table algebras of degree 3 with a faithful real element

Let $(A, \mathbf{B})$ be a table algebra and $\mathbf{C}$ a table subset of $\mathbf{B}$. There is an idempotent, denoted $e_{\mathbf{C}}$, which equals a positive real scalar times $\sum_{b_i \in \mathbf{C}}(f(b_i)/\beta_{i\bar{i}1})b_i$ [8, Corollary 3.13]. Also, $\{\mathrm{Supp}_{\mathbf{B}}(e_{\mathbf{C}}b_i) | b_i \in \mathbf{B}\}$ partitions $\mathbf{B}$ into disjoint classes and in fact, $b_j \in \mathrm{Supp}_{\mathbf{B}}(e_{\mathbf{C}}b_i)$ iff $e_{\mathbf{C}}b_j = (f(b_j)/f(b_i))e_{\mathbf{C}}b_i$ [8, Theorem 1, (1*)]. We call $\mathrm{Supp}_{\mathbf{B}}(e_{\mathbf{C}}b_i)$ the $\mathbf{C}$-*class* of b_i [11, Definition 2.1].

Definition 1.8 **([8])** Let $A, \mathbf{B}, \mathbf{C}$ and $e := e_{\mathbf{C}}$ be as above. Define

$$\mathbf{B}/\mathbf{C} := \{e\} \cup \{eb_i | b_i \in \mathbf{B} \setminus \mathbf{C} \text{ and } f(b_i) \leq f(b_j) \text{ for all } b_j \in \mathrm{Supp}_{\mathbf{B}}(eb_i)\}.$$

It follows from [8, Theorems 1,2] that $(Ae, \mathbf{B}/\mathbf{C})$ is a table algebra, is an epimorphic image of $(A, \mathbf{B})$ (as defined in [8, Definition 1.8]), and up to isomorphism is the unique such image where $\mathbf{C}$ comprises exactly those elements of $\mathbf{B}$ which map to positive scalar multiples of the identity. Then $(Ae, \mathbf{B}/\mathbf{C})$ is called the *quotient table algebra determined by* $\mathbf{C}$. The positive-valued homomorphism on $\mathbf{B}/\mathbf{C}$ is just the restriction of f. If $(A, \mathbf{B})$ is an ITA which is homogeneous of degree λ, then it follows from [8, Lemma 5.4] that $(Ae_{\mathbf{C}}, \mathbf{B}/\mathbf{C})$ is again an ITA which is homogeneous of degree λ.

Composition series and *composition factors* are defined in the obvious way for any table basis, and a Jordan-Holder theorem holds for such chains of table subsets [8, Theorem 5]. A table algebra $(A, \mathbf{B})$ is called *nilpotent* iff every composition factor of $\mathbf{B}$ is abelian [8, Definition 1.16].

Definition 1.9 The *upper central series* of a table algebra $(A, \mathbf{B})$ is the chain of table subsets $\mathbf{L}^{(i)}(\mathbf{B})$ for all $i \geq 0$, defined as follows: $\mathbf{L}^{(0)}(\mathbf{B}) := \{1\}$, $\mathbf{L}^{(1)}(\mathbf{B}) := \mathbf{L}(\mathbf{B})$, and recursively, $\mathbf{L}^{(i+1)}(\mathbf{B})$ is the table subset of $\mathbf{B}$ such that $\mathbf{L}^{(i+1)}(\mathbf{B})/\mathbf{L}^{(i)}(\mathbf{B}) = \mathbf{L}(\mathbf{B}/\mathbf{L}^{(i)}(\mathbf{B}))$ (see [8, Theorem 3]).

Thus, each $\mathbf{L}^{(i)}(\mathbf{B})$ is nilpotent, the $\mathbf{L}^{(i)}(\mathbf{B})$ form an increasing chain, and for some n, and all $j \geq 0$, $\mathbf{L}^{(n)}(\mathbf{B}) = \mathbf{L}^{(n+j)}(\mathbf{B})$. If there is no ambiguity, we abbreviate $\mathbf{L}^{(i)}(\mathbf{B})$ as $\mathbf{L}^{(i)}$.

For any table algebra $(A, \mathbf{B})$, it is easy to verify that the central series term $\mathbf{L}^{(n)}$ with $\mathbf{L}^{(n)} = \mathbf{L}^{(n+1)}$ coincides with the *nilpotent radical* $\mathbf{B}^{\mathrm{nil}}$, the unique maximal nilpotent table subset of $\mathbf{B}$ (as in [8, Theorem 7]). In particular, $\mathbf{L}(\mathbf{B}/\mathbf{B}^{\mathrm{nil}}) \cong_x \{1\}$.

In [14] and [15] one can find a description of, and notation for, some homogeneous table algebras. These will include all which occur in the conclusion of Theorem 2 below. When a basis vector is a product of a nonlinear basis element and a (scalar multiple of a) linear basis element, we list the vector as the product itself, and do not introduce another letter for it. We also omit the definition of products which are determined uniquely by the requirements of associativity and commutativity, and which can be obtained by multiplying some given equation through by some linear element.

Example 1.10 Example 1.6 or [14, Example 1.16].

Example 1.11 ([15]) Let $(A, \mathbf{B})$ be a homogeneous table algebra of degree $\lambda > 0$. Assume that $\mathbf{L(B)} \cong_x \mathbb{Z}'_m$ for some integer $m > 0$. Thus, for $\mathbb{Z}_m = \langle u \rangle$,

$$\mathbf{L(B)} = \{1\} \dot{\cup} \{\lambda u^j |\, 0 < j < m\}.$$

Let $\mathbb{Z}_{2m} = \langle w \rangle$ be another cyclic group of order $2m$. Then, as in the previous example, $\mathbf{C} := \mathbf{B} \otimes \mathbb{Z}'_{2m}$ is the basis of a homogeneous table algebra of degree λ, with $|\mathbf{C}| = 2m|\mathbf{B}|$. Let $x = \lambda u \otimes w^2 \in \mathbf{C}$, $\mathbf{X} = \mathbf{C}_x$, and idempotent $e = e_{\mathbf{X}}$. Then $\mathbf{X} \cong_x \mathbb{Z}'_m$ and $(1 \otimes w^2)e = (u^{-1} \otimes 1)e$. Define

$$\mathbf{B} * \mathbb{Z}_{2m} := (\mathbf{B} \otimes \mathbb{Z}_{2m})'/\mathbf{X}.$$

So $\mathbf{B} * \mathbb{Z}_{2m}$ is homogeneous of degree λ, and is integral if $\mathbf{B}$ is. Note that

$$\mathbf{B} * \mathbb{Z}_{2m} = \{(b_i \otimes 1)e \,|b_i \in \mathbf{B}\} \dot{\cup} \{(b_i \otimes w)e |\, b_i \in \mathbf{B} \backslash \{1\}\} \dot{\cup} \{(\lambda 1 \otimes w)e\}$$

where for all i, j,

$$(b_i \otimes w)e \cdot (b_j \otimes w)e = (b_i b_j u^{-1} \otimes 1)e.$$

In particular, the isomorphism class of $\mathbf{B} * \mathbb{Z}_{2m}$ is independent of the choice of generator w for $\mathbb{Z}_{2m}$. Also, $\mathbf{L}(\mathbf{B} * \mathbb{Z}_{2m}) \cong_x \mathbb{Z}'_{2m}$. Since $\mathbf{L}((\mathbf{B} \otimes \mathbb{Z}_2)') \cong \mathbb{Z}_m \times \mathbb{Z}_2$, it follows that $(\mathbf{B} \otimes \mathbb{Z}_2)' \ncong \mathbf{B} * \mathbb{Z}_{2m}$ when m is even. But if m is odd, then $(b_i u^{-(m-1)/2} \otimes w)e = (b_i \otimes w^m)e$ for all $b_i \in \mathbf{B}$ yields that $(\mathbf{B} \otimes \mathbb{Z}_2)' \cong_x \mathbf{B} * \mathbb{Z}_{2m}$. It is easy to see that $\mathbf{B} * \mathbb{Z}_{2m}/\mathbf{L}(\mathbf{B} * \mathbb{Z}_{2m}) \cong_x \mathbf{B}/\mathbf{L(B)}$.

Example 1.12 ([15]) Fix positive real λ, μ with $0 < \mu \leq \lambda$. Define the basis $\mathbf{V}(2, \lambda, \mu) := \{1, v\}$, where $v^2 = \mu\lambda 1 + (\lambda - \mu)v$. It is easy to check that $\mathbf{V}(2, \lambda, \mu)$ is a distinguished basis for a homogeneous table algebra of degree λ, and that the algebra is integral when λ, μ are integers. Of course, v is faithful and real. Define $\mathbf{V}_2 := \mathbf{V}(2, 3, 2)$, so that

$$\mathbf{V}_2 = \{1, v\}, \quad \text{where } v^2 = 6 \cdot 1 + v.$$

Example 1.13 ([14, Example 4.2]) Fix positive real λ, μ, β with $\lambda \geq \mu\beta$, $\lambda \geq \beta + 1$, and $\lambda + \mu^2\beta \geq \mu(\lambda + 1)$. Define $\mathbf{V}(3, \lambda, \mu, \beta) := \{1, b, c\}$ as a basis for a 3-dimensional $\mathbb{C}$-vector space A, and define multiplication on A so that

$$\begin{aligned}
b^2 &= \lambda 1 + (\lambda - \beta - 1)b + \beta c, \\
bc &= cb = \mu\beta b + (\lambda - \mu\beta)c, \\
c^2 &= \mu\lambda 1 + \mu(\lambda - \mu\beta)b + (\lambda + \mu^2\beta - \mu - \mu\lambda)c.
\end{aligned}$$

Then, as shown in [14], $(A, \mathbf{V}(3,\lambda,\mu,\beta))$ is a homogeneous table algebra of degree λ, with $\overline{} = id_A$. If λ, μ, β are integers then clearly $\mathbf{V}(3,\lambda,\mu,\beta)$ is integral. Define $\mathbf{V}_3 := \mathbf{V}(3,3,1,2)$. Thus,

$$\mathbf{V}_3 = \{1, v_1, v_2\}, \text{ where } v_1^2 = 3\cdot 1 + 2v_2,\ v_1v_2 = 2v_1 + v_2,\ v_2^2 = 3\cdot 1 + v_1 + v_2.$$

Example 1.14 ([14, Example 1.26]) Fix a positive integer λ. They define two families $\mathbf{N}((2),\lambda)$ and $\mathbf{N}((3),\lambda)$, of nilpotent, homogeneous ITA's of degree λ. First, define as the basis for a $\mathbb{C}$-vector space A,

$$\mathbf{N}((2),\lambda) := \{1, \lambda u,\ \lambda u^2, \cdots, \lambda u^{\lambda-1}, b\},$$

where $\langle u\rangle = \mathbb{Z}_\lambda$, a cyclic group of order λ, $\lambda u^i b = \lambda b = b\cdot\lambda u^i$ for all i, and $b^2 = \lambda(1 + u + u^2 + \cdots + u^{\lambda-1})$. It is easy to see that these products extend to a bilinear multiplication on A which is associative and commutative. Then $(A, \mathbf{N}((2),\lambda))$ is an ITA, with $\overline{u^i} = u^{-i}$ and $\bar{b} = b$; and $\mathbf{N}((2),\lambda)$ is homogeneous of degree λ. Note that $\mathbf{L}(\mathbf{N}((2),\lambda)) = \{1, \lambda u, \cdots \lambda u^{\lambda-1}\}$, and $\mathbf{L}^{(2)}(\mathbf{N}((2),\lambda)) = \mathbf{N}((2),\lambda)$. Next, define

$$\mathbf{N}((3),\lambda) := \{1, \lambda u, \lambda u^2, \cdots, \lambda u^{\lambda-1}, v_1, v_2, \cdots, v_{\lambda-1}, b, ub, u^2b, \cdots, u^{\lambda-1}b\},$$

as a basis for a complex vector space A of dimension $3\lambda - 1$. Define multiplication (with u^ib already given) so that $\langle u\rangle = \mathbb{Z}_\lambda$, and

$$v_iv_j = \begin{cases} \lambda v_{i+j} \text{ (read } i+j \bmod \lambda) & \text{if } i+j \neq \lambda; \\ \lambda\cdot\sum_{i=0}^{\lambda-1} u^i & \text{if } i+j = \lambda; \end{cases}$$

$$u^iv_j = v_ju^i = v_j \quad \text{for all } i, j;$$

$$b^2 = \lambda 1 + \sum_{j=1}^{\lambda-1} v_j;$$

$$v_jb = bv_j = b\cdot\sum_{i=0}^{\lambda-1} u^i \quad \text{for all } j.$$

This extends to a multiplication in A which is associative and commutative, $(A, \mathbf{N}((3),\lambda))$ is an ITA which is homogeneous of degree λ, where $\overline{u^i} = u^{-i}$, $\overline{v_i} = v_{\lambda-i}$, $\bar{b} = b$ [14, Proposition 4.7]. Note that $\mathbf{L}(\mathbf{N}((3),\lambda)) = \{1, \lambda u, \cdots, \lambda u^{\lambda-1}\}$, $\mathbf{L}^{(2)}(\mathbf{N}((3),\lambda)) = \{1, \lambda u, \cdots, \lambda u^{\lambda-1},\ v_1, v_2, \cdots, v_{\lambda-1}\}$, and $\mathbf{L}^{(3)}(\mathbf{N}((3),\lambda)) = \mathbf{N}((3),\lambda)$. Observe that b is real and faithful in each case. Define

$$\mathbf{N}_4 := \mathbf{N}((2),3) \text{ and } \mathbf{N}_8 := \mathbf{N}((3),3),$$

so that $|\mathbf{N}_4| = 4$, $|\mathbf{N}_8| = 8$.

Example 1.15 ([14, Example 4.3]) Fix integers n, m with $n \geq 0$, $m \geq 1$. Fix real $\lambda > 1$. Let $\alpha = (\lambda - 1)/2$, $\beta = (\lambda + 1)/2$. Let A be a $\mathbb{C}$-vector space of dimension $m(n+2)$, and consider a basis

$$\mathbf{B} = \mathbf{T}(n,\lambda,m) := \{1\} \dot\cup \{\lambda u^j | 0 < j < m\} \dot\cup \{v_iu^j | 0 \leq i \leq n,\ 0 \leq j < m\},$$

with a multiplication defined so that $\{1\} \cup \{u^j | 0 < j < m\} = \langle u \rangle \cong \mathbb{Z}_m$ is a cyclic group of order m, and for all $0 \leq i, j \leq n$,

$$v_i v_j = \begin{cases} \alpha v_{i+j} + \beta v_{i+j+1}, & i+j < n \\ \lambda u + \alpha u v_0 + \alpha v_n, & i+j = n \\ \alpha u v_{i+j-n} + \beta v_{i+j-n-1}, & i+j > n. \end{cases}$$

We extend this multiplication to all of $\mathbf{B}$, according to the convention above, and bilinearly to all of A. Then $(A, \mathbf{B})$ is a homogeneous table algebra of degree λ, and is integral if λ is an odd integer [14, Proposition 4.5]. Furthermore, $\overline{u^i} = u^{-i}$ and $\bar{v}_j = v_{n-j}u^{-1}$ for all $0 \leq i < m$, $0 \leq j \leq n$. Thus if n is even and m is odd, $v_{n/2}u^{(m-1)/2}$ is a real, faithful element. Also, $\mathbf{L}(\mathbf{B}) = \mathbf{B}^{\text{nil}} = \{1\} \cup \{\lambda u^j | 0 < j < m\} \cong_x \mathbb{Z}'_m$, and $\mathbf{B}/\mathbf{L}(\mathbf{B}) \cong_x \mathbf{T}(n, \lambda, 1)$. Define $\mathbf{T}_n(\lambda) := \mathbf{T}(n, \lambda, 1)$, so that

$$\mathbf{T}(n, \lambda, m)/\mathbf{T}(n, \lambda, m)^{\text{nil}} \cong_x \mathbf{T}_n(\lambda).$$

Note that if m is even, then $\mathbf{T}(n, \lambda, m)$ has no real, faithful element. But if n is even, and for any m, consider $\mathbf{T}(n, \lambda, m) * \mathbb{Z}'_{2m}$ as in Example 1.11 Here, the basis element $(v_{n/2} \otimes w)e$ is real and faithful.

Finally, observe that $\mathbf{T}_0(\lambda) = \{1, b_0\}$ with $b_0^2 = \lambda 1 + (\lambda - 1)b_0$.

Example 1.16 We present here the bases for six specific homogeneous ITA's of degree 3. Verification of their existence, and of the other claims, is straightforward.

$$\mathbf{V}_4 := \{1, v_1, v_2, v_3\}, \text{ where } v_1^2 = 3\cdot 1 + v_1 + v_2,\ v_1 v_2 = v_1 + 2v_3,$$
$$v_1 v_3 = v_3 + 2v_2,\ v_2^2 = 3\cdot 1 + 2v_2,\ v_2 v_3 = v_3 + 2v_1,\ v_3^2 = 3\cdot 1 + v_1 + v_2.$$

In $\mathbf{V}_4$, all elements are real, and v_1 and v_3 are faithful.

$$\mathbf{W}_3 := \{1, b, 3z\}, \text{ where } z^2 = 1,\ bz = b,\ b^2 = 3\cdot 1 + b + 3z.$$

All the elements of $\mathbf{W}_3$ are real, and b is faithful. Furthermore, $\mathbf{L}(\mathbf{W}_3) = \mathbf{W}_3^{\text{nil}} = \{1, 3z\}$, and $\mathbf{W}_3/\mathbf{W}_3^{\text{nil}} \cong_x \mathbf{V}_2$.

$$\mathbf{W}_4 := \{1, b, 3z, bz\}, \text{ where } z^2 = 1,\ b^2 = 3\cdot 1 + b + bz.$$

All the elements of $\mathbf{W}_4$ are real; b, bz are faithful; $\mathbf{L}(\mathbf{W}_4) = \mathbf{W}_4^{\text{nil}} = \{1, 3z\}$; $\mathbf{W}_4/\mathbf{W}_4^{\text{nil}} \cong_x \mathbf{T}_0(3)$.

$$\mathbf{W}_6 := \{1, b, bu, 3u, 3u^2, 3\bar{u}\}, \text{ where } u^3 = \bar{u},\ u\bar{u} = 1, bu^2 = b,\ b^2 = 3\cdot 1 + 3u^2 + bu.$$

In $\mathbf{W}_6$, b is real and faithful; $\mathbf{L}(\mathbf{W}_6) = \mathbf{W}_6^{\text{nil}} = \{1, 3u, 3u^2, 3\bar{u}\} \cong_x \mathbb{Z}'_4$; $\mathbf{W}_6/\mathbf{W}_6^{\text{nil}} \cong_x \mathbf{V}_2$.

$$\mathbf{W}_7 := \{1, b_1, b_1 w, b_1 \bar{w}, b_2, 3w, 3\bar{w}\}, \text{ where } w^2 = \bar{w},\ w\bar{w} = 1,\ b_2 w = b_2 \bar{w} = b_2,$$
$$b_1^2 = 3\cdot 1 + b_1 + b_2,\ b_1 b_2 = b_1 + b_1 w + b_1 \bar{w}, b_2^2 = 3\cdot 1 + 3w + 3\bar{w}.$$

In $\mathbf{W}_7$, b_1 is real and faithful; $\mathbf{L}(\mathbf{W}_7) = \{1, 3w, 3\bar{w}\}$; $\mathbf{W}_7^{\text{nil}} = \mathbf{L}(\mathbf{W}_7) \cup \{b_2\}$, $\mathbf{W}_7/\mathbf{W}_7^{\text{nil}} \cong_x \mathbf{V}_2$.

$$\mathbf{W}_9 := \{b_i w^j | i = 1,2; j = 0,1,2\}\dot{\cup}\{1, 3w, 3\bar{w}\}, \text{ where } w^2 = \bar{w},\ w\bar{w} = 1,$$
$$b_1^2 = 3\cdot 1 + b_2 w + b_2\bar{w},\ b_2^2 = 3\cdot 1 + b_1 + b_2,\ b_1 b_2 = b_2 + b_1 w + b_1\bar{w}.$$

In $\mathbf{W}_9$, b_1 and b_2 are real and faithful; $\mathbf{L}(\mathbf{W}_9) = \mathbf{W}_9^{\text{nil}} = \{1, 3w, 3\bar{w}\}$; $\mathbf{W}_9/\mathbf{W}_9^{\text{nil}} \cong_x \mathbf{V}_3$.

The central result of this article may now be stated.

Theorem 2 *Let $(A, \mathbf{B})$ be a homogeneous integral table algebra of degree three such that $\mathbf{B}$ contains a real and faithful element. Then one of the following must hold:*

(i) $\mathbf{B} = \mathbf{B}^{nil} \cong_x \mathbf{Z}_2', \mathbf{N}_4$, *or* $\mathbf{N}_8$;

(ii) $\mathbf{B}/\mathbf{B}^{nil} \cong_x \mathbf{V}_2$, *and* $\mathbf{B} \cong_x \mathbf{V}_2, (\mathbf{V}_2 \otimes \mathbf{Z}_2)', \mathbf{W}_3, (\mathbf{W}_3 \otimes \mathbf{Z}_2)', \mathbf{W}_3 * \mathbf{Z}_4, \mathbf{W}_6, \mathbf{W}_7$, *or* $(\mathbf{W}_7 \otimes \mathbf{Z}_2)'$;

(iii) $\mathbf{B}/\mathbf{B}^{nil} \cong_x \mathbf{V}_3$, *and* $\mathbf{B} \cong_x \mathbf{V}_3, (\mathbf{V}_3 \otimes \mathbf{Z}_2)', \mathbf{W}_9$, *or* $(\mathbf{W}_9 \otimes \mathbf{Z}_2)'$;

(iv) $\mathbf{B}/\mathbf{B}^{nil} \cong_x \mathbf{V}_4$, *and* $\mathbf{B} \cong_x \mathbf{V}_4$ *or* $(\mathbf{V}_4 \otimes \mathbf{Z}_2)'$;

(v) $\mathbf{B}/\mathbf{B}^{nil} \cong_x \mathbf{T}_n(3)$ *for some even* $n \geq 0$, *and* $\mathbf{B} \cong_x \mathbf{W}_4$, $(\mathbf{W}_4 \otimes \mathbf{Z}_2)'$, *either* $\mathbf{T}(n,3,m)$ *or* $(\mathbf{T}(n,3,m) \otimes \mathbf{Z}_2)'$ *for some odd* $m > 0$, *or* $\mathbf{T}(n,3,m) * \mathbf{Z}_{2m}$ *for some even* $m > 0$.

On antisymmetric homogeneous integral table algebras of degree 3

Throughout this section $(A, \mathbf{B})$ will denote a homogeneous integral table algebra (ITA) of degree 3. If $(A, \mathbf{B})$ contains a real element, say b, then the structure of $\mathbf{B}_b$ was given in Theorem 2. In this case, the quotient $\mathbf{B}/\mathbf{B}_b$ is once more an integral homogeneous table algebra of degree 3 (Lemma 5.4 of [11]). If $\mathbf{B}/\mathbf{B}_b$ again contains a real element, then we can factor out this algebra once more. Continuing this process, we may build a composition series $\mathbf{B}_0 = \{1\} \leq \mathbf{B}_1 \leq \ldots \leq \mathbf{B}_l \leq \mathbf{B}$ of table subsets such that the structure of $\mathbf{B}_i/\mathbf{B}_{i-l}$, $i \leq l$ is given by Theorem 2. If $\mathbf{B} \neq \mathbf{B}_l$, then $\mathbf{B}/\mathbf{B}_l$ has no non-trivial real element. We call such algebras *antisymmetric*. The main purpose of this section is to state a classification theorem about antisymmetric homogeneous integral table algebras of degree 3.

Definition 1.17 ([15, Definition 2.7]) Let $(A, \mathbf{B})$ be an integral table algebra. An element $b \in \mathbf{B}$ is called standard iff $(b\bar{b}, b_1) = f(b)$. $\mathbf{B}$ is standard iff every element of $\mathbf{B}$ is standard.

It is easy to see that in this case every element $b \in \mathbf{B}$ is either linear or standard. Replacing, if necessary, $\mathbf{B}$ by $\mathbf{B}/\mathbf{B}^{\text{nil}}$, we may always assume that all elements of $\mathbf{B}$ are standard.

So, from now on, $(A, \mathbf{B})$ will be a homogeneous antisymmetric integral C-algebra of degree 3 (C-algebra is our abbreviation for standard table algebra). If $\mathbf{B}$ contains an element b with $b\bar{b} = 3 + b + \bar{b}$, then the structure of $\mathbf{B}_b$ is given by Theorem 4.3 of [15].

Using this definition we can state the following theorem:

Theorem 3 ([15, Theorem 3]) *Suppose that* $(A, \mathbf{B})$ *is a homogeneous ITA of degree* 3 *and that* $\mathbf{B}$ *is standard. Let* $b \in \mathbf{B}$ *with* $b \neq \bar{b}$ *and* $b\bar{b} = 3 \cdot b_1 + b + \bar{b}$. *Then* $\mathbf{B} \simeq T_n(3)$ *for some odd* $n > 0$.

So we shall assume that $\mathrm{supp}(b\bar{b}\} \cap \{b, \bar{b}\} = \emptyset$ for every $b \in \mathbf{B}\{1\}$.

A natural example of homogeneous antisymmetric C-algebras arises from group theory.

Example 1.18 Let G be an abelian group of odd order, whose automorphism group contains an element ϕ of order m without non-trivial fixed points. Let O_i, $i \in I$ be a complete set of nontrivial orbits of ϕ. Then the $\mathbb{Z}$-submodule of $\mathbb{Z}G$ spanned by $\underline{O}_i$, $i \in I$, is a Homogeneous Antisymmetric Integral C-Algebra (HAICA) of degree m. We shall call any AICA obtained in this way as *group-like* of degree m. ·

Theorem 4 ([6]) Let $(A, \mathbf{B})$ be a homogeneous AICA of degree 3. Assume, in addition, that it satisfies the following condition:

$$\forall b \in \mathbf{B}^{\#}(\mathrm{supp}(b\bar{b}) \cap \{b, \bar{b}\} = \emptyset).$$

Then $\mathbf{B}_b$ is group-like of degree 3 for all $b \in \mathbf{B}^{\#}$.

Recently, homogeneous integral table algebras $(A, \mathbf{B})$ of degree three which also have a faithful element and with $L(\mathbf{B}) = 1$ were determined to exact isomorphism by H. Blau in [12].

Homogeneous integral standard table algebras of degree 4 with a faithful element

Work on completing the following theorem is in progress by the authors:

Theorem 5 ([3]) Let $(A, \mathbf{B})$ be an integral standard table algebra which is homogeneous of degree 4 and assume that $\mathbf{B}$ contains a faithful element b. Then either $(A, \mathbf{B})$ is *group-like* of degree 4 and all basis elements of $\mathbf{B}$ are real or $(A, \mathbf{B})$ is one of the following:

(i) $\mathbf{B} = \{b_1, b_2, b_3\}$ where $b_2^2 = 4b_1 + 2b_2 + b_3$, $b_3^2 = 4b_1 + 3b_2$ and $b_2b_3 = b_2 + 3b_3$ or

(ii) $\mathbf{B} = \{b_1, b_2, b_3, b_4\}$ where $b_2^2 = 4b_1 + 2b_2 + b_3$, $b_2b_3 = b_2 + b_3 + 2b_4$, $b_2b_4 = 2b_3 + 2b_4$, $b_3^2 = 4b_1 + b_2 + 2b_4$, $b_3b_4 = 2b_2 + 2b_3$ and $b_4^2 = 4b_1 + 2b_2 + b_4$.

Recently, a classification of integral standard table algebras $(A, \mathbf{B})$ with $L(\mathbf{B}) = 1$ and with a faithful nonreal element of degree 3 was given in [4] and [13].

A large part of the classification of integral standard table algebras with $L(\mathbf{B}) = 1$ and a faithful nonreal element $b \in \mathbf{B}$ of degree 4 and without elements in $\mathbf{B}$ of degree 2 was done in [5]. Work on completing this classification is in progress by the authors. The last result [5] implies that standard homogeneous ITA's $(A, \mathbf{B})$ of degree 4 are real (i.e., $b = \bar{b}$ for every $b \in \mathbf{B}$), and this fact is a key factor in classifying such algebras [3]

Open Problems 1) To classify homogeneous standard integral table algebras of degree 5 with a faithful element. 2) To classify standard ITA's $(A, \mathbf{B})$ with a faithful nonreal element of degree 5 with $L(\mathbf{B}) = 1$ and without elements in $\mathbf{B}$ of degrees 2 and 3.

References

[1] Z. Arad and H. Blau, "On table algebras and applications to finite group theory", *J. Algebra* **138** (1991) 137-185.

[2] Z. Arad, H.I. Blau, J. Erez and E. Fisman, "Real table algebras and applications to finite groups of extended Camina-Frobenius type", *J. Algebra* **168** (1994) 615-647.

[3] Z. Arad, E. Fisman, A. Husam and M. Muzychuk, "Homogeneous integral table algebras of degree four", in preparation.

[4] Z. Arad, E. Fisman, A. Husam and M. Muzychuk, "Integral table algebras with a faithful nonreal element of degree three I", manuscript.

[5] Z. Arad, E. Fisman, A. Husam and M. Muzychuk, "Integral table algebras with a faithful nonreal element of degree four", in preparation.

[6] Z. Arad, E. Fisman, V. Miloslavsky and M. Muzychuk, "On antisymmetric homogeneous integral table algebras of degree three", manuscript (see [16] below).

[7] E. Bannai and T. Ito, *Algebraic Combinatorics I: Association Schemes*, Benjamin Cummings, Menlo Park, 1984.

[8] H.I. Blau, "Integral table algebras, affine diagrams, and the analysis of degree two", *J. Algebra* **178** (1995) 872-918.

[9] H.I. Blau, "Homogeneous integral table algebras of degree two", submitted.

[10] H.I. Blau "C-algebras and table algebras", Lecture notes of Theory Workshop, Wuhan, Oct. 1993, J. Hubei University **16** (1994) 119-34.

[11] H.I. Blau, "Quotient structures in C-algebras", *J. Algebra* **175** (1995) 24-64.

[12] H.I. Blau, "Homogeneous integral table algebras of degree three with no nontrivial linear elements", manuscript (see [16] below).

[13] H.I. Blau, "Integral table algebras with a faithful nonreal element of degree three II", manuscript.

[14] H.I. Blau and B. Xu, "On homogeneous integral table algebras", *J. Algebra*, to appear.

[15] H.I. Blau and B. Xu "Homogeneous integral table algebras of degree three with a faithful real element", manuscript (see [16] below).

[16] *Homogeneous Integral Table Algebras of Degree Three: A Trilogy*, A volume containing the above three papers: [4], [9] and [15], submitted.

A POLYNOMIAL-TIME THEORY OF BLACK BOX GROUPS I

LÁSZLÓ BABAI* and ROBERT BEALS†

*Department of Computer Science, University of Chicago, 1100 E. 58th St., Chicago, IL 60637-1504 U.S.A.

†Department of Mathematics, University of Arizona, Tucson AZ 85721-0001 U.S.A.

Abstract

We consider the asymptotic complexity of algorithms to manipulate matrix groups over finite fields. Groups are given by a list of generators. Some of the rudimentary tasks such as membership testing and computing the order are not expected to admit polynomial-time solutions due to number theoretic obstacles such as factoring integers and discrete logarithm. While these and other "abelian obstacles" persist, we demonstrate that the "nonabelian normal structure" of matrix groups over finite fields can be mapped out in great detail by polynomial-time randomized (Monte Carlo) algorithms.

The methods are based on statistical results on finite simple groups. We indicate the elements of a project under way towards a more complete "recognition" of such groups in polynomial time. In particular, under a now plausible hypothesis, we are able to determine the names of all nonabelian composition factors of a matrix group over a finite field.

Our context is actually far more general than matrix groups: most of the algorithms work for "black-box groups" under minimal assumptions. In a black-box group, the group elements are encoded by strings of uniform length, and the group operations are performed by a "black box."

1 Introduction

1.1 Outline of objectives

Let G be a finite group given by a list of generators. Our aim is to design asymptotically efficient algorithms to obtain structural information about G.

Typically G will be a matrix group over a finite field, but most of our algorithms work in the more general context of "black-box groups." In a *black-box group*, a "black box" performs the group operations on codewords representing the group elements. Each codeword has the same length n, the *encoding length*. (See Section 3.1 for details). This generalization, introduced in [19], is critical to our work. Work in progress on statistical recognition of *simple* black-box groups will amplify the significance of the black-box approach (see Section 9). In particular, assuming success of that project, we shall be able to determine the names of all nonabelian composition factors of matrix groups over finite fields.

Our goal is to explore the power of *Monte Carlo* (randomized) *polynomial-time algorithms*. The rigorously proven performance guarantees accompanying the algorithms will include a user-prescribed bound ε on the probability of error and guaranteed polynomial time bounds. The complexity of such an algorithm will always be of the form $O(n^c \log(1/\varepsilon))$ where n is the encoding length and c is a constant, so in polynomial time we can achieve an exponentially small probability of error. (A more detailed upper bound on the complexity may be available as a function of specific parameters of the input.)

Unfortunately some of the rudimentary tasks such as membership testing and determining the order of the group are not likely to be solvable in polynomial time for matrix groups over finite fields, due to their close association with hard problems in computational number theory such as factoring integers and discrete logarithm.

In spite of this significant handicap, we show that a great deal of structural information about black-box groups can be obtained in polynomial time. We emphasize, that, in contrast to previous work [57, 24], in this paper we do not assume an "oracle" (black box) for discrete logarithms, and our estimates do not involve quantities such as the largest prime divisor of the order [57] or the minimum degree of permutation representations of composition factors [24]. Our algorithms are genuinely polynomial time (in the length of the input). The only concession we make is that for the most general black-box group results we assume that a superset of the primes dividing $|G|$ is available (preprocessing). However, for the most important applications (matrix groups over finite fields and their quotients), we get rid of this assumption and remain entirely within polynomial time.

In this paper we do not address questions of implementation. We believe, that, as has been the case in the past, the mathematical insights forced by the rigor of guaranteed polynomial time will lead, through further theoretical work as well as through the use of heuristic shortcuts, to algorithms with the potential of practical implementation[1].

1.2 Basic normal structure

We consider the following chain of characteristic subgroups of the finite group G:

$$1 \leq \mathrm{Sol}(G) \leq \mathrm{Soc}^*(G) \leq \mathrm{Pker}(G) \leq G. \tag{1}$$

Here $\mathrm{Sol}(G)$ is the *solvable radical* of G (largest solvable normal subgroup). We define $\mathrm{Soc}^*(G)$ by

$$\mathrm{Soc}^*(G)/\mathrm{Sol}(G) = \mathrm{Soc}(G/\mathrm{Sol}(G)) \tag{2}$$

where $\mathrm{Soc}(H)$ is the socle of H (product of the minimal normal subgroups of H). We say that a group H is *semisimple* if H is the product of *nonabelian* simple groups. We refer to $\mathrm{Soc}^*(G)/\mathrm{Sol}(G)$ as the *semisimple socle* of G. (Note that in general this is *not* the semisimple part of $\mathrm{Soc}(G)$.) Let

$$\mathrm{Soc}^*(G)/\mathrm{Sol}(G) = T_1 \times \cdots \times T_k \tag{3}$$

[1] A prime example is Luks's polynomial time algorithm for constructing the composition chain of a permutation group [56]. A sequence of asymptotic improvements culminated in a nearly linear time algorithm (still based on Luks's basic outline) by Beals and Seress [26]. With appropriate heuristic shortcuts, Seress implemented the [26] algorithm in GAP [66, 67].

where the T_i are nonabelian simple groups. Let $\varphi : G \to \Sigma_k$ be the permutation representation of G via conjugation action on the set $\{T_1, \ldots, T_k\}$. (By Σ_k we denote the symmetric group of degree k.) We define the *permutation kernel* $\mathrm{Pker}(G)$ of G by

$$\mathrm{Pker}(G) = \ker(\varphi). \tag{4}$$

We make the following observations regarding the quotients of the characteristic chain (1).

(i) $\mathrm{Sol}(G)$ is solvable;

(ii) $\mathrm{Soc}^*(G)/\mathrm{Sol}(G)$ is semisimple (eqn. (3));

(iii) $\mathrm{Pker}(G)/\mathrm{Soc}^*(G)$ is solvable; in fact $(\mathrm{Pker}(G)/\mathrm{Soc}^*(G))''' = 1$;

(iv) $G/\mathrm{Pker}(G) \leq \Sigma_k$, and $k \leq \log|G|/\log 60$.

The reason for item (iii) is the observation that

$$\mathrm{Pker}(G)/\mathrm{Soc}^*(G) \leq \mathrm{Out}(T_1) \times \cdots \times \mathrm{Out}(T_k). \tag{5}$$

This implies that $\mathrm{Pker}(G)/\mathrm{Soc}^*(G)$ is solvable according to Schreier's conjecture; in fact $(\mathrm{Pker}(G)/\mathrm{Soc}^*(G))''' = 1$ (cf. [33]). Item (iv) is immediate from equation (3).

1.3 The main results

Let us now return to our basic object, a black-box group G of encoding length n. To simplify the presentation, we assume that a superset $\mathcal{P} \supseteq \pi(G)$ is also given, where $\pi(G)$ denotes the set of primes dividing $|G|$. This assumption is partly justified by requiring a *preprocessing* only, rather than the availability of an "oracle" to solve problems like discrete logarithm on-line.

Nonetheless, we consider the assumption of prime factors to be an unpleasant one. In Section 8 we shall see how to dispense with this assumption in the most important subcase of *black-box groups of characteristic p*. This subcase includes the subgroups of $GL(d, p^k)$ and their quotients by recognizable normal subgroups, assuming the group elements are encoded as $d \times d$ matrices over $GF(p^k)$.

Terminology To *construct* (or synonymously, to *find*) a subgroup H of a black-box group G means to construct generators for H. To *recognize* H means to test, for any $g \in G$, membership of g in H. The *cost* of constructing H is the time or other measure of complexity (such as the number of group operations) required to construct a set of generators for H. The cost of recognizing H is the maximum over $g \in G$ of the cost of the decision whether or not $g \in H$.

Unfortunately, efficient construction is not known to imply efficient recognition or vice versa. This is a major obstacle we have to face all along. We note that for *permutation groups*, construction does imply recognition in polynomial time (membership testing is available in polynomial time [70, 38]), but even for permutation groups, recognition does not seem to imply polynomial-time construction: the automorphism group of a graph is *recognizable* in polynomial time (membership testing is straightforward), yet we are unable to construct a set of generators in polynomial time. The latter task is equivalent to the *graph isomorphism problem*, not known to be solvable in polynomial time (cf. [8, Section 6] for a survey of the graph isomorphism problem).

In this paper we shall try to *construct* the members of the chain (1), refine the chain to a composition chain, and identify the composition factors. Ideally, we would also want to *recognize* these subgroups. Certain "abelian obstacles" (Section 3.5) prevent us from completing this plan in polynomial time, but we claim considerable degree of success in the "nonabelian part" of the project.

By a *quasi-composition chain* of a group H we mean a subnormal chain $1 = H_m \triangleleft H_{m-1} \triangleleft \cdots \triangleleft H_0 = H$ such that all quotients H_{i-1}/H_i are abelian or simple nonabelian. In the case of an abelian quotient we allow the case $H_{i-1} = H_i$ since we do not have membership testing. If we say that a quasi-composition chain is *known*, we mean that generators of each H_i are known, and furthermore we know which quotients are abelian and which are simple nonabelian.

First we note that Sol(G) is *recognizable* in Monte Carlo polynomial time, and therefore we can treat $G/\text{Sol}(G)$ as a black-box group (Corollary 3.3). On the other hand, we are not able to *construct* Sol(G).

Theorem 1.1 *Given a black-box group G along with a set $\mathcal{P} \supseteq \pi(G)$, one can construct, in Monte Carlo polynomial time, the permutation kernel Pker(G), the semisimple socle Soc*(G)/Sol(G), the decomposition (3) of Soc*(G)/Sol(G) into a product of k simple groups, and the permutation representation $\varphi : G \to \Sigma_k$ with kernel Pker(G). Consequently one can construct, in Monte Carlo polynomial time, a quasi-composition chain of $G/\text{Sol}(G)$ as well as a black-box representation of each nonabelian composition factor of G.*

(The last statement includes exhibiting a correspondence between the nonabelian composition factors of G and the corresponding members of the quasi-composition chain of $G/\text{Sol}(G)$.)

We note that observation (iv) above (Section 1.2) implies[2] that $k \leq n/\log 60 < n/5$; therefore once φ has been computed, the extensive library of polynomial time algorithms for permutation groups becomes available for $G/\text{Pker}(G)$ [15, 58]. In particular, we can construct a composition chain of $G/\text{Pker}(G)$ as well as permutation representations of its composition factors; moreover we can construct a presentation of $G/\text{Pker}(G)$ in terms of generators and relations.

"Constructing" φ would mean defining the φ-images of generators. However, we "construct" φ in a stronger sense: we set up a data structure which allows $\varphi(g)$ to be computed in polynomial time, for arbitrary $g \in G$. We do not assume here that we know how to construct g from the generators of G. This makes Pker(G) *recognizable within G:* given $g \in G$, we have $g \in \text{Pker}(G)$ exactly if $\varphi(g) = 1$, and the latter is decidable in polynomial time.

We are, however, unable to recognize Soc*(G), determine the order of the solvable group Pker(G)/Soc*(G), or even decide whether or not Pker(G) = Soc*(G).

The most elusive part of the project is the determination of the solvable radical Sol(G); we are unable to find a set of generators for Sol(G), or even to decide

[2]Throughout this paper, "log" refers to base 2 logarithms.

whether or not $\mathrm{Sol}(G) = 1$.

For *black-box groups of characteristic p* (Section 8.2), we eliminate the assumption that a superset $\mathcal{P}$ of primes is given.

Theorem 1.2 *Given a black-box group G of characteristic p, one can construct, in Monte Carlo polynomial time, all the objects listed in Theorem 1.1.*

We remark that for black-box groups of characteristic p we are able to make considerable inroads into $\mathrm{Sol}(G)$ and the only really elusive part that remains is $O_p(G)$, the largest normal p-subgroup. We give the details in a subsequent paper [10].

1.4 Overview of the algorithm

We indicate the overall structure of the algorithm announced in Theorem 1.1.

The first phase consist of the construction of the semisimple socle. For this phase, we work in the black-box group $\bar{G} = G/\mathrm{Sol}(G)$. We note that $\mathrm{Sol}(\bar{G}) = 1$.

First we need to find a minimal subnormal subgroup of $\bar{G}$. This is accomplished by a process of "blind descent" along a subnormal chain (Sections 6.3, 7.1).

The normal closure of a minimal subnormal subgroup is semisimple (because $\mathrm{Sol}(\bar{G}) = 1$). Our key auxiliary process splits a semisimple group into its simple factors (Section 5). This decomposition helps to extend our semisimple normal subgroup of $\bar{G}$ until the entire socle of $\bar{G}$ is found (Section 7).

The decomposition of $\mathrm{Soc}(\bar{G}) = \mathrm{Soc}^*(G)/\mathrm{Sol}(G)$ makes it easy to construct the permutation representation φ (Corollary 5.2). Given φ, standard permutation group techniques yield generators for $\mathrm{Pker}(G)$ (Proposition 3.4). Finally, a quasi-composition chain of $\bar{G}$ is easily constructed: we apply Luks's composition chain algorithm [56] to $G/\mathrm{Pker}(G)$, followed by three steps of the derived series of $\mathrm{Pker}(G)/\mathrm{Sol}(G)$ and ending with a composition chain of $\mathrm{Soc}^*(G)/\mathrm{Sol}(G)$ which is obtained from the decomposition of $\mathrm{Soc}^*(G)/\mathrm{Soc}(G)$ (Section 5).

The proof of Theorem 1.2 requires the introduction of a set of "pretend-primes" to be used in determining the "pseudo-order" of elements (Section 8) to be used in our algorithms wherever computation of the order would be required. (Computing the order would require prime factorization, a problem not known to be solvable in polynomial time, cf. Section 8.1.)

Two results on finite simple groups, stated in Section 4, will play a critical role in the analysis of our algorithms. One is a bound by Landazuri and Seitz [54] stating that the degree of a representation of a Lie-type simple group in the wrong characteristic must be very large. We apply this result in several different contexts: in the analysis of "blind descent" (Lemma 6.3), in the selection of the "pretend-primes" for the calculation of "pseudo-order," and for handling the p'-part of the solvable radical [10].

The other critical result states that for any prime r, elements of order not divisible by r occur with non-negligible frequency in every finite simple group (Theorems 4.2, 8.7). This statistical result is the key to splitting a direct product of simple groups into its factors and finding the semisimple socle (Sections 5, 7). Another application is to finding the center of a quasisimple group (see [10]).

2 Brief history

2.1 Permutation groups

Charles Sims pioneered the design of efficient algorithms for the rudimentary tasks of managing permutation groups (membership, order, normal closure) based on his fundamental data structure of "strong generators" [69, 70]. A polynomial time analysis of closely related algorithms was first given in [6] in the context of an application to the Graph Isomorphism problem. [6] employed randomization and introduced the term "Las Vegas." Motivated by [6], a seminal paper by Furst, Hopcroft, and Luks [38] designed a variant of Sims's algorithm and proved that it runs in polynomial time. Other polynomial-time versions of Sims's algorithm were described by Knuth [50] and Jerrum [44]. Seress found that Sims's original version, too, runs in polynomial time [67].

[38] was followed by a succession of powerful polynomial-time algorithms for permutation groups. We highlight two results: Luks's algorithm to construct a composition chain along with permutation representations of the composition factors [56], and Kantor's algorithm to construct the Sylow subgroups [46].

2.2 Complexity theory in black-box groups

The first paper addressing the complexity of basic computational problems for finite matrix groups was [19]. That paper introduced the framework of black-box groups and proved that membership in subgroups of black-box groups is in the complexity class NP (has polynomial length verification) by proving that any element of a finite group G can be reached by at most $(1 + \log |G|)^2$ group operations from any set of generators. The combinatorial idea of the proof became later the basis of the polynomial-time Monte Carlo algorithm for selecting nearly uniformly distributed random elements in a black-box group ([7], see Theorem 3.2).

Another result of [19] asserts that determining the order of a black-box group is in the complexity class NP assuming the *"Short Presentation Conjecture"* which states that every finite simple group T has a presentation (in terms of generators and relations) of total bitlength $\leq (\log |T|)^c$, and that such a presentation can be constructed efficiently (in Las Vegas time $\leq (\log |T|)^c$), given the standard name of T. This conjecture has since been verified for all finite simple groups except for the three families of rank-1 twisted groups: the unitary groups $PSU(3, q) = {}^2A_2(q)$, the Suzuki groups $Sz(q) = {}^2B_2(q)$, and the Ree groups $R(q) = {}^2G_2(q)$ [14]. A full proof of this conjecture will play an essential role in turning Monte Carlo algorithms for matrix groups into the more desirable Las Vegas variety (guaranteed no error), a significant conceptual and practical leap because of the use of unproven heuristics in random sampling from black-box groups and matrix groups (Section 3.3, cf. [9]).

2.3 Polynomial-time algorithms in matrix groups

The first polynomial-time algorithms for a host of problems on matrix groups appear in Luks's paper [57]. In contrast to virtually all subsequent work, Luks's algorithms are deterministic. Luks showed that deciding solvability of matrix groups over finite fields is in polynomial time. Moreover he gave algorithms for membership testing and order, computing presentations, composition chain, as well as many other problems, for solvable matrix groups. Many of these algorithms are not polynomial time (in the bit-length of the input) but depend polynomially on the largest prime divisor of $|G|$, other than the characteristic. Some limitation of this kind seems indispensable for the membership and order problems because of the number theoretic obstacles to be discussed in Section 3.5.

Algorithms for membership testing, order, computing presentations, composition chain, and other problems for (not necessarily solvable) finite matrix groups are given by Beals and Babai [24]. Unlike the algorithms of [57], these algorithms are randomized (Las Vegas). As with Luks's algorithms, the timing is polynomial (in the bit-length of the input) in characteristic 0 (algebraic number fields), but depends polynomially on an additional parameter ν in characteristic p. The parameter $\nu = \nu(H)$ is defined as the largest integer k such that H has a composition factor $L \neq \mathbb{Z}_p$ which has no permutation representation of degree $< k$.

The large abelian composition factors represent a genuine obstacle (see Section 3.5). We survey recent progress on how to get around some of the nonabelian obstacles in Section 9.

2.4 The Aschbacher classification

In this paper we use "polynomial time" as the criterion of efficiency. A more traditional approach measures efficiency by practical performance. We note here some important work in this direction. Many of the algorithms referred to below have not been shown to run in polynomial time. However, implementations in GAP and MAGMA are available with impressive test data; they work even in dimensions over 100.

Aschbacher has classified the maximal subgroups of the finite classical groups [5]. We give a summary of the Aschbacher classification: Let $G \leq GL(d,q)$, and let $Z = Z(GL(d,q)) \cap G$. Let V denote the vector space F_q^d. Then at least one of the following holds:

C1 G is reducible.

C2 G is imprimitive.

C3 G fixes a tensor decomposition $V = U \otimes W$.

C4 G preserves a symmetric tensor decomposition: $V = V_1 \otimes V_2 \otimes \cdots \otimes V_m$, where each V_i has the same dimension r, and $d = r^m$.

C5 G acts semi-linearly over an extension field F_{q^e}, so $V = F_{q^e}^{d/e}$ and $G \leq \Gamma L(d/e, q^e)$.

C6 Modulo Z, G is conjugate to a subgroup of $GL(d, q')$, for some $q' \mid q$.

C7 $d = r^m$ for some prime R, G normalizes an r-group R, of order r^{2m+1} or 2^{2m+2}. In the first case, R is extraspecial; in the second case, R is symplectic.

C8 $K' \leq G/Z \leq K$, where K is a projective general linear, symplectic, orthogonal, or unitary group in dimension d over F_q.

C9 $T \leq G/Z \leq \mathrm{Aut}(T)$, where T is nonabelian simple.

The "recognition project," led by C. R. Leedham-Green, is guided by the following computational problem related to the Aschbacher classification: given a list of generators for a subgroup H of $GL(n,q)$, find a maximal subgroup $G \leq GL(n,q)$ containing H. For several of the classes C1–C9, algorithms exist which seem to perform well in practice. The classes C1 and C5 are handled by the Meataxe [64, 45]. Since the Meataxe, the first significant step in the recognition project was the Neumann–Praeger algorithm [61, 43] for case C8 in the case K is $PGL(d,q)$. The other classical groups that can arise in C8 are handled by Niemeyer and Praeger [63, 62].

In many cases, H has a normal subgroup N preserving some structure which is acted on by H. For example, in case C4, H permutes the V_i, and N is the kernel of this action. The SMASH algorithm [42], when given an element $x \in N \setminus Z$, attempts to find H in classes C2, C3, C4, and C7. SMASH does not solve, in general, the problem of finding such an x, though the SMASH authors have efficient heuristics for C2 [41]. And of course, G may intersect N trivially (or only in Z), in which case SMASH does not apply. Leedham-Green and O'Brien [51] have given another approach to C3.

Any implementation of our ideas should make use of the practical successes of the recognition project. The SMASH algorithm seems particularly suited to our needs, since we have, in many instances, algorithms for finding elements of nontrivial proper normal subgroups. Such an element, supplied to SMASH, may yield a permutation representation of G (in cases C2, C4, or C7), which is an important subgoal of our algorithm. Conversely, ideas from the theory of black-box group algorithms should be useful in the recognition project. In the case G is in class C9, what we have is, for practical purposes, a black-box group. In cases where SMASH is applicable, we present a well developed theory of how to find elements that are likely to lie in nontrivial proper normal subgroups (Section 6).

3 Algorithmic preliminaries

3.1 Black-box groups

A *black-box group* G is a finite group whose elements are encoded by $(0,1)$-strings ("codewords") of uniform length n [19]. We call n the *encoding length* of the black-box group. This convention implies $|G| \leq 2^n$.

Group operations on the codewords are performed by a "black box" at unit cost. The operations are *multiplication, inversion, and identity testing* (decision whether or not a given string encodes the identity). A black-box group is *given* by a list of generators.

Significantly, we do not require the encoding to be unique. Our most important models of black-box groups are *quotients of matrix groups* over finite rings: $G = H/N$ where $N \triangleleft H \leq GL(d, R)$ for some finite ring R (usually a finite field). The elements of G are encoded as matrices (elements of H). Identity testing requires membership testing in N (magically, or by some algorithm).

We say that a subgroup H of the black-box group G is *recognizable at cost t* if for every $g \in G$, membership of g in H can be tested at cost t. (The membership algorithm for H must not depend on how g is constructed from the generators of G. If the sequence of operations leading to g is required for the decision, we say that H is *weakly recognizable* at cost t. However, we shall not need this concept here.) The following evident fact will be very useful and demonstrates the power of the black box model.

Proposition 3.1 *Let G be a black-box group and $N \triangleleft G$ a normal subgroup recognizable at cost t. Then the black box for G can simulate a black box for G/N at cost t per operation.*

In fact, the cost of the simulation is one step (group operation) per multiplication/inversion, and t cost units per identity query. Note that identity queries typically make up only a small fraction of the operations; they do not occur at all in the process of random sampling, one of the expensive routines.

If a black-box group of encoding length n is given by a list of k generators then the *bit-length of the input* is kn and therefore a polynomial-time algorithm makes $\leq (kn)^c$ steps for some constant $c > 0$.

Convention. We shall use the letter c to denote positive constants; separate occurrences of c may refer to *different* constants. We also use $c_1, c_2, \ldots$ to denote positive constants; repeated occurrences of c_i refer to the *same* constant. The letter ε denotes a positive quantity, not necessarily a constant, and separate occurrences of ε may refer to different values.

3.2 Monte Carlo and Las Vegas algorithms

Our algorithms will use randomization; therefore the output will not necessarily be correct. However, the probability of error will be guaranteed to be less than ε, regardless of the input. The parameter ε is chosen by the user. The cost increases proportionally to $\log(1/\varepsilon)$, so exponentially small error probability is achieved at polynomial cost. We should emphasize that probabilities are over the coin tosses made in the course of the execution of the algorithm. We do not assume any probability distribution over the input space: we perform "worst case" analysis (as opposed to "average case analysis").

We use the term "Monte Carlo algorithm" to refer to any randomized algorithm with a user-specified arbitrarily small error bound ε (which may be proven or heuristic). A *polynomial time Monte Carlo algorithm* is guaranteed, for any requested error margin $\varepsilon > 0$, to stop in time $O(n^c \log(1/\varepsilon))$ and have error probability $\leq \varepsilon$.

"Las Vegas algorithms" form an important subclass of Monte Carlo algorithms: they never make an erroneous output, but with probability $\leq \varepsilon$, they are allowed to report failure.

3.3 Random sampling from groups

By a *random element* of a finite set S we mean a nearly uniformly distributed random member of S. *Near-uniformity* means that every element of S has $(1 \pm \varepsilon)/|S|$ chance to be selected; here $\varepsilon > 0$ is a parameter chosen by the user. When we speak of selecting *several random elements*, we always assume *independent* random choices.

A critical tool in all algorithms under discussion is a method of selecting random elements from a black-box group. Our polynomial-time claims heavily depend on the following result.

Theorem 3.2 ([7]) *An ε-uniformly distributed random element of a black-box group can be selected in Monte Carlo polynomial time at a cost of $O(n^c \log(1/\varepsilon))$.*

This algorithm uses randomization but it makes no error in the sense that it is guaranteed to produce an ε-uniformly distributed element. Several *independent* elements from the same distribution can be obtained by repeating the experiment with a separate set of coin tosses. The actual result in [7] gives a separate cost estimate for preprocessing and a considerably lower cost per random element generated (preprocessing costs $O(n^5)$ group operations and yields $\Theta(n)$ random elements; additional random elements cost $O(n)$ group operations each).

While this algorithm does not produce uniformly distributed random group elements, the slight deviation from uniformity has no significant effect on the analysis of our algorithms.

The "product replacement algorithm" [32], a popular heuristic introduced by Charles Leedham-Green and Leonard Soicher, is another candidate for a polynomial-time method for nearly uniform selection from a finite group. However, very little has yet been proven in this direction (cf. [9]). Of course in practical implementations of either of the above algorithms one uses much fewer operations than required for proven performance, and hopes for the best.

When efficient but unproven heuristics are invoked for sampling in groups under a randomized algorithm, Las Vegas algorithms are greatly superior to Monte Carlo since they never err, even if the distribution of the "random" group elements is not as nearly uniform as desired. The worst that can happen in such a case is that the algorithm reports failure more frequently then expected and thus a weakness of the sampling method is discovered. However, with Monte Carlo methods, the inadequacy of the sampling method may mean frequent erroneous outputs which may remain undetected.

Unfortunately, most of the algorithms in this area are not Las Vegas, and there is little hope for turning them into Las Vegas as long as some cases of the "Short Presentation Conjecture" (Section 2.2) remain open (cf. [9]).

For a detailed discussion of the concepts involved in and possible pitfalls of Monte Carlo algorithms in finite groups, we recommend [9].

3.4 Further rudiments: normal closure and applications

Recall our standard notation: G is a black-box group of encoding length n and therefore $|G| \leq 2^n$. It follows that any set of more than n generators is redundant. In fact, in Monte Carlo polynomial time one can replace any set of generators by a list of $O(n)$ generators [12], i. e., the number of generators is $\leq cn$ for some constant c. Therefore we shall assume that all groups considered are given by $O(n)$ generators.

A fundamental polynomial-time Monte Carlo algorithm given in [12] is constructing the *normal closure* $\langle S^G \rangle$ of a set $S \subseteq G$ (i. e., finding a set of $O(n)$ generators for $\langle S^G \rangle$). We note that this claim does not depend on nearly uniform random selection from G and uses the more elementary tool of "random subproducts."

As an immediate consequence, the following subgroups are constructible in polynomial Monte Carlo time: the *commutator subgroup* G', all members $G^{(i)}$ of the derived series, the *stable derivative* $G^{(\infty)} = G^{(n)}$ (the smallest normal subgroup such that $G/G^{(\infty)}$ is solvable), the lower central series. It follows that in polynomial Monte Carlo time one can decide solvability and nilpotence of a black-box group, as well as decide whether or not G is a p-group for a given prime p.

It follows that the following subgroups are *recognizable* in Monte Carlo polynomial time:

- $O_p(G)$ – the largest normal p-subgroup of G,
- $F(G)$ – the largest nilpotent normal subgroup (Fitting subgroup),
- $\text{Sol}(G)$ – the solvable radical of G (largest solvable normal subgroup),
- $Z(G)$ – the center.

Indeed, to test membership in $\text{Sol}(G)$, for instance, we test, given $g \in G$, whether or not the normal closure $\langle g^G \rangle$ is solvable. – We can also combine these operators; e. g., the inverse image of $Z(G/O_p(G))$ is also recognizable in Monte Carlo polynomial time.

We note an immediate consequence of particular importance:

Corollary 3.3 *If G is a black-box group, then a black box for $G/Sol(G)$ can be simulated in Monte Carlo polynomial time.*

Proof This follows by Proposition 3.1 in light of the recognizability of $\text{Sol}(G)$. □

We note that this corollary works in spite of the fact that we are not able to find any elements of $\text{Sol}(G)$ or even to tell whether or not $|\text{Sol}(G)| = 1$.

Another important consequence concerns the kernel of a permutation representation. The following is folklore.

Proposition 3.4 *Let G be a black-box group and $\varphi : G \to \Sigma_k$ be a permutation representation. Suppose φ is given by the images of the generators of G. Then* $\ker(\varphi)$ *can be constructed in Monte Carlo polynomial time.*

Proof Let $N = \ker(\varphi)$ and $H = G/N \leq \Sigma_k$. Let S be the given set of generators of G and let $\bar{S}$ be the image of S under φ. Starting from $\bar{S}$, standard permutation group techniques produce a presentation of H in terms of a new set $S' \supseteq \bar{S}$ of generators and relations in terms of S' ([70, 38, 50, 44]). Lift the same sequence of group operations to G, starting from S rather than from $\bar{S}$. Create a set R of elements of G by starting from the empty set and adding each $g \in G$ computed in the above process satisfying $\varphi(g) = 1$. (This procedure is deterministic.) It is easy to see that $N = \langle R^G \rangle$. □

3.5 Abelian obstacles

Abelian groups and abelian quotients represent obstacles to answering even the simplest rudimentary questions about black-box groups.

Assume the black-box group G is known to be elementary abelian of order p^k where p is known but k is unknown. Then it takes an exponential number of black-box operations to determine k, or, for large primes, even to decide whether or not $k = 1$ [19]. Therefore linear algebra cannot be addressed strictly within the context of black-box groups.

For matrix groups of characteristic p, linear algebra of elementary abelian p-subgroups is straightforward. However, this is not the case for elementary abelian r-subgroups, where $r \neq p$.

The *constructive membership problem* asks not only whether some $g \in G$ belongs to a given subgroup $H \leq G$ or not, but if so, it asks to construct g from the generators of H. If we want to use Gaussian elimination for linear algebra in elementary abelian r-groups, we need constructive membership. However, constructive membership in a cyclic subgroup of order r in $\mathbb{F}_q^\times$ $(r|q-1)$ is precisely the *discrete logarithm* problem which is not known (and not believed) to be solvable in (Monte Carlo) polynomial time. (For the best current algorithms, see [53], [2], [1].) Therefore, the order of an abelian matrix group over $\mathbb{F}_q$, consisting of diagonal matrices only, cannot be determined by known methods in polynomial time, even if the group is elementary abelian of known exponent.

Even determining the order of an element may not be feasible in polynomial time because of its close relationship to the problem of factoring integers, another problem of computational number theory not believed to be solvable in polynomial time. (We analyse the complexity of determining the order of elements in Sections 3.6 and 8.1). However, for most purposes, one can avoid using the exact order of an element, and be content with its "pseudo-order," based on a set of "pretend-primes" (Section 8.4).

Other "abelian obstacles" concern the discovery of certain abelian (or solvable) subgroups and quotients and may be unrelated to computational number theory. The two most important questions in this direction are stated in the "Open problems" section; see Problems 10.2 and 10.3.

The moral of the present paper is that, in a sense, the *only obstacles* to Monte Carlo polynomial-time computation in black-box groups are the "abelian obstacles."

3.6 Order, prime factors

The "primes of a group G" are the prime divisors of $|G|$. For our results about general black-box groups (as opposed to "black-box groups of characteristic p," Sec. 8.2), we shall assume that together with the black-box group G, a superset $\mathcal{P}$ of the primes of G is given. The combined bit-lengths of the primes in $\mathcal{P}$ is then part of the input length with respect to which our algorithms run in polynomial time. The following is well known.

Claim 3.5 *Given a black-box group G with a superset $\mathcal{P}$ of its primes, we can determine the order of any element $g \in G$ in polynomial time.*

Proof First we construct an integer $m = \prod_{p \in \mathcal{P}} p^{k_p}$ such that $|g|$ divides m. For instance, $|G|$ divides m if we we choose $k_p = \lfloor n/\log p \rfloor$. (Smaller exponents will suffice if more information about G is available, cf. [30].) Given m, the following algorithm determines the order of g.

Algorithm ORDER($G, \mathcal{P}, g, m$)

Initialize: $\ell := m$ (Note: $g^{\ell} = 1$)
for $r \in \mathcal{P}$
  find largest t such that $r^t | \ell$ and $g^{\ell/r^t} = 1$
  $\ell := \ell/r^t$
end
return ℓ

Using repeated squaring for calculating powers, the **for** loop takes $O(\log m)$ multiplications, adding up to a total of $O(|\mathcal{P}| \cdot \log m) \leq O(|\mathcal{P}|^2 \cdot n)$. This justifies the polynomial time claim. A divide-and-conquer trick described in [30] reduces the number of operations to $O(\log(1 + |\mathcal{P}|) \cdot \log m)$. □

4 Two results on simple groups

In this section we state two group-theoretic results which will be crucial for the analysis of our algorithms.

4.1 Smallest degree of representations in the wrong characteristic

Landazuri and Seitz [54] proved that if a simple group of T Lie type of characteristic r is represented in characteristic $p \neq r$ then the dimension of the representation must be very large: it is polynomially related to the *size* of the natural module on which T acts (projectively). A result of Feit and Tits [36] implies that the same conclusion holds for representations of extensions of T (i.e., groups with T as a homomorphic image). We state a consequence of their combined result.

Theorem 4.1 ([54, 36]) *Let T be a finite simple group of Lie type of characteristic r. Let V be the natural module (in characteristic r) on which T acts projectively. Let H be a finite group which involves T (as a quotient of a subgroup). Suppose*

H has a faithful projective representation of dimension d in characteristic $p \neq r$. Then $|V| \leq d^{c_1}$ for some absolute constant c_1.

4.2 Some statistical group theory

Let r be a prime number. A group element is called *r-regular* if its order is not divisible by r. We shall make extensive use of the following statistical result on r-regular elements.

Theorem 4.2 ([17]) *Let G be a finite simple group and r a prime number. Then at least a c/d fraction of the elements of G is r-regular where $c > 0$ is an absolute constant and d is the dimension of G, defined as follows. The dimension of the alternating group A_k is k, the dimension of a classical simple group is the dimension of the projective space on which the group acts; all other simple groups have dimension* 1.

We shall require a more detailed version of this result, stated as Theorem 8.7.

5 Splitting a semisimple group

We call a group G *semisimple* if G is the direct product of nonabelian simple groups.

If $G_1, \ldots, G_k$ is a family of black-box groups of total encoding length n then the direct product $G = G_1 \times \cdots \times G_k$ can be represented in the natural way as a black-box group of encoding length n. The cost of a group operation in G will be the sum of the costs of group operations in each G_i.

A *black-box decomposition* of a black-box group G into a direct product $G = G_1 \times \cdots \times G_k$ means the construction of black-box representations of each G_i.

Theorem 5.1 *Let G be a black-box group given together with a set $\mathcal{P} \supseteq \pi(G)$. Assume that G is known to be semisimple. Then a black-box decomposition of G into its simple factors can be constructed in Monte Carlo polynomial time. Each simple factor will be represented as a black-box group.*

Note that such a decomposition immediately implies that a composition series of G can be constructed in polynomial time: namely, if $G = T_1 \times \cdots \times T_k$ is a black-box decomposition then the the family of prefixes $H_i = T_1 \times \cdots \times T_i$ is also constructed in polynomial time.

We state an important corollary.

Corollary 5.2 *Let G be a black-box group given together with a set $\mathcal{P} \supseteq \pi(G)$. Let $N = T_1 \times \cdots \times T_k$ be a given semisimple normal subgroup of G and let $\varphi : G \to \Sigma_k$ be the permutation representation of G defined by conjugation action on the set $\{T_1, \ldots, T_k\}$. Then φ can be constructed in the following strong sense: in Monte Carlo polynomial time we can set up a data structure which allows, for any $g \in G$, to compute $\varphi(g)$ in (deterministic) polynomial time.*

Proof By Theorem 5.1 we find generators for each T_i in Monte Carlo polynomial time. Let now $g \in G$ and $\sigma := \varphi(g) \in \Sigma_k$. Then $\sigma(i) = j$ exactly if T_i^g does not centralize T_j; this circumstance is verified by comparing the generators of each. □

The rest of this section is devoted to the proof of Theorem 5.1. Versions of this process and its analysis lead to our key algorithm: constructing the semisimple socle (Section 7).

Proof of Theorem 5.1 Let m be a known multiple of $|G|$, constructed as in the proof of Claim 3.5; all primes dividing m are from $\mathcal{P}$.

Let $m = \prod_{r \in \mathcal{P}} r^{k_r}$ and let $m_r = m/r^{k_r}$, i.e., m_r is the maximal r-free divisor of m.

Assuming G is semisimple, first we describe how to construct one of its simple factors.

Algorithm CONSTRUCT_SIMPLE_FACTOR(G, ν)

```
Initialize: R := G
repeat ν times
    choose random g ∈ R
    for r ∈ P
        h := g^{m_r}
        if h ≠ 1 then R := ⟨h^G⟩
    end
end
return R
```

Claim 5.3 If $G \neq 1$ is semisimple then Algorithm CONSTRUCT_SIMPLE_FACTOR (G) returns a normal subgroup R of G which is simple with probability $\geq 1 - \varepsilon$ assuming $\nu \geq c_2\sqrt{n}\log(1/\varepsilon)$ where c_2 is an explicit constant.

Note that these are proven worst-case guarantees; much fewer rounds may suffice in practice, and the theoretical estimates can be improved if information about the T_i is available.

Proof Let $G = T_1 \times \cdots \times T_k$, T_i simple, nonabelian. For $I \subseteq \{1, \ldots, k\}$, set $T_I = \prod_{i \in I} T_i$. We use the fact that the groups T_I are the only normal subgroups of G.

We refer to each iteration of the "repeat" loop as a "round." Let R_j denote the group R at the end of round j; $R_0 = G$. Observe that for all j we have $1 \neq R_j \leq R_{j-1}$ and $R_j \triangleleft G$.

We say that *component i is successful in round j* if either $R_j = T_i$ (which means we are done) or $R_j \cap T_i = 1$. Clearly, the output R_ν is simple exactly if each component is successful in some round.

Let δ_i denote the lower bound provided by Theorem 4.2 for the probability that for a prime r, a random element of T_i is r-regular. By the estimate given in

Theorem 4.2 we have $\delta_i^{-2} \leq c_3 \log |T_i|$ and therefore

$$\sum_{i=1}^{k} \delta_i^{-2} \leq c_3 \log |G| \leq c_3 n \tag{6}$$

for an explicit constant c_3.

We claim, that, given any history of outcomes of earlier rounds, the probability that component i is not successful in round j is at most $1 - (59/60)\delta_i$.

Indeed, in order for this event to occur it is necessary that $R_{j-1} \geq T_i \times T_\ell$ for some $\ell \neq i$. Let us consider the random element $g = (g_1, g_2, \ldots)$ selected in round j, where the $g_u \in T_u$ are the components of g with respect to our direct decomposition. We may model the random choice of g by first selecting g_ℓ at random, then g_i, and then the rest. With probability $\geq 59/60$, $g_\ell \neq 1$. In this case, let r be a prime dividing the order of g_ℓ. Then with probability $\geq \delta_i$ the order of g_i is not divisible by r. If this is the case, then $h = (h_1, h_2, \ldots)$ satisfies $h_\ell \neq 1$ and $h_i = 1$. But this means $T_i \cap R_j = 1$ and component was successful in this round.

The probability that component i remains unsuccessful after ν rounds is therefore

$$\leq \left(1 - \frac{59\delta_i}{60}\right)^\nu < \exp\left(-\frac{59\delta_i \nu}{60}\right) = \exp\left(-\delta_i\sqrt{2c_3 n}\cdot 2\mu\right) < \left(\frac{1}{2\delta_i^2 c_3 n}\right)^\mu \tag{7}$$

where $\mu = 59\nu/(120\sqrt{2c_3 n}) > \nu/3\sqrt{c_3 n}$. Finally the probability that R_ν is not simple is less than

$$\sum_{i=1}^{k} \left(\frac{1}{2\delta_i^2 c_3 n}\right)^\mu < 2^{-\mu} < 2^{-\nu/3\sqrt{c_3 n}} \leq \varepsilon \tag{8}$$

by inequality (6), assuming $c_2 = 3\sqrt{c_3}$. □

Suppose now that R is a given simple factor of the semisimple group G. Our next task is to construct the complement H of R in G defined by $G = R \times H$.

Algorithm CONSTRUCT_COMPLEMENT(G, R, ν)

```
Initialize: H := 1
repeat ν times
    choose random g ∈ G
    for r ∈ P
        h := g^{m_r}
        if [h, R] = 1 then H := H · ⟨h^G⟩
    end
end
return H
```

Claim 5.4 *If G is semisimple and $R \triangleleft G$ is simple then Algorithm* CONSTRUCT_COMPLEMENT(G, R) *returns a normal subgroup H of G such that $G = R \times H$ with probability $\geq 1 - \varepsilon$ assuming $\nu \geq c_2\sqrt{n}\log(1/\varepsilon)$ where c_2 is the same explicit constant as in Claim 5.3.*

Proof As before, let $G = T_1 \times \cdots \times T_k$. We may assume $R = T_1$. Let $T = T_2 \times \cdots \times T_k$. Let H_j denote the group H at the end of round j; $H_0 = 1$. Observe that for all j we have $H_{j-1} \le H_j \le T$ and $H_j \triangleleft G$.

We say that *component $i \ne 1$ is successful in round j* if $H_j \ge T_i$. Clearly, the output H_ν is T exactly if each component $i = 2, \ldots, k$ is successful in some round.

We use the notation δ_i introduced in the proof of Claim 5.3. As before, we claim, that, given any history of outcomes of earlier rounds, the probability that component $i \ne 1$ is not successful in round j is at most $1 - (59/60)\delta_i$.

Indeed, as before, let us consider the random element $g = (g_1, g_2, \ldots)$ selected in round j ($g_i \in T_i$). We imagine this time that we first select g_i at random, then g_1, then the rest of the components. The rest of the proof proceeds exactly as the proof of Claim 5.3. □

Finally, to obtain the desired decomposition of the semisimple group G we separate a simple factor T_1, compute its complement H_1, and repeat the process with H_1 in the role of G. This completes the proof of Theorem 5.1. □

6 Finding a proper normal subgroup

Our first main goal is to construct a simple subnormal subgroup in $G/\mathrm{Sol}(G)$. This will be accomplished by a gradual descent along a subnormal chain. The steps of the process are given by the Monte Carlo algorithm PERM(G, m) [24] which takes as inputs a black box group G and a positive integer m and produces one of the following outputs using $(m+n)^c$ group operations:

(a) a faithful permutation representation of G;

(b) a nontrivial normal subgroup $1 \ne H \triangleleft G$.

In case (b), H is not necessarily proper. However, the algorithm comes with the following guarantee:

Theorem 6.1 *If G has a proper subgroup of index $\le m$ and case (b) is produced by Algorithm* PERM(G, m) *then $H \ne G$ holds with large probability. If case (a) is produced then the degree of the permutation representation obtained is always $\le m^c$.*

Note that the condition in the Theorem is equivalent to requiring that G has a nontrivial permutation representation of degree $\le m$. An important corollary states that if a simple group G of Lie type is represented in the wrong characteristic then a faithful permutation representation of G can be found in Monte Carlo polynomial time (see Theorem 8.6).

The name of the algorithm owes to its attempts at finding a (not necessarily faithful) permutation representation of G. Although in general it will not find such a representation, it is likely that it will find a nonidentity element of the kernel of a non-faithful permutation representation of small degree if such a representation exists.

In this section we describe the algorithm PERM and outline the proof of Theorem 6.1.

6.1 The normal still

Since normal closures can be calculated in Monte Carlo polynomial time, a case (b) output will be produced once we found *any* nontrivial element of a proper normal subgroup.

In fact it suffices to produce an element which has a non-negligible chance ($\geq 1/m^c$) of belonging to a proper normal subgroup. Then after $O(m^c)$ trials, we expect to have encountered an element that actually does belong to a proper normal subgroup. We cannot decide which one, since we do not have membership testing, so we have no direct way of deciding whether a subgroup is proper.

We overcome this difficulty by "distilling" any list of candidate elements into a single nonidentity element guaranteed to belong to a proper a proper normal subgroup if any member of the list does.

It suffices to show how to do this with a list of length 2. The following is from [24]:

Lemma 6.2 *Given nonidentity elements a, b of a nonabelian black-box group G, we can in Monte Carlo polynomial time calculate a nonidentity element c of G such that if either a or b lies in a proper normal subgroup of G, then so does c.*

Proof If $[a,b] \neq 1$ then we may take $c = [a,b]$, so assume a and b commute. Let T be a generating set for $\langle b^G \rangle$. If $[a,T] \neq 1$ then we may take $c = [a,t]$ for some $t \in T$, so assume a centralizes the normal closure of b. If a lies in the center of G we may take $c = a$. Otherwise, since a centralizes $\langle b^G \rangle$ but not all of G, we know $\langle b^G \rangle \neq G$ so we may take $c = b$. □

We remark that for matrix groups, Lemma 6.2 can be strengthened: the only randomized part of the algorithm is testing if a centralizes $\langle b^G \rangle$. If G is a matrix group, then we can compute in deterministic polynomial time a basis consisting of elements of $\langle b^G \rangle$ for the matrix algebra generated by b^G. Then a centralizes $\langle b^G \rangle$ iff it commutes with every element of this basis.

6.2 Conjugacy classes large and small

We need the following useful statistical property of simple groups: certain powers of elements of some large conjugacy classes belong to small conjugacy classes.

Lemma 6.3 *Let T be a nonabelian simple group admitting a faithful permutation representation of degree $\leq m$. Let x be a random element of T, and let r be a randomly selected prime divisor of $|x|$. Let $y = x^{|x|/r}$. Then with probability $\geq 1/m^c$, y will lie in a conjugacy class of $Aut(T)$ of cardinality $\leq m^{c_4}$, where c_4 is an absolute constant.*

Proof If T is not an alternating or classical group, then Theorem 4.1 implies that $|T| \leq m^{c_4}$, so any conjugacy class is small enough.

For alternating groups, let us consider the case $T = A_k$ where k is even and not divisible by 3. In this case the cycle structure of x is $\{3, k-3\}$ with probability $1/(3k-9)$. If x is such an element, then with probability at least $1/k$, the prime

3 will be chosen for r, and y will be a 3-cycle (with conjugacy class size $O(k^3)$). Now $k \leq m$, justifying the claim. In the other cases, look for elements x with cycle structure $\{3, k-s\}$ for an appropriate value $3 \leq s \leq 6$.

For classical groups acting projectively on $\mathbb{F}_q^k$, Theorem 4.1 implies that $q^k \leq m^{c_1}$ where c_1 is the constant appearing in Theorem 4.1. So it suffices to show that with probability $\geq q^{-ck}$, the element y acts trivially on a subspace of bounded codimension.

We may assume $k \geq c$. (Each occurrence of c represents a different constant.) Now G contains a subgroup $H \cong G_1 \times G_2$ where G_1 is the same kind of classical group (or its central extension) acting on a space of bounded codimension, and G_2 is nontrivial; $|G : H| < q^{ck}$. Now Theorem 4.2 implies that if $x \in H$ then $y \in G_2$ with probability $\geq c/(kt)$ where t is the number of prime divisors of $|x|$. □

6.3 Blind descent: the PERM algorithm

We now give the algorithm for finding an element of a proper normal subgroup. This algorithm represents one step in the "blind descent" with the goal to reach a simple subnormal subgroup (Section 7.1).

Algorithm PERM1(G, m)

Choose random $x, z \in G$
choose random prime r dividing $|x|$
$y := x^{|x|/r}, \quad w := [y, z]$
if $w \neq 1$ **then return** w
else partially enumerate y^G
  if $|y^G| \leq m^{c_4}$ **then**
    return permutation representation of G on y^G
  else report failure

Theorem 6.4 *Assume $G = G'$ and G has a nontrivial permutation representation of degree $\leq m$. Then with probability $\geq 1/m^{c_5}$, Algorithm* PERM1(G,m) *either returns an element $w \neq 1$ which belongs to a proper normal subgroup of G, or returns a nontrivial permutation representation of G of degree $\leq m^{c_4}$.*

Proof It follows from the assumptions that G has a nonabelian simple quotient $T \leq \Sigma_m$. Let $\varphi : G \to T$ be an epimorphism. Note that φ, and indeed the isomorphism type of T, are not known to the algorithm.

By Lemma 6.3, we have a reasonable hope ($\geq m^{-c}$) that the conjugacy class of $\varphi(y)^T$ has at most $m^{c_4}/2$ elements. Now the probability that $w \in \ker(\varphi)$ is $1/|T : C_T(\varphi(y))| = 1/|\varphi(y)^T| \geq 2m^{-c_4}$.

Moreover, if $|y^G| > m^{c_4}$ then the probability that $w = 1$ is $1/|G : C_G(y)| = 1/|y^G| < m^{c_4}$. So overall the probability that the algorithm succeeds in the sense stated in the Theorem is $> m^{-c-c_4}$. □

Algorithm PERM(G, m)

```
Flip coin
if heads then return G'
else repeat  m^c log(1/ε) times
        W := ∅
        PERM1(G, m)
        if nontrivial permutation representation φ found then
            if φ faithful then return φ, exit
            else let u ∈ ker(φ), u ≠ 1
            return ⟨u^G⟩, exit
        else if w ≠ 1 then add w to W
    end
    if W ≠ ∅ then
        u := element distilled from W (Section 6.1)
        return ⟨u^G⟩
    else return G
```

The proof of Theorem 6.1 is now immediate.

The algorithm described is a very simple one and serves to justify the polynomial-time claim. There is a lot of room for improvement, both theoretical and heuristic. We comment on some of these in Section 6.4.

6.4 Speedups

It is possible to extend the PERM algorithm to work in some cases where no permutation representation of small degree exists. Suppose that, for some simple group T, we can test in t steps (group operations) if a given black-box group G is isomorphic to T. Then, essentially in t steps, we can construct a nontrivial normal subgroup of G if G is not isomorphic to T but has T as a homomorphic image. This was observed in [23]. Note that there are many black-box recognition algorithms which work for entire classes of groups; we do not have to treat each possible homomorphic image T separately.

Theorem 6.5 ([23]) *Suppose that the black-box group G has a nontrivial normal subgroup N with G/N isomorphic to the group T. Then from the transcript of a black-box algorithm which decides that G is not itself isomorphic to T, we can in Monte Carlo polynomial time compute an element of a nontrivial proper normal subgroup of G.*

By the *transcript* of the algorithm we simply mean the sequence of queries to the black-box, together with the responses, from a run of the algorithm.

This means that any improvements to the small/large conjugacy class algorithm for black-box simple groups automatically yield corresponding improvements to the algorithm for the problem of obtaining proper normal subgroups. For alternating groups, the small/large conjugacy class algorithm takes $O(k^3)$ steps, where k is the degree of the group, to achieve success probability $1/k$. This algorithm thus takes $O(k^4)$ steps to have a high $(1-\varepsilon)$ success probability. However, a recent algorithm by Beals, Leedham-Green, Niemeyer, Praeger, and Seress [25], constructs a permutation representation of black box alternating groups in $O(k \log k)$ steps. This is an improvement by a factor of nearly k^3, hence the algorithm for finding a nontrivial normal subgroups of groups with an alternating homomorphic image is sped up by the same factor.

In the terminology of Section 9.2, the above result means that black-box alternating groups A_k admit weak as well as strong *constructive recognition* in $O(k \log k)$ steps.

Weak constructive recognition of T is clearly sufficient for the condition of Theorem 6.5. Therefore the striking results to be discussed in Section 9.2 become relevant: if T is a classical group of dimension d over a field of order q then T, as a black-box group, admits (weak and strong) constructive recognition in time $(dq)^c$ [47]. (Compare this with the time q^{cd} required by the large/small conjugacy class method.)

7 Constructing the semisimple socle

In this section, we assume $\mathrm{Sol}(G) = 1$.

7.1 Blind descent: constructing a semisimple normal subgroup

The term "blind descent" [24] refers to the fact that we shall keep descending along a subnormal chain without ever being able to verify progress. Algorithm PERM represents one step of the blind descent.

Note that the permutation representation obtained if item (a) is produced has degree $\le m^c$ for some constant c.

Recall that by $G^{(\infty)}$ we denote the stable derivative of G (cf. 3.4). The following algorithm performs the "blind descent."

Algorithm CONSTRUCT_SEMISIMPLE(G)

```
Initialize: H := G
repeat n times
    H := H^(∞)
    PERM(H, n/5)
    if nontrivial normal subgroup found then
        H := such a normal subgroup
    else (faithful permutation representation of H found)
        construct minimal subnormal subgroup of H, exit
end
return H
```

In the "**else**" branch we use Luks's composition chain algorithm for permutation groups [56] (as improved by Beals–Seress [26]). We then exit the "**repeat**" loop and return the minimal subnormal subgroup found.

Claim 7.1 *Assume G is a black-box group satisfying $Sol(G) = 1$ and $G \neq 1$. Then with large probability, the output of Algorithm* CONSTRUCT_SEMISIMPLE(G) *is a nontrivial semisimple subnormal subgroup of G.*

Proof Assume that the PERM subroutine does not make any errors throughout the execution of the algorithm.

It is then clear that throughout the process, H is subnormal in G. Moreover $H \neq 1$ because G has no solvable subnormal subgroups. Let now H be the output. We need to prove that H is semisimple.

If the "**else**" branch was invoked at any time then H is in fact simple. So we may assume this did not occur.

Since no subgroup chain of G is longer than $\log|G| \leq n$, H is stable under the loop of the algorithm. Therefore H is perfect and H has no nontrivial permutation representation of degree $\leq n$.

Let $N = T_1 \times \ldots \times T_k$ be the socle of G and let $\varphi : G \to \Sigma_k$ be the permutation representation defined by the conjugation action of G on the set $\{T_1, \ldots, T_k\}$. Observe that $H \leq \ker(\varphi)$ since otherwise H had a nontrivial permutation representation of degree $\leq k < n/5$. So $H \leq M := \mathrm{Aut}(T_1) \times \cdots \times \mathrm{Aut}(T_k)$. But $M^{(\infty)} = N$; therefore $H \leq N$. Any subnormal subgroup of N is normal in N and is of the form $\prod_{i \in I} T_i$ for some $I \subseteq \{1, \ldots, k\}$; therefore H is semisimple.

We conclude that our algorithm may err only if PERM errs; therefore the probability of error is at most $n\varepsilon$ where ε is the error probability of the PERM runs. We have to choose $\varepsilon < 1/n^2$, say, which is accomplished at a cost of $O(\log n)$-fold repetition. □

Now in order to construct a semisimple *normal* subgroup of G, take the normal closure of H.

7.2 Complementing a semisimple direct factor

Next we consider the situation that $G = N \times M$ where G is a black-box group and N is a semisimple normal subgroup of G. We continue to assume $\mathrm{Sol}(G) = 1$. Moreover, we assume that G and N are given (but M is not). We wish to find a nontrivial element of M or else decide that $M = 1$.

Using the algorithms of the preceding section, we may assume that we are given the decomposition $N = T_1 \times \cdots \times T_k$ where the T_i are simple.

Algorithm DISCOVER_COMPLEMENT(G, N, ν)

Initialize: $J := G$
for $i = 1$ **to** k **do**
  if $[J, T_i] \neq 1$ **then**
    $J :=$ CONSTRUCT_COMPLEMENT(J, T_i, ν)
    $J := \langle J^G \rangle$
end
return J

Claim 7.2 *If $G = N \times M$ where $N = T_1 \times \cdots \times T_k$, the T_i are simple, and $Sol(G) = 1$, then Algorithm* DISCOVER_COMPLEMENT(G, N, ν) *returns a subgroup $J \triangleleft M$. If $M \neq 1$ then $J \neq 1$ with probability $\geq 1 - \varepsilon$ assuming $\nu \geq c_2\sqrt{n}\log(1/\varepsilon)$ where c_2 is the same explicit constant as in Claim 5.3.*

Proof First we observe that if N is semisimple and M is an arbitrary group then any normal subgroup of $N \times M$ is of the form $K \times L$ where $K \triangleleft N$ and $L \triangleleft M$. Therefore J as well as the groups H arising in the inner (**repeat**) loop of CONSTRUCT_COMPLEMENT have the form $K \times L$ throughout the algorithm.

We claim that with high probability, $L \neq 1$ throughout the algorithm.

The reasoning is analogous to the proof of Claim 5.4. Consider round i of the **for**-loop, and within it, round j of the **repeat**-loop of the algorithm CONSTRUCT_COMPLEMENT(J, T_i, ν). Let $H_{i,j} = K_{i,j} \times L_{i,j}$ denote the group H of Algorithm CONSTRUCT_COMPLEMENT(J_i, T_i, ν) at the end of this round.

We say that *component i is successful in round (i, j)* if $T_i \cap K_{i,j} = 1$ and $L_{i,j} \neq 1$.

Let us assume that $L_{i,j-1} \neq 1$. This time we choose a random element $(x, y) \in K_{i,j-1} \times L_{i,j-1}$ by selecting $y \in L_{i,j-1}$ first. The probability that $y \neq 1$ is at least 59/60. Let, then, r be a prime dividing the order of y. Let $x = (g_1, \ldots, g_k)$. Now the probability that the order of g_i is not divisible by r is $\geq \delta_i$. If this is the case, component i is successful in this round. Therefore the probability that component i remains unsuccessful through ν rounds is bounded by the left-hand side of equation (7). Therefore the probability that at least one component remains unsuccessful during the entire algorithm is bounded by the left-hand side of equation (8). The result follows as before. □

7.3 Extending a semisimple normal subgroup

Suppose now that we have a black-box group G satisfying $\mathrm{Sol}(G) = 1$ and a semisimple normal subgroup N. We want either to increase N or decide that $N = \mathrm{Soc}(G)$ (and therefore no increase is possible).

Let $N = T_1 \times \cdots \times T_k$ where the T_i are simple. Let $\varphi : G \to \Sigma_k$ be the usual permutation representation on the set of simple factors. Let $K = \ker(\varphi)$ and let $L = K^{(\infty)}$. Note that K can be constructed in Monte Carlo polynomial time using the sifting technique for permutation groups [69, 70, 38] and L is constructed in [12].

Claim 7.3 *$Soc(G) \leq L$ and $L = N \times M$ for some $M \triangleleft G$.*

Proof Clearly, $\mathrm{Soc}(G) \leq K$. Moreover, the image of K/N in $\mathrm{Aut}(N)/N$ under the conjugation action is $\leq \mathrm{Out}(T_1) \times \cdots \times \mathrm{Out}(T_k)$; the right hand side being solvable, the image of L is trivial. Therefore $L \leq N \cdot C_G(N) = N \times C_G(N)$. Any normal subgroup $L \geq N$ of a group of the form $N \times T$, N semisimple, has the form $N \times M$. □

Next we apply Algorithm DISCOVER_COMPLEMENT(L, N, ν). Assuming $N \neq \mathrm{Soc}(G)$ and therefore $M \neq 1$, by Claim 7.3 we obtain a nontrivial element $u \in M$. Let $H = \langle u^G \rangle$. Construct a nontrivial semisimple G-normal subgroup T in H; replace N by $N \times T$. Repeat $\leq n$ times to obtain $\mathrm{Soc}(G)$.

8 How to do without prime factorization

8.1 The complexity of determining the order of elements

Proposition 8.1 *Without any knowledge about our black-box group other than its encoding length n, the determination of the order of an element may require exponential Monte Carlo time.*

Proof Let p be a random prime number between 1 and 2^n and assume $G = \langle g \rangle$ is cyclic of order p, (p is unknown to us). Suppose our polynomial-time Monte Carlo algorithm makes t steps, including $\leq t$ queries regarding whether an element so far constructed is the identity. All such questions will be of the form "$g^j = 1$?" where $|j| \leq 2^t$. For $j = 0$ the answer of course is "yes," but the probability that p divides any other j that arises is less than $\nu(j)/\pi(2^n)$ where $\nu(j) = O(\log|j|/\log\log|j|) = O(t/\log t)$ is the number of distinct prime divisors of j and $\pi(2^n) \sim 2^n/(n\ln 2)$ is the number of primes $\leq 2^n$. Therefore the probability that a query with $p|j \neq 0$ will ever be asked is $< c(t/\log t)/(2^n/n)$. For this probability to be non-negligible, t must be exponentially large. □

On the other hand, the assumption that $\mathcal{P}$ is known is unpleasant since for linear groups it assumes factoring integers, a problem generally believed not to be solvable in polynomial time [1]. Even if $G \leq GL(1,p)$, we need to factor $p-1$ (p prime), and this task alone is equivalent to factoring arbitrary integers, as seen from the following result.

Theorem 8.2 (E. Bach, J. Shallit) *Factoring any integer can be reduced, in Monte Carlo polynomial time, to factoring integers of the form $p-1$ (p prime) into their prime factors, assuming the Extended Riemann Hypothesis (ERH).*

Proof Suppose we wish to factor the n-digit integer N into prime factors. Linnik's celebrated theorem asserts that there exists a prime $p \equiv 1 \pmod N$ such that $p < N^c$. Bach and Shallit show that such a prime p can be found in Monte Carlo polynomial time, assuming ERH [21, p. 241, Ex. 30]. Now factoring $p-1$ into its prime factors will split N. □

We used the primes to determine the order of elements. We are unable to prove that determining the order elements of $GL(d,q)$ accessible to a Monte Carlo polynomial-time algorithm does actually require factoring integers of the form $p^i - 1$. But if we broaden our scope and consider linear groups over the rings $\mathbb{Z}/N\mathbb{Z}$, then factoring integers in fact becomes equivalent to finding the order of elements in such a group. Indeed, even determining the order of 1×1 invertible matrices over $\mathbb{Z}/N\mathbb{Z}$ is as hard as factoring N:

Proposition 8.3 (G. L. Miller [59]) *Factoring the composite integer N is reducible in Monte Carlo polynomial time to determining the order mod N of random integers k, $1 \leq k \leq N$, g.c.d.$(k, N) = 1$.*

This fact is the starting point of Peter Shor's polynomial-time algorithm for factoring integers on *quantum-mechanical* Turing-machines. Shor gives a nice explanation of Miller's result and provides additional background and literature on factoring [68, p. 1498].

These observations suggest that it is not possible to determine the order of elements of matrix groups in polynomial time. Nor is it necessary, however, at least for the class of black-box groups of greatest interest to us: "black-box groups of a known characteristic p."

8.2 Black-box groups of characteristic p

We say that G is a black-box group *of characteristic* p if G is a black-box group of some encoding length n and G is a section (quotient of a subgroup) of $GL(d,p)$ where $d = \lfloor n/\log p \rfloor$. When we say that such a group G is *given*, we tacitly assume that p is known.

Proposition 8.4 *For q a power of the prime p, the subgroups of $GL(f,q)$ as well as their quotients are black-box groups of characteristic p as long as their elements are encoded as matrices over $\mathbb{F}_q$ or over a subfield of $\mathbb{F}_q$.*

Proof If $q = p^k$ then $GL(f,q) \leq GL(fk,p)$ and therefore $|G|$ divides $|GL(fk,p)|$. Moreover, $n \geq f^2 \log q = f^2 k \log p \geq fk \log p$ and therefore $fk \leq d$. □

Remark 8.5 Note that any black-box group G becomes a black-box group of characteristic p (for any p) if we pad the encoding with a sufficiently long string of dummy symbols. However, for the purposes of polynomial-time algorithms, "padding" is limited to increasing the encoding length polynomially. In particular, the fact that G is a Lie-type simple group of characteristic p does not exclude the possibility of G being represented as a black-box group of characteristic $r \neq p$. However, in this case we have powerful tools to deals with G, as the following result shows.

Theorem 8.6 *Let G be a black-box group of characteristic r and suppose G is isomorphic to a simple group of Lie type of characteristic $p \neq r$. Then we can find a faithful permutation representation of G in polynomial time.*

This result then implies that we can learn virtually everything about G in polynomial time: by a theorem of Kantor [46], we can find the standard name of G and its representation in its projective action on the natural module.

Proof By the Landazuri–Seitz theorem (Theorem 4.1), G has a permutation representation of polynomially bounded degree, and therefore by Theorem 6.1 a faithful permutation representation of G can be computed in Monte Carlo polynomial time. □

8.3 More statistical group theory

We require a more complete version of Theorem 4.2, our main statistical tool about simple groups. For a set $S \subseteq G$ we write $S^G = \bigcup_{g \in G} S^g$ (the union of conjugates).

Theorem 8.7 ([17]) *Let G be a finite simple group of Lie type. Then G has two cyclic maximal tori T_1, T_2 of relatively prime orders such that for each i, the set T_i^G has non-negligible density: $|T_i^G|/|G| \geq c/d$, where d is the dimension of the linear space on which G acts projectively and $c > 0$ is a constant.*

(For classical groups, the result holds with $c = 1/2$.) This result clearly implies Theorem 4.2 for simple groups of Lie type: if r divides $|T_1|$ then all elements of T_2^G are r-regular; otherwise all elements of T_1^G are r-regular.

The orders of the maximal tori in question are listed in [17]. For a classical group $X_d(q)$ ($X \in \{PSL, PSp, PSU, P\Omega^{\pm}\}$; for the unitary groups, the field has order q^2) the orders of these tori are of the form $(k_1k_2)/(k_3k_4)$, where $k_1 = q^j \pm 1$ for $j \leq d$, $k_2 = 1$ or $k_2 = q \pm 1$, $k_3 = 1$ or $k_3 = q \pm 1$, and $k_4 | d$. For exceptional groups $Y(q)$ we have numbers of similar form with $j \leq 7$, $k_3 \leq 3$, and $k_4 = 1$; moreover, $\Phi_j(q^s)$ for $s \leq 3$ and $j = 6, 3$ (where $\Phi_j(x)$ denotes the j-th cyclotomic polynomial), furthermore, $\Phi_j(q)$ for $j = 30, 15$ (these arise for $E_8(q)$), and finally the following integer factors of $\Phi_j(q^s)$ which we shall refer to as *semicyclotomic* factors of $q^j - 1$: $q \pm \sqrt{2q} + 1$ (factors of $\Phi_4(q)$ for $q = 2^{2t+1}$), $q \pm \sqrt{3q} + 1$ (factors of $\Phi_6(q)$ for $q = 3^{2t+1}$), and $q^2 + q + 1 \pm \sqrt{2q}(q+1)$ (factors of $\Phi_6(q^2)$ for $q = 2^{2t+1}$).

8.4 Pseudo-order of elements in black-box groups of characteristic p

Let $\mathcal{P}$ be a set of pairwise relatively prime integers. Assume that $m = \prod_{p \in \mathcal{P}} p^{k_p}$ is a multiple of $|G|$. (We may take $k_p = \lfloor n/\log p \rfloor$ where n is the encoding length of G.) Let us call the members of $\mathcal{P}$ *pretend-primes*. Let us define the *pseudo-order* of $g \in G$ with respect to the set $\mathcal{P}$ of pretend-primes the smallest positive integer ℓ such that $g^\ell = 1$ and ℓ is a product of pretend-primes. Let $|g|_{\mathcal{P}}$ denote this quantity. Note that $|g|_{\mathcal{P}}$ is computed in polynomial time by Algorithm ORDER$(G, \mathcal{P}, g)$ (Section 3.6).

Celler and Leedham-Green [30] suggest that factoring the order of $GL(d,p)$ into easily computed pretend-primes will suffice in lieu of actual prime factorization for most applications. We turn this idea into a rigorous statement regarding the algorithms of this paper. Our pretend-primes will have to go slightly beyond the small primes and the cyclotomic factors $\Phi_j(p)$ of the integers $p^i - 1 = \prod_{j|i} \Phi_j(p)$ recommended by [30].

Let $\mathcal{L}$ be a set of positive integers. We define the *relatively prime refinement* of $\mathcal{L}$ as the smallest set $\mathcal{P}$ of pairwise relatively prime integers such that each member of $\mathcal{L}$ is a product of members of $\mathcal{P}$; and the sum of the elements of $\mathcal{P}$ should be maximal subject to this condition. We shall see that this set is unique and denote it by $\mathcal{P}(\mathcal{L})$.

For a set $\mathcal{M}$ of positive integers and an integer m, let us write $\mathcal{M} \vdash m$ if m is either the g.c.d. or the quotient of two members of $\mathcal{M}$. We say that $\mathcal{M}$ is closed under $\vdash$ if $\mathcal{M} \vdash m$ implies $m \in \mathcal{M}$. Let $\bar{\mathcal{M}}$ denote the closure of $\mathcal{M}$ (the smallest closed set containing $\mathcal{M}$).

Claim 8.8 *Let $\mathcal{L}$ be a set of positive integers. Then $\mathcal{P}(\mathcal{L})$ is the set of minimal elements with respect to divisibility of $\bar{\mathcal{L}} \setminus \{1\}$.*

(An element m of a set $\mathcal{N}$ of positive integers is *minimal* with respect to divisibility if no other element of $\mathcal{N}$ divides m.) The Claim includes the statement that $\mathcal{P}(\mathcal{L})$ is unique. We leave the easy proof to the reader. □

The following algorithm is folklore; it constructs $\mathcal{P}(\mathcal{L})$.

Algorithm REFINE($\mathcal{L}$)

Initialize: $\mathcal{P} := \mathcal{L}$
while not all pairs in $\mathcal{P}$ are relatively prime **do**
  pick $a, b \in \mathcal{P}, a \neq b$ such that $f :=$g.c.d.$(a, b) \neq 1$
  delete a, b from $\mathcal{P}$, add $f, a/f, b/f$ to $\mathcal{P}$
end
return $\mathcal{P}$

We note that each round reduces the product of the elements of $\mathcal{P}$ by a factor of $d \geq 2$; therefore the process terminates in $\leq \sum_{m \in \mathcal{L}} \log m$ rounds. This is less than the length of the input (total bit-length of the integers $m \in \mathcal{L}$), so the algorithm runs in polynomial time. We leave the easy proof of correctness to the reader. □

Remark 8.9 The "factor refinement problem" asks not only to produce the set $\mathcal{P}(\mathcal{L})$ but also to express the product $\prod_{m \in \mathcal{L}}$ as a product of elements of $\mathcal{P}(\mathcal{L})$. The REFINE algorithm can easily be adapted for this purpose. The "factor refinement problem," and specifically the idea of the REFINE algorithm, have a long history, going back to Stieltjes (1890). Bach, Driscoll, and Shallit [20] give a detailed account of early as well as recent work and a multitude of applications. They also present a definitive result on the complexity of the problem which we state.

Let k be the total bit-length of the integers in $\mathcal{L}$. Then a naive implementation of the REFINE procedure runs in $O(k^3)$ time (bit-operations). Note that the above description of REFINE does not specify how to keep track of which pairs remain not relatively prime and in what order to process them. By appropriately organizing this process, [20] reduces the running time to $O(k^2)$.

Now here comes our **recipe for creating the set of pretend-primes** for our algorithms.

Given that our groups are sections of $GL(d, p)$ (quotients of subgroups), first we create a list $\mathcal{L}(d, p)$ of positive integers as follows.

Include in $\mathcal{L}(d, p)$ all primes ≤ 47 and the primes 59, 67, 71 (these are the primes occurring in sporadic simple groups). Include all primes $\leq d^{c_1}$ where c_1 is the constant in Theorem 4.1. (This takes care of many things, see below.) Further, include all cyclotomic factors $\Phi_i(p^j)$ for $ij \leq d$.

Finally, include the following semicyclotomic factors: If $p = 2$, include $2^{2t+1} \pm 2^{t+1}+1, 0 \le t \le (d-4)/8$, and $2^{4t+2}+2^{2t+1}+1 \pm 2^{t+1}(2^{2t+1}+1), 0 \le t \le (d-12)/24$. If $p = 3$, include $3^{2t+1} \pm 3^{t+1}+1, 0 \le t \le (d-6)/12$. This completes the list $\mathcal{L}(d,p)$.

Let now $\mathcal{P}(d,p)$ denote the relatively prime refinement of $\mathcal{L}(d,p)$. This will be our set of pretend-primes.

Corollary 8.10 *Let S be a nonabelian simple section of $GL(d,p)$. Then for any prime r, at least a c/d fraction of the elements of $g \in S$ has pseudo-order $|g|_{\mathcal{P}}$ relatively prime to r with respect to the set $\mathcal{P} = \mathcal{P}(d,p)$ of pretend-primes.*

Proof If S is sporadic then $\pi(S) \subseteq \mathcal{P}$. The same holds if S is alternating of degree k since $k \le d+1$, and also if S is of Lie type of characteristic $s \ne p$ since in that case the natural module for S has order s^t for some t and $s^t \le d^{c_1}$ by Theorem 4.1. If S is Lie type of characteristic p then let T_1 and T_2 be the maximal tori of S discussed in Theorem 8.7. But then, $|T_i|$ is a product of the cyclotomic and semicyclotomic factors included in $\mathcal{L}(d,p)$. (We note in this connection that $\Phi_i(x^j) = \prod_{h:h|j,gcd(h,t)=1} \Phi_{tj/h}(x)$.) Therefore if the prime r does not divide $|T_i|$ (which is true for at least one of T_1 and T_2) then the pseudo-order of elements of T_i^G with respect to $\mathcal{P}(d,p)$ is not divisible by r either (since the pseudo-order will divide $|T_i|$). □

It follows that our algorithms work correctly using the set $\mathcal{P}(d,p)$ of pseudo-primes.

The following observation speeds up the refinement algorithm for the case $\mathcal{L} = \mathcal{L}(d,p)$.

Proposition 8.11 *Let r be a prime, q an arbitrary integer, $i > j$ positive integers. If $r|\Phi_i(q)$ and $r|\Phi_j(q)$ then $r|i$.*

Proof The conditions imply that r divides g.c.d.$(q^i-1, q^j-1) = q^k - 1$, where k =g.c.d.(i,j). Now $q^i - 1$ is divisible by $\Phi_i(q)(q^k-1)$, therefore q is a multiple root of the polynomial $x^i - 1$ over $\mathbb{F}_r$. This implies that q is a root of the derivative, i. e., $r|iq^{i-1}$, and therefore $r|i$. □

As a consequence, the following version of the REFINE procedure will succeed. For a prime r, the *r-free part* of an integer $a \ne 1$ is the integer $a' = a/r^k$ where r^k is the largest power of r which divides a.

Algorithm SIMPLE-REFINE(d,p)

> Initialize: $\mathcal{P} := \mathcal{L}(d,p)$
> **for** all primes $r \le d$ **do**
>   **for** all $a \in \mathcal{P}$ **do**
>   replace a by its r-free part
>   **end**
> **end**
> **return** $\mathcal{P}$

Claim 8.12 *Algorithm* SIMPLE-REFINE(d,p) *returns the set* $\mathcal{P}(d,p)$.

Indeed, this is immediate from Proposition 8.11 and the fact that all primes $\le d$ are included in $\mathcal{L}(d,p)$, with the additional observation that the pairs of semicyclotomic factors included in $\mathcal{L}(d,p)$ are trivially relatively prime. □

Remark 8.13 Given the set $\mathcal{L}(d,p)$, the Algorithm SIMPLE-REFINE(d,p) runs in $O(d^3 \log p)$ bit operations. Note that the total number of bits of the input $\mathcal{L}(d,p)$ is $\Theta(d^2 \log p)$.

Remark 8.14 (P.P. Pálfy kindly communicated the following result to us.) *Under the conditions of Proposition 8.11, the g.c.d. of* $\Phi_i(q)$ *and* $\Phi_j(q)$ *is either* 1 *or a prime number* r; *in the latter case,* $i = sr^k$ *and* $j = sr^\ell$ *for some nonnegative integers* s, k, ℓ *where* $s|r-1$.

9 Statistical recognition of black box simple groups: a project

In this section we briefly review major progress on the following question.

Problem 9.1 *Given a black-box* simple *group* G *of characteristic* p, *what can we say about* G *in Monte Carlo polynomial time?*

This is a *promise problem:* we expect a correct answer only if G is indeed simple, but the fact of simplicity need not be verified. (See the Problem 10.2.)

9.1 Name-recognition

The simplest type of answer is *name-recognition:* we wish to tell which simple group G is isomorphic to (print the standard name of G). It now seems reasonable to expect that the following conjecture holds:

Conjecture 9.2 Let G be a simple group of Lie type of characteristic p given as a black-box group. Assume p is known. Then one can compute the standard name of G in Monte Carlo polynomial time.

Combined with Theorem 1.2 this conjecture implies that we can *list the names of all nonabelian composition factors (with multiplicity) of a black-box group of given characteristic in Monte Carlo polynomial time.* (If G is given as a black-box group of the wrong characteristic, then we can deal with G in a very strong sense, see Theorem 8.6.)

Conjecture 9.2 is addressed in [16]. In 1955, Artin introduced four basic invariants of simple groups of Lie type to distinguish simple groups by their orders [4]. For an update to include the classes discovered after Artin's paper, see [48]. Two of Artin's invariants, called α and β in [4], turn out to be computable in Monte Carlo polynomial time for black-box simple groups of a given characteristic. (The other two do not seem to, since their computation would depend on finding p-singular elements in a group of Lie type of characteristic p, an elusive quest.) However, it turns out that α and β already "almost determine" G in the sense that at most 7 groups of Lie type of characteristic p share the same pair (α, β). Results of Niemeyer and Praeger [63] on ppd-elements in classical groups and their extension to exceptional groups [17] imply further breakup of these classes.

While the determination of the Artin invariants depends only on sampling the orders of the elements of the given black-box group, such a simple-minded approach will not separate the two infinite classes of simple groups of equal orders: $PSp(2k, q)$ and $P\Omega(2k+1, q)$, q odd. However, a breakthrough by Altseimer and Borovik eliminated this obstacle: *these two classes are distinguishable in Monte Carlo polynomial time* [3]. This result gives reason for optimism regarding the completion of the proof of Conjecture 9.2.

9.2 Constructive recognition

A much more ambitious goal is *strong constructive recognition*: given a black-box simple group T, find an isomorphism to a "natural representation," with the isomorphism efficiently computable in both directions. (The "natural representation" of Lie-type simple groups is its projective action on the natural module. The natural representation of the alternating groups is self-explanatory. For the purposes of the polynomial-time theory we are not concerned about sporadic groups.)

Weak constructive recognition requires finding a presentation in terms of generators and relations.

We note that strong constructive recognition implies weak constructive recognition, assuming the *Short Presentation Conjecture* mentioned in Section 2.2. Therefore this implication is known to hold for all finite simple groups except for the rank-1 twisted types [14].

Sims's "strong generators" yield weak constructive recognition of permutation groups in polynomial time. This observation, combined with Theorem 6.4, yields weak constructive recognition of simple black-box groups in Monte Carlo time polynomial in $n + m$ where n is the input length and m is the smallest degree of a permutation representation. For classical groups, Kantor [46] turns the permutation representation into "strong constructive recognition" in polynomial time.

A breakthrough in strong constructive recognition of black-box simple groups came with the recent paper by Cooperman, Finkelstein, and Linton [35] who solved this problem in Monte Carlo polynomial time for $PSL(d, 2)$. Following work by Prabhav Morje on nearly linear time algorithms for Sylow subgroups in permutation groups [60], the result of [35] was extended by Bratus, Cooperman, Finkelstein, and Linton [28] to $PSL(d, q)$ for *tiny* q (see below) and finally in a monumental paper by Kantor and Seress [47] to all classical groups over *tiny fields.*

Following accepted terminology in the theory of computing, we say that an input parameter q is *tiny* if it is input in *unary* (rather than in binary), i. e., it contributes q rather than $\log q$ to the length of the input. So a polynomial-time algorithm for a $d \times d$ matrix group over a *tiny field* runs in time $O((dq)^c)$ rather than $O((d \log q)^c)$, as required if the field is not tiny (q is the order of the field).

(We note that *over tiny fields, strong recognition automatically implies weak recognition* since the Steinberg presentations [71] are of polynomial length as a function of dq (but not of $d \log q$).)

We mention that weak constructive recognition is provably hard (exponential time) for *elementary abelian black-box groups* [19].

10 Open problems

We conclude with a list of open problems. All these problems represent bottlenecks in the polynomial-time attempts to obtain a more complete description of the normal structure of a black-box group G. For a further explanation of the connections we refer to [10].

Problem 10.1 (The p-singular element problem) *Given a black-box group known to be isomorphic to $PSL(2,p^k)$ (or any other simple group of Lie type of characteristic p), find an element of order p.)*

The difficulty is that when p^k is large then p-singular elements (elements of order divisible by p) become rare in these groups, so simple sampling is unlikely to find a p-singular element. This problem may hold the key to the *constructive recognition* (strong or weak) of $PSL(2,p^k)$ (and other simple groups), the most significant open problem in the area.

Problem 10.2 (The p-core problem) *Given a black-box group G of characteristic p, decide (in Monte Carlo polynomial time) whether or not $O_p(G) = 1$.*

This problem is open even in the case when $G = G'$, $O_p(G)$ is known to be an elementary abelian minimal normal subgroup, and $G/O_p(G) = PSL(2,p^k)$. A Monte Carlo polynomial-time solution exists for $k = 1$, but the question is open even for $k = 2$ [18]. Again, Problem 10.1 seems relevant. This problem indicates the difficulty of recognizing simplicity of a (nonabelian) black-box group. (For abelian black-box groups, deciding simplicity requires exponential time [19], but for nonabelian black-box groups the problem could be easier.)

In contrast to this (affine) case, the case of central extensions is easy: *if G is a black-box group known to satisfy $G = G'$ and $G/Z(G)$ is known to be simple then $Z(G)$ can be found in Monte Carlo polynomial time.* (This is yet another simple algorithmic consequence of Theorem 4.2, cf. [10].)

We note that for matrix groups $G \leq GL(d,p)$ (p prime), the quotient $G/O_p(G)$is itself a subgroup of $GL(d,p)$ and can be found in deterministic polynomial time. Indeed a composition chain of the G-module $\mathbb{F}_p^d$ can be constructed in polynomial time via Rónyai's algorithm for the radical of an algebra [65]. We restrict the G-action to the direct sum of the quotients of the composition chain. $O_p(G)$ is the kernel of this action. (A practical implementation would use the Holt–Rees generalization of the Meataxe [45], available in MAGMA [27].)

The preceding problem considered the key question in the case when an abelian group was sitting "under" a nonabelian simple group. The next problem concerns the converse: an abelian (or solvable) group on the "top."

Problem 10.3 (Outer automorphism problem) *Given a black-box group of characteristic p known to satisfy $T \leq G \leq AutT$, recognize T within G in Monte Carlo polynomial time. (I. e., we ask for membership testing in T for any $g \in G$.) Here T is a simple group of Lie-type of characteristic p and we assume that the name of T is known.*

References

[1] L. M. Adleman: Algorithmic Number Theory: The Complexity Contribution. *Proc. 35th IEEE FOCS*[3], 1994, 88-113.

[2] L. M. Adleman, J. DeMarrais: A subexponential algorithm for discrete logarithms over all finite fields. *Proc. CRYPTO'93*, 1993.

[3] Christine Altseimer, A. Borovik: On the recognition of $PSp(2n, q)$ and $P\Omega(2n+1, q)$ for q odd. Manuscript, 1997.

[4] E. Artin: The orders of the classical simple groups. *Comm. Pure Appl. Math.* **8** (1955), 455-472.

[5] M. Aschbacher: On the maximal subgroups of the finite classical groups. *Invent. Math.* **76** (1984), 469–514.

[6] L. Babai: Monte Carlo algorithms in graph isomorphism testing. Université de Montréal Tech. Rep. DMS 79-10, 1979 (pp. 42).

[7] L. Babai: Local expansion of vertex-transitive graphs and random generation in finite groups. *Proc. 23rd ACM STOC*[4], 1991, 164-174.

[8] L. Babai: Automorphism groups, isomorphism, reconstruction. Chapter 27 of the *Handbook of Combinatorics*, R. L. Graham, M. Grötschel, L. Lovász, eds., North-Holland – Elsevier, 1995, 1447-1540.

[9] L. Babai: Randomization in group algorithms: conceptual questions. In: [40], 1-16.

[10] L. Babai, R. Beals: A polynomial-time theory of black box groups 2. In preparation.

[11] L. Babai, R. Beals, D. Rockmore: Deciding finiteness of matrix groups in deterministic polynomial time. *Proc. ISSAC'93*, Kiev, ACM Press, 1993, 117-126.

[12] L. Babai, G. Cooperman, L. Finkelstein, E. M. Luks, Á. Seress: Fast Monte Carlo algorithms for permutation groups. *J. Comput. System Sci.* **50** (1995), 296-307. (Preliminary version appeared in *Proc. 23rd ACM STOC*, 1991, 90-100.)

[13] L. Babai, G. Cooperman, L. Finkelstein, Á. Seress: Nearly linear time algorithms for permutation groups with a small base. *Proc. ISSAC'91 (Internat. Symp. on Symbolic and Algebraic Computation)*, Bonn 1991, 200-209.

[14] L. Babai, A. J. Goodman, W. M. Kantor, E. M. Luks, P. P. Pálfy: Short presentations for finite groups. *J. Algebra* **194** (1997), 79-112.

[15] L. Babai, E. M. Luks, Á. Seress: Computing composition series in primitive groups. In: [39], 1–16.

[16] L. Babai, P. P. Pálfy: Recognizing finite simple groups by order statistics. Manuscript, 1998.

[17] L. Babai, P. P. Pálfy, J. Saxl: On the number of p-regular elements in simple groups. Manuscript, 1998.

[18] L. Babai, A. Shalev: Finding p-singular elements in affine groups. In preparation.

[19] L. Babai, E. Szemerédi: On the complexity of matrix group problems I. *Proc. 25th IEEE FOCS*, 1984, 229–240.

[3]FOCS: IEEE Symposium on Foundations of Computer Science. IEEE Computer Society Press.

[4]STOC: ACM Symposium on Theory of Computing. ACM Press.

[20] E. Bach, J. Driscoll, J.O. Shallit: Factor refinement. *J. Algorithms* **15** (1993), 199–222.
[21] E. Bach, J.O. Shallit: *Algorithmic Number Theory, Vol. I: Efficient Algorithms.* MIT Press, 1996.
[22] R. Beals: Algorithms for matrix groups and the Tits alternative. *Proc. 36th IEEE FOCS,* 1995, 593–602.
[23] R. Beals: Towards polynomial time algorithms for matrix groups. In: [40], 31–54.
[24] R. Beals, L. Babai: Las Vegas algorithms for matrix groups. *Proc. 34th IEEE FOCS,* 1993, 427–436.
[25] R. Beals, C. Leedham-Green, Alice Niemeyer, Cheryl Praeger, Á. Seress: A *mélange* of algorithms for recognising black-box alternating and symmetric groups, in prep.
[26] R. Beals, Á. Seress: Structure forest and composition factors for small base groups in nearly linear time. *Proc. 24th ACM STOC,* 1992, 116–125.
[27] W. Bosma, J. Cannon: MAGMA *Handbook.* Sydney, 1993.
[28] S. Bratus, G. Cooperman, L. Finkelstein, S. A. Linton: Constructive recognition of a black box group isomorphic to $GL(n,q)$. In preparation.
[29] R. W. Carter: *Simple Groups of Lie Type*, Wiley, New York, 1989.
[30] F. Celler, C. Leedham-Green: Calculating the order of an invertible matrix. In: [40], 55–60.
[31] F. Celler, C. Leedham-Green: A non-constructive recognition algorithm for the special linear and other classical groups. In: [40], 61–68.
[32] F. Celler, C.R. Leedham-Green, S.H. Murray, A.C. Niemeyer, E.A. O'Brien: Generating random elements of a finite group. *Comm. Algebra* **23** (1995), 4931–4948.
[33] J.H. Conway, R.T. Curtis, S.P. Norton, R.A. Parker, R.A. Wilson: *ATLAS of Finite Groups.* Clarendon Press, Oxford, 1985.
[34] G. Cooperman, L. Finkelstein: Combinatorial tools for computational group theory. In: [39], 53–86.
[35] G. Cooperman, L. Finkelstein, S. Linton: Constructive recognition of a black box group isomorphic to $GL(n,2)$. In: [40], 85–100.
[36] W. Feit, J. Tits: Projective representations of minimum degree of group extensions. *Canad. J. Math.* **30** (1978), 1092–1102.
[37] K. Friedl, L. Rónyai: Polynomial time solutions of some problems in abstract algebra. *Proc. 17th ACM STOC,* 1985, 153–162.
[38] M.L. Furst, J. Hopcroft, E.M. Luks: Polynomial-time algorithms for permutation groups. *Proc. 21st IEEE FOCS,* 1980, 36–41.
[39] *Groups and Computation.* Proc. 1991 DIMACS Workshop. (L. Finkelstein and W.M. Kantor, eds.) DIMACS Ser. Discr. Math. Theoret. Comput. Sci. 11, Amer. Math. Soc. 1993.
[40] *Groups and Computation II.* Proc. 1995 DIMACS Workshop. (L. Finkelstein and W.M. Kantor, eds.) DIMACS Ser. in Discr. Math. and Theor. Comput. Sci. Vol 28, A.M.S. 1997.
[41] D.F. Holt, C.R. Leedham-Green, E.A. O'Brien, Sarah Rees: Testing matrix groups for primitivity. *J. Algebra* **184** (1996), 795–817.
[42] D.F. Holt, C.R. Leedham-Green, E.A. O'Brien, Sarah Rees: Computing matrix group decompositions with respect to a normal subgroup. *J. Algebra* **184** (1996), 818–838.
[43] D.F. Holt, Sarah Rees: An implementation of the Neumann–Praeger algorithm for the recognition of special linear groups. *J. Experimental Math.* **1** (1992), 237–242.
[44] M. Jerrum: A compact representation for permutation groups. *J. Algorithms* **7** (1986), 60–78.
[45] D.F. Holt, Sarah Rees: Testing modules for irreducibility. *J. Austral. Math. Soc.*

57 (1994), 1–16.

[46] W.M. Kantor: Sylow's theorem in polynomial time. *J. Comput. System Sci.* **30** (1985), 359–394.

[47] W.M. Kantor, Á. Seress: Black box classical groups. Manuscript, 1997. pp. 152. *Trans. Amer. Math. Soc.*, to appear.

[48] W. Kimmerle, R. Lyons, R. Sandling, D.N. Teague: Composition factors from the group ring and Artin's theorem on orders of simple groups. *Proc. London Math. Soc.* **60** (1990), 89–122.

[49] P. Kleidman, M. Liebeck: *The Subgroup Structure of the Finite Classical Groups.* LMS Lecture Notes 129, Cambridge Univ. Press, 1990.

[50] D.E. Knuth: Efficient representation of perm groups. *Combinatorica* **11** (1991), 33–44. (Preliminary version circulated since 1981.)

[51] C.R. Leedham-Green, E.A. O'Brien: Tensor products are projective geometries. *J. Algebra* **189** (1997), 514–528.

[52] J.S. Leon: On an algorithm for finding a base and strong generating set for a group given by generating permutations. *Math. Comput.* **35** (1980), 941–974.

[53] R. Lovorn: *Rigorous, subexponential algorithms for discrete logarithms over finite fields.* Ph.D. thesis, University of Georgia, 1992.

[54] V. Landazuri, G.M. Seitz: On the minimal degrees of projective representations of the finite Chevalley groups. *J. Algebra* **32** (1974), 418–443.

[55] E.M. Luks: Isomorphism of graphs of bounded valence can be tested in polynomial time. *J. Comput. System Sci.* **25** (1982), 42–65.

[56] E.M. Luks: Computing the composition factors of a permutation group in polynomial time. *Combinatorica* **7** (1987), 87–99.

[57] E.M. Luks: Computing in solvable matrix groups. *Proc. 33rd IEEE FOCS*, 1992, 111–120.

[58] E. M. Luks: Permutation groups and polynomial-time computation. In: [39], 139–175.

[59] G.L. Miller: Riemann's hypothesis and tests for primality. *J. Comput. System Sci.* **13** (1976), 300–317.

[60] P. Morje: *A nearly linear algorithm for Sylow subgroups of permutation groups.* Ph.D. Thesis. Ohio State University, 1996.

[61] P.M. Neumann, Cheryl E. Praeger: A recognition algorithm for special linear groups. *Proc. London Math. Soc.* **65** (1992), 555–603.

[62] Alice Niemeyer, Cheryl E. Praeger: Implementing a recognition algorithm for classical groups. In: [40], pp. 273-296.

[63] Alice C. Niemeyer, Cheryl E. Praeger: A recognition algorithm for classical groups over finite fields. *Proc. London Math. Soc.* **77** (1998), 117-169.

[64] R.A. Parker: The computer calculation of modular characters (the meat-axe). *In: Computational Group Theory* (Durham, 1982), (M.D. Atkinson, ed.), Academic Press, 1984, 267-274.

[65] L. Rónyai: Computing the structure of finite algebras. *J. Symbolic Comput.* **9** (1990), 355–373.

[66] M. Schönert *et al.*: GAP - *Groups, Algorithms, and Programming.* Lehrstuhl D für Mathematik, RWTH Aachen, Germany, 1994.

[67] Á. Seress: *Permutation Group Algorithms.* Book, in prep.

[68] P.W. Shor: Polynomial-time algorithms for prime factorization and discrete logarithms on a quantum computer. *SIAM J. Comput.* **26** (1997), 1484–1509.

[69] C.C. Sims: Computation with permutation groups. *Proc. Second Symp. on Symbolic and Algebraic Manipulation* (New York) (S.R. Petrick, ed.), ACM, 1971, 23–28.

[70] C. C. Sims: Some group-theoretic algorithms. *In: Springer Lecture Notes in Math.*

697 (1978), 108–124.
[71] R. Steinberg: *Lectures on Chevalley groups.* Mimeographed notes. Yale University, 1967.

TOTALLY AND MUTUALLY PERMUTABLE PRODUCTS OF FINITE GROUPS

A. BALLESTER-BOLINCHES*, M. D. PÉREZ-RAMOS* and M. C. PEDRAZA-AGUILERA†[1]

*Departamento de Algebra, Universidad de València, C/ Doctor Moliner 50, 46100 Burjassot, València, Spain
†Departamento de Matemática Aplicada, E.U.I., Universidad Politécnica de Valencia, Camino de Vera, s/n 46071 Valencia, Spain

Only finite groups are considered here. The well-known fact that the product of two normal supersoluble subgroups of a group is not necessarily supersoluble shows that formations, even saturated, need not be closed under the product of normal subgroups. Clearly, any formation is, however, closed under direct products and even under central products. Therefore it is interesting to study factorized groups whose subgroup factors are connected by certain permutability properties. In fact, the following question can be formulated. Let the group $G = HK$ be the product of subgroups H and K of G which lie in a formation $\mathcal{F}$. What is a relationship between the factors H and K -weaker than their elementwise permutability in the case of a direct product- which will guarantee $G \in \mathcal{F}$? In [1] the following results are proved:

Theorem 1 *Let $G = HK$ be a group which is the product of two supersoluble subgroups H and K. If every subgroup of H permutes with every subgroup of K (we say in this case that H and K are totally permutable subgroups of G), then G is supersoluble.*

Theorem 2 *Let $G = HK$ be the product of two supersoluble subgroups H and K. Suppose that H permutes with every subgroup of K and K permutes with every subgroup of H (we say then that H and K are mutually permutable subgroups of G). Then G is supersoluble provided that either G' is nilpotent or H is nilpotent or K is nilpotent.*

Direct products and more generally, central products are typical examples of totally permutable products. It is also clear that totally permutable products are mutually permutable products. However the converse does not hold. Let G be the symmetric group of degree 4. If H is a Sylow 2-subgroup of G and K is the alternating group of degree 4, then H and K are mutually permutable subgroups of G. It is clear that H and K are not totally permutable.

In [8] Maier proved that Theorem 1 is not accidental and could be obtained owing to a general completeness property of all saturated formations which contain the class $\mathcal{U}$ of supersoluble groups. More precisely, he proved:

[1]The research of the first and second authors is supported by Proyecto PB 94-0965 of DGICYT, MEC, Spain. Part of the research of the third author has been done in the Johannes Gutenberg-Universität Mainz under the support of Beca Bancaixa-Europa III ciclo (Doctorado). Also part of the work of this author was supported by Beca Predoctoral de formación del Personal Investigador de la Consellería de Educación y Ciencia of Valencia (Spain).

Theorem 3 *Let the group $G = HK$ be the product of the totally permutable subgroups H and K. Let $\mathcal{F}$ be a saturated formation such that $\mathcal{L} \subseteq \mathcal{F}$, where $\mathcal{L}$ denotes the class of groups all of whose Sylow subgroups are cyclic. If H and K lie in $\mathcal{F}$, then so does G.*

Later, Carocca [6] presented a generalization of Maier's result to an arbitrary finite number of factors. In the same paper, Maier proposed the following question: *Does Theorem 3 extend to non-saturated formations which contain all supersoluble groups?* This question was answered affirmatively by Ballester-Bolinches and Pérez-Ramos in [2]. They obtained the following result:

Theorem 4 *Let $\mathcal{F}$ be a formation containing the class $\mathcal{U}$ of all supersoluble groups. Suppose that the group $G = HK$ is the product of the totally permutable subgroups H and K. If H and K belong to $\mathcal{F}$, then G belongs to $\mathcal{F}$.*

Our main goal is to take the above studies further. We present here the results contained in the references [3, 4, 5]. We draw attention to a range of interesting phenomena associated with totally and mutually permutable products. We prove that Carocca's result [6] can be extended to non-saturated formations containing $\mathcal{U}$ and we study the behaviour of $\mathcal{F}$-residuals, $\mathcal{F}$-projectors and $\mathcal{F}$-normalizers, where $\mathcal{F}$ is a saturated formation, in totally permutable products. We finish by proving extensions of Asaad and Shaalan's results on mutually permutable products. Let us begin by presenting a non-trivial example of a totally permutable product:

Example 1 ([2]) Let $P =< u, v, z \mid u^5 = v^5 = z^5 = 1, [u, v] = z >$ be the extraspecial group of order 5^3. Let $A =< a >$ be the cyclic group of order 5^2. Consider the group $C =< u, v, z, a \mid u^5 = v^5 = z^5 = a^{25} = 1, [u, v] = z, ua = au, va = av, z = a^5 >$ constructed in [7, A; 19.5]. C is a central product of P and A with amalgamated subgroup $< a^5 >\simeq Z(P)$. By [7, A; 20.8], the elements of the symplectic group $\mathrm{Sp}(2, 5)$ can be regarded as automorphisms of P centralizing $Z(P)$. Moreover, since $\mathrm{Sp}(2, 5) = \mathrm{SL}(2, 5)$ by [7, A; 20.10], the elements

$$\beta = \begin{pmatrix} 0 & 1 \\ -1 & -1 \end{pmatrix} \text{ and } \alpha = \begin{pmatrix} 0 & 1 \\ 1 & 0 \end{pmatrix} \in \mathrm{GL}(2, 5)$$

have the properties that β induces an automorphism of P centralizing $Z(P)$ and α can be regarded as an automorphism of P which inverts $Z(P)$. A simple calculation allows us to see that the group $< \alpha, \beta >\simeq \Sigma_3$, the symmetric group of degree 3, acts on P. On the other hand, it is rather easy to see that the action of α on $A =< a >$ inverting a and the trivial action of β on A, induces an action of $< \alpha, \beta >$ on A. So the group Σ_3 acts on C in a natural way. Consider $G = [C]\Sigma_3$, the semidirect product with respect to this action. Then $G = HK$ is the product of the totally permutable subgroups H and K, where H is isomorphic to A and K is isomorphic to the semidirect product $[P]\Sigma_3$.

Our next result shows that the converse of Maier's Theorem holds.

Lemma *Let $\mathcal{F}$ be a formation containing $\mathcal{U}$ such that either $\mathcal{F}$ is saturated or $\mathcal{F} \subseteq \mathcal{S}$, the class of all soluble groups. Assume that $G = HK$ is the product of the totally permutable subgroups H and K. If $G \in \mathcal{F}$, then $H \in \mathcal{F}$ and $K \in \mathcal{F}$.*

A well-known theorem of Doerk and Hawkes states that for a formation $\mathcal{F}$ of soluble groups the $\mathcal{F}$-residual respects the operation of forming direct products. This result has been generalized in [2]: if $G = HK$ is the product of the totally permutable subgroups H and K, then $G^{\mathcal{F}} = H^{\mathcal{F}}K^{\mathcal{F}}$. Here, $\mathcal{F}$ is a saturated formation of soluble groups containing the class $\mathcal{U}$ of all supersoluble groups. Recall that $G^{\mathcal{F}}$ is the $\mathcal{F}$-residual of G, that is, the smallest normal subgroup of G such that $G/G^{\mathcal{F}}$ belongs to $\mathcal{F}$.

Our first theorem extends this result to arbitrary formations of soluble groups containing $\mathcal{U}$.

Theorem 5 *Let $\mathcal{F}$ be a formation of soluble groups such that $\mathcal{U} \subseteq \mathcal{F}$. If $G = HK$ is the product of the totally permutable subgroups H and K, then $G^{\mathcal{F}} = H^{\mathcal{F}}K^{\mathcal{F}}$.*

We mention that the corresponding result for saturated formations is used in the proof of Theorem 5.

The above result is not valid for formations which do not contain the class $\mathcal{U}$, for example, the saturated formation $\mathcal{N}$ of nilpotent groups. Let G be the symmetric group of degree 3. G is the product of the totally permutable subgroups H and K, where H is a cyclic group of order 3 and K is a cyclic group of order 2. However $G^{\mathcal{N}} \neq H^{\mathcal{N}}K^{\mathcal{N}} = 1$.

From Theorem 5, we have that totally permutable factors satisfy, as well-placed subgroups, the property of lying in the formation generated by the group.

Let $\mathcal{X}$ be a Schunck class. It is quite clear that knowledge of the $\mathcal{X}$-projectors of a group usually reveals little about the $\mathcal{X}$-subgroup structure of the group. In fact there is no connection in general between the projectors of a group and those of a proper subgroup. We show here that totally permutable subgroups are exceptions to this general rule when $\mathcal{X}$ is a saturated formation containing the formation of supersoluble groups.

Theorem 6 *Let $\mathcal{F}$ be a saturated formation such that $\mathcal{U} \subseteq \mathcal{F}$. Let $G = HK$ be the product of the totally permutable subgroups H and K. If A is an $\mathcal{F}$-projector of H and B is an $\mathcal{F}$-projector of K, then AB is an $\mathcal{F}$-projector of G.*

Our next theorem shows that Theorem 6 also holds for $\mathcal{F}$-normalizers in the soluble universe.

Theorem 7 *Let the soluble group $G = HK$ be the product of the totally permutable subgroups H and K. If $\mathcal{F}$ is a saturated formation such that $\mathcal{U} \subseteq \mathcal{F}$, A is an $\mathcal{F}$-normalizer of H and B is an $\mathcal{F}$-normalizer of K, then AB is an $\mathcal{F}$-normalizer of G.*

The above results become false if the formation $\mathcal{F}$ does not contain the class $\mathcal{U}$, as the formation of all nilpotent groups and the symmetric group of degree 3 show.

Other different extension to the result of Maier [8] is made in [6] by Carocca. He generalized Maier's result to an arbitrary finite number of factors in the following way.

Theorem 8 *Let $G = G_1 G_2 \dots G_r$ be a group such that $G_1, G_2, \dots, G_r$ are pairwise totally permutable subgroups of G. Let $\mathcal{F}$ be a saturated formation which contains $\mathcal{U}$. If, for all $i \in \{1, 2, \dots, r\}$, the subgroups G_i are in $\mathcal{F}$, then $G \in \mathcal{F}$.*

Extensions of Theorems 5, 6 and 7 to an arbitrary finite number of factors are also possible.

Finally we focused the attention on the study of finite mutually permutable products. This can be considered as the continuation of Theorem 2. We show here that supersoluble residual respects the operation of forming mutually permutable products under some restrictions.

Theorem 9 *Let the finite group $G = HK$ be the mutually permutable product of the subgroups H and K. Assume that one of the following conditions is satisfied:*

(a) *K is supersoluble and G' is nilpotent,*

(b) *K is nilpotent,*

then $G^{\mathcal{U}} = H^{\mathcal{U}}$.

References

[1] M. Asaad and A. Shaalan. On the supersolvability of finite groups, Arch. Math., 53 (1989), 318-326.

[2] A. Ballester-Bolinches and M. D. Pérez-Ramos. A question of R. Maier concerning formations, J. Algebra 182 (1996), 738-747.

[3] A. Ballester-Bolinches M. C. Pedraza-Aguilera and M. D. Pérez-Ramos. On finite products of totally permutable groups, Bull. Austral. Math. Soc. 53 (1996), 441-445.

[4] A. Ballester-Bolinches M. C. Pedraza-Aguilera and M. D. Pérez-Ramos. Finite groups which are products of pairwise totally permutable subgroups, To appear in Proc. Edinburgh Math. Soc.

[5] A. Ballester-Bolinches M. C. Pedraza-Aguilera and M. D. Pérez-Ramos. Mutually permutable products of finite groups. Preprint.

[6] A. Carocca. A note on the product of $\mathcal{F}$-subgroups in a finite group. Proc. Edinburgh Math. Soc. 39 (1996), 37-42.

[7] K. Doerk and T. Hawkes, Finite Soluble Groups, Walter De Gruyter, Berlin / New York, (1992).

[8] R. Maier. A completeness property of certain formations. Bull. London Math. Soc. 24 (1992), 540-544.

ENDS AND ALGEBRAIC DIRECTIONS OF PSEUDOGROUPS

ANDRZEJ BIŚ

Faculty of Mathematics, University of Łódź, ul. S. Banacha 22, 90-238 Łódź, Poland

Abstract

The notion of an end of a group is generalised to the case of finitely generated pseudogroups. The notion of a direction of a pseudogroup, subtler than the end of a pseudogroup, is introduced in this paper. The ends and directions of the Markov pseudogroup are described.

Introduction

The notion of an end of a group was originated in the 40's by H. Freudenthal [Fre] and H. Hopf [Hop]. A classical result of H. Hopf [Hop] says that the end space $E(X)$ of a connected topological space X is an invariant of the group G of covering transformations. Thus it becomes meaningful to define the end space of the finitely generated group G as $E(G) := E(X)$.

The notion of the end of a group was studied again by J. Stallings [Sta] and D. Cohen [Coh1], [Coh2] in the 70's. D. Cohen has used a combinatorial method to define the number of ends of an arbitrary group G. The end spaces of finitely generated groups were completely classified (see [Coh2], [Hop], [Sta]).

The theory of ends of groups was generalised. A more general concept, introduced by C. Hougton [Hou] and further explored by G. Scott [Sco] and M. Sageev [Sag], is the number of a group G relative to a subgroup H.

This paper arose from an attempt to generalise the notion of an end to pseudogroups. It was shown by Hopf that the number of ends of a group is 0, 1, 2 or ∞, but in the case of ends of pseudogroups there are no such limitations (see Example 2.3). A slight more general concept of a direction of a pseudogroup is introduced in this paper. The space of ends of a pseudogroup G is described in terms of sequences of generators. Using a properly defined relation, we obtain that the space of ends of a pseudogroup splits into equivalence - directions of the pseudogroup. The notion of a direction of a group is a suitable tool allowing us to examine the action of a group on a topological space in a more subtler way than by applying the theory of ends. This notion can be also applied to the study of the ergodicity of a group (see [Biś1]) or the structure of its limit set (see [Biś2]).

1 Ends and directions

Definition 1.1 A *graph* Γ consists of two sets, called, respectively, *vertices* and *edges* of Γ, and a map from the set of edges to the set of unordered pairs of vertices.

If the image of the edge e is a pair $\{v, w\}$, then we say that e has vertices v and w, or that e joins v and w.

We shall usually use the symbol Γ to denote both the graph and its set of vertices.

Definition 1.2 We say that Γ is locally finite, for each vertex v of Γ, there are only finite many edges having v as a vertex.

Definition 1.3 By a *path* from v to w in a set $A \subset \Gamma$ we mean a sequence $v = w_1, w_2, \ldots, w_n = w$ of vertices of A, such that there exists an edge joining w_{k-1} and w_k for $k = 2, \ldots, n$.

Definition 1.4 We call a set A *connected* if any two elements of A can be joined by a path.

By a *connected component* of A including the vertex v we mean the union of all paths in A including the vertex v.

Definition 1.5 The coboundary of a set A, denoted by δA, is the set of edges having exactly one vertex in A.

Remark 1.6 An example of locally finite graph is a finitely generated group G with a set of generators $G_1 = \{g_1, g_2, \ldots, g_m\}$. Then the pairs $\{g, g_i g\}$, $g \in G$, $g_i \in G_1$, are edges.

Let us recall (see [Sac], [Wal]) the definition of a pseudogroup.

Definition 1.7 Let M be a topological space. The family $H := \{h_i : D_i \to R_i\}$ of local homeomorphisms of M is called a pseudogroup if and only if:

1) for any $h_i \in H$, the element h_i^{-1} belongs to H,
2) for any $h_i, h_j \in H$ such that $D_i \cap R_j \neq \emptyset$, the composition $h_i \circ h_j : h_j^{-1}(D_i \cap R_j) \to h_i(D_i \cap R_j)$ belongs to H,
3) $\mathrm{id}_M \in H$,
4) for any $h_i \in H$ and any open set U , $h|U \in H$,
5) for any $h : D \to R$ and open cover W of D such that $h|U \in G$ for any $U \in W$, then $h \in G$.

Definition 1.8 The smallest pseudogroup of local homeomorphisms of M which contains a set A of homeomorphisms between open subsets of M exists, is said to be generated by A. It is denoted by $G(A)$. A pseudogroup G is finitely generated iff there exists a finite set A such that $G = G(A)$. The map $g : D_g \to R_g$ belongs to $G(A)$ if for any $x \in D_g$ there exists an open neighbourhood U of x, maps $g_1, \ldots, g_n \in A$ and exponents $e_1, \ldots, e_n \in \{-1, 1\}$ for which $g = g_1^{e_1} \circ \cdots \circ g_n^{e_n}$ on $D_g \cap U$.

Remark 1.9 Any finitely generated pseudogroup G of local homeomorphisms defined on a topological space M is an example of a graph. Indeed, denote by A a set of generators. Then a vertex of the graph is an element of $V := \{(g_{i_1}, ..., g_{i_n}) : g_{i_j} \in A$ *or* $g_{i_j}^{-1} \in A$, domain $(g_{i_1} \circ \cdots \circ g_{i_n}) \neq \emptyset$, $n \in \mathbb{N}\}$, the pairs $(g, g_i g)$ are edges provided $g, g_i g \in V$ and $g_i \in A$ or $g_i^{-1} \in A$.

Definition 1.10 Define a *filter* F on $P := \{E \subset \Gamma : \#\delta E < \infty\}$ as a subset of P satisfying the following conditions:

1) $\emptyset \in F$,
2) if $E_1, E_2 \in F$, then $E_1 \cap E_2 \in F$,
3) $E_1, E_2 \in P$, $E_1 \subseteq E_2$ and $E_1 \in F$, then $E_2 \in F$.

Definition 1.11 The maximal filter defined on P (with respect to the inclusion) is called an *end* of the graph Γ.

Proposition 1.12 *Let (F_i) be an ascending sequence of finite subsets of Γ, such that $\bigcup_{n=1}^{\infty} F_n = \Gamma$. Denote by $C_1, C_2 \ldots$, the infinite connected components of the sets $\Gamma \backslash F_1, \Gamma \backslash F_2, \ldots$. Then there exists exactly one end including all sets C_n, $n \in \mathbb{N}$.*

For distinct sequences (C_n), $C_1 \supseteq C_2 \supseteq \ldots$, and (C'_n), $C'_1 \supseteq C'_2 \supseteq, \ldots$, there exist distinct ends.

Proof See [Coh2] p.37. □

Definition 1.13 A subset E of G is called left-invariant if symmetric difference $gE \stackrel{\bullet}{-} E$ is a finite set for all $g \in G$.

Proposition 1.14 *If there exist two distinct left-invariant subsets E_1, E_2 of a finitely generated group G, such that the symmetric difference $E_1 \stackrel{\bullet}{-} E_2$ is infinite, then there exist at least two ends in G.*

Proof Let $\{t_1, ..., , t_k\}$ be a set of generators of G. From the definition of the left-invariant subset E_1 we obtain that, for any $x \in G$, $E_1 \stackrel{\bullet}{-} xE_1$ is finite. Note that

$$\begin{aligned} \delta E_1 &= \{(t,g) : g \in E_1 \wedge t \in G_1 \wedge tg \notin E_1\} \\ &\subset \bigcup_{i=1}^{k} \{(t_i, g) : g \in E_1 \backslash t_i E_1\}. \end{aligned}$$

So, δE_1 is a finite set. We can also write

$$\delta(E_1 \backslash E_2) \subset \delta E_1 \cup \delta E_2.$$

The set $E_1 \backslash E_2$ has a finite coboundary and belongs to some end K_1. Similarly, the set $E_2 \backslash E_1$ has a finite coboundary and belongs to some end K_2. The intersection of the sets $E_1 \backslash E_2$ and $E_2 \backslash E_1$ is empty; that is why they do not belong to the same end. In consequence, we obtain that there exist at least two distinct ends in G. □

Remark 1.15 Let $G \approx \mathbb{Z}^2$ be a group of transformations of $\mathbb{R}^2$, generated by $G_1 = \{v_1, v_2\}$ where v_1 (respectively v_2) is a translation of $\mathbb{R}^2$ determined by the vector $(1,0)$ (resp. $(0,1)$). Let

$$G_n = \{w_1 + w_2 + \cdots + w_n : w_i \in \{v_1, v_2, -v_1, -v_2, \mathrm{id}_{\mathbb{R}^2}\}\}.$$

Then $G \backslash G_n$ has exactly one connected component. So, the group G has exactly one end.

Remark 1.16 Let G be a finitely generated free group on generating set $G_1 = \{a_1, a_2, \ldots, a_n\}$, $n \geq 2$. For brevity, we assume that $G_1 = G_1^{-1}$. We assume that there exists only one relation in G such that we identify by it the word

$$a_{i_1} \cdot \ldots \cdot a_{i_k}^p a_{i_k}^q \cdot \ldots \cdot a_{i_n}, \qquad p, q \in \mathbb{Z},$$

with the word

$$a_{i_1}^p \cdot \ldots \cdot a_{i_k}^{p+q} \cdot \ldots \cdot a_{i_n}.$$

Denote the neutral element in G by e. Put

$$H_b = \{h \in G : \exists_{h \in \mathbb{N} \cup \{0\}} \exists_{g_{i_1}, \ldots, g_{i_n} \in G_1} \; h = g_{i_1} \cdot \ldots \cdot g_{i_n} b \text{ and } g_{i_1} \cdot \ldots \cdot g_{i_n} \neq b^{-1}\}$$

for any $b \in G_m \backslash G_{m-1}$.

The set H_m defined above is an infinite and left-invariant set. Indeed, if $g \in G$, then there exist $g_{j_1} \cdot \ldots \cdot g_{j_k} \in G_1$, $k \in \mathbb{N}$, such that

$$g = g_{j_1} \cdot \ldots \cdot g_{j_k}.$$

Therefore we get

$$\begin{aligned} gH_b &= \{gh \in G : \exists_{h \in \mathbb{N} \cup \{0\}} \exists_{g_{i_1}, \ldots, g_{i_n} \in G_1} \; h = g_{i_1} \cdot \ldots \cdot g_{i_n} b \neq e\} \\ &= \{d \in G : \exists_{h \in \mathbb{N} \cup \{0\}} \exists_{g_{i_1}, \ldots, g_{i_n} \in G_1} \; d = g_{j_1} \cdot \ldots \cdot g_{j_k} g_{i_1} \cdot \ldots \cdot g_{i_n} b \\ &\qquad \text{and } g_{i_1} \cdot \ldots \cdot g_{i_n} \neq b^{-1}\} \\ &= \begin{cases} H_b \\ H_b \cup \{e\} \end{cases} \end{aligned}$$

So, $H_b \stackrel{\bullet}{-} gH_b$ is an at most one-element set. That is why H_b is an invariant set. Moreover,

$$\begin{aligned} \delta H_b &= \{(t, g) : g \in H_b \text{ and } t \in G_1 \text{ and } tg \notin H_b\} \\ &\subset \bigcup_{i=1}^{n} \{(a_i, g) : g \in H_b \text{ and } g \notin a_i^{-1} H_b\} \\ &= \bigcup_{i=1}^{n} \{(a_i, g) : g \in H_b \backslash a_i^{-1} H_b\}. \end{aligned}$$

Thus δH_b is a finite set. Using Proposition 1.14, we obtain that the subsets $H_{b_1}, H_{b_2} \in G$, $b_1 \neq b_2$, $b_1, b_2 \in G_m \backslash G_{m-1}$, determine distinct ends in G. Thus there are infinitely many ends in G.

Further, we shall consider only a finitely generated pseudogroup G of transformations defined on a topological space X. Denote by G_1 a set of generators of G such that $G_1^{-1} \subset G_1$ and $\mathrm{id}_X \in G_1$.

Let $G_n = \{g \in G : g = g_1 \cdot \ldots \cdot g_n,\ \ g_i \in G_1\}$, then

$$G_\infty = \bigcup_{i=1}^{\infty} G_n \text{ and } G_1 \subset G_2 \subset G_3 \subset \ldots.$$

Put $G_0 = \{\mathrm{id}\,|D_g : g \in G_\infty\}$

Definition 1.17 We say that $f, g \in G_\infty \backslash G_n$ can be connected in $G_\infty \backslash G_n$ and denote this by $f\mathrm{con}\,(n)g$ if

$$\exists_{k \in \mathrm{N}}\ \exists_{g_1,\ldots,g_k \in G_\infty \backslash G_n}\ \forall_{1 \le i \le k-1}\ f = g_1 \text{ and } g_k = g \text{ and } g_{i+1}g_i^{-1} \in G_1.$$

Definition 1.18 By a connected component of the set $G_\infty \backslash G_n$ including $a \in G_\infty \backslash G_n$ we mean the set

$$\{f \in G_\infty : f\mathrm{con}\,(n)a\}.$$

Consider a sequence (C_i) of subsets of G_∞ satisfying the following conditions:

1) the set C_i is an infinite connected component of $G_\infty \backslash G_i$ for each $i \in \mathrm{N}$,

2) $C_1 \supset C_2 \supset C_3 \supset \ldots$.

Definition 1.19 The sequence (C_i) of subsets of G_∞ satisfying 1) and 2) is called an algebraic end of the pseudogroup G.

In the set S of paths in the Cayley graph $Cay(G, G_1)$ i.e. sequences (f_n), where $f_n \in G_\infty$, satisfying the conditions:

3) $\forall_{n \in \mathrm{N}} f_n \in G_{n+1} \backslash G_n$,

4) $\forall_{n \in \mathrm{N}} f_{n+1} f_n^{-1} \in G_1$,

we introduce a relation $\tilde{e}$ in the following way: $(f_n)\tilde{e}(g_n)$ if and only if $f_n\mathrm{con}\,(n)g_n$ for each $n \in \mathrm{N}$. The relation defined above is an equivalent relation. Put $K = S/\tilde{e}$.

We shall show that the algebraic end of a pseudogroup can be described in an equivalent way in the language of equivalent paths, which is stated by the theorem below.

Theorem 1.20 *There exists a bijective correspondence between algebraic ends of the pseudogroup G and elements of K.*

Proof Consider an algebraic end (C_i) of G. Then

$$C_i = \bigcup_{n=i+1}^{\infty} (C_i \cap G_n)$$

where each $C_i \cap G_n$, $n = i+1, i+2, \ldots$, is a finite and nonempty set. Indeed, C_i is an infinite set and, therefore, infinitely many sets $C_i \cap G_n$ are nonempty. Assume that there exists $j \geq i+1$ such that $C_i \cap G_j = \emptyset$ and $C_i \cap G_{j+1} \neq \emptyset$. Then there exists an element $g = g_{j+1}g_j \cdot \ldots \cdot g_1 \in C_i \cap G_{j+1}$ where $g_1, \ldots, g_{i+1} \in G_1$. Notice that the element $g' = g_j \cdot \ldots \cdot g_1 \in G_j$ can be connected by g_{j+1} with the infinite set C_i. From the definition of a connected component we obtain that $g' \in G_j \cap C_i$, which contradicts our assumption that $G_j \cap C_i = \emptyset$.

Put $A_n = C_n \cap G_{n+1}$. We shall show that any $f_n \in A_n$ can be obtained from some element of A_{n-1}. Fix $f_n \in A_n$; then there exist $f_{n-1} \in G_n$ and $f \in G_1$ such that $f_n = ff_{n-1}$. So, f_{n-1} belongs to some connected component of $G_\infty \backslash G_{n-1}$. Since $f_n \in C_n \subset C_{n-1}$ and f_n can be connected with f_{n-1} in $G_\infty \backslash G_{n-1}$ by the element f, therefore $f_{n-1} \in C_{n-1} \cap G_n = A_{n-1}$.

Define a mapping $L : A_1 \to \mathbf{N} \cup \{\infty\}$ in the following way:

$$L(g) = \max\{l \in \mathbf{N} \cup \{\infty\} : \exists_{g_1,\ldots,g_l \in G_1} \ \forall_{i \leq l} \ g_i \cdot \ldots \cdot g_2 g_1 g \in A_{i+1}\}.$$

If, for each g from the finite set A_1, we have $L(g) < \infty$, then, for a sufficiently large $n \in \mathbf{N}$, we obtain $A_n = \emptyset$, which contradicts the above reasoning. So, there exists $g_0 \in A_1$ such that $L(g_0) = \infty$ and this implies that there exists a sequence (f_n), $f_n \in G_\infty$, such that

5) $\forall_{n \in \mathbf{N}} f_n \in C_{n+1}$,

6) $\forall_{n \in \mathbf{N}} f_{n+1} f_n^{-1} \in G_1$.

Note that if the sequences (f_n) and (g_n) satisfy conditions 5) and 6), then they are $\widetilde{e}$-equivalent. So, the mapping ψ, sending an algebraic end (C_i) to an equivalent class determined by the sequence (f_n) fulfilling 5) and 6), is well defined.

The mapping ψ is one-to-one. Indeed, denote by $[(f_i)]_{\widetilde{e}}$ and $[(f_i')]_{\widetilde{e}}$ the equivalent classes determined by the mapping ψ via the algebraic ends (C_i) and (C_i'). Assume that $(C_i) \neq (C_i')$; then there exists $j \in \mathbf{N}$ such that $C_j \cap C_j' \neq \emptyset$. The elements $f_{j-1} \in C_j$ and $f_{j-1}' \in C_j'$ belong to different connected components of $G_\infty \backslash G_j$, therefore f_j and f_j' cannot be connected in $G_\infty \backslash G_j$. The sequences (f_n) and (f_n') determine different equivalent classes in the relation $\widetilde{e}$, which proves that ψ is a one-to-one mapping.

Consider a sequence (f_n) satisfying 5) and 6). Denote by D_i the connected component of $G \backslash G_i$ including f_i. Then $\psi((D_i)) = [(f_i)]_{\widetilde{e}}$ and we obtain that ψ is a surjective mapping. □

Consider an algebraic end (C_n) of G and two sequences (f_n) and (f_n'), such that

$$\psi((C_n)) = [(f_n)]_{\widetilde{e}} = [(f_n')]_{\widetilde{e}}.$$

Then $f_n \mathrm{con}\,(n) f_n'$ for each $n \in \mathbf{N}$ and there exist $g_{n,1}, g_{n,2}, \ldots, g_{n,s(n)} \in G_1$ such that

7) $f_n = g_{n,1}, g_{n,2} \cdots g_{n,s(n)} f_n'$.

Definition 1.21 We say that sequences (f_n) and (f_n') determining the same algebraic end define the same algebraic direction if and only if one can choose a connection of (f_n) and (f_n') such that the word distance $d(f_n, g_n)$ (i.e. sequence $s(n)$ defined by Condition 7)) is bounded. Then we write $(f_n) \sim (f_n')$.

Lemma 1.22 *The relation $\sim$ described above is an equivalent relation.*

Proposition 1.23 *Let G be a finitely generated pseudogroup of local diffeomorphisms of a manifold M, such that*

$$\forall_{g \in G_1} \|g_*\| < \infty.$$

If the sequences (f_n) and (f'_n) determine the same algebraic direction, then, for any point p belong to the intersection of all domains of f'_n and f_n and for a vector $0 \neq v \in TpM$, we have

$$\limsup_{n\to\infty} \frac{1}{n} \log \|f_{n*}v\| = \limsup_{n\to\infty} \frac{1}{n} \log \|f'_{n*}v\|.$$

Proof Let (f_n) and (f'_n) satisfy the assumption; then

$$[(f_n)]_\sim = [(f'_n)]_\sim.$$

There exist $g_{n,1}, \ldots, g_{n,s(n)} \in G_1$, such that

$$f_n = g_{n,1}, \ldots, g_{n,s(n)} f'_n.$$

Thus

$$\begin{aligned}
\limsup_{n\to\infty} \ \tfrac{1}{n} \ \log \|f_{n*}v\| &= \limsup_{n\to\infty} \frac{1}{n} \log \|g_{n,1*} \cdot \ldots \cdot g_{n,s(n)*} f'_{n*} v\| \\
&\leq \limsup_{n\to\infty} \frac{1}{n} \log \|g_{n,1*} \cdot \ldots \cdot g_{n,s(n)*}\| + \limsup_{n\to\infty} \frac{1}{n} \log \|f'_{n*}v\| \\
&\leq \limsup_{n\to\infty} \frac{1}{n} \log(\max\{\|g_{i*}\| : g_i \in G_1\})^{s(n)} \\
&+ \limsup_{n\to\infty} \frac{1}{n} \log \|f'_{n*}v\| \leq \limsup_{n\to\infty} \frac{1}{n} \log \|f'_{n*}v\|
\end{aligned}$$

because the sequence $s(n)$ is bounded and $\max\{\|g_{i*}\| : g_i \in G_1\}$ is finite.

The second inequality is obtained in a similar way, by using the equality $f'_n = g^{-1}_{n,s(n)} \cdot \ldots \cdot g^{-1}_{n,1} f_n$. □

Definition 1.24 With the notation as above, the term

$$\limsup_{n\to\infty} \frac{1}{n} \log \|f_{n*}v\|$$

is called the Lapunov exponent for the algebraic direction $[(f_n)]_\sim$ along a vector v.

The example mentioned below describes a group with one end and infinitely many directions.

Remark 1.25 Let $a > b > 2$. Define mappings $g, h : \mathbb{R}^2 \to \mathbb{R}^2$ in the following way:

$$g(x_1, x_2) = (x_1, ax_2), \qquad h(x_1, x_2) = (x_1, bx_2), \quad \text{for } (x_1, x_2) \in \mathbb{R}^2.$$

Consider a group of transformations of $\mathbb{R}^2$ generated by

$$G_1 = \{g, h, g^{-1}, h^{-1}, \mathrm{id}_{\mathbb{R}^2}\}.$$

Fix $l \in \mathrm{N}$ and define a sequence $(f_n^{(k,l)})$ in the following way:

$$f_1^{(k,l)} = g,$$

$$f_n^{(k,l)} = \begin{cases} gf_{n-1}^{(k,l)} & \text{if } 1 \le n - [\frac{n}{k+l}] \le k, \\ hf_{n-1}^{(k,l)} & \text{if } k+1 \le n - [\frac{n}{k+l}] \le k+l. \end{cases}$$

Denoting by S_n the number of elements g_* which appear in the term $f_{n*}^{(k,l)}$, we get

$$f_{n*}^{(k,l)} = \begin{bmatrix} 1 & 0 \\ 0 & a^{S_n} b^{n-S_n} \end{bmatrix}$$

and

$$\frac{k\frac{n}{k+l}}{n} \le \frac{S_n}{n} \le \frac{k([\frac{n}{k+l}]+1)}{n}.$$

Counting the Lapunov exponent for the sequence $(f_n^{(k,l)})$ along the vector $(1,1) \in T_pM$, we obtain

$$\begin{aligned} \limsup_{n\to\infty} \frac{1}{n} \log \|f_{n*}^{(k,l)}(1,1)\| &= \limsup_{n\to\infty} \frac{1}{n} \log \left\| \begin{bmatrix} 1 & 0 \\ 0 & a^{S_n} b^{n-S_n} \end{bmatrix} \right\| \\ &= \limsup_{n\to\infty} \frac{1}{n} \log (a^{2S_n} b^{2(n-S_n)} + 1)^{\frac{1}{2}}. \end{aligned}$$

Since

$$\begin{aligned} \limsup_{n\to\infty} \frac{1}{n} \log(a^{S_n} b^{n-S_n}) &= \limsup_{n\to\infty} \left(\frac{S_n}{n} \log a + \frac{n - S_n}{n} \log b \right) \\ &= \frac{k}{l+k} \log a + \frac{l}{l+k} \log b, \end{aligned}$$

therefore

$$\limsup_{n\to\infty} \frac{1}{n} \log \|f_{n*}^{(k,l)}(1,1)\| = \frac{k}{l+k} \log a + \frac{l}{l+k} \log b.$$

This means that there are infinitely many different algebraic directions in the group generated by G_1 because directions determined by distinct pairs (k, l) are distinct.

In the case where a and b are chosen in such a way that the equalities $a^m = b^n$ do not hold for any integers m and n, any element $f \in G$ can be represented in a unique way in the form

$$f = g^l h^k \text{ where } l, k \in \mathbb{Z}.$$

Therefore G is isomorphic to $\mathbb{Z} \times \mathbb{Z}$ and we conclude that the group G has only one end.

2 Markov pseudogroups

Following [CC] we define a Markov pseudogroup as a pseudogroup G of local homeomorphisms of the real line $\mathbb{R}$ generated by finite set $A = \{g_1, ..., g_m\}$ of maps between open bounded intervals D_i and R_i, $g_i : D_i \to R_i$, satisfying the following conditions:

(i) $R_i \cap R_j = \emptyset$ when $i \neq j$,

(ii) either $R_i \subset D_j$ or $R_i \cap D_j = \emptyset$.

The Markov pseudogroup generated by A determines a $m \times m$ matrix $P = (p_{ij})$ with entries $p_{ij} \in \{0, 1\}$, where $p_{ij} = 1$ if and only if $R_i \subset D_j$.

Example 2.1 According to J. Cantwell and L. Conlon [CC] any Markov pseudogroup of local C^2- diffeomorphisms of the real line $\mathbb{R}$ can be realized as the holonomy pseudogroup of codimension-one C^2–foliation of a compact manifold (compare [Ina]).

Example 2.2 A group G has only one, two or infinitely many ends [Coh2], but there is no such limitations in the case of pseudogroups. For any even positive integer k , there exists a pseudogroup with exactly k ends.

Remark 2.3 For a fixed $k \in \mathbf{N}$, there exists a Markov pseudogroup possessing exactly $2k$ directions.

Indeed, let $f_i : (i, i+1) \to (i, i+1)$, $i = 1, 2, ..., k$, be a homeomorphism. The pseudogroup G generated by $G_1 = \{\mathrm{id}_{\mathbb{R}}, f_1, f_1^{-1}, ..., f_k, f_k^{-1}\}$ satisfies the above condition.

Proposition 2.4 *Let G be a Markov pseudogroup generated by a set A. Then, any two sequences (f_n) and (g_n) determining the same end in G are equal.*

Proof Assume that the sequences (f_n) and (f_n') determining the same end are different. Then, for any $n \in \mathbf{N}$ there exist $h_1, ..., h_k \in G_1$,for some positive integer k, such that

$$f_n = h_k...h_1 f_n'$$

So,

$$\mathrm{id}\,|D_{f_1} = h_l...h_1 f_n' \in G_1, \text{ or } \mathrm{id}\,|D_{f_1} = f_m \in G_0$$

for some l or $m > 1$. In both cases $\mathrm{id}\,|D_{f_1} \in G_0$, but the right sides of above equalities, by the definition of (f_n), do not belong to G_0. □

Corollary 2.5 *Any two sequences determining the same direction in a Markov pseudogroup G are equal. The numbers of ends and directions in G are equal.*

Proposition 2.6 *Let $A = \{g_1, ..., g_m\}$ generates a Markov pseudogroup G. Denote by $\lambda_1, ..., \lambda_m$ the complex roots of the characteristic polynomial of the matrix P. Then,*

(i) *if all the roots satisfy the inequality $|\lambda_i| < 1$, then G has finite numbers of ends (resp., directions),*

(ii) *if there exists λ_i, such that $|\lambda_i| > 1$, then G has infinitely many ends (resp., directions),*

(iii) *if all the roots have multiplicity one and satisfy $|\lambda_i| = 1$, then G has finitely many ends (resp., directions),*

(iv) *if all the roots satisfy $|\lambda_i| = 1$ and the multiplicity λ_j is greater than one, for some j, then G has infinitely many ends (resp., directions).*

Proof Using the Jordan normal form of the matrix $P = (p_{ij})$ we get $m \times m$ matrices B and J such that $P = B^{-1}JB$, where $J =$ diag$(J_1, J_2, \ldots, J_k)$ and

$$J_l = \begin{pmatrix} \lambda_l & 1 & 0 & \cdots & 0 \\ 0 & \lambda_l & 1 & \cdots & 0 \\ \cdots & \cdots & \cdots & \cdots & \cdots \\ 0 & 0 & 0 & \cdots & \lambda_l \end{pmatrix}, \; l = 1, 2, ..., k$$

So, for any $n \in \mathbf{N}$ we obtain $P^n = B^{-1}J^nB$ where $J^n =$ diag$(J_1^n, J_2^n, \ldots, J_k^n)$ and

$$J_l^n = \begin{pmatrix} \lambda_l^n & \frac{n}{1!}\lambda_l^{n-1} & \frac{n(n-1)}{2!}\lambda_l^{n-2} & \ldots & \frac{n(n-1)...(n-p+2)}{(p-1)!}\lambda_l^{n-p+1} \\ 0 & \lambda_l^n & \frac{n}{1!}\lambda_l^{n-1} & \ldots & \frac{n(n-1)...(n-p+3)}{(p-2)!}\lambda_l^{n-p+2} \\ \ldots & \ldots\ldots & \ldots\ldots\ldots & \ldots & \ldots\ldots\ldots \\ 0 & 0 & 0 & \ldots & \lambda_l^n \end{pmatrix}$$

On the other hand, $P^n = (p_{ij}^{(n)})$, where

$$p_{ij}^{(n)} = \sum_{1 \le i_1, ..., i_n \le n} p_{ii_1} \cdot p_{i_1 i_2} \cdot \ldots \cdot p_{i_n j}$$

The entry $p_{ij}^{(n)}$ determines the number of paths, consisted of $n-1$ elements, joining g_i and g_j of A.

In the case (i) and (iii), there exists a constant C such that for every $n \in \mathbf{N}$ the entries of J^n are bounded by C. Similary, there exists a positive constant C_1 such that for the all entries of the matrix P^n we obtain inequalities:

$$p_{ij}^{(n)} \le C_1$$

So, the number of infinite distinct paths in the Cayley graph of G is finite. It yields, compare Proposition 2.4, that the number of ends of G is finite.

In the case (ii) and (iv), let $J^n = (a_{ij}^{(n)})$, then

$$\lim_{n\to\infty} \sum_{i,j} a_{ij}^{(n)} = \infty \text{ so } \lim_{n\to\infty} \sum_{i,j} p_{ij}^{(n)} = \infty.$$

It means that there are infinitely many infinitely many infinite distinct paths in the Cayley graph of G. Proposition 2.4 completes the proof for the end space of G. □

Respective statements for the directions are directly implied by Corollary 2.5.

3 Equivalent algebraic directions and frequency of generator

Consider a finitely generated pseudogroup G of mappings of a topological space X, generated by G_1. Let sequences (f_n) and (g_n) determine the same algebraic direction in G.

Definition 3.1 Denote by $N(g_i, f_n)$ the minimal number of repetitions of the generator $g_i \in G_1$ in the term f_n. Then

$$F(g_i, (f_n)) := \liminf_{n\to\infty} \frac{N(g_i, f_n)}{n}$$

denotes the frequency of appearance of g_i in the sequence (f_n).

Proposition 3.2 *Let G be a finitely generated pseudogroup G generated by G_1. Then, for any sequences (f_n), (g_n) determining the same algebraic direction in G and for each $g_i \in G_1$, the equality*

$$F(g_i, (f_n)) = F(g_i, (g_n))$$

holds.

Proof Assume that for some $g_i \in G_1$ we have $F(g_i, (f_n)) > F(g_i, (g_n))$. Put $\delta := F(g_i, (f_n)) - F(g_i, (g_n))$. From the definition of the lower limit we conclude that only a finite number of terms of the sequence $\frac{N(g_i, f_n)}{n}$ is less than $F(g_i, (f_n)) - \frac{\delta}{3}$, and that there exists sequence (n_k) such that

$$\lim_{k\to\infty} \frac{N(g_i, g_{n_k})}{n_k} = F(g_i, (g_n)).$$

Thus, for a sufficiently large k, we obtain

$$\frac{N(g_i, f_{n_k})}{n_k} > F(g_i, (f_n)) - \frac{\delta}{3},$$

$$F(g_i,(g_n)) - \frac{\delta}{3} < \frac{N(g_i, g_{n_k})}{n_k} < F(g_i,(g_n)) + \frac{\delta}{3} \text{ and } \frac{s(n_k)}{n_k} < \frac{\delta}{3},$$

where the sequence $s(n)$ was defined in Definition 1.21.

The above inequalities yield, for a sufficiently large k, the following inequality:

$$N(g_i, f_{n_k}) > N(g_i, g_{n_k}) + s(n_k).$$

There exist $g_{n_k,1}, \ldots, g_{n_k,s(n_k)} \in G_1$ such that

$$f_{n_k} = g_{n_k,1} \cdot \ldots \cdot g_{n_k,s(n_k)} g_{n_k}$$

thanks to the definition of an algebraic direction. So, the generator g_i appears more times on the left-hand side of the above equality than on the right-hand one, which contradicts the fact that $N(g_i, f_{n_k})$ is the minimal number of repetitions of g_i in the term f_{n_k}. □

Corollary 3.3 *Let G be a finitely generated pseudogroup and let sequences (f_n) and (g_n) determine the same algebraic direction in G. Then the sets of generators with positive frequencies are equal in both sequences.*

References

[Biś1] A.Biś, *Entropy of topological directions*, to appear in Ann. Faculte Sci. Toulouse.

[Biś2] A.Biś, *Geometrical directions and ends of a manifold...*, Supplemento ai Rendiconti del Circolo Matematico di Palermo, **39** (1996), 37–56.

[CC] J.Cantwell and L.Conlon, *Foliations and subshifts*, Tohoku Math. J., **40** (1988), 165–187.

[Coh1] D. Cohen, *Ends and Free Products of Groups*, Math. Z. **114** (1970), 9–18.

[Coh2] D. Cohen, *Groups of Cohomological Dimension One*, Springer-Verlag, Lecture Notes in Math. 245, (1972).

[Fre] H. Freudenthal, *Über die Enden diskreter Räume und Gruppen*, Comment. Math. Helvet., **17** (1944), 1–38.

[Hop] H. Hopf, *Enden offener Raume und unendliche diskontinuierliche Groupen*, Comment. Math. Helv., **16** (1943), 81–100.

[Hou] C.H. Houghton, *Ends of locally compact groups and their quotient spaces*, J. Austral. Math. Soc., **17** (1974), 274–284.

[Ina] T.Inaba, *Examples of exceptional minimal sets*, in A Fete of Topology, Papers Dedicated to Itiro Tamura, (1988).

[Sac] R. Sacksteder, *Foliations and pseudogroups*, Amer. J. Math., **87** (1965), 79–102.

[Sag] M.Sageev, *Ends of group pairs and nonpositively curved cube complexes*, J. London Math. Soc., **71** (1995), 585–617.

[Sco] G.P. Scott, *Ends of pairs of groups*, J. Pure Appl. Algebra, **11** (1977), 179–198.

[Sta] J. Stallings, *Group Theory and Three-dimensional Manifolds*, New Haven Yale Univ. Press, Yale Mathematical Monographs 4, (1971).

[Wal] P. Walczak, *Losing Hausdorff dimension while generating pseudogroups*, Fundamenta Mathematicae, **149** (1996), 211–237.

ON LOCALLY NILPOTENT GROUPS WITH THE MINIMAL CONDITION ON CENTRALIZERS

VASILIY BLUDOV[1]

Department of Mathematics, Irkutsk State University, Irkutsk, 664003, Russia

Introduction

In this report we prove that locally nilpotent groups satisfying the minimal condition on centralizers (or groups with the min-c condition) are hypercentral. This result is a generalization of well-known results for linear locally nilpotent groups (see [6], [7]) and periodic locally nilpotent groups [1], and gives a positive answer for a question of F. Wagner ([3], question 13.16).

We use the following notation: $\langle A\rangle$ or $\langle a_i \mid i \in I\rangle$ is the subgroup, generated by $A = \{a_i \mid i \in I\}$, $a^b = b^{-1}ab$, H^G is the normal closure of a subgroup H in a group G, $[a,b] = a^{-1}b^{-1}ab$, $[a,b_1,\ldots,b_{n-1},b_n] = [[a,b_1,\ldots,b_{n-1}],b_n]$, $[A,B] = \langle [a,b] \mid a \in A, b \in B\rangle$, $G' = [G,G]$, $\mathrm{C}_H(A)$ ($\mathrm{C}_H(a)$) is the centralizer of a set A (element a) in H, $\zeta_1(G) = \zeta(G)$ is the centre, $\zeta_\alpha(G)$ is the α-th hypercentre of a group G. $A \leq B$ ($A < B$) denotes that A is a (proper) subgroup of B. The identity element of any group is denoted by e.

1 Preliminary results

It is well-known that the minimal and maximal conditions on centralizers are equivalent and inherited by subgroups. Moreover, if a group G satisfies the minimal conditions on centralizers then every nonempty subset A of G contains a finite subset S such that $\mathrm{C}_G(A) = \mathrm{C}_G(S)$ (see [1]). We will use these facts later without special mention.

We also need the following results:

Assertion 1.1 (S.N. Chernikov [2], see also [4]) 1. *A group G is hypercentral if and only if for any element $a \in G$ and any sequence $x_1, x_2, \ldots, x_n, \ldots \in G$ there is an integer k such that $[a, x_1, x_2, \ldots, x_k] = e$.*

2. *If all countable subgroups of a group G are hypercentral then G is hypercentral.*

Assertion 1.2 (T. Yen, see [1]) *Every locally nilpotent group with the min-c condition is soluble.*

Assertion 1.3 *Let G be a group and $H \leq G$. If $[g,H] \leq \mathrm{C}_G(f)$ for some elements $g, f \in G$ then $[g,H] \leq \mathrm{C}_G(\langle f\rangle^H)$.*

Proof The result follows from the equality $[g,H]^H = [g,H]$. □

[1]Partially supported by RFBR (grant 96-01-00358)

Assertion 1.4 *Let G be a group and $g \in C_G(G')$. Then for any elements $x_1, \ldots, x_n \in G$ and for any permutation σ of $\{1, \ldots, n\}$*

$$[g, x_1, \ldots, x_n] = [g, x_{\sigma(1)}, \ldots, x_{\sigma(n)}].$$

Proof The result follows from the Witt-Hall commutator identities (see [5], Theorem 5.1) by induction on the number of elements in the sequence $x_1, \ldots, x_n$. □

Corollary 1.5 *Let G be a locally nilpotent group with the min-c conditions and $g \in C_G(G')$. Then there is an integer $n = n(g)$ such that $g \in \zeta_n(G)$.*

Proof There is a finite subset $S \subset G$ such that $C_G(S) = \zeta(G)$. The subgroup $\langle g, S\rangle$ is finitely generated, hence it is nilpotent of some class n. Now we show that $g \in \zeta_n(G)$. If $[g, x_1, \ldots, x_n] \neq e$ for some sequence $x_1, \ldots, x_n \in G$ then $[g, x_1, \ldots, x_{n-1}] \notin \zeta(G) = C_G(S)$ and so there exists an element $y_1 \in S$ such that $[g, x_1, \ldots, x_{n-1}, y_1] \neq e$. By Assertion 1.4 $[g, x_1, \ldots, x_{n-1}, y_1] = [g, y_1, x_1, \ldots, x_{n-1}]$. Continuing in this way n times we obtain $[g, y_1, \ldots, y_n] \neq e$ for some elements $y_1, \ldots, y_n \in S$, a contradiction with the nilpotent class of $\langle g, S\rangle$. □

Assertion 1.6 *Let G be a group with the min-c conditions and let A be a countable subset of G, $A = \{a_1, a_2, \ldots, a_n, \ldots\}$. Then there exists an integer $n = n(A)$ such that $C_G(\{a_i \in A \mid i \geq n\}) = C_G(\{a_i \in A \mid i \geq k\})$ for any $k \geq n$.*

Proof For any integer j we put $A_j = \{a_i \in A \mid i \geq j\}$ and construct the nondecreasing chain of centralizers

$$C_G(A_1) \leq C_G(A_2) \leq \ldots \leq C_G(A_j) \leq \ldots.$$

Since G satisfies the maximal condition on centralizers, this chain stabilizes at a finite step $n = n(A)$. □

Assertion 1.7 *Let G be a group, $g, h \in G$ and $[g, G] \leq C_G(h)$. Then for any sequence $x_1, \ldots, x_n \in G$*

$$[g, h, x_1, x_2, \ldots, x_n] = [g, [h, x_1, x_2, \ldots, x_n]].$$

Proof We use induction on n. Let $n = 1$ and $x \in G$ be an arbitrary element. Putting $H = G$ in Assertion 1.3, we obtain $[x, g, h^x] = e$, $[g, h, x^g] = [g, h, [g, x^{-1}] \cdot x] = [g, h, x] \cdot [g, h, [g, x^{-1}]]^x = [g, h, x]$. In a similar way we obtain $[g^h, [h, x]] = [g, [h, x]]$. Now we use the Witt-Hall commutator identity: $[g, h, x^g] \cdot [x, g, h^x] \cdot [h, x, g^h] = e$ (see [5], Theorem 5.1) and obtain $[g, h, x] = [g, [h, x]]$. Let $n > 1$. We put $h_1 = [h, x_1, \ldots, x_{n-1}]$. By inductive hypothesis $[g, h, x_1, \ldots, x_{n-1}] = [g, h_1]$. By Assertion 1.3, the conditions of Assertion 1.7 are valid for h_1. We repeat the calculations of the first step for g, h_1 and x_n, and obtain $[g, h_1, x_n] = [g, [h_1, x_n]]$. □

2 The main result

Lemma 2.1 *Let G be a countable locally nilpotent group with the min-c conditions and let a, b be non-commuting elements in G. Then there exists an element $c \in \langle b \rangle^G$ such that $a \in (C_G([c,G]) \setminus C_G(c))$.*

Proof We index the elements of the group G with the positive integers, $G = \{g_1, g_2, \ldots, g_k, \ldots\}$, and for every $n \geq 1$ put $G_n = \langle a, b, g_1, g_2, \ldots, g_n \rangle$. Then we find an integer s such that $C_G(G) = C_G(G_s)$. Since the groups G_n are nilpotent, $a, b \in G_n$ and $[a,b] \neq e$, one can construct a sequence $S = \{b_0, b_1, \ldots, b_m, \ldots\}$ by the rule: $b_0 = b$, and if $i > 0$ then b_i is a commutator of maximal length which does not commute with a and has the form $[b_{i-1}, x_1, \ldots, x_{k_i}]$, where $x_1, \ldots, x_{k_i}$ is a suitable (possible empty) sequence of elements of G_i. Such b_i exists in G_i because G_i is nilpotent. Now we show that the sequence S has finitely many different elements. By Assertion 1.6 we find an integer $r = r(S)$ such that for any $m \geq r$

$$C_G(\{b_i \in S \mid i \geq m\}) = C_G(\{b_i \in S \mid i \geq r\}. \tag{1}$$

Then for any $k \geq \max(s,r)$ we construct a commutator of maximal length in the group G_s of the form $a_k = [a, h_1, \ldots, h_{k_s}]$ with $h_1, \ldots, h_{k_s} \in G_s$ and such that $[a_k, b_k] \neq e$. Since lengths of commutators a_k are bounded by the nilpotent class of G_s, there exists an integer $q \geq \max(s,r)$ such that the length of a_q is maximal. In this case the definition of a_q implies that

$$[a_q, G_s] \leq C_G(b_k) \quad \text{for any } k \geq \max(s,r). \tag{2}$$

Since $a_q \in \langle a \rangle^{G_s} \leq \langle a \rangle^{G_k}$ when $k \geq s$, the definition of b_k and Assertion 1.3 imply that

$$[b_k, G_k] \leq C_G(a_q) \quad \text{for any } k \geq q. \tag{3}$$

Putting $G = G_k$, $g = b_k$, $h = a_q$ in Assertion 1.7, we obtain from (3) for $k \geq q$:

$$[b_k, a_q, x] = [b_k, [a_q, x]] \quad \text{for any } x \in G_k. \tag{4}$$

Since, in this case, $G_s \leq G_k$ and $C_G(G_s) = \zeta(G)$, we obtain from (2) and (4) that

$$[a_q, b_k] \leq \zeta(G) \quad \text{for any } k \geq q. \tag{5}$$

Taking into account (5), we obtain from (4) that

$$[a_q, G_k] \leq C_G(b_k) \quad \text{for any } k \geq q. \tag{6}$$

Fixing an arbitrary $k \geq q$, we obtain $[a_q, G_k] \leq [a_q, G_t] \leq C_G(b_t)$ for any $t \geq q$. In view of (1), this gives $[a_q, G_k] \leq C_G(b_q)$ and, since the group G is the join of its subgroups G_k, we have $[a_q, G] \leq C_G(b_q)$. We apply again Assertion 1.7 with $g = a_q$, $h = b_q$ and, taking into account (5), obtain that $[b_q, G] \leq C_G(a_q)$. If for some $k > q$ there is $b_k \neq b_q$ then for any $t \geq k$, $b_t \in [b_q, G]$, $[a_q, b_t] = e$ and, in view of (1), $[a_q, b_q] = e$, a contradiction with the choice of elements a_q, b_q. Therefore $b_q = b_k$ for for any $k \geq q$ and, by definition of the sequence S, $a \in C_G([b_q, G])$. And so the element b_q can be chosen as the required element c. □

Theorem 2.2 *Every locally nilpotent group satisfying the minimal condition on centralizers is hypercentral.*

Proof By Assertion 1.1 we can restrict to a countable group G. By Assertion 1.2 all such groups are soluble. If there exists non-hypercentral locally nilpotent group with the min-c condition then we choose a group with the minimal derived length. Then G is nonabelian and its derived subgroup is hypercentral. By Assertion 1.1, there is an infinite sequence $a_1, a_2, \ldots, a_n \ldots$, satisfying the condition: $a_i = [a_{i-1}, x_i]$ for a suitable $x_i \in G$ for all $i > 1$ and $a_i \neq e$ for all i. We denote the set of all such sequences by S. We remark that if a sequence $A = \{a_i \mid i \geq 1\}$ belongs to S then it is true for its subsequence $\{a_i \in A \mid i \geq k\}$ for every integer k. Therefore G' contains at least one sequence from S, and one can find a minimal number α such that some sequence $A \in S$ belongs to $\zeta_\alpha(G')$. We fix elements of this sequence $A = \{a_i \mid i \geq 1\}$ and elements $x_i \in G$ such that $a_i = [a_{i-1}, x_i]$. By Corollary 1.5 $\alpha > 1$. Moreover, in view of the choice of α, for every $\beta < \alpha$, $A \bigcap \zeta_\beta(G') = \emptyset$. By Assertion 1.6 we find an integer r such that $C_G(\{a_i \mid i \geq r\}) = C_G(\{a_i \mid i \geq k\})$ for any $k \geq r$. Since $\alpha > 1$ and $a_r \in \zeta_\alpha(G') \setminus \zeta_\beta(G')$ for any $\beta < \alpha$ then there exists $b \in G'$ such that $[a_r, b] \neq e$. By Lemma 2.1 we find an element $c \in \langle b \rangle^G \leq G'$ such that

$$[c, a_r] \neq e \text{ and } [c, x, a_r] = e \text{ for any } x \in G. \tag{7}$$

By definition of hypercentre $[c, a_r] \in \zeta_\beta(G')$ for some $\beta < \alpha$. Since no sequence from S belongs to $\zeta_\beta(G')$ then there exists an integer s such that for any $t \geq s$

$$[c, a_r, x_{r+1} \ldots, x_{r+t}] = e. \tag{8}$$

Now we use Assertion 1.7 with $g = c$, $h = a_r$ and obtain for any t:

$$[c, a_r, x_{r+1}, \ldots, x_{r+t}] = [c, [a_r, x_{r+1}, \ldots, x_{r+t}]] = [c, a_{r+t}].$$

Together with (8) this gives

$$c \in C_G(\{a_i \mid i \geq r+s\}) = C_G(\{a_i \mid i \geq r\}) \leq C_G(a_r),$$

a contradiction with (7). □

References

[1] R.M. Bryant, Groups with the minimal condition on centralizers. J. Algebra **60** (1979), 371–383.
[2] S.N. Chernikov, On special p-groups. Mat. Sb. **27** (1950), 185–200 (Russian).
[3] Unsolved problems in group theory. The Kourovka notebook. Novosibirsk, 1995.
[4] A.G. Kurosh, "Group theory." Moscow, Nauka, 1967 (Russian).
[5] W. Magnus, A. Karrass, D. Solitar, "Combinatorial group theory." New York/London/Sydney, 1966.
[6] D.A. Suprunenko, "Groups of matrix." Moscow, Nauka, 1972 (Russian).
[7] B.A.F. Wehrfritz, "Infinite linear groups." Springer–Verlag, Berlin/Heidelberg/New York, 1973.

INFINITE GROUPS IN PROJECTIVE AND SYMPLECTIC GEOMETRY

F. BOGOMOLOV*[1]and L. KATZARKOV† [2]

* Courant Institute, N.Y.U., 251 Mercer Street, New York, NY10012-1185, U.S.A.
Steklov Institute, ul. Vavilova 42 GSP-1, Moscow 117966, Russia
† U.C. Irvine, Irvine, CA92717, U.S.A.

1 Introduction

For a long time it was not known if the groups which appear as fundamental groups of projective manifolds are similar to linear groups. First J. P. Serre asked if there are smooth complex projective manifolds with nonresidually finite fundamental groups. D. Toledo [22] found examples of nonresidually finite groups which appear as fundamental groups of complex projective manifolds. Further examples were given by Catanese, Kollár and Nori [6]. We shall call a group Kähler if it can be realized as a fundamental group of a compact Kähler manifold. We assume (this is a popular belief) that Kähler groups can be realized as fundamental groups of projective manifolds and hence by the Lefschetz hyperplane section theorem they can also be realized as fundamental groups of complex projective surfaces.

The problem, which arises naturally is to find some "natural borders" for the class of groups which appear as fundamental groups of projective (Kähler manifolds) inside the class of finitely presented groups. In this paper we are going to present a construction which shows that fundamental groups of projective surfaces (and hence Kähler groups) are rather densely distributed among all finitely presented groups.

Consider for any finitely presented group Γ the family R_Γ of all finitely presented groups G_i which contain Γ as a normal subgroup with G_i/Γ being equal to the fundamental group of a compact Riemann surface of genus $g(i)$. Any group $G_s \in R_\Gamma$ defines also a sequence of group $G_j(s)$ which are obtained from G_i by increasing the genus $g(s)$ of the base which does not change the action of the fundamental group of a Riemann surface on Γ.

We show that the family R_Γ for any finitely presented group Γ contains sequences $G_j(s)$ which can be "approximated" by the fundamental groups of an algebraic surface. Assuming that the group Γ is given together with a surjective homomorphism $f : \pi_g \to \Gamma$ we construct a surface of genus $h \geq g$ with a related surjective map $f' : \pi_h \to \Gamma$ where the $\mathrm{Ker} f'$ is generated as a normal subgroup by a special finite configuration of simple cycles. Dehn twists corresponding to the subset of cycles in $\mathrm{Ker} f'$ generate a subgroup $M \subset \mathrm{Map}(h)$. We construct then families of groups depending on two discrete parameters. First one is a natural number N. The other one parameterizes some family of open subgroups $M_t \subset \mathrm{Map}(h)$ containing M. We

[1]Partially supported by DMS Grant- 9500774
[2]Partially supported by DMS Grant- 9700605

assume that the parameter N converges to infinity and the intersection of all the groups M_t is equal to M.

If we let N go to infinity and $M_t = M$ then the resulting group will be equal to some $G_j(s) \subset R_\Gamma$. However for all finite values of N and t we obtain a fundamental group of some projective algebraic surface.

Thus by increasing the genus of the base we can implant into a fundamental group of a projective surface an "approximation" (in the above sense) of any finitely presented group as a normal subgroup with a rather standard quotient. We still don't quite understand the relation between the group and its "approximations" but we think this problem is related not to geometry but to the theory of infinite groups.

We also formulate several precise questions, which hopefully will be addressed by group theorists. We show that the answers of these questions have far going geometric consequences. These questions came out from our attempts to approach a problem formulated by Shafarevich. He conjectured that the universal covering of a smooth projective variety X, $\widetilde{X}$ is holomorphically convex. Recall that a complex manifold M is holomorphically convex if for every infinite discrete subset S of M there exists a holomorphic function on M that is unbounded on S. T. Napier has studied the question when (unramified, infinite) coverings of a complex surface are holomorphically convex and has shown [19] that one of the basic obstructions to the holomorphic convexity is the existence of connected noncompact analytic sets all of whose irreducible components are compact. We will call such an analytic set an infinite chain.

Let X be an algebraic surface and D be a compact connected curve consisting of components D_i. Assume that the image of the fundamental group $\pi_1(D)$ in $\pi_1(X)$ is infinite but the image of the fundamental group of the normalization of every component D_i is finite. Then the preimage of D in the universal covering $\tilde{X}$ of X is an infinite connected graph of compact curves and therefore $\tilde{X}$ is not holomorphically convex.

Using the results of Lassel and Ramachandran [13] and of Simpson (see e.g. [20]) one can prove the following:

Theorem 1.1 *Let $\rho : \pi_1(X) \to GL(N, \mathbb{C})$ be a linear representation of the fundamental group of an algebraic surface X and let $D = \cup D_i$ is a divisor in X. If the restriction of ρ on $\pi_1(D_i)$ is trivial then the restriction of ρ on $\pi_1(D)$ is a finite group.*

This theorem implies (see e.g. [14]):

Theorem 1.2 *Let X be a smooth projective surface. If $\pi_1(X)$ can be realized as a subgroup of a linear group $GL(N, \mathbb{C})$ for some N, then the universal covering of X is holomorphically convex.*

Simpson's technique of mixed twistor structures (MTS) [21] gives stronger results which is a subject of a recent work by Katzarkov, Pantev and Ramachandran [15]. It is plausible that (MTS) technique will give a proof for X an algebraic surface with residually finite torsion free fundamental group.

The fact that the properties of the fundamental group enforce some analytic properties of the universal covering of the surface makes it very tempting to use analytic notions and concepts in the investigation of the properties of abstract infinite groups. It is worth noticing that since the pioneering work of Gromov the analogues of the basic concepts of differential geometry became standard concepts in the theory of infinite groups.

The article contains six sections and two appendices.

In the first section we describe a local version of the base change construction which allows to enlarge the fundamental group of a local fibration. The second section is devoted to its globalization and in the third section we consider results which clarify the relation between fundamental groups and universal coverings of projective and quasiprojetcive surfaces. In the fourth section we discuss the groups which appear as fundamental groups after applying our construction to some specially constructed initial surfaces. In Section 5 we consider a special case of our construction and produce a series of potential counterexamples to the Shafarevich's conjecture discussed above. The results and the constructions of this section are heavily based on the results of Appendix B which are related to the structure of groups with relations $x^3 = 1$ for some sets of elements. They generalize classical results on the structure of Burnside groups of exponent 3.

In Section 6 we consider some application to symplectic fourfolds. Using recent work of Donaldson [10] we define new invariants of symplectic fourfolds. To illustrate the motivic nature of our considerations we suggest an arithmetics version of Shafarevich's conjecture. Appendix A contains some results on the topology of projective surfaces and maps of the fundamental groups which we use in the article.

2 The construction

2.1 Vanishing cycles - a local construction

In this subsection we explain the local computation with the vanishing cycles on which the whole construction is based. We begin with some classical results on degeneration of curves that can be found in [7]. For more details see also [3].

Let X_D be a smooth complex surface fibered over a disc D. We assume that fibers over a punctured disc $D^* = D - 0$ are smooth curves of genus g and the projection $t : X \to D$ is a complex Morse function. In particular the fiber X_0 over $0 \in D$ has only quadratic singular points and it has no multiple components. Denote by P the set of singular points of X_0 and by $T : X_t \to X_t$ the monodromy transformation acting on the fundamental group of the general fiber X_t. This action can be described in terms of Dehn twists. Obviously this action defines an action on the first homology group of the general fiber. The following proposition describes completely the topology of X_D and the projection $t : X_D \to D$.

Proposition 2.1 1) *There is a natural topological contraction* $cr : X_D \to X_0$.

2) *The restriction of* cr *to* X_t *is an isomorphism outside singular points* $P_i \in X_0$. *It contracts the circle* $S_i \in X_t$ *into* P_i. *The monodromy transformation* T *is*

the identity outside small band B_i around S_i and in B_i the transformation T coincides with a standard Dehn twist.

Proof See [7]. □

The above contraction is an isomorphism from X_t minus preimage of P on $X_0 - P$. The preimage of any singular point P_i is a smooth, homotopically nontrivial curve $S_i \subset X_t$.

Definition 2.1 We will call the free homotopy class of S_i in the fundamental group of X_t a geometric vanishing cycle.

Remark 2.1 The direction of the standard Dehn twist is defined by the orientation of the neighborhood of S_i which in turn is defined by the complex structure of the neighborhood of the corresponding singular point of the singular fiber.

We denote by De_i the topological Dehn transformation of X_t and by $De_{i,H}$ its action on the homology of X_t. In a neighborhood of X_0 the monodromy transformation T is a product of Dehn transformations De_i with non-intersecting support.

Lemma 2.1 1) *The monodromy transformation T acts via unipotent transformation T_H on the homology group $H_1(X_t, \mathbb{Z})$.*

2) $(1 - T_H)^2 = 0$.

3) $(1 - T_H^N) = 0 (mod\ N)$ *for any N.*

Proof It is enough to prove (2) for $De_{i,H}$. The topological description of De_i implies that $(1 - De_{i,H})x = (x, s_i)s_i$, where s_i is a homology class of the vanishing cycle s_i and (x, s_i) is the intersection number.

Since $(s_i, s_i) = 0$ we obtain $(1 - De_{i,H})^2 = 0$. The image of $(1 - De_{i,H})$ consists of the elements proportional to s_i. We also have $(1 - D^N e_{i,H}) = (1 - N De_{i,H}) = 0(\text{mod } N)$.

The topological transformations De_i commute for different i since they have disjoint supports. Therefore the same holds for $De_{i,H}$. We also have $De_{i,H} s_j = 0$ for any vanishing cycle s_j since the corresponding circles S_i, S_j do not intersect in X_t. We now see that

$$(1 - T_H)^2 = (1 - \prod De_{i,H})^2 = \prod (1 - De_{i,H})^2 = 0$$

where the last equality follows from the above formulas. Similarly we obtain 1) and 3). □

The geometric vanishing cycles consist of two different types:
1) The first type includes homologically nontrivial vanishing classes. They are all equivalent under the mapping class group Map(g). The latter can be described either as a group of connected components of the orientation preserving homeomorphisms of the Riemann surface of genus g or as the group of exterior automorphisms of the group π_g. The vanishing cycle from this class is a primitive element

of π_g. Throughout the whole paper primitive element means an element that can be included into a set of generators of π_g satisfying standard relation defining the fundamental group of the curve. We shall denote the vanishing cycles of the first type as NZ-cycles.

2) The second type consists of elements in π_g which are homologous to zero. Any vanishing cycle of this type cuts the Riemann surface X_t into two pieces and the number of handles in these pieces is the only invariant which distinguishes the type of a cycle under the action of Map(g). We shall denote the vanishing cycles of the second type by Z-cycles.

Vanishing Z-cycles correspond to the singular points of the singular fiber which divide this fiber into two components and NZ-cycles to the ones which do not. Each Z-cycle defines a primitive element in the center of $\pi_g/[[\pi_g, \pi_g], \pi_g]$, while each NZ-cycle defines a primitive element in the abelian quotient $\pi_g/[\pi_g, \pi_g] = \mathbb{Z}^{2g}$. Assume now that we made a change of the variable $t = u^N$ and consider the induced family of curves over D with a coordinate u. Denote the resulting family as X_U. It is a singular surface and the singular set can be identified with the set P o f the singular points of the fiber X_0. For the new monodromy transformation we have $T_U = T^N$. Therefore by Lemma 2.3 it acts trivially on $H_1(X_t, \mathbb{Z}_N), t \neq 0$.

Lemma 2.2 *The surface X_U contracts to the central fiber X_0.*

Indeed the fiberwise contraction of X_D to X_0 can be lifted into a contraction of X_U.

Remark 2.2 The fundamental group $\pi_1(X_D - P) = \pi_1(X_D) = \pi_1(X_0)$ since the singular points of the fiber X_0 are nonsingular points of X_D. The analogous statement is not true however for X_U.

Theorem 2.1 *The fundamental group $\pi_1(X_U - P)$ is equal to the quotient of $\pi_1(X_t) = \pi_g$ by a normal subgroup generated by the elements s_i^N.*

Proof The fundamental group of a complex coincides with the fundamental group of any two-skeleton of the complex. There is a natural two-dimensional complex with a fundamental group as in the theorem. Namely let us take a curve X_t and attach two-dimensional disks D_i via the boundary maps $f_i : dD_i \to S_i$ of degree N. The resulting two dimensional complex X_t^a evidently has a fundamental group isomorphic to the group described in the theorem. We are going to show that X_t^a can be realized as a two skeleton of $X_U - P$. We prove first the following statement:

Lemma 2.3 *The surface $X_U - P$ retracts onto a three dimensional complex X^3 which is a union of generic curve X_t and a set of three-dimensional lens spaces L_N^i. Each lens space corresponds to the singular point P_i of the singular fiber and L_N^i intersects X_t along the band B_i.*

Proof Locally near P_i the surface X_U is described by the equation $t^N = z_1 z_2$. Hence a neighborhood U_i of P_i is a cone over the three-dimensional lens space $L_N^i = S^3/\mathbb{Z}_N$. Here S^3 is a three dimensional sphere and $\mathbb{Z}_N$ is generated by the matrix with eigenvalues χ, χ^{-1}, where χ is a primitive root of unity of order N. We can assume that the surface X_t intersects L_N^i along a two-dimensional band B_i with a central circle S_i defining a generator of $\pi_1(L_N^i)$. Though the band B_i is a direct product of S_i by an interval its embedding into L_N^i is nontrivial : the boundary circles have nonzero linking number. We use a fiberwise contraction to contract $X_U - P$ to the union to X^3. It coincides with a standard contraction outside of the cones over L_N^i. Therefore we obtain a contraction of $X_U - P$ onto X^3. □

The two-skeleton of X^3 can be obtained as a union of X_t and a two-skeleton of L_N^i. The latter can be seen as a retract of a complimentary set to the point in L_N^i. Here is the topological picture we are looking at.

Lemma 2.4 *Let L_N^i be a lens space as above. Then the complimentary set to a point retracts on complex L_i^2 obtained by attaching a disc to the circle S_i via the boundary map of degree N.*

Proof The sphere S^3 can be represented as a joint of two circles S_i and S'. In other words it consists of intervals connecting different points of S_i, S'. The action of $\mathbb{Z}_N$ with $L_N^3 = S^3/\mathbb{Z}_N$ rotates both circle. Define the disc D_x to be a cone over S_i. Different discs D_x, $x \in S'$, don't intersect and the image of D_x in L_N^3 is L_i^2. The fundamental domain of $\mathbb{Z}_N$-action lies between discs D_x, D_{gx}, where g is a generator of $\mathbb{Z}_N$. Hence this domain is isomorphic to D^3 and coincides with complimentary of L_i^2 in L_N^i. The last proves the lemma. □

Corollary 2.1 *There is an embedding of X_t^a into $X_U - P$ which induces an isomorphism of the fundamental groups.*

Indeed we obtain the two skeleton of X^3 gluing X_t and L_i^2 along S_i, but the resulting two-complex coincides with X_t^a. Since X^3 is a retract of $X_U - P$ we obtain the corollary and finish the proof of the theorem. □

Let us denote by G_X the fundamental group of $X_U - P$ and by $\widetilde{X}$ its universal covering. The description of the group G_X can be obtained in pure geometric terms. Namely the vanishing cycles S_i don't intersect and therefore we can first contract Z-cycles to obtain the union of smooth Riemann surfaces X_i with normal intersections. The graph corresponding to this system of surfaces is a tree since every point corresponding to Z-cycle splits it into two components. The remaining NZ- cycles lie on different surfaces X_j and constitute a finite isotropic subset of primitive elements in $\pi_1(X_j)$. The vanishing cycle S_i which contracts to a point of X_j defines a map of a free group onto $\pi_1(X_j)$ with S_i corresponding to standard relations in $\pi_1(X_j)$. We also have the following lemma.

Lemma 2.5 *If N is odd or divisible by four then the group G_X has a natural surjective projection on the quotient group $\pi_g/[[\pi_g, \pi_g], \pi_g]$ with additional relation $x^N = 1$. If $N = 2$ then G_X maps surjectively on a central extension of $\mathbb{Z}_2^{2g}$ by $\mathbb{Z}_2$ and the images of all s_i have order 2.*

Proof The case when N is odd or divisible by four is clear since all the elements s_i have order N. In the case $N = 2$ we use the fact that NZ cycles s_i contained in the component X_j lie in the isotropic subspace. Hence there exists a standard $\mathbb{Z}_2$ central extension of $\mathbb{Z}_2^{2g_j}$ where the images of all $s_i \in X_j$ are exactly of order 2. Denote this group by G_j and the generator of its center by c_j. Now consider the product of G_j for all j and factor it by a central subgroup generated by $c_k - c_j$ if X_k and X_j intersect. Since the graph is a tree we obtain that the quotient of the center by the above group is equal to $\mathbb{Z}_2$ which is identified with the zero homology with coefficients $\mathbb{Z}_2$. The image of $\mathbb{Z}$-cycle s_i coincide with c_j if $s_i \in X_j$ and hence is never zero. Vanishing NZ-cycles project into nonzero elements of the abelian quotient of the group. □

Remark 2.3 In the case of N is odd or divisible by four we obtain a canonical quotient of G_X which is invariant under $\mathrm{Aut}(\pi_g)$. We denote this group by UC_g^N. In the case of $N = 2$ our construction is less canonical, it depends on the choice of isotropic subspace in $H_1(X_t, \mathbb{Z}_2)$.

Here we start to develop an idea that will be constantly used through out the paper. Namely we show that we can work with open surfaces and get results concerning the universal coverings of the closed surfaces. The advantage is that we get bigger variety of fundamental groups.

Theorem 2.2 *There is a natural G_X-invariant embedding of $\widetilde{X}$ into a smooth surface $\widetilde{X}_U$ with $\widetilde{X}_U/G_X = X_U$. The complimentary of $\widetilde{X}$ in $\widetilde{X}_U$ consists of a discrete subset of points.*

Proof Let P_i is a point that corresponds to a NZ vanishing cycle. The preimage of a local neighborhood U_i of P_i in $\widetilde{X}$ consists of a number of nonramified coverings of $U_i - P_i$. Since $\pi_1(U_i - P_i) = \pi_1(L_N^i) = \mathbb{Z}_N$ we obtain that the nonramified covering we get is finite. Therefore by local pointwise completion we obtain $\widetilde{X}_U$ with the action of G_X extending on it.

Now the universal nonramified covering of $U_i - P_i$ is a punctured unit disk in $\mathbb{C}^2$. Therefore if the map $\pi_1(U_i - P_i) \to G_X$ is injective then the surface $\widetilde{X}_U$ is nonsingular. Since the group $\pi_1(U_i - P_i) = \mathbb{Z}_N$ and is generated ed by the vanishing cycle s_i we can easily get the result. Indeed the groups $H_1(G_X, \mathbb{Z}_N)$ and $H^1(X_t, \mathbb{Z}_N)$ are isomorphic. The general result follows from theorem 2.1 since we have constructed finite quotients for G_X which map every s_i into an element of order N. □

We also have the following:

Lemma 2.6 *Let G_X^0 be the kernel of projection of G_X into one of the finite groups defined in Lemma 1.13. Then G_X^0 acts freely on $\widetilde{X}_U$ and the quotient is a family of compact curves without multiple fibers and with a fundamental group G_X^0.*

Proof Indeed the surface $\widetilde{X}_U$ was obtained from $\widetilde{X}$ by adding points and since $\widetilde{X}$ was simplyconnected the former is symplyconnected either. Any element of G_X

which has invariant point in $\widetilde{X}_U$ is conjugated to the power of s_i, but the latter are not contained in G^0_X. □

We are done with our description of the local computations. By taking inite coverings and taking away the singularities of the covering we were able to make all local geometric vanishing cycles torsion elements.

2.2 The global construction

In this subsection we globalize the construction. We will show that we can do the construction described in the previous section globally in a compact surface.

Let X be smooth surface with a proper map to a smooth projective curve C. We assume that the map $f : X \to C$ is described locally by a set of holomorphic Morse functions and hence satisfies locally the conditions of the previous section. Thus the generic fiber is a smooth curve $X_t, t \in C$ of genus g. As in the previous section we denote the fundamental group of X_t by π_g.

Abusing the notation now we denote by P the set of all singular points of the fibers and by P_C the set of points in C corresponding to the singular fibers. Thus $f(P) = P_C$ and all points in P are singular double points. The main difference of the global situation lies in the presence of the global monodromy group which is the image of $\pi_1(C - P_C)$ in the mapping class group Map(g). We denote this group by M_X.

Let us select now an integer N and consider a base change $h : R \to C$ where R such that the map h is N-ramified at all the preimages of the points from P_C in R. Consider a surface S obtained via a base change $h : R \to C$. We have a finite map $h' : S \to X$ defined via h and the projection $g : S \to R$ with a generic fiber $S_t = X_{h(t)}$. The surface S is singular with the set of singular points equal to $h^{-1}(P) = Q$ and the set of singular fibers over the points of $h^{-1}P_C = P_R$. The monodromy group M_S of the family S is subgroup of finite index of the group M_X.

Remark 2.4 The fact that the monodromy group M_S of the family S is a subgroup of a finite index of the group M_X gives us some flexibility in choosing the corresponding families.

Theorem 2.3 *The fundamental group $\pi_1(S - Q)$ is an extension of $\pi_1(R)$ by the quotient of π_g by a normal subgroup generated by the orbits $M_S(s_i^N)$.*

Proof Indeed we have a natural surjection of π_g on the kernel of projection onto $\pi_1(R)$ since there are no multiple fibers in the projection g. A standard argument reduces all relations to the local ones and the local relations were described in the previous section (see Theorem 2.1). □

Now we move to the second step of our construction getting out of $S - Q$ a smooth compact surface S^N with almost the same fundamental group as $S - Q$. Let us assume that $N \geq 2$.

Lemma 2.7 *There exists a smooth projective surface S^N with a finite map $f : S^N \to S$ such that the image of the homomorphism $f_* : \pi_1(S^N) \to \pi_1(S - Q)$ is a subgroup of finite index in $\pi_1(S - Q)$.*

Proof As it was already shown in Lemma 2.5 there exists a projection of $\pi_1(X_g)$ to a finite group UC_g^N which is invariant under $\mathrm{Aut}(\pi_g)$ and factors through the fundamental group of a small neighborhood of a singular fiber in $S - Q$. Therefore, if the map $h : S \to R$ has a topological section we obtain a finite fiberwise covering of S which is ramified only over the singular set Q. The resulting surface coincides locally with the smooth surface described in Theorem 2. It is also smooth and the map f is finite. If there is not a topological section we first make a base change nonramified at P_R in order to obtain such a section (see Theorem 7.2 Appendix A) and then apply the previous argument. Since the map f is finite the preimage $f^{-1}(Q)$ consists of a finite number of smooth points and therefore $\pi_1(S^N) = \pi_1(S^N - f^{-1}(Q))$ (see Theorem 7.1 Appendix A). This also proves that the image of the homomorphism $f_* : \pi_1(S^N) \to \pi_1(S - Q)$ is a subgroup of finite index in $\pi_1(S - Q)$. □

As we have said in the introduction to find a counterexample to the Shafarevich's conjecture we try to construct an infinite connected chain of compact curves in the universal covering of S^N. The above construction shows that we can do it by controlling the image of the fundamental groups of the open irreducible components of the reducible singular fibers in the fundamental group of the open surface $S - Q$. In the next sections we introduce a variety of constructions and groups respectively that suggest ways of constructing an infinite connected chain of compact curves.

3 Classes of groups

The construction discussed in the previous section can be applied to both projective and quasiprojective fibered surfaces. This indicates that the universal coverings and the fundamental groups for these two classes of surfaces have a similar structure. In this section we illustrate another flavor of the same principle by describing a procedure, comparing the fundamental groups of surfaces with quotient singularities to the fundamental groups of certain smooth surfaces. We will set up this transition in a slightly bigger generality.

The global construction described in Section 2.2 treats separately the part of the fundamental group of the fibered surface which lies in the image of the fundamental group of the fiber. Let V be normal projective complex surface and Q be a set of its singular points. Suppose that there is a projection of V on a smooth curve C which has no multiple fibers and the generic fiber of the projection is a curve of genus $g > 1$.

Definition 3.1 Denote by $\pi_{1,f}(V - Q)$ the image of the fundamental group π_g of the general fiber in $\pi_1(V - Q)$. We will call this group a general fiber group.

In this article we mostly consider the case when the set of singular points in V includes only singularities with finite local fundamental groups. It is well known that these are exactly the quotient singularities.

Definition 3.2 We will call the group $\pi_{1,f}(V-Q)$ above a fiber group if Q consists of the quotient singularities only.

We shall also give a special notation for the case when Q is empty.

Definition 3.3 We will call the group $\pi_{1,f}(V)$ projective fiber group if V is a projective fibered surface over C without multiple fibers.

Remark 3.1 Though we don't allow multiple fibers in the above definition we allow some multiple components in the singular fibers. We need that at least one component of each singular fiber has multiplicity one.

Thus we have defined three classes of groups. These groups are equipped with a surjective map from the group π_g. The principal difference between these three classes of groups lies in the nontriviality of the local fundamental groups of normal surface singularity.

Remark 3.2 It follows that the general fiber group occurs also as a fiber group of a projective surface if the images of all local fundamental groups are finite. In particular this is true if all singular points have finite local fundamental groups.

The following theorem is a generalization of a theorem by J. Kollár (see [17]).

Theorem 3.1 *Let V be a normal projective surface and let Q be the finite set of its singular points. Consider for any $q \in Q$ a normal subgroup of finite index $K_q \triangleleft L_q$ which contains the kernel of the natural map $L_q \to \pi_1(V-Q)$. Then there exists a smooth projective surface F and a surjective finite map $r : F \to V$ which induces an isomorphism between $\pi_1(F)$ and the quotient of $\pi_1(V-Q)$ by the normal subgroup generated by the images of $K_q \in \pi_1(V-Q), q \in Q$.*

Proof Denote by K_Q the normal subgroup of $\pi_1(V-Q)$ generated by the subgroups $K_q \subset L_q$. We obtain the surface N as a generic hyperplane section of a singular projective variety W with the following property: W contains a subvariety S of codimension at least three and such that $\pi_1(W-S) = \pi_1(V-Q)/K_Q$. We may assume that F does not intersect S since the latter has codimension at least three in W. The fundamental group $\pi_1(F) = \pi_1(W-S)$ since F is a generic complete intersection in W. We are going to construct W as a union of two quasiprojective subvarieties. Denote by G_q the finite quotient L_q/K_q and by G the direct product of all the groups G_q. Denote by g_q the coordinate projection of G onto G_q and by i_q the coordinate embedding of G_q into G. For any q there is a natural finite covering M_q of U_q corresponding to the projection $L_q \to G_q$. The preimage of q in M_q consists of a single point and the projection $M_q \to U_q$ is nonramified outside q. In the next lemma we prove the existence of an algebraic extension of this local covering.

Lemma 3.1 *There exists an open affine subvariety $V_q \subset V$ containing q and an affine variety B_q which is a G_q-Galois covering of V_q ramified only at q so that $B_q \times_{V_q} U_q$ and M_q are isomorphic.*

Proof Let $\hat{A}_q$ be the completed local ring of $q \in V$. A local G_q-covering defines a finite algebraic extension $\hat{B}_q$ of $\hat{A}_q$. By Artin's approximation theorem [1] there exists an affine ring $A \subset \mathbb{C}(V)$ and a finite algebraic extension B of A which locally at q corresponds to the extension $\hat{B}_q$ over $\hat{A}_q$. Explicitly the extension $\hat{B}_q$ is described by a monic polynomial $f(x)$ with coefficients in $\hat{A}_q$. If we now consider any monic polynomial $g(x)$ over the ring A with $g(x) = f(x) \bmod \mathfrak{m}_q^N$ for a big enough N then the resulting algebraic extension B will be the one we need. The ring A defines an open algebraic subvariety spec$(A) \subset V$ containing q. Similarly B defines an affine variety spec(B) with a finite projection p_q : spec$(B) \to$ spec(A). This projection is unramified outside q in the formal neighborhood of q. Since we have the freedom to impose any finite number of extra conditions on $g(x)$ we can choose p_q to be unramified at any finite number of points. In particular we may assume that the projection p_q is nonramified over $Q - \{q\}$. That means that the divisorial part $D \subset$ spec(A) of the ramification of p_q does not intersect Q. Now we can take spec$(A)/D$ as V_q. Let B_q denote spec$(B) \times_{\text{spec}(A)} V_q$. It is an affine variety with affine G_q-action since it extends the local nonramified Galois covering $U_q - \{q\}$ and has the same degree. □

Let B_0 be the product of all B_q's over V. This is an affine variety with the action of G. The quotient $B_0/G = V_0'$ is an open affine subvariety of V which contains Q. The action of G on B_0 is free outside of the preimage of Q. Let $G \to \mathrm{GL}(E)$ be a (not-necessarily irreducible) faithful linear representation of G of dimension e with the property that only $1 \in G$ is represented by scalar matrix. Consider the diagonal action of G on the product $B_0 \times E$. There exists a natural $\mathbb{C}^*$ action on E - multiplication by scalars. It extends to a $\mathbb{C}^*$-action on the product which commutes with the G-action.

Let $F_0' = (B_0 \times E)/G$ be the quotient variety. It is an affine variety with induced $\mathbb{C}^*$ action which has a natural projection $\pi_0 : F_0' \to V_0'$ and a zero section $i(V_0') = (B_0 \times 0)/G$. For any $s \in V_0' - Q$ the preimage $\pi_0^{-1}(s)$ is a vector space isomorphic to E. Moreover F_0' contains a natural vector bundle I over $V_0' - Q$. Its sheaf of sections coincides with the sheaf of G-equivariant sections of the constant sheaf $\mathcal{O} \otimes E$ over B_0. Let us choose a smaller affine variety $V_0 \subset V_0', Q \subset V_0$ with the property that I is constant on $V_0 - Q$. We define F_0 as the preimage of V_0 in F_0'. Let V_1 be an open subvariety of V which does not contain Q and such that the union of V_1 and V_0 is equal to V. Let J be the trivial bundle of rank e on V_1. Choose a linear algebraic isomorphism of F_0 and J over the intersection of V_0 and V_1. Use this isomorphism to glue the projectivization $\mathbb{P}(J)$ with the singular variety $X = (N_0 - i(V_0'))/\mathbb{C}^*$. The resulting proper variety W has a natural projection $p : W \to V$ with all the fibers outside Q isomorphic to projective space $\mathbb{P}^{e-1}$. Moreover the preimage of $V - Q$ in W coincides with the projectivization of a vector bundle according to the construction of W. The fiber W_q over q coincides with $\mathbb{P}^{e-1}/i_q(G_q)$.

The action of $i(G_q)$ on $\mathbb{P}^{e-1}$ is effective because of our assumption on the representation of $G \to \mathrm{GL}(E)$. We denote by S_q the singular subset of W_q. It lies in the image of the fixed sets $\mathrm{Fix}_g(\mathbb{P}^{e-1}) \subset \mathbb{P}^{e-1}$ for different elements $g \in i_q(G_q), g \neq 1$. Define a subvariety S as the union of the varieties $S_q, q \in Q$. The set S has codimension ≥ 3 in W since the codimension of S_q in W_q is at least 1.

Lemma 3.2 *The variety $W - S$ has a fundamental group isomorphic to $\pi_1(V - Q)/K_Q$.*

Proof The fundamental group of $W-S$ is the quotient of the fundamental group of $\pi_1(V-Q)$ since W contains an open subvariety which is $\mathbb{P}^{e-1}$ fibration over $V-Q$ and therefore has the same fundamental group. The group K_q maps into zero under the surjective map $\pi_1(V-Q) \to \pi_1(W-S)$ since the image of a neighborhood of q via the zero section i has K_q as a local fundamental group. All the relations are local and concentrated near special fibers. A formal neighborhood of $W_q - S_q$ in $W - S$ is topologically isomorphic to a fibration over $W_q - S_q$ with M_q as a fiber. Therefore all local relations follow from $K_q = 1$. It finishes the proof that $\pi_1(W-S) = \pi_1(V-Q)/K_Q$. □

Finally we prove the projectivity of W by constructing an ample line bundle on it. Start with a line bundle L on W whose sections give an embedding of X into a projective space. To see that such an L exists consider first the G-invariant and $\mathbb{C}^*$ homogeneous sections of the trivial bundle $\mathcal{O} \otimes E$ over B_0. Assume that the degree of homogeneity is big enough and divisible by the order of G. It is known that such sections separate the points in the quotient variety X and we obtain an embedding of X into a projective space. Thus the induced bundle $\mathcal{O}(1)$ is defined on X. Denote by L some extension of $\mathcal{O}(1)$ to W. Such extension exists since the complement of X in W is smooth.

Next by choosing a polarization H on V appropriately we may assume that the global sections of $L \otimes p^*H$ on W separate all the points of X. The restriction of L on $\mathbb{P}(J)$ coincides with $\mathcal{O}_{\mathbb{P}(J)}(m)$ for some positive integer m. By replacing H with a high power of H if necessary we can produce enough sections of $L \otimes p^*H$ to separate the points of $\mathbb{P}(J)$. Therefore $L \otimes p^*H$ gives an embedding of W into a projective space. □

As a consequences of the above theorem we get:

Corollary 3.1 *The classes of fundamental groups and fiber groups are the same for projective smooth surfaces and projective surfaces minus quotient singularities.*

The above results suggest that finding examples of smooth projective surfaces with pathological behavior of fundamental groups and universal coverings can be reduced to a similar problem for normal projective surfaces minus singular points. The latter seems to be an easier task.

4 Relations to group theory

Now we suggest a construction that could lead to new examples of surfaces with nonresidually finite groups. We thank V. Alexeev, S. Keel and M. Nori, for fruitful discussions in this direction.

Let M_g^L be the compactified moduli space of curves of genus g corresponding to a subgroup of finite index M_L in the group Map(g). (In other words M_g^L is a covering of the Deligne-Mumford compactification of the moduli space of curves of genus g that corresponds to the subgroup M_L in the group Map(g).) It is an algebraic variety with quotient singularities only which contains a family of similar type irreducible divisors S_I with normal crossings corresponding to different type of stable degenerations. Singular points of M_g^L correspond to stable curves with automorphisms and constitute a subset of codimension more than 1 if genus $g > 2$. Let M_g^{Lo} be an open nonsingular subvariety in M_g^L which corresponds to smooth curves of genus g without extra automorphisms. Then $\pi_1(M_g^{Lo}) = M_L$.

The space $M_g^0, g > 3$ is far from being affine. In fact there is a natural map of M_g^0 into a Satake compactification of the moduli of abelian varieties of dimension g with a principal polarization. It maps each stable curve to the corresponding the Jacobian of its normalization. Thus all divisors corresponding to degenerate curve have images of codimension at least 2 if $g > 2$ under this map. In particular generic hyperplane sections of Satake compactification produce complete curves which lie in M_g^0.

We are interested in constructing holomorphic families of curves with a given set of singularities. The following construction shows that there are almost no restrictions in constructing such families over a disc.

Lemma 4.1 *Let X_0 be curve of genus g with a given set of smooth noncontarctible cycles s_i^k on it. Assume that for a given k all cycles s_i^k, s_j^k do not intersect and correspond to different conjugated classes in the fundamental group of X_0. Then there is a holomorphic family of curves over a disc D which contains X_0 as nonsingular fiber, the families s_i^k correspond to the vanishing cycles for degenerate fibers and monodromy group is generated by the products of Dehn twists over s_i^k for each k.*

Proof The condition on s_i^k means that each set s_i^k corresponds to some type of stable degeneration $I(k)$ modulo the action of Map(g). Consider a small complex disc D_k around a generic point of $S_{I(k)}$ in M_g. There is a local family of stable curves over D_k which consists of smooth curves outside the point of intersection of D_k and $S_{I(k)}$. Let p_k be a point on the boundary circle dD_k. We can find a path t_0^k connecting 0 and D_k inside M_g^0 which provides with a diffeomorphisms $X_0 \to X_{p_k}$ which maps s_i^k into a family of vanishing cycles on X_{p_k}. Indeed different paths provide with maps which differ by the elements of $\text{Map}(g) = \pi_1(M_g^0)$.

Let us take an extension of t_0^k into smooth real analytic curves which ends up transversally at the intersection point of D_k and $S_{I(k)}$. We can assume that all the curves t_k^0 meet at 0 being tangent to some one-dimensional complex subspace. Thus we constructed a one dimensional "octopus" W consisting of the extended curves t_k^0. We can assume that W in the neighborhood of 0 belong to an analytic disc inside M_g^0. A small analytic neighborhood of $U(W)$ is a Stein variety. The subset W is totally real which means that complex valued functions on W are approximated by holomorphic functions on $U(W)$. After a small variation of W we can complexify the resulting one-dimensional real set into a complex disc D which contains 0 and intersects a given set of divisors $S_{I(k)}$ with a prescribed monodromy corresponding to the connecting path t_k^0.

The family of stable curves induced on D satisfies the lemma. □

Lemma 4.2 *For any finitely presented group Γ we can construct a relative projective family of curves X_t over a holomorphic disc which has Γ as a fundamental group.*

Proof For any group Γ above we can find a surjective homomorphism $r : \pi_g \to \Gamma$ for some g. Let N be the kernel of r. It has a finite number of generators k_i as normal subgroup of π_g. The elements k_i can be realized as cycles with normal intersections (including selfintersections) only on the curve X_g. Let us add a handle at each intersection. We can lift k_i in a new Riemann surface X_h into a family of cycles $\tilde{k}_i$ without selfintersections. If we add to this family of cycles the generating cycles a_j, b_j of the additional small handles we obtain the set which generates the kernel of the projection $\pi_h \to \Gamma$ and is realized by simple cycles without selfintersections.

We introduce a complex structure on X_h and consider the Dehn twists corresponding to $\tilde{k}_i, a_j, b_j$. All these cycles correspond to the conjugation classes which belong to the kernel of the projection $r_h : \pi_h \to \Gamma$. Note that the Dehn twist along the cycle s acts trivially on the quotient of π_h by a normal subgroup generated by s. Therefore we can apply the above construction to the family $X_h, \tilde{k}_i, a_j, b_j$ and obtain a relatively projective family X of curves of genus h over a disc with $\pi_1(X) = \Gamma$. □

Though the holomorphic families of curves look very different from algebraic ones we can rather easily embed a small deformation of such a family into an algebraic one.

Denote by M the monodromy group of the family of curves over disc described in the lemma. Note that D is a Stein subvariety in M_g. We can lift it into any covering of M_g which is unramified along D. In particular we can lift D into any variety M_g^L for a subgroup of finite index $M^L \subset \mathrm{Map}(g)$ containing M. Since D lies in an affine subset of M_g we can find an algebraic curve $C \subset M_g$ (respectively M_g^L) which contains a small variation of D. The family over C induced from M_g or M_g^L) extends globally a small variation of the family over D without changing its topological data. By taking generic C we can assume that the resulting family Y of curves over a normalization of C is smooth and the projection $p : Y \to C$ is a Morse type map. Thus having a family over disc with arbitrary data g, Ms_i where M is generated by the Dehn twists over s_i we obtain the following geometric data $g, M_L s_i, M_L s_j$ for any subgroup of finite index in M_g containing M. Considering accordingly any data g, Ms_i we can obtain a Burnside type approximation $g, M_L s_i^{N_1}, M_L s_j^{N_2}$ for any N_1, N_2. By taking $N_1 = N$ and $N_2 = N^S$ for increasing S and decreasing M_L we obtain series which approximate the group given by g, Ms_i^N.

Remark 4.1 It is expected (see Conjecture 4.1) that most of the groups in such a series are nonresidually finite and violate any other good properties of linear groups if of course the group g, Ms_i is nonresidually finite. Since the latter can be any group and the set s_i can be chosen rather arbitrarily the above construction seems to indicate that almost any hereditary property which breaks for finitely presented groups can be broken for the fundamental groups of projective complex surfaces.

Consider the map f of the moduli space M_g^L into Satake compactification S_g of the moduli space of abelian varieties of dimension g with a principal polarization. The latter is a projective variety which has a representation as a union of $A_g, A_{g-1}, \ldots$ where A_g is a quotient of the space of positively defined hermitian matrices of rank g by the action of $\mathrm{Sp}(2g, \mathbb{Z})$. If $x \in M_g^L$ corresponds to a stable curve X then $f(x) \in S_g$ is a point corresponding to the Jacobian of the normalization of the curve X. Denote by SM_g the closure of the image $f(M_g)$ in S_g. If $g > 3$ the map f contracts analytic subvarieties corresponding to the degenerate curves and curves with nontrivial automorphisms into proper analytic subvarieties in SM_g of codimension at least two. Denote by Δ_0 the divisor in M_g^L corresponding to irreducible stable curves with one node. The image of $f(\Delta_0)$ consists of all the points of $S_{g-1} \subset S_g$ which correspond to the Jacobian varieties of dimension $g-1$. On the other hand the images of divisors corresponding to different type of stable degeneration intersect S_{g-1} in proper subvarieties.

Consider the surface V in SM_g obtained by a set of hyperplane sections. We can assume that:

1) The fundamental group of an open part of V surjects onto $\mathrm{Map}(g)$.
2) The surface V intersects the images of different divisors S_I at points only.
3) The intersection of V and $f(\Delta_0)$ does not include points from the images of other divisors S_I or subsets corresponding to curves with automorphisms.

Denote the latter as $R \subset V$. By resolving V only at R we again obtain a projective surface V'. General hyperplane section C of V' will lift into a curve C' in M_g^L which intersects only Δ_0 and $C' - \Delta_0$ is contained in $M_g^{L,0}$. Thus we have a family of curves X_g over C' which has singular fibers of one type only and the monodromy group of it coincides with M^L. We introduce the following definitions:

Definition 4.1 Let us take a free group $\mathbb{F}^l$ on l generators. Let Ξ be a normal subgroup generated by Ms^N where s is a primitive element of $\mathbb{F}^l$ and M is the group of all automorphisms of $\mathbb{F}^l$. We will denote $\mathbb{F}^l/\Xi$ by $BT(l, N)$.

Definition 4.2 Denote by $\pi_g/(x^N = 1)$ the quotient of the fundamental group of a Riemann surface of genus g by the group generated by the N-th powers of all primitive elements x in π_g.

We would like to relate our construction to the following:

Conjecture (Zelmanov) For big l, g, N the groups $BT(l, N)$ and $\pi_g/(x^N = 1)$ are nonresidually finite groups.

Now we apply our construction to the above surface V' and get a surface V^N whose fundamental group $\pi_1(V^N)$ is a of a finite index in an extension of the fundamental group of a Riemann surface by $\pi^g/(x^N = 1)$. Therefore Zelmanov's conjecture implies that the group $\pi_1(V^N)$ is nonresidually finite. Hence we obtain a series of simple potential examples of surfaces with nonresidually finite fundamental groups.

The considerations from the previous subsection allow us to get even bigger variety of examples.

Definition 4.3 Let x be a primitive element in the fundamental group $\pi_1(g)$ of a Riemann surface of genus $g > 1$ and M^L be a subgroup of finite index in Map(g). Consider the orbit of x under M^L, $(M^L x)$ and take the N-th powers of all these elements. Consider the normal closure of this powers in $\pi_1(g)$. Let us denote this normal closure by $(M^L x)^N = 1$. We will denote the quotient group by $\pi_1(g)/(M^L x)^N = 1$. (Observe that definition depends on the choice of x.)

Now the following generalization of the conjecture of Zelmanov's gives us a way of constructing more examples of surfaces with nonresidually finite fundamental groups.

Question 4.1 (Zelmanov) For big g and N the groups $\pi_1(g)/(M^L x)^N = 1$ are nonresidually finite for any primitive x and M^L a subgroup of finite index in Map(g).

5 Shafarevich's conjecture

In this section we consider potential counterexamples to the Shafarevich's conjecture based on our construction. We also show that on the other hand a positive solution to the Shafarevich's conjecture has strong implications to the group theory.

We begin with the general setting. Let $f : X \to R$ be a Morse type fibration with X_t as generic fiber. Suppose that the fiber X_0 is singular and has more than one component $X_0 = \cup C_i$. We also assume that all components C_i are smooth and without selfintersection. Denote the intersection graph of X_0 by Γ_0. Consider the retraction $cr : X_t \to X_0$ of generic fiber on special fiber.

The preimage $cr^{-1}(C_i)$ in X_t is an open Riemann surface with a boundary consisting of geometric vanishing cycles corresponding to the intersection points of C_i with other components of X_0. The fundamental group $\pi_1(cr^{-1}(C_i))$ is free. The natural embedding $cr^{-1}(C_i)$ into generic fiber X_g defines an embedding of the fundamental groups $\pi_1(cr^{-1}(C_i)) = \mathbf{F}_i \to \pi_g$. Similar construction holds for any proper subgraph of curves in X_0.

Definition 5.1 For any proper subgraph $K \subset \Gamma_0$ define a subgroup $\mathbf{F}_K \subset \pi_g$ as a fundamental group of the preimage $cr^{-1}(\cup C_i), i \in K$.

Remark 5.1 If the graph K is connected then its preimage in X_t has only one component and vice versa.

Let $\pi_{1,f}$ be a fiber group obtained from π_g by our base change construction for some N.

Lemma 5.1 *Suppose that there is a decomposition of a connected subgraph $K \subset \Gamma_0$ into a union $K_1 \cup K_2$ so that the image of $\mathbb{F}_K$ in $\pi_{1,f}$ is infinite, but the image of both $\mathbb{F}_{K_1}, \mathbb{F}_{K_2}$ is finite then the Shafarevich conjecture is not true.*

Proof Indeed under the conditions of the lemma we obtain an infinite connected graph of compact curves in the universal covering of the surface S^N. □

Now we choose $N = 3$. In this case we can apply the above lemma due to the following group theoretic result which concerns the quotients of the free groups.

Definition 5.2 Let $\mathbb{F}_k^g$ be a free group with k generators with a realization as a fundamental group of curve minus one or two points (depending on k). Define $P^g(k,3)$ as the quotient of $\mathbb{F}_k^g$ by the set of relations $x^3 = 1$ for all primitive elements of $\mathbb{F}_k$ which can be realized by smooth nonintersecting cycles in the above geometric realization $\mathbb{F}_k^g$ of $\mathbb{F}_k$.

Recall that $\pi_g/(x^N = 1)$ was defined as a quotient of the fundamental group of a Riemann surface of genus g by the group generated by the N-th powers of all primitive elements x in π_g.

Here are some results about the groups defined above (for the proofs see Appendix B).

Theorem 5.1 *The group $P^g(k,3)$ is equal to the Burnside group $B(k,3)$ and hence finite.*

Corollary 5.1 *The group $\pi_g/(x^3 = 1)$ is a quotient of $B(2g,3)$ by one additional relation.*

Let X_0 be a graph of smooth curves C_i with each curve intersecting at most two others (chain or ring). Suppose that there exists a family of curves $X \to R$ satisfying the following properties:
1) The monodromy group and the set of vanishing cycles are big enough to make the image of $\pi_1(cr^{-1}(C_i))$ in π_g/Ms_j^3 finite for any i.
2) The quotient group $\pi_g/(Ms_j^3)$ is infinite.

If such a family exists we get a counterexample to Shafarevich's conjecture. The first condition is easy to achieve. Suppose that the monodromy group contains a subgroup M_D whose action commutes with the contraction cr of the generic fiber onto X_0. Suppose that M_D is transitive on the images of the primitive elements of each open surface $cr^{-1}(C_i)$ inside the corresponding Burnside group. Suppose also that we have a set of NZ-vanishing cycles s_i which are equivalent to the primitive elements of $\pi_1(cr^{-1}(C_i))$. Then the first condition is satisfied.

As we said the action of the group M_D commutes with cr. Therefore if we can arrange that $M = M_D$ and s_i are the only vanishing cycles outside X_0 then second condition will be also satisfied. Indeed π_g/Ms_j^3 in this case would surject onto free product of Burnside 3-groups.

We construct our potential counterexamples using the family of curves X of genus $g > 1$ over the interval $I = [0,1]$ with the property that X_0 is curve of geometric genus $g-1$ with one normal self intersection and X_1 has a geometric genus $g-2$ with two self-intersections.

The corresponding vanishing cycles for the projection of generic fiber are NZ cycles s_0 for X_0 and s_0, s_1 for X_1. This is the core of our construction. Consider a $\mathbb{Z}_l$-cyclic covering of the above family ($l > 1$) which transforms the curve X_0 into l-ring of smooth curves of genus $g-1$ and X_1 into a ring of singular curves of geometric genus $g-2$. Denote by Y the resulting family of curves over an interval with the action of $\mathbb{Z}_l$. Consider now the family of curves V which has the following properties:

1) It has a fiberwise $\mathbb{Z}_l$ action and coincides with Y with the $\mathbb{Z}_l$ action defined above. (A simple transversality argument shows that we can always do that over an $\mathbb{Z}_l$ of the corresponding moduli space of curves of genus g.)
2) It is a Morse type family. We shall also add an assumption on the monodromy group. It is clear that the monodromy group of V maps into the monodromy group M of the quotient family which stabilizes the kernel of the corresponding $\mathbb{Z}_l$- character above. We assume that M is big enough to ensure that the image in $\pi_1(V)$ of the fundamental group of X_0 without singular points $\pi_1(X_0 - sing)/Ms_1^3$ is a finite group (Burnside group). On the other hand we assume that Ms_0 coincides with s_0 modulo a subgroup $R \subset \pi_1(X_t)$. Such a family can be easily constructed for example for any subgroup R of finite index in Map(g). Indeed we just need to take a $\mathbb{Z}_l$ covering of the corresponding moduli space of curves of genus g.

Let us take some number I and apply the construction described in Section 2. As it follows from Lemma 1.1 we obtain a new family V_I with $N_0 = N_1 = 3$ and $N_p = 3^I$ for any singular fiber over a point $p \neq 0,1$. The image of the fundamental group of the general fiber of V_I in $\pi_1(V_I)$ will be

$$\pi_{1,f} = \pi_1(V_t)/(\mathbb{Z}_l M s_0^3, \mathbb{Z}_l M s_1^3, \mathbb{Z}_l M s_i^{3^I}).$$

Question 5.1 Is it true that for big enough I and a small enough R the corresponding group $\pi_{1,f}$ is infinite?

If the answer of the above question is positive then the Shafarevich conjecture is not true for the family constructed above since the subgroup of $\pi_{1,f}$ corresponding to any component of the preimage of V_0 gives a finite Burnside group $B(2g-1,3)$.

Remark 5.2 The construction of the group above can be divided into two steps. First we obtain a quotient group by the element $Ms_i^3 = 1, i = 0,1$. This group is infinite if we choose M close enough to the diagonal group M_D inside Map(g). This follows from the computation used in the proof of lemma B.1 in Appendix B. The group we get is closed to a free product of Burnside 3-groups corresponding to different components. On the second step we factor the resulting group by additional relations $Ms_i^{3^I}, i > 1$ where s_i runs through the vanishing cycles of the family. Though we cannot control the elements s_i but by increasing I we can hope to achieve infiniteness of the corresponding quotient.

A free product of Burnside 3-groups has infinite linear representations but the fiber group constructed above has only finite linear representations ([14]). It is worth noticing that the resulting quotient group indeed changes substantially if we change the mondoromy group to a subgroup of finite index. A simple example can be seen in the following lemma, which is a direct consequence of Lemma 8.1.

Lemma 5.2 *Let $x \in \pi_g$ be a primitive element, $g > 1$ and $\chi : \pi_g \to \mathbb{Z}_3, \chi(x) \neq 1$ be any character. Let $M = Map(g)$ and $M_\chi \subset Map(g)$ be a subgroup of finite index which stabilizes a subgroup $Ker\chi$. Then*

1) $(\pi_g/(Map(g)x = 1)) = (\pi_g/(M_\chi x = 1)) = 1.$

2) $\pi_g/(Map(g)x^3 = 1)$ *is a quotient of the finite Burnside group* $B(2g, 3)$.

3) $\pi_g/(M_\chi x^3 = 1)$ *is an infinite group.*

Proof There is a projection $h : \pi_g \to \mathbf{F}_2$ which induces the character χ on π_g from the character on $\mathbf{F}_2$ denoted in the same way. The statement 1) is evident and 2) is proved in Section 8, Theorem 8.1 . Now we can use Lemma 8.1. According to this lemma we have a surjective map $g : \mathbf{F}_2 \to G_1$ with the property that $g(x)^3 = 1$ if $\chi(x) \neq 1$. The group G_1 is an extension of $\mathbb{Z}_3$ by $\mathbb{Z} + \mathbb{Z}$ and hence infinite. The composition map $gh : \pi_g \to G_1$ has the same property and hence it can be represented as a combination of the projection on $\pi_g/(M_\chi x^3 = 1)$ and a surjective map of the latter onto G_1. It proves 3) since the corresponding group surjects on G_1. □

There is also another possibility to satisfy Condition 2 . We can easily construct a family of curves of genus $g = 2k, k > 1$ such that there is a fiber in this family which consists of two components each of genus k that give us a tree of components. Applying the base change construction for a given N we obtain a surface $S - Q$ with the image of the fundamental group of every component in $\pi_1(S - Q)$ being equal to $P^{2k}(2k, N)$. The fiber group of $S - Q$ is equal to $\pi_{2k}/(x^N = 1)$. Now as we have shown we have the same behavior on S^N for the closed curves and surfaces. We can formulate the following question:

Question 5.2 Are there such a k and N such that $P^{2k}(2k, N)$ is a finite group and $\pi_{2k}/(x^N = 1)$ is an infinite group?

If the answer of the above question is affirmative for some N, k we get a counterexample to the Shafarevich conjecture. We should point out that if the groups obtained from the components are finite then $\pi_{2g}/(x^N = 1)$ does not have infinite linear representations whose image is virtually not equal to $\mathbb{Z}$ coming from linear representations of $\pi_1(S^N)$. If the Shafarevich conjecture is correct the answer of the above question is negative. It also implies the answer to many similar group theoretic questions. The most basic question seems to be the following:

Question 5.3 Is there such an N and such $2 \leq m_1 < m_2$ for which $B(m_1, N)$ is finite and $B(m_2, N)$ is infinite?

Recently we were informed by Zelmanov that he can show that there exists an integer $d(0)$ so that for every prime number p and an integer d the group $B(d_0, p)$ is finite if and only if the group $B(d, p)$ is finite. This result suggests the existence of an abstract group theoretic version of holomorphic convexity. The above considerations indicate a possibility for analysis of infinite groups by analytic methods.

6 Some arithmetic and symplectic reflections

6.1 Symplectic Lefschetz pencils

Observe that the constructions described in the previous sections can be performed in the symplectic category. It leads to interesting examples of symplectic four-dimensional manifolds. Using the base change construction we have defined in [4] an obstruction to a symplectic Lefschetz pencil being a Kähler Lefschetz pencil.

Remark 6.1 Here by Kähler Lefschetz pencil we mean a Kähler surface with a pencil structure on it. Since we are working with Kähler surfaces, Kähler Lefschetz pencil is a small deformation of a projective Lefschetz pencil. In higher dimensions this is not known. In this subsection we will also consider a little bit more general situation allowing reducible fibers in our Lefschetz pencils.

Let X be a symplectic fourfold with a given structure of a Lefschetz pencil. Apply our base change construction for every $N > 2$. It is shown in [4] that the construction goes through and for a fixed integer N we get a symplectic fourfold S^N. Denote by Y_i the components of reducible fibers in S^N and denote by F the general fiber of S^N. Consider all linear representations ρ of $\pi_1(S^N) \to \mathrm{GL}(n, \mathbb{C})$ which are infinite on $\pi_1(F)$. Consider for a given ρ all components Y_i with a finite image of $\rho(\pi_1(Y_i))$. Let G be a connected graph of such components Y_i. If $\rho(\pi_1(\bigcup Y_i)), i \in G$ is also finite for any such G then we set $O(X)^{N,n} = 0$.

If on the other hand there exists ρ and G with $\rho(\pi_1(\bigcup Y_i)), i \in G$ being infinite then we set $O(X)^{N,n} = 1$. Thus we assign to every symplectic Lefschetz pencil X a function on the set of pairs of integers $N > 2, n \geq 1$ with values $0, 1$.

Proposition 6.1 *If X is a Kähler surface then $O(X)^{N,n}$ is identically zero.*

Indeed otherwise some covering of S^N with linear Galois group contains infinite chain of compact curves. So it is not holomorphically convex but this contradicts Theorems 1.1 , 1.2 mentioned in the introduction.

Remark 6.2 So far the above obstruction does not distinguish in general Kähler surfaces from symplectic fourfolds. It distinguishes only symplectic from Kähler Lefschetz pencils.

We formulate some properties of the monodromy of a Kähler Lefschetz pencil. Let X be a Kähler Lefschetz pencil and we denote by X_t its generic fiber. Let $\rho : \pi_1(\mathbb{P}^1 - (p_1, \dots, p_k)) \to \mathbb{GL}(H_1(X_t, \mathbb{Q}))$ be the monodromy representation of the family X_t. As it follows from [9]:

Proposition 6.2 $\rho : \pi_1(\mathbb{P}^1 - (p_1, \dots, p_k)) \to \mathbb{GL}(H_1(X_t, \mathbb{Q}))$ *is a direct sum of irreducible representations.*

Here we are going to describe a general construction of symplectic Morse fibrations with rather arbitrary topological properties. In particular we construct a symplectic Lefschetz pencil with any finitely presented group as a fundamental group. We begin with the following general construction which allows to diminish the intersections of cycles on the Riemann surface.

Definition 6.1 Let X^g be a Riemann surface of genus g. We define the Wiler transform X^h as Riemann surface obtained by adding $h - g$ handles to X^g.

Remark 6.3 We denote the above procedure - adding local handles- as Wiler transform since Wiler invented a similar procedure of the addition of local topological handles in attempt to fit quantum particles (particularly electrons) into the geometrical framework of general relativity.

Let s_1, s_2 be two local smooth curves in the Riemann surface X^g which intersect transversally at point p lying in the middle of the interval $I \subset s_1$. After adding a handle H which joins the surface X^g at the ends of the interval I we can change s_1 by removing the interval I and inserting instead the interval I' lying in H. The curve s_1' thus constructed won't intersect s_2. We have a natural projection $p : X^h \to X^g$ which maps a new handle H into the interval I. The kernel of $p_* : \pi_1(X^h) \to \pi_1(X^g)$ is generated by the subgroups corresponding to the handle H and hence it is generated by cycle represented by smooth homologically nontrivial curve inside X^h. Moreover there are two smooth cycles a, b in the new handle H with the intersections: $(a, b) = 1, (a, s_1') = 1, (a, s_2) = 0, (b, s_1') = 0, (b, s_2) = 1$. Geometrically a corresponds to the "short" (latitude) cycle on the handle and b to the "long" meridian cycle. The image $p(a)$ is an interval which is perpendicular to I and $p(b)$ coincides with I.

Let $f : \pi_g \to \Gamma$ be any surjective homomorphism with a nontrivial kernel from the fundamental group of a Riemann surface of genus g, π_g to an arbitrary finitely presented group Γ. Assume that:

1) Ker$f$ of the map $f : \pi_g \to \Gamma$ is generated by cycles $s_i, s = s_1$ and cycles $d_1, \dots, d_j$ so that $T^{n_1} T_1^{m_1}, \dots, T_j^{m_j} = 1, n_1 > 0, m_k > 0$ for any k. Here $T, T_1, \dots, T_j$ are positive Dehn twists corresponding to the cycles s and $d_1, \dots, d_j$. The cycles s_i, d_k satisfy the following conditions:
2) s_i, d_k have no self-intersections.
3) s_i, s_j intersect transversally in at most one point.
4) The intersection graph G corresponding to a subset s_i is connected.

The graph G is obtained by assigning a point to each loop s_i and we join i, j if s_i, s_j intersect. Any two vertices of G are connected by at most one edge.

Definition 6.2 We shall denote the above set of conditions as LS-conditions.

Remark 6.4 The abbreviation LS stands for Lefschetz symplectic and we are going to show that the system of cycles satisfying LS-conditions can be realized as a subset of vanishing cycles for a symplectic Lefschetz pencil with the group Γ as a fundamental group. It is worth noticing that if we have a topological Lefshetz pencil then the system of its vanishing cycles can be drawn on any smooth curve X_0 in the pencil. If we join the point 0 in the base with points corresponding to the singular fiber then the vanishing cycles can be moved into cycles on X_0 using the monodromy transformation along the paths.

Remark 6.5 The cycles s_i constitute the most essential part of the system of relations. We need additional cycles d_i in order to complete a symplectic family into a family over S^2. They provide a positive relation in the semigroup generated by T_{s_i}, T_{d_k} and hence a possibility to complete the family over S^2.

Theorem 6.1 *Let $f : \pi_g \to \Gamma$ be any surjective homomorphism with a nontrivial kernel from the fundamental group of a Riemann surface of genus g, π_g. Assume that Kerf is generated by the system of cycles $s, d_1, ..., d_j, s_i, 1 < i < l$ satisfies LS-conditions 1), 2), 3), 4) and $g \geq 2$. Then there exists a symplectic fourmanifold V with a symplectic Morse type fibration over S^2 and such that all s_i are among the vanishing cycles of this fibration and $\pi_1(V) = \Gamma$.*

Proof Let us construct first the corresponding symplectic family over a disc with the cycles s_i, d_j being the vanishing cycles. We can do it inside the universal family of curves over the moduli space of curves of genus g. The boundary of this family is a fibration over a circle S^1 with a boundary map being the product of the diffeomorphisms $h_i T_{s_i} h_i^{-1}$ and $h_k T(d_k) h_k^{-1}$. Here h_i, h_k are some diffeomorphisms of X_t which are in turn the products of the Dehn twists T_{s_i}, T_k.

The completion of the argument is in two steps:

Step 1: We add first similar transformations in order to obtain a new boundary transformation which includes only the products of $h_j T h_j^{-1}, T = T_{s_1}$ and T_k.

We can do it using the connectedness of G. Namely consider any shortest path P_1^i inside G which is joining s_i and s_1. If two elements $s_k, s_l \in P_1$ then they intersect only of they are next to each other in P_1. Otherwise it would be possible to shorten the path. Hence we can put the corresponding chain of cycles onto an open Riemann surface $X^{g'}$ which lies inside X^g with the border $dX^{g'} = C$ where C is a smooth cycle. The image of $f(\pi_1(X^{g'})) = 0$. Thus we can find a family of positive Dehn twists T_{l_j} corresponding to the cycles l_j inside $X^{g'}$ with the following property $T_{s_i} T_{l_1}^{n_1} ... T_{l_m}^{n_m} = T^{r_i}$.

Therefore after adding singular fibers corresponding to vanishing cycles from $X^{g'}$ we obtain a new symplectic family over a disc. The boundary transformation of this new family contains $h_l T h_l^{-1}$ instead of $h_i T_{s_i}^{h_i^{-1}}$ in the product. The vanishing cycles which we add are also contained in Ker$f$. After performing a similar transform for any s_i we obtain a boundary transform which is a product of $h_l T h_l^{-1}$ and T_k only.

Step 2: We can homotopically trivialize the boundary transform by adding vanishing cycles corresponding to s and d_i since $T_{s_1} = T$ is included into a positive relation containing d_i (Condition 1).

Thus after adding vanishing cycles we obtain a symplectic family which has a border transformation homotopic to identity.

It can be completed then into a symplectic family X over S^2. □

Lemma 6.1 *The fundamental group of the above family X is equal to Γ.*

Proof The cycles s_i, d_j generate Ker$f$. The normal subgroup Ker$f \subset \pi_g$ is invariant under Dehn twist T_x for any element $x \in \pi_g$ which is contained in Ker$f$. The above construction is inductively adding new singular fibers. The vanishing cycle s_v on each step belongs as to a subset $M_p s$ where $s \in$ Ker$f$ and M_p is a monodromy group corresponding in the symplectic family existing on the previous step. Since the group $\pi_1 X = \pi_g / M s_v$ where s_v runs through the set of vanishing cycles we obtain that the fundamental group $\pi_1(X) = \pi_g/\text{Ker}f = \Gamma$ □

Any symplectic Lefschetz pencil (even with mild singularities and multiple fibers) provides with some relation between positive Dehn twists which involves only their positive powers. We shall use one special positive relation between Dehn twists which can be obtained from a special family of curves of genus two.

There is a relation between five positive Dehn twists $(T^2.T_1^2 \ldots T_4^2)^2 = 1$ in Map_2 according to [23]. This relation corresponds to the projective family X of curves of genus two which is a two sheeted covering of the direct product $S^2 \times S^2$. The latter has a structure of a complex projective surface which is a direct product $\mathbb{P}^1 \times \mathbb{P}^1$. Denote the classes of divisors corresponding to $\mathbb{P}^1_i, i = 1, 2$ by l_i. The ramification divisor $D_r \subset \mathbb{P}^1 \times \mathbb{P}^1$ is equal to the union of five different lines equivalent to l_1, the diagonal line equivalent to $l_1 + l_2$ and a line l_2. We assume that there are only pairwise intersections between components. The two sheeted covering X is a surface which is fibered over $\mathbb{P}^1 \times \mathbb{P}^1$ with five singular fibers containing a singular double point each and a rational fiber of multiplicity 2. The monodromy group is generated by positive Dehn twists T, $T_i, i = 1, 2, 3, 4$ and they satisfy the relation $(T_0^2 T_1^2 \ldots T_4^2)^2 = 1$.

The following general statement allows us to obtain similar positive relation for the Riemann surface of any genus $g \geq 2$.

Let X be a Riemann surface with a boundary and T_i be a number of Dehn twists with central circles d_i which satisfy a relation P in the group Map(X), the group of connected components of the group of diffeomorphisms of X. Let $f : Y \to X$ be a finite unramified covering of X. Then the preimage of T_i is defined as a product T_i' of the Dehn twists corresponding to the cycles which are in the preimage of d_i. We have the equality $hT_i' = T_i$.

Lemma 6.2 *The elements T_i' satisfy the same relation P in the group Map(Y).*

Proof There is natural map $r : \text{Map}(X) \to \text{Out}(\pi_1(X))$ and the latter is an embedding. There is an equality $hP(T_i') = P(T_i)$ which follows from $hT_i' = T_i$. Hence if h is a nonramified covering then $r(P(T_i)) = id$ implies $r(P(T_i')) = id$ since $\pi_1(Y) \subset \pi_1(X)$ is a subgroup of finite index. □

Example 6.1 Suppose that $d_i, i = 0,1,2,3,4$ are cycles on the surface X of genus 2 corresponding to the vanishing cycles of above family. Let $\chi : H_1(X,\mathbb{Z}) \to \mathbb{Z}_N$ with the property $\chi(d_i) = 1$. Consider the corresponding cyclic nonramified covering $X_\chi \to X$. The preimage of every cycle s_i is a cycle $s_i' \subset X_\chi$. The positive Dehn twists T' $T_i', i = 1,2,3,4$ corresponding to $d_i' \subset X_\chi$ satisfy the relation $(T'^2.T_1'^2 \ldots T_4'^2)^2 = 1$ The genus of X_χ is equal to $N+1$ and hence we obtain a positive relation which includes five cycles only on a surface of any genus g.

Let us look at the geometry of the surface X . It has two separated handles: H_1 which contains cycles d_0, d_1 and H_2 which contains d_3, d_4. The surface X is a connected sum of the handles H_1, H_2 by the handle Q with two ends. The covering $X_\chi \to X$ induces cyclic covering on each of the handles $H_j, j = 1,2$ and in addition increases the number the connecting handles. Therefore X_χ has a natural representation as a union of two handles H_1', H_2' (tori) joint by N handles Q_i. Thus we have a natural representation of X_χ as a union of handles.

The preimages of d_0, d_1 are two transversal cycles in H_1' with the intersection index N. Denote the cycles which are preimages of d_i in X_χ similarly. Consider now a new surface X' which is obtained from X_χ by a contraction of the handles H_1', H_2', $p : X_\chi \to X'$. It is a smooth surface of genus $N-1$ with one special element $p(d_2)$ corresponding to the cycle $d_2 \in X_\chi$.

Lemma 6.3 *The cycle $p(d_2)$ is equivalent to the product of generators q_i corresponding to the cycles on the handles defined by Q_i.*

Proof Indeed the cycle d_2 projects into a cycle which is a product of generators corresponding to different handles. □

Theorem 6.2 *Let $f : \pi_1(X_g) \to \Gamma$ be any surjective homomorphism. Then there exists a surface X_h which is obtained from X_g by adding handles with the following properties:*

1. *The surface X_h contains $X_g' = X_g$ minus a finite number of discs as an open subsurface.*
2. *The surface X_h is a generic fiber of a symplectic Lefschetz pencil X over S^2.*
3. *The fundamental group $\pi_1(X) = \Gamma$.*
4. *A natural surjective map $r : \pi_1(X_h) \to \Gamma$ induces on $\pi_1(X_g')$ the composition map $f.i_*$ where $i_* : \pi_1(X_g') \to \pi_1(X_g)$ is induced by the natural embedding and f is the map above.*

Proof Let l_i be a finite number of elements in π_g which generate $\text{Ker} f$ as a normal subgroup of π_g. We can realize them by a family of cycles with normal pairwise intersections and selfintersections. Without changing their free homotopy type (modulo conjugation) we make their intersection graph connected. Next we apply a sequence of Wiler transforms which keeps the graph connected and transforms it into a connected graph G of cycles s_i with intersections $(s_i, s_j) = 0, 1$ on $X_{h'}$. Assume that G has at least two distant ends s_1, s_k. That means each of them intersect only one cycle from the $(s_i, s_1) = (s_j, s_k) = 1$ and $s_i \neq s_j$

We can easily achieve it by adding to the initial set of l_i some homotopically trivial elements. Now we identify $s_1 = d_0, d_1 = s_i, d_3 = s_j, d_4 = s_k$. The cycles $s_1 = d_0, d_1 = s_i$ and $d_3' = s_j, d_4' = s_k$ are contained in the nonintersecting handles H_1, H_2 inside $X_{h'}$.

We can decompose X_h' as a connected some of H_1, H_2 by means of $h' - 1$ handles and therefore as a cyclic covering of X_2. We select now d_1 as a smooth cycle inside H_1 which is equivalent to $d_0 + (h' + 1)d_1'$ and $d_3 = d_4 + (h' + 1)d_3$ inside H_2.

The only cycle we need to construct now is d_2. We obtain first d_2' from the identification of $X_{h'}$ with $h' - 1$ covering of X_2. The map $f : \pi_g \to \Gamma$ defines a natural map $f' : \pi_{h'} \to \Gamma$ which is identical on every handle which was added. Consider $g = f'(d_2)$. If $g' = 1$ then the construction is finished. Suppose that $g' \neq 1$ By adding one more connecting handle between H_1, H_2 we obtain a surface $X_h, h = h' + 1$ and two new elements $a, b \in \pi_1(X_h)$ where a corresponds to the meridianal circle and b to the latitude circle (cutting cycle) on the additional handle. The surface X_h has a representation as $h - 1$ cyclic covering of X_2 which slightly modifies similar representation for the surface $X_{h'}$. Now we can define d_2 as a product of a^{-1}, d_2' and define $r(a) = g, r(b) = 1$ and $r = f'$ on the subgroup generated corresponding to $X_{h'}$. The resulting homomorphism $r : \pi_h \to \Gamma$ satisfies LS-properties from Theorem 6.1 and hence there is a symplectic Lefschetz pencil X satisfying the conditions 1), 2), 3), 4). This finishes the proof of the theorem. □

Example 6.2 (1) Let $f : \pi_g \to \Gamma$ in the above construction be a composition $f = t.h$ with $h : \pi_g \to \mathbf{F}_g$ where $\mathbf{F}_g$ is a free group with g generators and $t : \mathbf{F}_g \to \Gamma$. We assume that $\mathbf{F}_g$ is realized as the fundamental group of the space formed by two spheres intersecting normally at $g + 1$ points and h corresponds to the contraction of $g + 1$ non-intersecting cycles s_i nonhomologous to zero in X^g. Thus X^g is split by s_i in two discs D_1, D_2 with g holes. Assume that the group Ker$f$ is generated by some connected finite subset of curves in X^g which include all s_i. Now we apply Wiler transform to realize the above set as the set of vanishing classes in some symplectic fibration over S^2. In this case we can realize a special fiber in the above construction as the union of two curves with the trivial images of the fundamental groups of the individual components plus may be more components but also with trivial images of their fundamental groups in Γ. In other words we assume that the new handles of the Wiler transform have a support which does not intersect s_i. The new handles correspond to the intersection of two curves. If neither of them is s_i then we add a handle with both ends inside one of the discs D_1 or D_2. If one of the curves is s_i we add a handle H which joins D_1, D_2. The cycle $s_H \subset H$ which separates D_1, D_2 is then added to the set of s_i. Since all nonintersecting cycles s_i can represent a vanishing cycles for a particular singular fiber of a fibration we obtain that the corresponding fiber consists of smooth normally intersecting curves and the image of the fundamental groups of the components is trivial in Γ. Hence if Γ is infinite the obstruction $O(X)^{N,n}$ will be one in this case for N, n big enough.

We can also easily construct a symplectic Lefschetz pencil X over S^2 with the following additional properties:

1) The fundamental group of the family Γ is a Kähler group (e.g. a lattice in a semisimple group).
2) The action of the monodromy group on the first homology group of the general fiber is absolutely irreducible.

Now we observe that for the symplectic pencil constructed above the obstruction $O(X)^{N,n}$ will be one and all other known obstructions for X to be a Kähler Lefschetz pencil seem to vanish. In a similar vein one can construct examples that violate other monodromy restrictions of Kähler Lefschetz pencils. Here we list some of them:

Let us denote by $\mathbb{C}(\pi_1(X_t))$ the Lie algebra of the Malcev completion of $\pi_1(X_t)$ (see e.g. [14]). The monodromy representation $\rho : \pi_1(\mathbb{P}^1-(p_1,\ldots,p_k)) \to H_1(X_t,\mathbb{Q})$ induces a representation: $\rho' : \pi_1(\mathbb{P}^1 - (p_1,\ldots,p_k)) \to \mathrm{Aut}(\mathbb{C}(\pi_1(X_t)))$. If X is a Kähler Lefschetz pencil then $\rho : \pi_1(\mathbb{P}^1-(p_1,\ldots,p_k)) \to \mathrm{Aut}(\mathbb{C}(\pi_1(X_t)))$ is a Hodge representation [21] (namely it satisfies the Griffiths transversality).

Proposition 6.3 *If $\rho' : \pi_1(\mathbb{P}^1 - (p_1,\ldots,p_k)) \to Aut(\mathbb{C}(\pi_1(X_t)))$ is a Hodge representation then it is an irreducible representation.*

Conjecture The representation $\rho' : \pi_1(\mathbb{P}^1 - (p_1,\ldots,p_k)) \to \mathrm{Aut}(\mathbb{C}(\pi_1(X_t)))$ is a Hodge representation iff X is a Kähler Lefschetz pencil.

Simillar conjecture should be correct for representations into the automorphisms group of the relative completions of $\pi_1(X_t)$ on which we have a MTS (mixed twistor structure), see [21].

Conjecture The representation $\rho' : \pi_1(\mathbb{P}^1 - (p_1,\ldots,p_k)) \to \mathrm{Aut}(\mathbb{C}(\pi_1(X_t)))$ is a MTS (see [21]) representation iff X is a Kähler Lefschetz pencil.

The technique described above (Theorem 6.2) allows us to construct examples of symplectic Lefschetz pencils such that $\rho' : \pi_1(\mathbb{P}^1 - (p_1,\ldots,p_k)) \to \mathrm{Aut}(\mathbb{C}(\pi_1(X_t)))$ is not a Hodge representation. Similarly we can construct examples of symplectic Lefschetz pencils so that $\rho' : \pi_1(\mathbb{P}^1 - (p_1,\ldots,p_k)) \to \mathrm{Aut}(\mathbb{C}(\pi_1(X_t)))$ is a not a MTS representation.

As it follows from the works of Kotschick , Morgan and Taubes [16] and Fintushel and Stern [11] the Seiberg-Witten invariants of symplectic fourfold X are invariants of the symplectic structure together with the Seiberg-Witten invariants of every finite nonramified covering of X. Our construction allows to look at the ramified coverings as well but instead of studying at the Seiberg Witten invariants of the coverings we try to measure how the local monodromies come together to generate the global monodromy of X.

We would like to indicate how our obstruction can be used to distinguish two symplectic Lefschetz pencils X_1 and X_2 with similar local monodromies as well as really define invariants of symplectic four manifolds. Let X_1 and X_2 be two symplectic Lefschetz pencils with a generic fiber a Riemann surface of genus greater than one. An invariant that distinguishes X_1 and X_2 as symplectic Lefschetz pencils could come from their global monodromies. The obstruction $O(X)^{N,n}$ can be used for defining such an invariant in the following way. Denote by $W(X_i), i = 1,2$ the smallest N for which $O(X_i)^{N,n}$ is not trivial. In case all $O(X_i)^{N,n}$ are trivial we will say that $W(X_i) = 0$.

Proposition 6.4 *If $W(X_1) \neq W(X_2)$ then there is no symplectomorphism from X_1 X_2 that respect the given symplectic Lefschetz pencil structures.*

Indeed as it follows from the construction above this means that the global monodromies of X_1 and X_2 are different. The following question arises naturally.

Question 6.1 Compute the Seiberg - Witten invariants of a symplectic fourmanifold X in terms of the local monodromies around the singular fibers of some Lefschetz pencil structure on X?

6.2 Invariants of the symplectic fourfolds

Now we show how to use the construction we have defined to get invariants of the symplectic fourfolds.

Remark 6.6 In this subsection we consider only generic Lefschetz pencils - no reducible fibers are allowed.

Recently Donaldson [10] has introduced a new invariant of the symplectic structure of a symplectic fourfold. He has shown that for every smooth symplectic manifold has a symplectic Lefschetz pencil structure that corresponds to the class of the symplectic form ω. Now by taking powers of ω Donaldson defines a series of Lefschetz pencil structures that are invariants of the symplectic structure on X. A different way to express this invariant is to consider the following inverse limit :

$$(\lim_k \rho_{j(k)})/B$$

where $\rho_{j(k)} : \pi_1(\mathbb{P}^1 \setminus p_1, \ldots, p_{l(k)}) \to \mathrm{Map}_{m(k)}$ are representations of the fundamental group of the base for the different Lefschetz pencil structures described above. $\mathrm{Map}_{m(k)}$ is the mapping class group of the generic fibers of the pencils and $j(k), l(k), m(k)$ are polynomials of k. B is the infinite braid group which acts on the limit of the points $p_1, \ldots, p_{l(k)}$.

In other words after taking higher enough power of ω the monodromy of the corresponding Lefschetz pencil stabilizes and becomes an invariant of symplectic fourfold X. Let us consider such a Lefschetz pencil L with monodromy $\rho_{j(k)}$: $\pi_1(\mathbb{P}^1 \setminus p_1, \ldots; p_{l(k)}) \to \mathrm{Map}_{m(k)}$ that is invariant of the symplectic fourfold X. Let us apply to L the construction from the beginning of the section. We get a new symplectic manifold S^N for every N. Observe that $\pi_1(S^N)$ is an invariant of X since it expresses the monodromy of L. We combine our construction with

Donaldson's theory [10] to get new invariants of the symplectic fourfolds. We give the following:

Example 6.3 (2) Let X_1 and X_2 are two symplectic fourmanifolds. Consider the sequences X_1^k and X_2^k constructed by Donaldson [10]. X_1^k and X_2^k are sequences corresponding to the $k-th$ power of the symplectic forms $\omega(X_1)$ and $\omega(X_2)$. Now we apply to X_1^k and X_2^k the construction described above. We get two sequences of manifolds $S_{1,k}^N$ and $S_{2,k}^N$ depending on two parameters k and N. Any of the topological invariants is an invariant of $S_{1,k}^N$ coming from the monodromy of the symplectic Lefschetz pencils X_1^k . The same is true for X_2^k. Therefore these invariants are invariant of the symplectic structures on X_1 and X_2.

In particular we have:

Proposition 6.5 *If there exists a pair of integers k and N such that $\pi_1(S_{1,k}^N)$ and $\pi_1(S_{2,k}^N)$ are different then X_1 and X_2 are not symplectomorphic.*

Indeed $\pi_1(S_{1,k}^N)$ and $\pi_1(S_{2,k}^N)$ measure the monodromy of $S_{1,k}$ and $S_{2,k}$ which as it follows from [10] is an invariant of the symplectic structure. For possible realizations of X_1 and X_2 see the appendix. We have a corollary:

Corollary 6.1 *If there exists a pair of integers k and N such that $W(S_{1,k}^N)$ and $W(S_{2,k}^N)$ are different then X_1 and X_2 are not symplectomorphic.*

The above construction gives a new tool of studying projective surfaces. Indeed to any projective surface X_1 we assign a sequence of canonically defined Lefschetz pencils X_1^k. To each X_1^k we assign a sequence of projective surfaces $S_{1,k}^N$. Any Hodge theoretic information of $S_{1,k}^N$ (e.g. the Hodge structure on the cohomology of $S_{1,k}^N$ or the Mixed Hodge structure on some relative completions of $\pi_1(S_{1,k}^N)$) is invariant of the Hodge structure namely of the motive of X_1 and the canonically defined Lefschetz pencils X_1^k. Of course we should say that as it follows from Zelmanov's conjecture the groups $\pi_1(S_{1,k}^N)$ are nonresidually finite. Therefore the Hodge theoretic tools will miss a lots of information. Analyzing the motive of X_1 using surfaces $S_{1,k}^N$ requires combinatorial group theory. After saying all that the combination of Donaldson theory [10] and our constructions should be seen as an attempt to extend some motivic virtues of the projective surfaces to the case of symplectic fourfolds.

We pose the following question:

Question 6.2 Let X_1 and X_2 be two algebraic surfaces with structures Lefschetz pencils of curves of the same genus but with different monodromies.

1) Is it true that X_1 and X_2 are not symplectomorphic?

2) Is it true that X_1 and X_2 are not deformation equivalent?

Finally we suggest the following group theoretic question coming from our considerations above. The answer of this question gives an interesting example of nonsymplectomorphic simplyconnected fourfolds. We will call monodromy that comes from a single Dehn twists corresponding to nonhomologous to zero vanishing cycle in any singular fiber a generic monodromy. As it follows from the results of Gompf [12] every monodromy $\rho_{j(k)} : \pi_1(\mathbb{P}^1 \setminus p_1, \ldots, p_{l(k)}) \to \mathrm{Map}_{m(k)}$ for a big k realizes an invariant symplectic Lefschetz pencil structure for some symplectic fourfold.

Question 6.3 Are there such a g, N, x and subgroups, generic monodromies, of Map_g M^{L_1} and M^{L_2} so that:

1) $\pi_1(g)/(M^{L_1}(x)) = 1 = \pi_1(g)/(M^{L_2}(x))$,
2) $\pi_1(g)/(M^{L_1}(x)^N)$ is finite, and
3) $\pi_1(g)/(M^{L_2}(x)^N)$ is infinite?

Indeed if such an example exists we will have two symplectic manifolds X_1 and X_2 so that corresponding $\pi_1(S_1^N)$ and $\pi_1(S_2^N)$ have different fundamental groups. (Suggestions for such examples are contained in Appendix B. One also needs to use the realization Lemma 6.3.)

Remark 6.7 It might be possible to define easier invariants of the symplectic structure for symplectic four manifolds. For this one should try to develop the the usual projective Lefschetz - Moisheson theory for symplectic four manifolds. Then instead on level of fundmental groups it might be possible to define invariants on level of the second cohomology groups. Some of this can be found in [5].

6.3 Arithmetics reflections

In this article we consider only surfaces with given projection on a curve. Algebraically this means that the field of rational functions on the surface is provided with a structure of a one-dimensional field over the field of rational functions of the base curve. This picture is parallel to that of a curve defined over number field. It suggests that there should be a natural arithmetic version of the Shafarevich's conjecture. The notion of holomorphic convexity is not well defined in the arithmetic case. Instead we can describe the analogue of the absence of infinite chains of compact curves in the universal covering in the arithmetic case. The absence of infinite chains of compact curves seems to be the only obstruction to the holomorphic convexity in the case of compact complex surfaces. For different arithmetics considerations and further results see [3].

Let C be a projective semistable curve over K. Since K is not algebraically closed we obtain a nontrivial map $C \to \mathrm{Spec}(O_K)$ where $O(_K)$ is the ring of integers. Extending K if necessary we can assume that C is a semistable curve. That means C is a normal variety with semi-stable fibers consisting of normally intersecting divisors. The function field $K(C)$ is regular and has dimension one over K. Any maximal ideal ν in O_K defines a subring O_ν of $K(C)$ which consists of the elements of $K(C)$ which are regular at the generic point of any component of the preimage of ρ in the scheme of C. The ring O_ν contains the ideal of elements which are trivial on the preimage of ν. Denote this ideal by M_ν. The quotient ring O_ν/M_ν is a finite sum of fields of rational functions on the components of the fiber of C over ν. Consider the maximal nonramified extension $K(C)^{nr}$ of $K(C)$. It is a Galois extension with a profinite Galois group $\mathrm{Gal}^{nr}(K(C)$. The field $K(C)^{nr}$ contains the maximal nonramified extension K^{nr} of the field K as a subfield. (Observe that the field $\mathbb{Q}^{nr} = \mathbb{Q}$ but for many other fields K the field K^{nr} is an infinite extension.) The group $\mathrm{Gal}^{nr}(K(C)$ maps surjectively onto $\mathrm{Gal}^{nr}(K)$. Denote the kernel of the corresponding projection as $\mathrm{Gal}_g^{nr} K(C)$.

Any maximal ideal ρ in the ring of integers $O_{K^{nr}}$ contains the unique maximal ideal ν of O_K. We can now define the subring A_ρ as an integral algebraic closure of the subring O_ν in $K^{nr}(C)$. Let $Res_\rho = A_\rho/I(\rho)$ be the quotient ring by the ideal generated by ρ in A_ρ. It is a semisimple ring of finite characteristics. We formulate a strong arithmetic analogue of the Shafarevich's conjecture.

Conjecture Let $K(C)$ be a field as above. For some finite extension F of K we can find a semistable model C' of the field $F(C)$ such that the ring Res_ρ is a direct sum of a finite number of fields for any ideal ρ of the field F^{nr}.

We can also formulate this conjecture in a more geometric language. Namely if C' is a semistable curve over F then for any finite nonramified extension L of $F(C')$ we have a model C_L with a finite map onto C'. The fibers of this new model C_L are uniquely defined by C'.

Conjecture There is a number $J(F(C'))$ such that the number of components of any fiber of C_L is bounded by $J(F(C'))$ for any finite extension L of $F(C')$ containing in $F(C')^{nr}$.

Remark 6.8 The result of the conjecture depends (at least formally) on the chosen model C'. If we blow up a generic point on a fiber the conjecture becomes false if the corresponding covering induces an infinite covering of the fiber. Thus the arithmetic version is sensitive to the change of the semistable model whereas the geometric conjecture is not.

Remark 6.9 The fields $F(C')^{nr}$ correspond to the factor of $\pi_1^{fin}(C')$ that is acted trivially by all inertia groups. (Here we denote by $\pi_1^{fin}(C')$ the profinite completion of the geometric fundamental group of C'). Geometrically this means that we consider coverings of C_L that are unramified over a generic point of every irreducible divisor in C_L.

There exists some evidence for the above arithmetic conjectures. Partial results in this direction were obtained by the first author several years ago (1982). Namely he proved that the torsion group of an abelian variety A is finite for any infinite algebraic extension of K which contains only finite abelian extensions of K.

In particular it is true for infinite nonramified extension of K where K is a finite extension of $\mathbb{Q}$. (This result was announced at Delange-Puiso seminar in Paris, May 1982 and later appeared in [8]). Yu. Zarhin proved that the same is true if the infinite extension of K contains only finite number of roots of unity under some conditions on the algebra of endomorphisms A.

The above result states that the group $\mathrm{Gal}_g^{nr} K(C)$ has a finite abelian quotient and hence Conjecture 6.2 is evidently true for the quotient of $\mathrm{Gal}^{nr} K(C)$ by the commutant of $\mathrm{Gal}_g^{nr} K(C)$. For the same reason it is true for the quotient of $\mathrm{Gal}_g^{nr} K(C)$ by any iterated commutant.

Remark 6.10 Any curve over arithmetic field can be obtained as a covering of $\mathbb{P}^1$ ramified at $(0, 1, \infty)$ according to the famous Bely's theorem. As it was pointed out by Yu. Manin any arithmetic curve has a ramified covering which is a nonramified covering of a modular curve. Therefore one might reduce the above conjecture to modular curves being considered over different number fields. In the complex case we don't have similar simple class of dominant manifolds (see the discussion in [2]).

If Conjecture 6.2 is correct for every curve over every finite extension of $\mathbb{Q}$ then we get that there isn't any infinite chains of compact curves on the universal covering of a projective surface with a residually finite fundamental group. According to a conjecture of M. Ramachandran this is the only obstruction to holomorphic convexity for the universal coverings of a projective surfaces.

7 Appendix A - Fiber groups and monodromy

This appendix includes several technical results from the theory of surfaces which we use in our article. First theorem generalizes the result we used in order to pass from quasiprojective surface to the projective smooth surface in Section 2.

Theorem 7.1 *Let V be a normal projective surface with Q being the set of singular points in V and $f : V' \to V$ be a finite surjective map from another normal projective surface V'. Assume that for any $q \in f^{-1}(Q)$ the map of the local fundamental groups $f_* : \pi_1(U_q - q) \to \pi_1(V - Q)$ is zero, where U_q is small topological neighborhood of q. Then there is a natural map $f_* : \pi_1(V') \to \pi_1(V - Q)$ which is a surjection on a subgroup of finite index bounded from above by the degree of f.*

Proof Indeed Q is a set of isolated singular points. Following the proof of the Lemma 2.5 we obtain a map $f_* : \pi_1(V' - f^{-1}(Q)) \to \pi_1(V - Q)$. In order to prove the theorem it is sufficient to show that the kernel of natural surjection $i_* : \pi_1(V' - f^{-1}(Q)) \to \pi_1(V')$ lies in the kernel of f_*. The preimage $f^{-1}(Q)$ also consists of a finite number of points and the kernel of i_* is generated as a normal subgroup in $\pi_1(V' - f^{-1}(Q)$ by the local subgroups $\pi_1(U_q - q), q \in f^{-1}(Q)$. Due to the condition of the theorem the images of these groups are trivial in $\pi_1(V - Q)$ and hence we obtain the map f_*. □

We shall also need the following general result which allows to compare the properties of fiber groups and fundamental groups of the smooth quasiprojective surfaces.

Theorem 7.2 *Let $f : X \to R$ be a quasiprojective surface which has a structure of a family of curves over a smooth curve R. Assume that generic fiber is an irreducible smooth projective curve and any fiber contains a component of multiplicity* *one. Let $h : \pi_{1,f}(X) \to G$ be M_X invariant homomorphism into a finite group G with M_X action.*

Let K be the kernel of h. Then there exists a subgroup of finite index $H \in \pi_1(X)$ so that the intersection of H with $\pi_{1,f}(X)$ is equal to K.

Remark 7.1 This is true if there exists a section $s : \pi_1(R) \to \pi_1(X)$, since we can define H as a subgroup of $\pi_1(X)$ generated by products of the elements of $s(\pi_1(R))$ and K. If the curve R is open then the group $\pi_1(R)$ is free and hence a section always exists. Indeed, the group K is a normal subgroup of $\pi_1(X)$ since it is invariant under the conjugations from both $\pi_1(R)$ and $\pi_{1,f}(X)$. Denote by Q the quotient $\pi_1(X)/K$. It is an extension of $\pi_1(R)$ by G. Hence there is an action of $\pi_1(R)$ over G. Since G is a finite group $\pi_1(R)$ contains a subgroup of finite index which acts trivially on G via interior endomorphisms. This subgroup corresponds to a finite nonramified covering $\phi : C \to R$ and will be denoted by $\pi_1(C)$. Its preimage Q' in Q is a subgroup of finite index. Consider a subgroup K_G of Q' consisting of the elements commuting with all the elements of G. The group K_G projects onto $\pi_1(C)$ by the definition of $\pi_1(C)$ and therefore its intersection with G is a cyclic subgroup of the center of G. The group $\pi_1(C)$ contains a subgroup of finite index K' where the corresponding central extension splits. Namely if the order of the corresponding cyclic extension is n then it splits over any cyclic covering of order n. The preimage of K' in Q' splits into a direct product of G and K'. Hence on the preimage of a subgroup $K' \subset \pi_1(C) \subset \pi_1(R)$ we have a natural extension of the projection from $\pi_{1,f}(X) \to G$.

8 Appendix B - Some group theoretic results

This appendix contains several group theoretic results. We present proofs though most of the results are presumably known to the experts in group theory.

Recall that we denote by $BT(n,m)$ the quotient of a free group $\mathbb{F}^n$ by a normal subgroup generated by the elements $x^m = 1$ where x runs through all primitive elements of $\mathbb{F}^n$.

Proposition 8.1 *If m is divisible by 4 and $n \geq 2$ then $BT(2,m)$ is infinite.*

Proof The group $BT(2,4)$ has an infinite representation. Namely let $\mathbb{Q}_8$ be the group of the unit quaternions of order 8. It acts on $H = \mathbb{R}^4$ by multiplications. Consider a group of the affine transformations of $\mathbb{R}^4$ generated by two generating rotations g_1, g_2 in $\mathbb{Q}_8$ but with different invariant points. The resulting group G will be infinite. It has two generators and G has $\mathbb{Q}_8$ as its quotient group. Any element of $g \in G$ which projects nontrivially into G has order 4. Indeed g is an affine transformation of $\mathbb{R}^4$ with a linear part of order 4 and without 1 as eigenvalue. Hence g in the conjugacy class of its linear part and has order 4. □

Remark 8.1 Similar argument can be applied to any finite subgroup of a skewfield instead of $\mathbb{Q}_8$ (see the description of such finite groups in J. Amitsur, Ann. Math., 1955, vol. 62, p. 8). The above result makes it plausible that the groups $P(n,m)$ are infinite if $m > 3, n > 1$.

The case $BT(n,3)$ is different. The following result gives a hint on the effects which occur with the exponent 3.

Lemma 8.1 *Let $G_i, i = 1,2$ be the groups generated by a, b with relations*

1) $G_1 : a^3 = b^3 = (ab)^3 = 1.$

2) $G_2 : a^3 = b^3 = (ab)^3 = (ab^2)^3 = 1.$

Then G_1 is infinite and G_2 coincides with the Burnside group $B(2,3)$ and hence it is finite.

Proof The group G_1 has a natural geometric realization. Namely let us take $\mathbf{P}^1$ minus three points $p_i, i = 1,2,3$. The fundamental group $\pi_1(\mathbf{P}^1 - p_i) = \mathbf{F}_2$. If we impose relations x^3 on the elements which can be realized by smooth curves in $\mathbf{P}^1 - p_i$ then we obtain the set 1). Let us take a $\mathbf{Z}_3$ character χ of $\mathbf{F}_2$ which is nontrivial on a, b, ab. We obtain a covering of $\mathbf{P}^1$ ramified over three points. Imposing the above relations corresponds to the completion of the curve. Hence the subgroup of G_1 which is the kernel of χ coincides with the fundamental group of the corresponding complete curve. Since the above curve is a torus the group G_1 is an extension of $\mathbf{Z}_3$ by a free abelian group $\mathbf{Z} + \mathbf{Z}$. The description of 2) immediately follows since it is the quotient group of the group in 1). The element ab^2 generates $\mathbf{Z} + \mathbf{Z}$ as a $\mathbf{Z}_3$-module. Hence $(ab^2)^3$ generates $3(\mathbf{Z} + \mathbf{Z})$ and the resulting group is an extension of $\mathbf{Z}_3$ by the group $\mathbf{Z}_3 + \mathbf{Z}_3$. □

Remark 8.2 The groups above can be used to answer some of the questions asked in the symplectic part of the paper.

We shall use the following standard notations. Denote by (a,b) the commutator of the elements a, b and we put a sequence of brackets to denote an element obtained by iteration of the procedure.

The group $B(n,3)$ has a rather simple description. It is a metabelian group with a central series of length three. The elements (a,x) where $x \in [B(n,3), B(n,3)]$ are in the center of $B(n,3)$. In fact $((a,b),c)$ varies under permutation according to standard $\mathbf{Z}_2$ character of the group S_3 for any $((a,b),c)$. In particular $((x,r),x) = 1$ for any x.

Lemma 8.2 *Let G be a finitely generated group with a given set S of generators. Assume that $((a,b),f)$ is invariant under any even permutation of a, b, f for any $a, b \in S, f \in S \bigcup (S,S)$ where the latter denotes the set of pairwise commutators of the elements from S. Then G is a metabelian group and it has a central series of length at most 3.*

Proof The proof closely follows the proof from [18]. We can write $((a,b),(c,d)) = (a,(c,d)),b)$. We deduce next that the above expression is invariant under even permutations. The latter implies that it is equal to 1 and hence $((a,b),c)$ commutes with any element from S and hence lies in the center of G. The quotient of G by the center $Z \in G$ is also generated by S with equality $((a,b),c) = 1$ for any $a, b, c \in S$ which means that (a,b) is in the center of G/Z for any $a, b \in S$. That means G/Z is a central extension of abelian group which finishes the proof. □

Corollary 8.1 *Under the above conditions the commutant $[G,G]$ is additively generated by the elements $(a,b),((a,b)c), a,b,c \in S$.*

Lemma 8.3 *Assume that S in the previous lemma consists of n elements and $[S,S]$ consists of elements of order 3 and there is a surjective map $p : G \to B(n,3)$. Then p is an isomorphism.*

Proof The commutant $[G,G]$ is additively generated as G^{ab}-module by the elements (a,b) under the above conditions. Hence it is of exponent 3 and the number of elements in G is not greater than the number of elements in the commutant of $B(n,3)$.

The abelian quotient G^{ab} is isomorphic to $B(n,3)^{ab}$. Therefore the number of elements in G is not greater than in $B(n,3)$ and a surjective map $p : G \to B(n,3)$ is an isomorphism. □

Proposition 8.2 *The group $BT(n,3)$ is finite and coincides with the Burnside group $B(n,3)$ if $BT(3,3) = B(3,3)$.*

Proof Indeed this is true for $n = 2$ as it was shown above (see e.g. [18]). The group $BT(n,3)$ has a natural surjective map onto $B(n,3)$ and hence we have to check the condition on $((a,b),c)$. The latter is enough to check for the group with three generators. □

Lemma 8.4 $BT(3,3) = B(3,3)$.

Proof The group $BT(n,3)$ is obtained as an extension of $B(3,2)$ by c and since cx is a generator for any $x \in B(3,2)$ we obtain that $(cx)^3 = 1$. Therefore $x^{-1}cx, c$ commute for any $x \in B(3,2)$ and the kernel K of the projection $p_c : BT(3,3) \to B(3,2)$ is an abelian group of exponent 3. The sum $x^{-1}cx + c + xcx^{-1} = 0$ and we can easily deduce that K has 4 generators as a $\mathbb{Z}_3$-space. Therefore the group $BT(3,3)$ has the same number of elements as $B(3,3)$ and since there is a natural surjection $BT(3,3) \to B(3,3)$ they are isomorphic. □

Lemma 8.5 *Let S be a finite set of generators of the group G. Assume that any subset of 4 elements in S generates a subgroup of exponent 3. Then G is of exponent 3.*

Proof The assumption implies that any three-commutator of $a,b,c \in S$ lies in the center C of the group G and the group satisfies Lemmas 8.2, 8.3. □

Lemma 8.6 *Assume that G has four generators a,b,c,d that G contains a set of subgroups of exponent 3 which includes groups generated by triples of generators a,b,c,d and also the groups $a,b,(cd)$, $c,d,(a,b)$. Then G is of exponent 3.*

Proof It follows from the above results that $((a,b)c)$ and similar combinations are invariant under cyclic permutations. We also have $((a,b)(c,d)) = (((c,d)a)b) = (((d,c)b)a) = (((a,b)c)d) = (((a,b)d)c)$. Thus the value of the above commutator does not depend on any even permutation of the symbols. On the other hand it transforms into opposite if we permute $(c,d)(a,b)$. Hence all of the above elements are equal to 1. This implies the lemma. □

We are interested in the groups which occur as the quotients of the fundamental groups of curves. Namely we will study the groups $P^g(n,m)$ defined in section four. We are going to use a representation of the fundamental group of a Riemann surface as a subgroup of index 2 in the group generated by involutions which was successfully used by J. Birman, W. Thurston, B. Wainrieb and many others. Let X be a curve of genus g. It can be represented as a double covering of $\mathbb{P}^1$ ramified over $2g+2$ points and π_g is a subgroup of index two in the group generated by $2g+2$ involutions x_i with additional relation that the product of all involutions is an involution again. Similarly the group $\pi_1(X_g - pt)$ is realized as subgroup of index two in the group generated by $2g+1$ involutions. The fundamental group of a curve minus two points is realized as a subgroup of index two in the group generated by $2g+2$ involutions without any additional relations.

The groups above are free groups, but they are provided with a special realization as the fundamental groups of open curves.

Definition 8.1 We shall denote by $\mathbb{F}_n^g$ a free group of n generators provided with a realization as a fundamental group of a curve of genus g minus one or two points. In case n is odd then $\mathbb{F}_n^g$ is realized as a fundamental group of a curve of genus g minus one point. In case n is even $\mathbb{F}_n^g$ is realized as a fundamental group of a curve of genus g minus two points.

Recall that $P^g(n,m)$ is a quotient of $\mathbb{F}_n^g$ by the relations $x^m = 1$ where x runs through the primitive elements of $\mathbb{F}_n^g$ which can be realized as a smooth loops in the above geometric realization.

The following lemma reduces a general case to the case $n \leq 4$.

Lemma 8.7 *If $P^g(4,3) = B(4,3)$ then $P^g(n,3) = B(n,3)$ for any $n \geq 4$.*

Proof The group $\mathbb{F}_n^g$ is represented as subgroup of index two generated by involutions $x_1, .., x_{n+1}$. The set of standard generators of $\mathbb{F}_n^g$ can be taken as $x_1x_i, = i \neq 1$. Any four elements x_1x_j generate a subgroup of $P^g(n,3)$ which is a quotient of $P^g(4,3)$ and by assumption of the lemma the latter is of exponent 3. Hence by lemma 6.5 $P^g(n,3)$ is of exponent 3. Since the set of generators includes only primitive elements $\mathbb{F}_n$ the group $P^g(n,3)$ coincides with $B(n,3)$. □

Lemma 8.8 *If $P^g(3,3) = B(3,3)$ then $P^g(4,3) = B(4,3)$.*

Proof The group $P^g(4,3)$ is obtained from a curve of genus 2 minus a point. Consider a standard decomposition of X^2 into a union of two handles corresponding to pairs of generators a,b and c,d of $\mathbb{F}_4^g$. There are topological embeddings of tori with two discs removed corresponding to the subgroups generated by any three symbols from the set (a,b,c,d). The group $P^g(3,3)$ is realized as the quotient of the fundamental group of torus minus two discs. The above tori are obtained from one of the handles by adjoining a neighborhood of a generator in another handle. By assumption of the lemma $((a,b),c)$ is transformed into itself or the opposite element under the permutation of symbols.

Similar groups correspond to the triples $(a, b(cd)), (c, d, (ab))$. Namely we can consider a corresponding handle minus a point. Thus we can apply Lemma 8.6 and obtain that $P^g(4,3)$ is of exponent 3 if the group $P^g(3,3)$ is of exponent 3. □

Theorem 8.1 *$P^g(3,3) = B(3,3)$ and hence $P^g(n,3) = B(n,3)$ for any n.*

Let us first describe the geometric picture. Consider the torus T^2 with a small embedded interval I. Let p_1, p_2 be two different points in the interval I and I_1 be the interval between p_1, p_2 inside I. We assume that p_0 is a point in $I - I_1$ and identify p_0 as an initial point for the fundamental group $\pi_1(T^2 - p_1 - p_2) = \mathbb{F}_3^g$. Assume that T^2, I are given with orientation. We consider smooth oriented loops through p_0 which are transversal to I.

Lemma 8.9 *Any element of $P^g(2,3)$ with p_0 as an initial point is represented by an oriented curve A without selfintersection which does not intersect I and such that A, I defines a standard orientation of T at p_0.*

Proof Let a, b be standard generators of $\mathbb{F}_2^g$ with a given orientation. The group $\mathrm{Out}(\mathbb{F}_2^g) = \mathrm{SL}(2,\mathbb{Z})$ and can be realized by linear periodic map of torus. In particular we can represent topologically the elements of $\mathrm{SL}(2,\mathbb{Z})$ by maps which stabilize the points of I. In this way we obtain any map of $\mathbb{F}_2^g$ into itself which transforms $aba^{-1}b^{-1} = C$ into itself. Any homomorphism of the free group with above property is induced by this action of $\mathrm{SL}(2,\mathbb{Z})$. Thus $g(a), g \in \mathrm{SL}(2,\mathbb{Z}), a \in B(3,2)$ can be any element such that there exists $b' \in B(3,2)$ with $g(a)b'g(a)^{-1}(b')^{-1} = C$ where C is a given generator of the center. But we can find such b' for any $g(a)$ which is not in the center. The curve $g(a) = A$ will be the image of a map which is linear outside a neighborhood of I and keeps orientation intact. □

Remark 8.3 Since $A^{-1}I$ represents the opposite orientation the lemma actually shows that the element A^{-1} can be represented by another simple closed curve B with orientation BI opposite to $A^{-1}I$.

The group $P^g(3,3)$ is generated by $B(2,3)$ realized as above and an element r realized by a curve R with one selfintersection. We have a natural representation of r as $x_1x_2^{-1}$ where x_1, x_2 are simple curves with the same orientation which move around p_1, p_2 respectively. The element $c = x_1x_2$ is a natural central loop in $P^g(2,3) = B(2,3)$.

Lemma 8.10 *Any element bx_i is realized by a simple loop if b is not in the center of $B(2,3)$.*

Proof We have to find a simple representative of b through p_0 with an appropriate orientation. The latter exists due to the previous lemma. □

Lemma 8.11 *Any element $br \in P^g(3,3), b \in B(2,3)$ can be represented by a simple curve in its conjugation class unless b is in the center of $B(2,3)$.*

Proof We have $br = bx_1x_2$. The element bx_1 is represented by a simple curve B. Let S be curve which contains I_1 and intersects B transversally at exactly one point inside I_1. The compliment $T - (S - I_1)$ defines another group $B(3,2)$. We assume that p_0 is not in S and hence bx_1 is equivalent to a simple curve with a desired orientation with respect to x_2. That means $(bx_1)x_2$ can be realized by a class of simple curve in $P^g(3,3)$. The classes x_i are also realized by simple curves. □

Corollary 8.2 *The elements* $(br)^3 = 1$ *for any* $b \neq c = x_1x_2$.

Indeed if b is not in the center then br is realized by a simple curve and we get the result. The element $c^{-1}r = x_2^{-1}x^{-1}x_1x_2^{-1} = x_2^{-2} = x_2$ in $P^g(3,3)$.

In dealing with the extension of $B(2,3)$ by an element r we will be using the following general argument. Let G be a finite group of exponent 3 and G' is obtained from G by adding r and some relations of type $(br)^3 = 1$. Then the kernel of the natural projection $p : G' \to G$ is generated by the elements $r^a = ara^{-1}, a \in G$. The group G acts on these set of elements by left translation $g : r^a \to r^{ga}$. Any relation $(br)^3 = 1$ implies the relation : $1 = brbrbr = brb^{-1}b^{-1}rbr = r^br^{b^{-1}}r = 1$ and similar relation for left translations of the orbits of the cyclic group $B = (1, b, b^{-1})$. If in addition $(b^{-1}r)^3 = 1$ all the elements $r^b, r^{b^{-1}}, r$ commute and any pair of them generate the same abelian group.

Lemma 8.12 *The group* $P^g(3,3)$ *is an abelian extension of* $B(2,3)$.

Proof The kernel of the projection $p_r : P^g(3,3) \to B(2,3)$ which maps r to 1 is generated by the elements $r^b = brb^{-1}$. All these elements commute with r unless $b = c$. We also have $r^brr^{b^{-1}} = 1$ and they all commute if b is not in the center. Therefore r^a commutes with r if it commutes with both $r^b, r^{b^{-1}}$. On the other hand $r^{ab}, r^a, r^{ab^{-1}}$ commute if $r^b, r, r^{b^{-1}}$ commute. Let a be a generator of $B(2,3)$. Then $r^{ca}r^cr^{ca^{-1}} = 1$ and all these elements commute, but $r^{ca}, r^{ca^{-1}}$ commute with r. This implies that all the elements $r^b, b \in B(2,3)$ commute with r. After translation by $B(2,3)$ we obtain that all the elements r^g commute. □

Lemma 8.13 *Let T be the group generated by $r^b, b \in B(2,3)$. Then T is an abelian group with* 4 *generators.*

Proof Indeed the set of cyclic subgroups which don't lie in the center generate a family of relations. Since we have established that T is an abelian group we shall write them in the additive form $r^x + r^{ax} + r^{a^2x} = 0$, $x \in B(2,3)a$ but not in the center C. Denote by T_S a subgroup of T generated by a subset $S \subset B(2,3)$. Let A be an abelian subgroup generated by a and c which generates the center of $B(2,3)$. The summation over orbits of cyclic noncentral subgroups gives zero. Thus if we consider T_A modulo a subgroup T_C corresponding to the center we obtain $r^g + r^{gh} = 0, g \notin C$. Hence the elements $r^g = -r^{g^{-1}}$ and $r^{gc} = r^g$ modulo subgroup r^c, r. We obtain that r^a generates T_A modulo the subgroup (r_c, r) and a sum over any orbit of C is also zero.

Thus we have $r(x^{-1}) = -r - r^x$ and $r^x + r^y + r^{y^{-1}x^{-1}} = 0$ for the elements x, y which lie in one abelian subgroup of $B(3,2)$. The same is true modulo r^c since we can apply the same argument to the quotient of $B(3,2)$ by the center. Hence $r^x + r^y - r = r^{xy}$ (modulo(r^c)) for any $x, y \in B(2,3)$. In particular r^a, r^b, r^c, r generate the group T. □

Lemma 8.14 *T is an elementary abelian group.*

Proof We have $r^{x^2} = 2r^x - r$ and $r = 3r^x - 2r$. Hence $3r = 3r^x$ for any x. Hence $3(r - r^x) = 0$ for any x. Thus the elements of zero degree in T constitute an elementary 3-group T_0 which is a normal subgroup of $P^g(3,3)$. The quotient T/T_0 is a cyclic group. The group $P^g(3,3)/T_0$ is a central extension of $B(2,3)$. Since $P^g(3,3)$ is generated by elements of order 3 we obtain that $T/T_0 = \mathbb{Z}_3$. □

Corollary 8.3 *The number of elements in T is 3^4.*

Hence the number of elements in $P^g(3,3)$ is equal to 3^7 and coincides with the number of elements in $B(3,3)$. Since there exists a surjective map $p : P^g(3,3) \to B(3,3)$ the groups coincide. Thus we have proved that group $P^g(n,3)$ coincide with $B(n,3)$ for all n.

Acknowledgments We would like to thank S. Donaldson and E. Zelmanov for various suggestions and remarks. We also thank the referee for useful comments and M. Fried about his comments on the arithmetics version of Shafarevich's conjecture.

References

[1] M. Artin *Algebraic approximation of structures over complete local rings.* Publ. Math. I.H.E.S. **36** (1969), 23-58.

[2] F. Bogomolov, D. Husemoller *Geometric properties of curves defined over number fields.* University of Maryland, preprint 1993.

[3] F. Bogomolov, L. Katzarkov *Complex projective surfaces and infinite groups.* to appear in GAFA.

[4] F. Bogomolov, L. Katzarkov *Symplectic fourfolds and projective surfaces,* to appear in Proceedings of Georgia topology conference, 1996.

[5] F. Bogomolov, L. Katzarkov *Symplectic Moisheson theory,* in preparation.

[6] E. Ballico, F. Catanese, C. Ciliberto (Eds.) *Trento examples,* Classification of higher dimensional varieties, Proceedings, Trento 1990, Springer LNM, 1515, 1992, 134-139.

[7] H. Clemens *Degeneration of Kähler varieties.* Duke Mathematical Journal, **44** (1977), 215 -290.

[8] R. Coleman *Ramified torsion points on curves.* Duke Mathemtical Journal **54** (1987), 615 - 640.

[9] P. Deligne, N. Katz *Groupes de monodromie en geometrie algbrique. II.* Seminaire de Geometrie Algebrique du Bois-Marie 1967–1969 (SGA 7 II). Dirige par P. Deligne et N. Katz. Lecture Notes in Mathematics, Vol. 340. Springer-Verlag, Berlin-New York (1973), 14-06.

[10] S. Donaldson, in preparartion.

[11] R. Fintushel, R. Stern *Isotopy classes of surfaces in 4-manifolds.*, in preparation.

[12] R. Gompf, in preparartion.

[13] B. Lasell, M. Ramachandran *Observations on harmonic maps and singular varieties.* Ann. Sci. École Norm. Sup. **29** (1996), 135-148.
[14] L. Katzarkov *On the Shafarevich maps.* to appear in Proc. Symp. in Pure Math., Proceedings of AMS Alg. Geometry conference Santa Cruz 1995.
[15] L. Katzarkov, M. Ramachandran, T. Pantev *Geometric factorization of linear fundamental groups*, in preparation.
[16] D. Kotschick, J. W. Morgan, C.H. Taubes *Four-manifolds without symplectic structures but with nontrivial Seiberg-Witten invariants.* Math. Res. Lett. **2** (1995), no. 2, 119–124.
[17] J. Kollár *Shafarevich maps and automorphic forms.* Princeton University Press, Princeton, NJ, (1995).
[18] W. Magnus, A. Karrass, D. Solitar *Combinatorial group theory. Presentations of groups in terms of generators and relations.* , second revised edition. Dover Publications, Inc., New York, 1976.
[19] T. Napier *Convexity properties of coverings of smooth projective varieties.* Math. Ann. **286** (1990), 433-479.
[20] C. T. Simpson *Constructing variations of Hodge structures using Yang-Mills theory and applications to uniformization.*, J. Amer. Math. Soc. **1** (1988), 867–918.
[21] C. T. Simpson, *Mixed twistor structures.* alg-geom /9705006.
[22] D. Toledo *Projective surfaces with nonresidually finite fundamental groups.* Publ. Math I.H.E.S., No. 77 (1993), 103–119.
[23] B. Wajnryb *Mapping class group of a surface is generated by two elements.* Topology **35** (1996), 377–383.

NON-POSITIVE CURVATURE IN GROUP THEORY

MARTIN R. BRIDSON[1]

Mathematical Institute, 24–29 St Giles', Oxford, OX1 3LB England

Abstract

This article is an edited account of the four lectures that I gave at the Groups St Andrews meeting in Bath during the summer of 1997. The aim of these lectures was to introduce an audience of group theorists, with varying backgrounds, to the role that non-positive curvature plays in the theory of discrete groups. A few new results are included, but basically this is an expository article aimed as non-experts. These notes do not constitute a comprehensive survey of curvature in group theory. Nevertheless, I hope that they give the reader a substantial and enticing taste of this active area of research.

Introduction

In the last fifteen years the close connection between geometry and combinatorial group theory, which was at the heart of the pioneering work of Dehn [52], has re-emerged as a central theme in the study of infinite groups. Thus combinatorial group theory has been joined by (and to a large extent has merged with) what has become known as *geometric group theory* — this is the broad context of these lectures. My basic goal will be to illustrate how various notions of non-positive curvature can be used to illuminate and solve a range of group theoretic problems. I shall also attempt to illustrate how group theory can serve as a potent tool for exploring the geometry of non-positively curved spaces. I wish to promote the idea that manifestations of non-positive curvature are deeply inherent in combinatorial group theory, often in contexts where there is little reason, *prima facie*, to suppose that the problems concerned have any geometric content whatsoever.

This last point is particularly emphasized in the first lecture, which begins with Dehn's formulation of the basic decision problems at the heart of combinatorial group theory: the word problem, the conjugacy problem and the isomorphism problem. Following a general discussion of the foundations of geometric group theory, I shall explain how Dehn's approach to the word problem leads us naturally to the theory of hyperbolic groups à la Gromov [68]. I shall briefly outline some elements of this theory. Through hyperbolicity we enter the world of non-positive curvature and Lecture 1 concludes with some illustrations of the way in which a qualitative understanding of the geometry of non-positively curved manifolds can lead us to striking group theoretic results.

Gromov's notion of hyperbolicity, which emerges in Lecture 1 from core considerations in combinatorial group theory, encapsulates many of the features of the

[1]The author is supported by an EPSRC Advanced Fellowship. His work on this subject has also been supported by grants from the NSF (USA) and the FNRS (Switzerland).

global geometry of simply-connected spaces of negative curvature, whereas classically one thinks of curvature as a local or infinitesimal concept. In Lecture 2 we will examine the geometry of spaces which are non-positively curved in a strict, local, sense first identified by A. D. Alexandrov (see [2]). Much of the importance of this more local approach to curvature rests on the existence of a local-to-global theorem (the Generalized Cartan-Hadamard Theorem): if a complete geodesic space X is non-positively curved (in a local sense), then its universal cover $\widetilde{X}$ is (a strong non-positive curvature condition that implies much about the large-scale geometry and topology of the space).

A central theme in Lecture 2 is the closeness of the relationship between the global geometry of CAT(0) spaces and the structure of the groups which act properly by isometries on them. Moreover, as the Generalized Cartan-Hadamard Theorem indicates, the existence of a metric of non-positive curvature on a space tells one a great deal about the topology of the space. Thus we are presented with an environment in which group theory, geometry and topology are deeply interconnected. Such an environment is obviously an exciting one, but before one can get too excited one needs to know that there is a substantial range of examples. Thus in Lecture 3 we turn our attention to the construction and identification of non-positively curved spaces, with particular emphasis on examples that are of group-theoretic interest.

Our point of departure in Lecture 1 will be Dehn's original formulation of the basic decision problems of combinatorial group theory. (This was also the point of departure for much of twentieth century combinatorial group theory.) In Lecture 4 I shall return to the study of these basic problems in the context of non-positively curved spaces and groups, and I shall describe the state of the art in this active area.

The origins of combinatorial group theory are deeply intertwined with those of low-dimensional topology. In both subjects geometry, in particular non-positive curvature, plays a central role that is not immediately evident[2]. For most of this century, though, curvature has played little role in the study of either subject. Indeed, in the case of combinatorial group theory, the whole geometric vein in the subject has lacked prominence. Its return to prominence in recent years is due largely to the deep insights of Mikhael Gromov [67, 68, 69], and I should make it clear that although his name and that of Max Dehn do not appear on every page of these notes, their influence runs throughout.

Acknowledgements

A more comprehensive account of much of the material in these lectures can be found in my book with André Haefliger [33]. Exploring the geometry of CAT(0) spaces is an adventure that Haefliger and I have shared over the last five years and I am deeply grateful to him for sharing his many insights. I am equally grateful to both him and his wife, Minouche, for welcoming me so warmly into their home during this time.

[2]In the case of 3-manifolds this insight is largely due to Thurston [111, 112].

Some of the material in Lecture 4 is based on joint work with Gilbert Baumslag, Chuck Miller and Hamish Short; I would like to take this opportunity to thank them for many hours of good-humoured conversation. Finally, I would like to thank the organizers of Groups St Andrews in Bath. I particularly wish to thank Geoff Smith for accommodating my disorganization both in Bath and in the process of getting this article finished within a (strenuous) stone's throw of the deadline.

The tone of the lectures

The level and tone of the lectures reflect both the background of the audience and my own tastes and prejudices. First the audience: each of them might acquiesce in the title "group theorist", but within that category one would find many whose daily thoughts revolve entirely around finite groups, while others might consider themselves to be topologists. From the point of view of these lectures, the importance of this diversity is that it would be unreasonable to assume that a majority of the audience were well-acquainted with ideas from topology and geometry (beyond such basic concepts as the fundamental group of a space). I shall assume no knowledge of differential geometry. Various comments in the text will only be transparent to readers who are familiar with the rudiments of hyperbolic geometry, but for the most part such a familiarity is not required. At some points in lectures 3 and 4 I shall need some low dimensional topology and the reader who finds this to be a problem may wish to omit the sections concerned.

With regard to my own tastes and prejudices: I came to group theory, more precisely combinatorial group theory, through problems in geometry and topology. Moreover, I am by nature inclined to seek geometry underlying any given piece of mathematics. I hope that admitting this subjectivity will not weaken my contention that geometry is inherent (though not always immediately apparent) in many aspects of group theory and that to ignore it can be debilitating.

Lecture 1: In the beginning there was Max Dehn

I shall begin be explaining what I mean by combinatorial and geometric group theory, and by framing some of the central issues in each subject. I particularly wish to emphasize the common origins of these subjects and the natural interdependence of them.

First, combinatorial group theory: roughly speaking, *combinatorial group theory* is the study of groups given by generators and defining relations. Thus a group Γ is described to us by means of a set $\mathcal{A}$ and a subset $\mathcal{R}$ of the free group F on $\mathcal{A}$; to say that $\langle \mathcal{A} \mid \mathcal{R} \rangle$ is a presentation of Γ means that Γ is isomorphic to the quotient of F by the smallest normal subgroup containing $\mathcal{R}$. This method of describing groups emerged from work on discrete groups of isometries of hyperbolic space, and was first articulated by von Dyck in [56], which is generally regarded as the first paper in combinatorial group theory. For a comprehensive history of the subject up to 1980, see [46].

1.1 Decision problems

It was some years after von Dyck's original paper that Max Dehn framed the central issues of combinatorial group theory that have occupied researchers in the field for most of this century. Building on work of Poincaré [97, 98], Dehn was working on the basic problems of recognition and classification for low-dimensional manifolds (see [52]). In that setting, one finds that for many purposes the key invariant is the fundamental group of the space at hand. Moreover, when one is presented with the space in a concrete way, the fundamental group often emerges in the form of a presentation. In the course of his attempts to recover knowledge about fundamental groups (and hence manifolds) from such presentations, Dehn came to realize that the problems that he was wrestling with were manifestations of fundamental problems in the theory of groups. Let us recall the manner in which Dehn originally formulated the three basic problems of this kind.

1.2 "Über unendliche diskontinuierliche Gruppen", (Dehn, 1912)

"The general discontinuous group is given by n generators and m relations between them, as defined by Dyck (Math. Ann., 20 and 22). The results of those works, however, relate essentially to finite groups. The general theory of groups defined in this way at present appears very undeveloped in the infinite case. Here there are above all three fundamental problems whose solution is very difficult and which will not be possible without a penetrating study of the subject.

1. The identity [word] problem: *An element of the group is given as a product of generators. One is required to give a method whereby it may be decided in a finite number of steps whether this element is the identity or not.*

2. The transformation [conjugacy] problem: *Any two elements S and T of the group are given. A method is sought for deciding the question whether S and T can be transformed into each other, i.e. whether there is an element U of the group satisfying the relation $S = UTU^{-1}$.*

3. The isomorphism problem: *Given two groups, one is to decide whether they are isomorphic or not (and further, whether a given correspondence between the generators of one group and elements of the other is an isomorphism or not).*

These three problems have very different degrees of difficulty. [...] One is already led to them by necessity with work in topology. Each knotted space curve, in order to be completely understood, demands the solution of the three above problems in a special case."

I particularly wish to emphasize Dehn's last two remarks: the degree of difficulty of these three problems varies considerably, and each arises naturally in low-dimensional topology. I would also like to add something to the second remark: Dehn was studying surface groups and knot groups, and surfaces of positive genus and knot-complements[3] all support metrics of non-positive curvature. Thus

[3] viewed as compact manifolds with boundary

the context in which Dehn was working was deeply imbued with non-positive curvature.

1.3 Dehn's algorithm for solving the word problem

This is perhaps the most direct approach that one can hope for whereby the information in a finite presentation is used directly to solve the word problem in the group presented. We shall see in subsequent sections that this direct algebraic approach to the word problem provides a remarkable bridge into the world of negative curvature (1.21).

Given a finite set of generators $\mathcal{A}$ for a group Γ, one would have a particularly efficient algorithm for solving the word problem if one could construct a finite list of words $u_1, v_1, u_2, v_2, \ldots, u_n, v_n$, with $u_i = v_i$ as elements of Γ, with lengths $|v_i| < |u_i|$, and with the property that if a word w represents the identity in Γ then at least one of the u_i is a subword of w.

If such a list of words exists, then given an arbitrary word w one looks for a subword of the form u_i; if there is no such subword, one stops and declares that w does not represent the identity; if u_i occurs as a subword of w, then one replaces it with v_i and repeats the search for subwords of the new (shorter) word — this new word represents the same element of the group as w. After at most $|w|$ steps one will either have reduced w to the empty word or else verified that w does not represent the identity.

This is the algorithm that Dehn used to solve the word problem for surface groups [53].

Definition 1.4 A finite presentation $\langle \mathcal{A} \mid \mathcal{R} \rangle$ of a group Γ is called a *Dehn presentation* if $\mathcal{R} = \{u_1v_1^{-1}, \ldots, u_nv_n^{-1}\}$, where the words $u_1, v_1, \ldots, u_n, v_n$ satisfy the conditions of Dehn's algorithm.

Readers should ask themselves how one might decide whether or not $\mathbb{Z}^2$ has a (possibly obscure) Dehn presentation. (In fact it does not.) As further food for thought, let me mention some consequences of having a Dehn presentation.

Theorem 1.5 *If a group Γ has a Dehn presentation, then:*

1. *Γ does not contain $\mathbb{Z}^2$;*
2. *Γ has a solvable conjugacy problem;*
3. *Γ has only finitely many conjugacy classes of finite subgroups;*
4. *$H^*(\Gamma, \mathbb{Q})$ is finitely generated and finite dimensional.*

I have chosen to mention these properties because in each case the hypothesis and conclusion are purely algebraic, whereas the proof of each assertion goes via the global geometry of the group, as we shall see later.

In order to tease out the geometry inherent in Dehn's algorithm, we need the basic vocabulary of geometric group theory.

Geometric group theory

This is a subject close to combinatorial group theory whose distinct identity has been forged in the last ten years. It is based to a large extent on two strands of thought.

The first is that by getting a group to act on a space (preferably by isometries) one can often use knowledge of the space to elucidate the nature of the group, or else use knowledge of the group to elucidate the nature of the space.

The second idea begins with the observation (due to Cayley (1878) and Dehn (1910)) that one can regard any finitely generated group as a metric space. As the result of efforts inspired largely by the writings of Gromov, we now know that one can glean a remarkable amount of algebraic information by studying the asymptotic geometry of groups regarded as metric spaces. Gromov outlined this idea in his article *"Infinite groups as geometric objects"* for the proceedings of the ICM in Warsaw [67], and he explored it in much greater detail in his two subsequent essays *"Hyperbolic groups"* [68] and *"Asymptotic invariants of infinite groups"* [69].

1.6 The word metric

Given a group Γ with finite generating set $\mathcal{A}$, one realizes the group as a geometric object by endowing it with the left-invariant metric:

$$d(g,h) = \inf\{|w| \mid w \in F(\mathcal{A}),\ w = g^{-1}h \text{ in } \Gamma\},$$

where $|w|$ is the length of the word w.

In many contexts it is easier to work with geodesic metric spaces[4] rather than simple metric spaces. With this in mind, one often replaces the metric space (Γ, d) with its *Cayley graph.*

Cayley graphs (Gruppenbilder)

Arthur Cayley introduced these graphs in 1878 *"to study the quasi-geometry"* of (finite) groups. Dehn, who was aware of Cayley's work, made extensive use of them in his work on Fuchsian groups twenty years later.

Definition 1.7 The *Cayley graph* $C_{\mathcal{A}}(\Gamma)$ of a group Γ with respect to a generating set $\mathcal{A}$ is the metric graph whose vertices are in 1-1 correspondence with Γ and which has an edge (labelled a) of length one joining γ to γa for each $\gamma \in \Gamma$ and $a \in \mathcal{A}$.

Example 1.8 The Cayley graph of $\mathbb{Z}^2$ with respect to the standard generators is (isometric to) the 1-skeleton of the tiling of the plane by unit squares, endowed with the ℓ_1-metric.

[4] A metric space X is called a geodesic space if every pair of points $x, y \in X$ can be joined by a geodesic, i.e. a map $c : [0, D] \to X$ such that $c(0) = x$, $c(D) = y$ and $d(c(t), c(t')) = |t - t'|$ for all $t, t' \in [0, D]$.

Remark The word metric $d_{\mathcal{A}}$ on Γ is the restriction to Γ (the vertex set) of the metric on the Cayley graph $\mathcal{C}_{\mathcal{A}}(\Gamma)$.

Note that the action of Γ on itself by left multiplication gives an embedding $\Gamma \to Isom(\Gamma, d_{\mathcal{A}})$, and this action extends to an action of Γ by isometries on the Cayley graph. (The action of $\gamma_0 \in \Gamma$ by right multiplication $\gamma \mapsto \gamma\gamma_0$ is an isometry only if γ_0 lies in the centre of Γ.)

The point of introducing word metrics and Cayley graphs is to realize finitely generated groups as geometric objects. Since we are interested in groups Γ rather than pairs $(\Gamma, \mathcal{A})$, where $\mathcal{A}$ is a choice of generators, the geometry that we introduce ought to be, as far as possible, independent of the choice of generators. The key observation in this regard is that the word metrics associated to different choices of finite generating sets are Lipschitz equivalent, and the corresponding Cayley graphs are quasi-isometric in the following sense.

Definition 1.9 (quasi-isometry) Let (X_1, d_1) and (X_2, d_2) be metric spaces. A (not necessarily continuous) map $f : X_1 \to X_2$ is called a (λ, ε)-*quasi-isometric embedding* if there exist constants $\lambda \geq 1$ and $\varepsilon \geq 0$ such that for all $x, y \in X_1$

$$\frac{1}{\lambda} d_1(x, y) - \varepsilon \leq d_2(f(x), f(y)) \leq \lambda d_1(x, y) + \varepsilon.$$

If, in addition, there exists a constant $C \geq 0$ such that every point of X_2 lies in the C–neighbourhood of the image of f, then f is called a (λ, ε)-*quasi-isometry.* When such a map exists, X_1 and X_2 are said to be *quasi-isometric.*

The following remarks show in particular that quasi-isometry is an equivalence relation on any set of metric spaces.

(1) If there exists a (λ, ε)–quasi-isometry $f : X_1 \to X_2$ then there exists a (λ', ε')–quasi-isometry $f' : X_2 \to X_1$ (for some λ' and ε') and a constant $k \geq 0$ such that $d(ff'(x'), x') \leq k$ and $d(f'f(x), x) \leq k$ for all $x \in X_1$ and all $x' \in X_2$. Such a map f' is called a *quasi-inverse* for f.

(2) The composition of a (λ, ε) quasi-isometric embedding and a (λ', ε') quasi-isometric embedding is a (μ, ν) quasi-isometric embedding with $\mu = \lambda\lambda'$ and $\nu = \lambda'\varepsilon + \varepsilon'$. The composition of two quasi-isometries is a quasi-isometry.

(3) Let X be a metric space. Say that two maps $f, g : X \to X$ are equivalent, $f \sim g$, if $\sup_x d(f(x), g(x))$ is finite. Let $[f]$ denote the $\sim$ equivalence class of f. The *quasi-isometry group of* X, denoted $\mathcal{QI}(X)$, is the set of $\sim$ equivalence classes of quasi-isometries $X \to X$. Composition of maps induces a group structure on X, and any quasi-isometry $\phi : X \to X'$ induces an isomorphism $\phi_* : \mathcal{QI}(X) \to \mathcal{QI}(X')$.

Example 1.10 The inclusion of a subspace Y into a metric space X is a quasi-isometry if and only if Y is *quasi-dense*, i.e. there exists a constant $C > 0$ such that every point of X lies in the C–neighbourhood of Y. For example, the natural inclusion $\mathbb{Z} \hookrightarrow \mathbb{R}$ is a quasi-isometry (for the metric $d(x, y) = |x - y|$). In fact, the natural inclusion $(\Gamma, d_{\mathcal{A}}) \hookrightarrow \mathcal{C}_{\mathcal{A}}(\Gamma)$ of any finitely generated group as the vertex set its Cayley graph is a $(1, \frac{1}{2})$ quasi-isometry.

Proposition 1.11 *A homomorphism $\varphi : \Gamma_1 \to \Gamma_2$ of finitely generated groups is a quasi-isometry if and only if* $\ker(\phi)$ *and* $\Gamma_2/\text{im}(\varphi)$ *are both finite.*

This proposition can be proved directly without difficulty, but it can also be viewed in terms of the natural action of Γ_2 by isometries on its Cayley graph $\mathcal{C}_{\mathcal{A}}(\Gamma_2)$: via ϕ there is an induced action of Γ_1 on $\mathcal{C}_{\mathcal{A}}(\Gamma_2)$; this induced action is proper and cocompact (in the sense defined below) if and only if $\ker(\phi)$ and $\Gamma_2/\text{im}(\varphi)$ are both finite. Thus (1.11) provides us with our first illustration of the fact that quasi-isometries arise naturally from proper cocompact group actions.

A bridge between the two aspects of geometric group theory

Definition 1.12 A metric space is called *proper* if all of its closed bounded subsets are compact. When a group Γ acts on a space X by isometries, one endows the quotient (set of orbits) $\Gamma\backslash X$ with the (pseudo)metric:

$$d(\Gamma.x, \Gamma.x') = \inf\{d(y, y') \mid y \in \Gamma.x,\ y' \in \Gamma.x'\}.$$

The action of Γ is called *proper* if for each bounded set $B \subset X$, the set $\{\gamma \in \Gamma \mid \gamma.B \cap B \neq \emptyset\}$ is finite. If $\Gamma\backslash X$ is a compact metric space then the action is said to be *cocompact.*

The following result was discovered by the Russian school in the nineteen fifties (see [58],[110]). It was rediscovered by John Milnor [87] some years later in the course of his investigations into the geometry of nilpotent groups.

Proposition 1.13 (Švarc-Milnor Lemma) *Let (X, d) be a geodesic space. If Γ acts properly and cocompactly by isometries on X, then Γ is finitely generated and for any choice of basepoint $x_0 \in X$ the map $\gamma \mapsto \gamma.x_0$ is a quasi-isometry.*

Proof The idea of the proof is as follows. First we fix a point $x_0 \in X$ and a ball B of radius D centred at x_0, where D is large enough to ensure that X is the union of the translates by Γ of the ball of radius $D/3$ about x_0. (B exists because the action is cocompact.) The set $\mathcal{A} := \{a \in \Gamma \mid B \cap a.B \neq \emptyset\}$ is finite because the action of Γ is proper, and in a moment we'll show that $\mathcal{A}$ generates Γ. If two balls $\gamma.B$ and $\gamma'.B$ intersect then $\gamma' = \gamma a$ for some $a \in \mathcal{A}$, and by drawing a line labelled a from $\gamma.x_0$ to $\gamma'.x_0$ for each such γ and γ' we obtain a copy of the Cayley graph $\mathcal{C}_{\mathcal{A}}(\Gamma)$ immersed in X. We then have to check that this version of $\mathcal{C}_{\mathcal{A}}(\Gamma)$ is not too distorted. In one direction we have:

Lemma 1.14 *Let (X, d) be a metric space. Let Γ be a group with finite generating set $\mathcal{A}$ and associated word metric $d_{\mathcal{A}}$. If Γ acts by isometries on X, then for every choice of basepoint $x_0 \in X$ there exists a constant $\mu > 0$ such that $d(\gamma.x_0, \gamma'.x_0) \leq \mu\, d_{\mathcal{A}}(\gamma, \gamma')$ for all $\gamma, \gamma' \in \Gamma$.*

Proof Let $\mu = \max\{d(x_0, a.x_0) \mid a \in \mathcal{A} \cup \mathcal{A}^{-1}\}$. If $d_{\mathcal{A}}(\gamma, \gamma') = n$ then $\gamma^{-1}\gamma' = a_1 a_2 \dots a_n$ for some $a_j \in \mathcal{A} \cup \mathcal{A}^{-1}$. Let $g_i = a_1 a_2 \dots a_i$. By the triangle inequality, $d(\gamma.x_0, \gamma'.x_0) = d(x_0, \gamma^{-1}\gamma'.x_0) \leq d(x_0, g_1.x_0) + d(g_1.x_0, g_2.x_0) + \cdots +$

$d(g_{n-1}.x_0, \gamma^{-1}\gamma'.x_0)$. And for each i we have $d(g_{i-1}.x_0, g_i.x_0) = d(x_0, g_{i-1}^{-1}g_i.x_0) = d(x_0, a_i.x_0) \leq \mu$. Thus the lemma is proved. □

To complete the proof of the proposition we must show that $\mathcal{A}$ generates Γ and we must bound $d_{\mathcal{A}}(\gamma, \gamma')$ in terms of $d(\gamma.x_0, \gamma'.x_0)$. Because both metrics are Γ-invariant, it is enough to compare $d_{\mathcal{A}}(1, \gamma)$ and $d(x_0, \gamma.x_0)$.

Given $\gamma \in \Gamma$, let $c : [0, l] \to X$ be a geodesic joining x_0 to $\gamma.x_0$. We choose a partition $0 = t_0 < t_1 < \ldots < t_n = l$ of $[0, l]$ such that $d(c(t_i), c(t_{i+1})) \leq D/3$ for all i. For each t_i we choose an element $\gamma_i \in \Gamma$ such that $d(c(t_i), \gamma_i.x_0) \leq D/3$; choose $\gamma_0 = 1$ and $\gamma_n = \gamma$. Then, for $i = 1, \ldots, n$ we have $d(\gamma_i.x_0, \gamma_{i-1}.x_0) \leq D$ and hence $a_i := \gamma_{i-1}^{-1}\gamma_i \in \mathcal{A}$. Thus $\mathcal{A}$ generates Γ:

$$\gamma = \gamma_0(\gamma_0^{-1}\gamma_1)\ldots(\gamma_{n-2}^{-1}\gamma_{n-1})(\gamma_{n-1}^{-1}\gamma_n) = a_1 \ldots a_{n-1}a_n.$$

If we take as coarse a partition $0 = t_0 < t_1 < \ldots < t_n = l$ as possible with $d(c(t_i), c(t_{i+1})) \leq D/3$, then $n \leq (3/D)d(x_0, \gamma.x_0) + 1$. Since γ can be expressed as a word of length n, we get $d_{\mathcal{A}}(1, \gamma) \leq (d(x_0, \gamma.x_0) + 1)(3/D) + 1$. □

1.15 Some consequences of the Švarc-Milnor Lemma

1. $\mathbb{Z}^n$ is quasi-isometric to Euclidean n-space $\mathbb{E}^n$.
2. If G is a Lie group that any cocompact lattice $\Gamma \subset G$ is quasi-isometric to G.
3. Let T_n denote the connected metric tree in which every vertex has valence n and every edge has length 1. If $n, m \geq 3$ then T_n is quasi-isometric to T_m.

To prove (3), note that the Cayley graph of the free group F_r of rank r is T_{2r}, so the case where n and m are even follows from the fact that every finitely generated free group of rank ≥ 2 is a subgroup of finite index in F_2. Hence all such groups (and their Cayley graphs) are quasi-isometric.

In the case where n is odd, T_n is quasi-isometric to the Cayley graph of $G_{2,n} = \mathbb{Z}_2 * \mathbb{Z}_n$, which has a finitely generated free subgroup of finite index, namely the kernel of the abelianization map $G_{2,n} \to \mathbb{Z}_2 \times \mathbb{Z}_n$.

A surprising number of group-theoretical properties that one might think had nothing to do with geometry turn out to be invariants of quasi-isometry, for example finite presentability, the existence of a nilpotent subgroup of finite index [70], the solvability of the word problem [3], the existence of an abelian subgroup of finite index ([93], [32]) and the existence of representations as lattices in various semi-simple Lie groups (see [60] for references). I quote these results to exemplify the theme that geometry is lying just under the surface of much of group theory.

Classical curvature

Let us now turn our attention to the main subject of these lectures *non-positive curvature*[5].

---

[5]Terminology: In English one normally uses the term "negative curvature" to mean curvature strictly less than zero and one says "non-positive curvature" when one wishes to include zero. The French term "courbure negative" should normally be translated as "non-positive curvature"

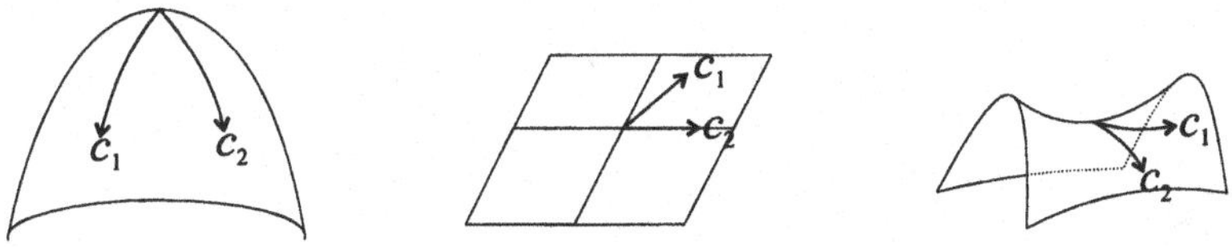

Figure 1.1. Sectional curvature κ

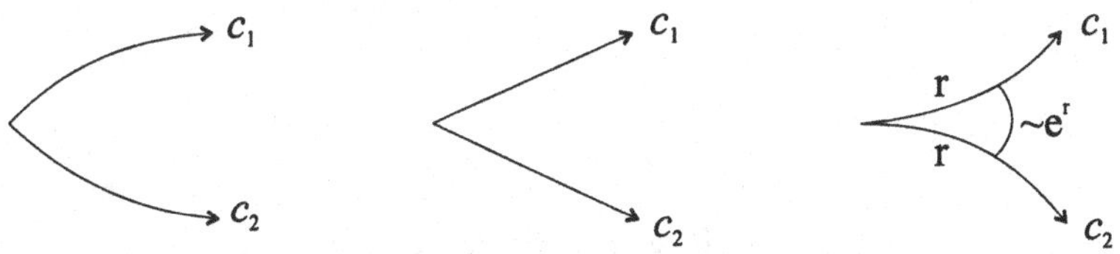

Figure 1.2. The divergence of geodesics in each case

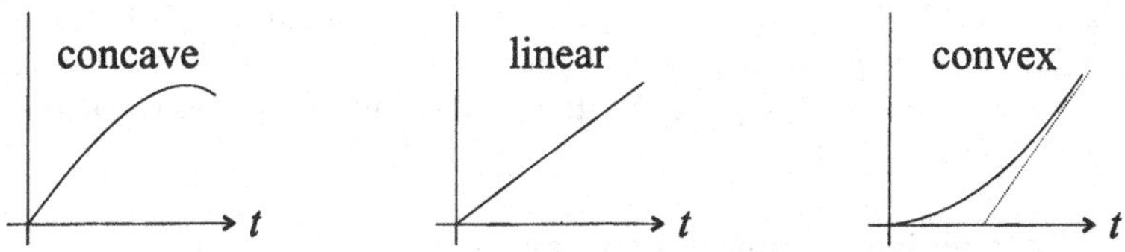

Figure 1.3. The function $t \mapsto d(c_1(t), c_2(t))$ in each case

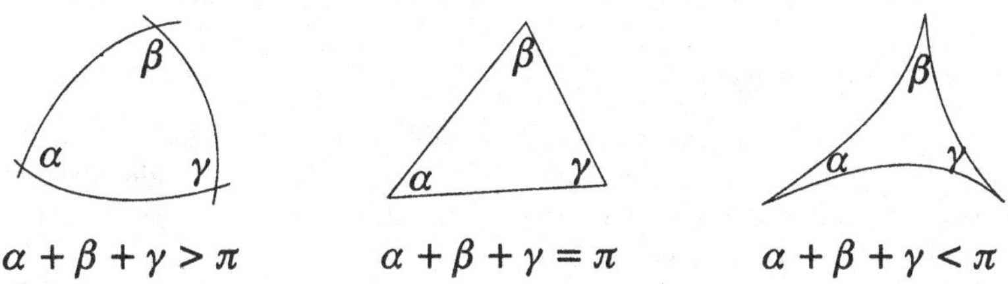

Figure 1.4. Triangles in each case

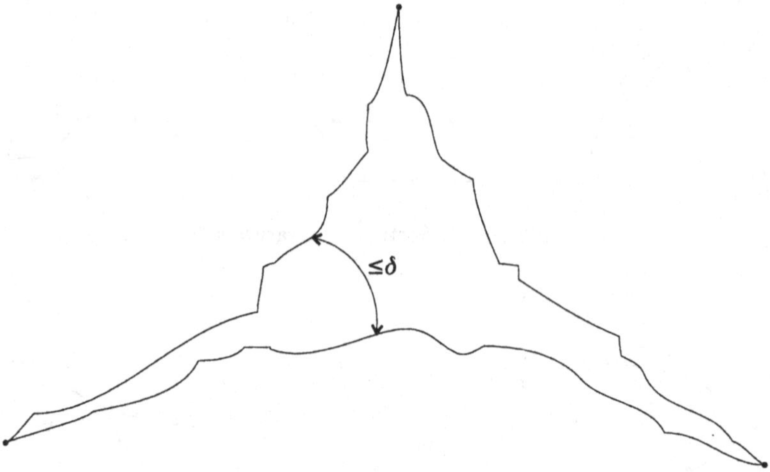

Figure 1.5. A triangle in a hyperbolic space

Figure 1.4 portrays the fact that classical (sectional) curvature can be quantified in terms of the geometry of triangles. With this in mind, in seeking generalized notions of non-positive curvature we concentrate our attention on the geometry of triangles. By definition, a *(geodesic) triangle* $\Delta = \Delta(p,q,r)$ in a metric space X consists of three points $p,q,r \in X$, its *vertices*, and a *choice* of three geodesic segments joining them, its *sides*.

Generalized notions of non-positive curvature

We shall be primarily concerned with two generalizations of non-positive curvature. The first generalization we consider is Gromov's notion of hyperbolicity, a global manifestation of strictly negative curvature. Gromov's approach is essentially to ignore the local structure of the space and to try instead to encapsulate the essence of negative curvature by requiring that triangles in the spaces under consideration behave on the large scale like geodesic triangles in real hyperbolic space $\mathbb{H}^n$.

1.16 Definition of hyperbolicity

Given $\delta > 0$, a geodesic metric space X is said to be δ-hyperbolic if for every geodesic triangle $\Delta \subseteq X$, each edge of Δ is contained in the δ-neighbourhood of the union of the other two sides (see Figure 1.5). X is said to be *hyperbolic* (in the sense of Gromov) if it is δ-hyperbolic for some δ.

A finitely generated group is said to be hyperbolic (or "word hyperbolic") if its Cayley graph is hyperbolic in the above sense. (We shall see in (1.27) that hyperbolicity is an invariant of quasi-isometry among geodesic spaces and therefore this definition of a hyperbolic group does not depend on a choice of generators.)

I should mention that there are a number of useful variations on the definition given above. (Gromov credits Rips with the definition that we have given, "the slim triangles condition".) For an account of these variations, and a proof of their equivalence, we refer to [22], [33, III.H], [49], [64] and [106]. In these lectures I shall give only a very brief introduction to the theory of hyperbolic spaces and groups. For more details see the references listed above. Gromov's original article [68] is not easy to read, but contains an abundance of fascinating ideas beyond those examined in the above references.

Exercise 1.17 Find δ_0 such that the hyperbolic plane $\mathbb{H}^2$ is δ_0-hyperbolic. (Hint: Because the area of a hyperbolic triangle is less than π, there is a bound on the size of semi-circles that one can inscribe in a triangle.)

The second notion of curvature that we shall consider in these notes is local in nature. This approach is due to A.D. Alexandrov [2] (and to a lesser extent Busemann [40]); it is the subject of my book with Haefliger [33] (see also [6]). The idea is to compare triangles in the given space to triangles in the Euclidean plane $\mathbb{E}^2$. If all small triangles in the given space are no fatter than their comparison triangles in $\mathbb{E}^2$, then the space is said to be non-positively curved. More precisely:

Let X be a metric space. A comparison triangle in $\mathbb{E}^2$ for a geodesic triangle $\Delta = \Delta(p,q,r)$ in X is a triangle $\overline{\Delta} = \Delta(\overline{p},\overline{q},\overline{r})$ in $\mathbb{E}^2$ such that $d(\overline{p},\overline{q}) = d(p,q)$, $d(\overline{q},\overline{r}) = d(q,r)$ and $d(\overline{p},\overline{r}) = d(p,r)$. Note that $\overline{\Delta}$ is unique up to isometry. A point $\overline{x} \in [\overline{q},\overline{r}]$ is called *a comparison point* for $x \in [q,r]$ if $d(q,x) = d(\overline{q},\overline{x})$. Comparison points on $[\overline{p},\overline{q}]$ and $[\overline{p},\overline{r}]$ are defined in the same way.

Definition 1.18 (CAT(0) space) Let X be a metric space. Let $\Delta \subset \mathbb{E}^2$ be a geodesic triangle in X. Let $\overline{\Delta} \subset M^2_\kappa$ be a comparison triangle for Δ. Then Δ is said to satisfy the *CAT*(0) *inequality* if for all $x,y \in \Delta$ and comparison points $\overline{x},\overline{y} \in \overline{\Delta}$,

$$d(x,y) \leq d(\overline{x},\overline{y}).$$

If X is a geodesic space and all geodesic triangles in X satisfy the CAT(0) inequality, then X is said to be a CAT(0) space.[6]

A metric space X is said to be *non-positively curved* (in the sense of Alexandrov) if it is locally a CAT(0) space, i.e. for every $x \in X$ there exists $r_x > 0$ such that the ball $B(x,r_x)$, endowed with the induced metric, is a CAT(0) space. Henceforth, whenever I say that a space is non-positively curved, I shall mean this in the sense of Alexandrov.

Remarks (1) Standard comparison theorems in Riemannian geometry show that a complete Riemannian manifold is non-positively curved in the above sense if and only if all of its sectional curvatures are non-positive (see [48] or [33, II.1A]).

(2) By taking comparison triangles in the hyperbolic plane instead of the Euclidean plane, one obtains the notion of a CAT(−1) space and *curvature* ≤ -1.

[6]This terminology is due to Gromov. The initials are in honour of Cartan, Alexandrov and Toponogov.

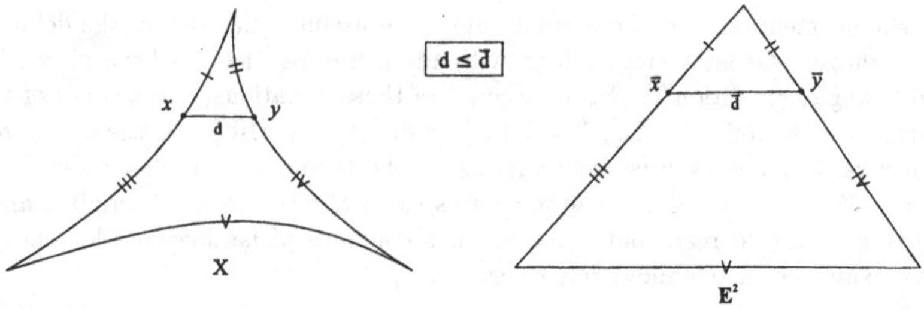

Figure 1.6. The CAT(0) inequality

More generally, by taking comparison triangles in the plane of constant sectional curvature κ one gets the notion of a CAT(κ) space and *curvature* $\leq \kappa$.

By reversing the direction of the inequality in (1.18) one gets the notion of *curvature* ≥ 0. The study of spaces with lower curvature bounds is also interesting (see [38]), but has less to do with group theory.

Lemma 1.19 *CAT*(-1) *spaces are* δ_0*-hyperbolic, where* δ_0 *is as in (1.17).*

In the remainder of this lecture I shall concentrate on δ-hyperbolic spaces. First though, to whet the reader's appetite for CAT(0) spaces, let me state a group-theoretic result that I will explain later in the context of CAT(0) spaces.

Let $\mathcal{P} = \langle \mathcal{A} \mid \mathcal{R} \rangle$ be a presentation for Γ, with $\mathcal{R} \neq \emptyset$. Regard the elements of $\mathcal{R}$ as cyclic words and consider the following graph $W(\mathcal{P})$: the vertex set of $W(\mathcal{P})$ is $\{a, \overline{a} \mid a \in \mathcal{A}\}$; the edges of $W(\mathcal{P})$ are in 1–1 correspondence with the subwords of length two in the cyclic words $r \in \mathcal{R}$ — for each occurrence of $(ab^{-1})^{\pm 1}$ there is an edge connecting a to b, and for each occurrence of $(ab)^{\pm 1}$ there is an edge connecting a to $\overline{b}$.

Theorem 1.20 *If each* $r \in \mathcal{R}$ *has length at least* 4 *and* $W(\mathcal{P})$ *contains no circuits of length less than* 4, *then*

1. Γ *is torsion-free;*
2. Γ *and all of its f.p. subgroups have solvable word and conjugacy problems;*
3. $H^n(\Gamma, \mathbb{Z}) = 0$ *for* $n \geq 3$*; and*
4. *if* Γ *contains a solvable subgroup of finite index, then* $|\mathcal{A}| = 2$, $|\mathcal{R}| = 1$ *and* $\Gamma \cong \mathbb{Z}^2$ *or* $\mathbb{Z} \rtimes \mathbb{Z}$.

Dehn's algorithm: the word problem and negative curvature

We now have a sufficient vocabulary to explain how negative curvature is hidden in the obvious solution to the word problem that one gets from a Dehn presentation.

Theorem 1.21 (Gromov, Cannon, Dehn) *A group is hyperbolic if and only if it admits a (finite) Dehn presentation.*

Max Dehn proved that Fuchsian groups admit Dehn presentations [53]. Jim Cannon extended Dehn's theorem to include the fundamental groups of all compact negatively curved manifolds [42, 43]. In the generality stated above, the theorem is due to Gromov [68]. (Cannon gave a proof in [43] and alternative proofs can be found elsewhere, e.g. [106].)

Proof [an outline] A detailed proof requires more elements of the geometry of hyperbolic spaces than we shall establish here, but we indicate the main points.

First suppose that the Cayley graph of Γ with respect to some finite generating set $\mathcal{A}$ is δ-hyperbolic. By studying the geometry of quasi-geodesics in hyperbolic spaces one sees (Corollary 1.30) that every closed loop in the Cayley graph $\mathcal{C}_{\mathcal{A}}(\Gamma)$ contains a subpath of length at most 8δ that is not a geodesic (and whose endpoints are at vertices). It follows that we obtain a Dehn presentation $\langle \mathcal{A} \mid \mathcal{R} \rangle$ for Γ by defining $\mathcal{R}$ to be the set of words $u_i v_i^{-1}$, where u_i varies over all words of length at most 8δ in the generators and their inverses and v_i is a word that is equal to u_i in Γ and whose length $|v_i|$ is less than $|u_i|$.

The proof of the converse is less direct; it goes via the linear isoperimetric inequality. □

Definition 1.22 (Isoperimetric inequalities) A finitely presented group $\Gamma = \langle \mathcal{A} \mid \mathcal{R} \rangle$ is said to satisfy a linear *isoperimetric inequality* if there exists a constant K such that every word w in the kernel of the natural map $F(\mathcal{A}) \to \Gamma$ can be expressed in the free group $F(\mathcal{A})$ as a product of at most $K|w|$ conjugates of relators $r \in \mathcal{R}^{\pm 1}$.

Quadratic, polynomial and (sub)exponential isoperimetric inequalities are all defined similarly. In the quadratic case, for example, the requirement is that w can be expressed as a product of at most $K|w|^2$ conjugates of relators.

One can show that this definition (but not the value of the constant K) is independent of the given finite presentation of Γ. Indeed the optimal type of isoperimetric inequality satisfied by a finitely presented group is an invariant of quasi-isometry [3].

It is obvious that if a finitely presented group has a Dehn presentation then the group satisfies a linear isoperimetric inequality. Thus the following result of Gromov completes the proof of (1.21).

Theorem 1.23 *If a group satisfies a linear isoperimetric inequality then it is hyperbolic.*

An elementary proof of this result, due to Hamish Short, can be found in an appendix to [63]. (See also Lysenok [82].)

Remark Gromov [67] found a useful companion for Theorem 1.23: if a group satisfies a sub-quadratic isoperimetric inequality then it satisfies a linear isoperimetric inequality. For a proof see [23], [90] or [95].

In Lecture 4 we shall discuss the conjugacy problem and the isomorphism problem in the class of hyperbolic groups.

Quasi-geodesics in δ-hyperbolic spaces

The essence of a geodesic space is encoded in the geometry of its geodesics. What do we wish to know about geodesics in hyperbolic spaces? Well, we are primarily interested in groups, and a finitely generated group is only well-defined as a geometric object up to quasi-isometry. Moreover, our main tool for relating the geometry of groups to the geometry of the spaces on which they act (1.13) also concerns only the quasi-isometry type of the group. Thus we are led to ask which properties of geodesics are preserved by quasi-isometries.

Definition 1.24 (quasi-geodesic) A (λ, ε) *quasi-geodesic* in a metric space X is a (λ, ε) quasi-isometric embedding $c : I \to X$, where I is an interval of the real line (bounded or unbounded) or $\mathbb{Z}$. More explicitly,

$$\frac{1}{\lambda}|t - t'| - \varepsilon \leq d(c(t), c(t')) \leq \lambda\,|t - t'| + \varepsilon$$

for all $t, t' \in I$. If $I = [a, b]$ then $c(a)$ and $c(b)$ are called the endpoints of c.

In general quasi-geodesics need not resemble geodesics.

Exercise 1.25 Consider the spiral $[0, \infty) \to \mathbb{E}^2$ given in polar coordinates by $t \mapsto (t, \log(1 + t))$. Prove that this is a quasi-geodesic.

In hyperbolic spaces quasi-geodesics approximate geodesics closely — this is a key difference between Euclidean and hyperbolic geometry.

Theorem 1.26 (Stability of quasi-geodesics) *For all $\delta > 0, \lambda \geq 1, \varepsilon \geq 0$ there exists a constant $R = R(\delta, \lambda, \varepsilon)$ such that the following statement holds. If X is a δ-hyperbolic geodesic space, c is a (λ, ε) quasi-geodesic in X and $[p, q]$ is a geodesic segment joining the endpoints of c, then $[p, q]$ lies in the R-neighbourhood of the image of c and vice versa.*

It follows from this theorem that a group is hyperbolic if and only if for every $\lambda, \varepsilon > 0$, the (λ, ε) quasi-geodesic triangles in its Cayley graph are uniformly slim. This property is obviously preserved under quasi-isometry, therefore:

Corollary 1.27 *Hyperbolicity is an invariant of quasi-isometry amongst geodesic spaces (in particular Cayley graphs, hence finitely generated groups). More generally, if X and Y are geodesic spaces, X is hyperbolic and $Y \hookrightarrow X$ is a quasi-isometric embedding, then Y is hyperbolic.*

The following result gives a local criterion for recognising quasi-geodesics.

Definition 1.28 Let X be a metric space and fix $k > 0$. A path $c : [a, b] \to X$ is said to be a *k-local geodesic* if $d(c(t), c(t')) = |t - t'|$ for all $t, t' \in [a, b]$ with $|t - t'| \leq k$.

Theorem 1.29 (8δ-local geodesics are q.g.) *Let X be a δ-hyperbolic geodesic space and let $c : [a, b] \to X$ be a 8δ-local geodesic. Then c is a (λ, ε)-quasi-geodesic, where λ depends only on δ and $\varepsilon < 8\delta$.*

For a proof see [49], [64], [106] or [33, III.H].

Corollary 1.30 *If X is a δ-hyperbolic geodesic space, then no path $c : [a, b] \to X$ with $a \neq b$ and $c(a) = c(b)$ is an 8δ-local geodesic.*

In Lecture 4 we shall discuss the conjugacy problem and the isomorphism problem in the class of hyperbolic groups.

Quasification and finite subgroups of hyperbolic groups

I wish to illustrate how profitable it can be to transport classical arguments from the theory of non-positively curved manifolds into the world of hyperbolic and related groups. The key to all such adaptations is that one must find an appropriate way to "quasify" the key role that non-positive (or negative) curvature is playing in the classical setting; one then attempts to encapsulate a robust form of the salient feature of curvature in the more relaxed world of hyperbolic spaces. The first example of this is the very definition of δ-hyperbolic space: instead of insisting that geodesic triangles satisfy the CAT(-1) inequality, which imposes tight control on the local structure of the space, one observes that many important aspects of the global geometry of CAT(-1) spaces rely only on the fact that the triangles are uniformly slim – thus one arrives at a notion of negative curvature (namely δ-hyperbolicity) that is much more robust, indeed quasi-isometry invariant.

To illustrate how this general philosophy of quasification can be useful, we prove the following:

Theorem 1.31 *If a finitely generated group Γ is hyperbolic, then it contains only finitely many conjugacy classes of finite subgroups.*

The proof is adapted from the following classical argument of Elie Cartan [44]. Suppose that we have a finite (more generally, compact) group of isometries G of a complete Riemannian manifold X of non-positive curvature. Fix $x \in X$ and consider $G.x \subset X$. This is a bounded set, and in a complete Riemannian manifold of non-positive curvature there are various intrinsic notions of "centre" that one can associate to bounded sets. We focus on the *circumcentre* (this is not the notion of centre used by Cartan, but it was used by Bruhat and Tits [37] in a later adaptation of Cartan's argument) — the circumcentre of a bounded set is the centre of the unique smallest ball that contains that set. (See [33, II.2] for a proof that there is indeed a unique such ball.)

Since $G.x$ is G-invariant, so is its centre c, and hence G fixes a point of X. If G is a subgroup of a group Γ that is acting cocompactly on X, say $X = \Gamma.K$ where K is compact, then $\gamma.c \in K$ for some $\gamma \in \Gamma$ and hence G is conjugate in Γ to a subgroup of the stabilizer of some point of K. If the action of Γ is proper, then the union of such stabilizers is finite and therefore Γ has only finitely many conjugacy classes of finite subgroups.

The existence of a circumcentre for bounded sets was the key feature of the above argument. By quasifying the usual proof of the existence of such centres in non-positively curved spaces we get:

Lemma 1.32 (quasi-centres) *Let X be a δ-hyperbolic geodesic space. Let $Y \subset X$ be a non-empty bounded subspace. Let $r_Y = \inf\{\rho \geq 0 \mid Y \subset B(x,\rho)$, some $x \in X\}$. For all $\varepsilon > 0$, the set $C_\varepsilon(Y) = \{x \in X \mid Y \subseteq B(x, r_Y + \varepsilon)\}$ has diameter less than $(4\delta + 2\varepsilon)$.*

Proof Given $x, x' \in C_\varepsilon(Y)$, let m be the midpoint of a geodesic segment $[x, x']$. For each $y \in Y$ we consider a geodesic triangle with vertices x, x', y and with $[x, x']$ as one of the sides. As X is δ-hyperbolic, m is within a distance δ of some $p \in [x,y] \cup [x',y]$; suppose that $p \in [x,y]$. Then, since $d(x,m) = d(x,x')/2$ and $d(p,x) \geq d(x,m) - \delta$, we have $d(y,p) = d(y,x) - d(p,x) \leq d(y,x) + \delta - d(x,x')/2$, and hence

$$d(y,m) \leq r_Y + \varepsilon + 2\delta - \frac{1}{2}d(x,x').$$

But $d(y,m) \geq r_Y$ for some $y \in Y$, hence $\varepsilon + 2\delta - \frac{1}{2}d(x,x') \geq 0$. □

Proof[of Theorem 1.31] Let Γ be a hyperbolic group and consider the natural action of Γ on its Cayley graph. Let $H \subset \Gamma$ be a finite subgroup, and let $C_1(H)$ be as in the lemma. $C_1(H)$ contains at least one vertex and the action of H leaves $C_1(H)$, and hence its vertex set, set-wise invariant. If x is one of the vertices of $C_1(H)$, then $x^{-1}Hx$ leaves $x^{-1}C_1(H)$ invariant. But $x^{-1}C_1(H)$ is a set of diameter less than $(4\delta + 2)$ containing the identity 1, and it contains $x^{-1}Hx = (x^{-1}Hx).1$. Thus every finite subgroup of Γ is conjugate to a subset of the ball of radius $(4\delta+2)$ about the identity. □

The proof given above does not appear in the literature but I believe that it is known to a number of researchers in the field, in particular Brian Bowditch and Ilya Kapovich. Alternative proofs were given by Ol'shanskii and Bogopolskii and Gerasimov [21].

Quasiconvexity and abelian subgroups of hyperbolic groups

A subspace C of a geodesic space X is said to be *convex* if for all $x, y \in C$, each geodesic joining x to y is contained in C. Following Gromov [68], one can quasify this notion: a subspace C of a geodesic metric space X is said to be *quasiconvex* if there exists a constant $k > 0$ such that for all $x, y \in C$, some geodesic joining x to y is contained in the k-neighbourhood of C.

Exercise 1.33 Let G be a group with finite generating set $\mathcal{A}$. Let $H \subset G$ be a subgroup. Prove that if H is a quasiconvex subset of the Cayley graph $\mathcal{C}_{\mathcal{A}}(G)$ then H is finitely generated, and $H \hookrightarrow G$ is a quasi-isometric embedding (with respect to any choice of word metrics).

In the circumstances of the above exercise one says that H is a *quasiconvex subgroup* of G. In the light of (1.27) we have:

Proposition 1.34 *The quasiconvex subgroups of hyperbolic groups are hyperbolic.*

The following results are due to Gromov [68]. The ideas underlying them have been used by other authors, particularly Hamish Short, to obtain similar results in wider contexts (see [59], [63], [4], [107]).

Proposition 1.35 *Let Γ be a hyperbolic group.*
1. *The centralizer $C(\gamma)$ of every $\gamma \in \Gamma$ is a quasiconvex subgroup.*
2. *If the subgroups $H_1, H_2 \subset \Gamma$ are quasiconvex then so is $H_1 \cap H_2$.*

Corollary 1.36 *Suppose that Γ is hyperbolic and that $\gamma \in \Gamma$ has infinite order. Then:*
1. *The map $\mathbb{Z} \to \Gamma$ given by $n \mapsto \gamma^n$ is a quasi-geodesic.*
2. *$\langle\gamma\rangle$ has finite index in $C(\gamma)$. In particular Γ does not contain $\mathbb{Z}^2$.*

Proof $C(\gamma)$ is quasiconvex, hence finitely generated and hyperbolic. By intersecting the centralizers of a generating set for $C(\gamma)$, we see that the centre $Z(C(\gamma))$ is also finitely generated and hyperbolic. It is easy to see that a finitely generated abelian group is hyperbolic if and only if it contains a cyclic subgroup of finite index. Hence $Z(C(\gamma))$ contains $\langle\gamma\rangle$ as a subgroup of finite index, and since $Z(C(\gamma)) \hookrightarrow C(\gamma)$ and $C(\gamma) \hookrightarrow \Gamma$ are quasi-isometric embeddings (1.33), so is $\langle\gamma\rangle \hookrightarrow \Gamma$. This proves (1).

Fix a finite generating set with respect to which Γ is δ-hyperbolic. Let d be the associated word metric. Define the *translation number* $\tau(\gamma)$ to be $\lim_{n\to\infty} d(1,\gamma^n)/n$. (This limit exists because the function $n \mapsto d(1,\gamma^n)$ is sub-additive.) It follows from (1) that $\tau(\gamma) > 0$.

$\tau(\gamma^r) = r\,\tau(\gamma)$ and $\tau(x^{-1}\gamma x) = \tau(\gamma)$ for all $x \in \Gamma$. Thus, replacing γ by a suitable power if necessary, we may assume that γ is not conjugate to any element a distance 4δ or less from the identity. I claim that it follows from this that if $[\gamma, g] = 1$ then g lies within a distance $K := d(1,\gamma) + 4\delta$ of $\langle\gamma\rangle$. For suppose not, choose g so that $d(g, \langle\gamma\rangle) = d(g,1) > K$, and consider a geodesic quadrilateral Q in the Cayley graph of Γ with vertices $\{1, \gamma, g, g\gamma = \gamma g\}$. The side of Q joining γ to γg is chosen to be the translate by γ of the side joining 1 to g; we write g_t to denote the point a distance t from 1 on this side.

δ-hyperbolicity (1.16) implies that g_t lies within a distance 2δ of a point on one of the other three sides of Q. If $d(1,g) - d(1,\gamma) - 2\delta > t > 2\delta$, then this other point must be on the side joining γ to γg; say it is $\gamma g_{t'}$. Since $t' = d(\gamma g_{t'}, \langle\gamma\rangle)$, we have $t' \le t + 2\delta$. Similarly $t \le t' + 2\delta$, and hence $d(g_t, \gamma g_t) \le 4\delta$. But this implies that $d(1, g_t^{-1}\gamma g_t) \le 4\delta$, contrary to our assumption on γ. This proves (2). □

1.37 Translation numbers are rational

A remarkable result of Gromov [68], an elegant proof of which has been given by Delzant [55], states that associated to each word metric on a hyperbolic group Γ

there is an integer M such that each translation number $\tau(\gamma)$, $\gamma \in \Gamma$ (as defined in the preceding proof) is a rational number whose denominator divides M.

1.38 Semihyperbolic groups

One obvious omission from these lectures is the theory of semihyperbolic groups as laid out by myself and Alonso in [4] (following the influence of [68], [59] and [63]). In this theory one seeks to encapsulate the global essence of non-positive curvature (as opposed to strictly negative curvature) by a suitable analogue of the δ-hyperbolic condition. In the theory of semihyperbolic groups quasi-convexity again plays a key role.

Lecture 2: The geometry of spaces

In this lecture I shall present those elements of the basic theory of CAT(0) spaces that are most useful from the point of view of combinatorial and geometric group theory. All of this material is taken from my book with André Haefliger [33]. My main aim is to illustrate the close connection between the geometry of CAT(0) spaces and the structure of the groups which act on them.

A key feature in this theory is the existence of a local-to-global theorem for non-positively curved[7] spaces (Theorem 2.2). This is important from the point of view of group theory because in many cases it provides a remedy to the problems caused by the fact that the rather dull local structure of Cayley graphs precludes any local-to-global analysis of their geometry: if one can realise a group Γ as the fundamental group of a compact non-positively curved space X, then one can analyze Γ via its action by deck-transformations on the universal covering $\widetilde{X}$, and we have the option of studying Γ laid out as the set of translates of a basepoint in $\widetilde{X}$ rather than as the set of vertices of the Cayley graph. Theorem 2.2 tells us that $\widetilde{X}$ is a CAT(0) space. The action of Γ on $\widetilde{X}$ is proper, cocompact and by isometries. Thus if by a study of CAT(0) spaces we can deduce facts about groups which act on them in this manner, then we will have an approach to understanding a (hopefully wide) class of groups.

In order to motivate such a study let me list some of the benefits of getting groups to act on CAT(0) spaces.

Theorem 2.1 (A summary of results) *Let Γ be a group that acts properly and cocompactly by isometries on a CAT(0) space X. Then:*

1. *Γ is finitely presented.*
2. *Γ satisfies a quadratic isoperimetric inequality (and hence has a solvable word problem).*
3. *Γ has a solvable conjugacy problem.*
4. *Γ has only finitely many conjugacy classes of finite subgroups.*
5. *Every solvable subgroup of Γ is virtually abelian.*

[7] In all that follows "non-positive curvature" will be meant in the sense of (1.18).

6. *Every abelian subgroup of* Γ *is finitely generated.*
7. *If* Γ *is torsion-free, then it is the fundamental group of a compact cell complex whose universal cover is contractible.*
8. Γ *does not contain subgroups of the form* $\langle x, y \mid x^{-1}y^p x = y^q\rangle$ *with* $|p| \neq |q|$.
9. *If* $A \cong \mathbb{Z}^n$ *is central in* Γ *then there exists a subgroup of finite index in* Γ *that contains* A *as a direct factor.*

The class of groups which act properly and cocompactly by isometries on CAT(0) spaces is closed under the following operations:

(a) *direct products,*

(b) *free products with amalgamation along virtually cyclic subgroups.*

All of these results are proved in [33]. I will give a sketch of the proof of parts (1) to (9) in the course this lecture and lecture 4. Part (a) is an easy exercise and (b) is proved in (3.11).

Local-to-global

The following theorem is a variation on a result of M. Gromov [68], following an idea of A. D. Alexandrov. A detailed proof was given by W. Ballmann in the locally compact case ([64], Chap.10). S. Alexander and R. L. Bishop [1] proved the stronger result stated below under the additional hypothesis that $\widetilde{X}$ is a geodesic space. The form of the result stated here is proved in [33, II.4].

The metric d on a geodesic space X is said to be *convex* if given any pair of geodesic paths $c, c' : [0,1] \to X$ parametrized proportional to the arc length, the following inequality holds for all $t \in [0,1]$:

$$d(c(t), c'(t)) \leq (1-t)\, d(c(0), c'(0)) + t\, d(c(1), c'(1)).$$

The metric is said to be *locally convex* if every point has a neighbourhood on which the induced metric is convex.

Theorem 2.2 (Generalized Cartan-Hadamard Theorem) *Let* X *be a complete connected metric space.*

1. *If the metric on* X *is locally convex, then the induced length metric on the universal covering* $\widetilde{X}$ *is (globally) convex. (In particular there is a unique geodesic segment joining each pair of points in* $\widetilde{X}$ *and geodesic segments in* $\widetilde{X}$ *vary continuously with their endpoints.)*
2. *If* X *is of curvature* $\leq \kappa$, *where* $\kappa \leq 0$, *then* $\widetilde{X}$ *(with the induced length metric) is a CAT*(κ) *space.*

The second part of Theorem 2.2 is deduced from the first by a 'patchwork' process indicated in the following picture. Following Alexandrov [2], one proves a gluing lemma: given a geodesic triangle $\Delta(p,q,r)$ and a point m on the side $[p,r]$, if the triangles $\Delta(p,m,r)$ and $\Delta(q,m,r)$ both satisfy the CAT(κ) inequality, then so does $\Delta(p,q,r)$. One applies the gluing lemma repeatedly, first to fill out the

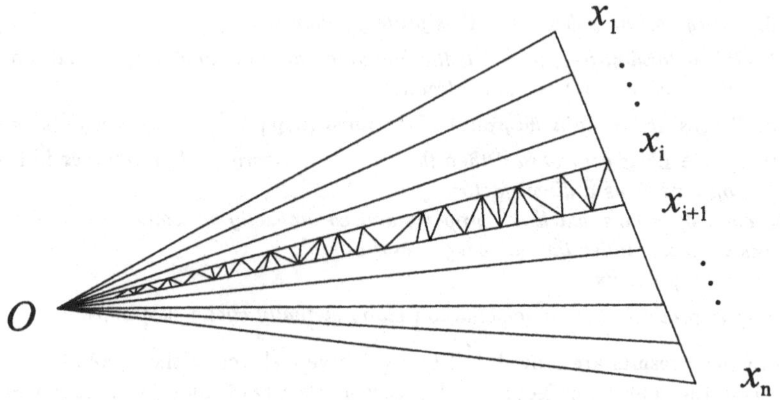

Figure 2.1. Alexandrov's Patchwork

narrow subtriangles in Figure 2.1, such as $\Delta_i = \Delta(0, x_i, x_{i+1})$, then to assemble the Δ_i into the given (large) triangle.

We note three consequences of Theorem 2.2 that are important from the point of view of group theory.

Corollary 2.3 *If X is a compact non-positively curved space, then:*

1. *The universal cover of X is contractible.*
2. *$\Gamma = \pi_1 X$ has a finite $K(\Gamma, 1)$.*
3. *If Y is a compact length space and $f : Y \to X$ is a local isometry, then $f_* : \pi_1 Y \to \pi_1 X$ is an injection.*

(1) is proved in (2.5). In order to deduce (2) from (1) one has to show that X has the homotopy type of a finite simplicial complex — see [33, II.5]. (3) is proved in [33, II.4].

Convexity and its consequences

One of the most basic properties determining the nature of CAT(0) spaces is the convexity of the metric.

Proposition 2.4 *If X is a CAT(0) space, then the metric on X is convex (in the sense defined above).*

Proof First assume that we are given geodesics $c, c' : [0,1] \to X$ with $c(0) = c'(0)$. Consider a comparison triangle $\overline{\Delta} \subset \mathbb{E}^2$ for $\Delta(c(0), c(1), c'(1))$. Given $t \in [0,1]$, elementary Euclidean geometry tells us that $d(\overline{c(t)}, \overline{c'(t)}) = t\, d(\overline{c(1)}, \overline{c'(1)}) = t\, d(c(1), c'(1))$. And by the CAT(0) inequality, $d(c(t), c'(t)) \le d(\overline{c(t)}, \overline{c'(t)})$. Hence we obtain $d(c(t), c'(t)) \le t\, d(c(1), c'(1))$.

In the general case, we introduce the linearly reparametrized geodesic $c'' : [0,1] \to X$ with $c''(0) = c(0)$ and $c''(1) = c'(1)$. By applying the preceding special case, first to c and c'' and then to c' and c'' with reversed orientation, we obtain: $d(c(t), c''(t)) \leq t\, d(c(1), c''(1))$ and $d(c''(t), c'(t)) \leq (1-t)\, d(c''(0), c'(0))$. Hence,

$$d(c(t), c'(t)) \leq d(c(t), c''(t)) + d(c''(t), c'(t)) \leq t\, d(c(1), c'(1)) + (1-t)\, d(c(0), c'(0)),$$

as required. □

Remark 2.5 Let X be a CAT(0) space, fix $x_0 \in X$ and define a homotopy $H : X \times [0,1] \to X$ by $H(x,t) = c_x(t)$, where $c_x : [0,1] \to X$ is a constant-speed parametrization of the geodesic connecting x_0 to x. It follows from the above proposition that H is a continuous map. Thus X is contractible.

I shall not include many proofs in this lecture (all relevant proofs can be found in Part II of [33]). The above proof is given so as to illustrate the nature of elementary arguments using comparison triangles. One of the hallmarks of the theory of CAT(0) spaces is that by concatenating many such elementary arguments one arrives at non-trivial results remarkably quickly. This remark applies in particular to the following results.

Recall that a subset C of a CAT(0) spaces X is said to be *convex* if the geodesic in X joining each pair of points $x, y \in C$ lies entirely in C.

Proposition 2.6 *Let X be a complete space.*

1. *If $C \subseteq X$ is closed and convex, then for every $x \in X$ there exists a unique point $\pi(x) \in C$ such that $d(x, \pi(x)) = d(x, C) := \inf_{y \in C} d(x,y)$. Moreover the map π does not increase distances.*
2. *Every bounded subset of X has a unique circumcentre.*

Following the discussion in (1.31), from 2.6(2) we deduce:

Corollary 2.7 1. *If a group Γ acts by isometries on a complete CAT(0) space X, then every finite subgroup of Γ fixes a point of X.*

2. *If the action of Γ is proper and cocompact, then Γ contains only finitely many conjugacy classes of finite subgroups.*
3. *If Y is a compact non-positively curved space, then $\pi_1 Y$ is torsion-free.*

The action of the fundamental group of any space on the universal cover is free, so part (3) of the above corollary follows from part (2) and Theorem 2.2.

Flat subspaces of CAT(0) spaces

The two most classical examples of CAT(0) spaces are Euclidean space $\mathbb{E}^n$ and hyperbolic space $\mathbb{H}^n$. As every undergraduate knows, the flavours of the geometry enjoyed by these two spaces are quite distinct. Given an arbitrary CAT(0) space, then, one might wonder whether it is essentially hyperbolic in nature, or more Euclidean. More precisely, one might ask whether it contains significant flat

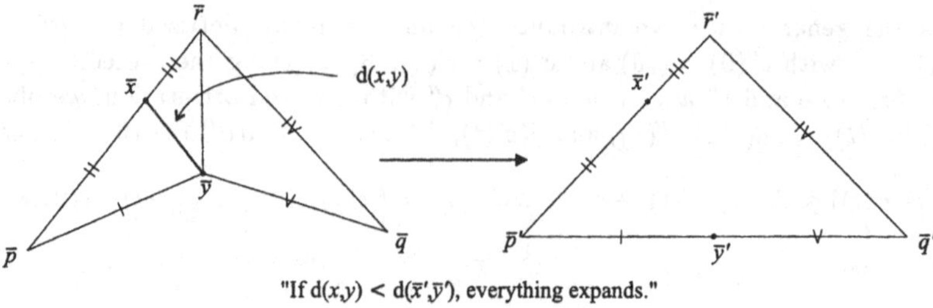

Figure 2.2. The idea behind the Flat triangle lemma

(Euclidean) subspaces, and if it does not then can one make precise the idea that it is morally speaking negatively curved? We shall see that in the case of cocompact CAT(0) spaces such a dichotomy does indeed exist (2.10).

Our search begins with a remarkable observation of Alexandrov: given a geodesic triangle Δ in a CAT(0) space X, compare Δ with its comparison triangle $\overline{\Delta} \subset \mathbb{E}^2$ as in (1.18); if there is *any* non-trivial equality $d(x, y) = d(\overline{x}, \overline{y})$, then one can fill Δ to get an isometrically embedded triangular Euclidean disc in X.

To state this more precisely, we define the *convex hull* of a subset A of a geodesic space X to be the intersection of all of the convex subspaces of X containing A. (Despite what our Euclidean intuition may suggest, the convex hull of a geodesic triangle in a CAT(0) space may not be 2-dimensional. For example, in the complex hyperbolic plane $\mathbb{C}\mathrm{H}^2$ the convex hull of three points in general position is 4-dimensional — see [33, II.10].)

Proposition 2.8 (Flat triangle lemma) *Let $\Delta(p, q, r)$ be a geodesic triangle in a CAT(0) space, and let $\Delta(\overline{p}, \overline{q}, \overline{r}) \subset \mathbb{E}^2$ be its comparison triangle. If there exist points $\overline{x}$ in the interior of $[\overline{p}, \overline{r}]$ and $\overline{y} \in [\overline{p}, \overline{q}]$ with $\overline{y} \neq \overline{p}$ such that $d(x, y) = d(\overline{x}, \overline{y})$, then the triangle $\Delta(p, q, r)$ is flat; more precisely, the convex hull of Δ in X is isometric to the convex hull of $\overline{\Delta}$ in $\mathbb{E}^2$.*

A *geodesic line* in a metric space X is, by definition, a map $c : \mathbb{R} \to X$ such that $d(c(t), c(t')) = |t - t'|$ for all $t, t' \in \mathbb{R}$. Two geodesic lines c, c' in X are said to be *asymptotic* if there exists a constant K such that $d(c(t), c'(t)) \leq K$ for all $t \in \mathbb{R}$. We remind the reader that if a function $\mathbb{R} \to \mathbb{R}$ is convex and bounded, then it is constant.

Theorem 2.9 (The flat strip theorem) *Let X be a CAT(0) space, and let $c : \mathbb{R} \to X$ and $c' : \mathbb{R} \to X$ be geodesic lines in X. If c and c' are asymptotic, then the convex hull of $c(\mathbb{R}) \cup c'(\mathbb{R})$ is isometric to a flat strip $\mathbb{R} \times [0, D] \subset \mathbb{E}^2$.*

Proof [33] Let π be the projection of X to the closed convex subspace $c(\mathbb{R})$. By reparametrizing if necessary, we may assume that $c(0) = \pi(c'(0))$.

The function $t \mapsto d(c(t), c'(t))$ is convex, non-negative and bounded, hence constant, equal to D say. Similarly, for all $a \in \mathbb{R}$, the function $t \mapsto d(c(t + a), c'(t))$ is constant. In particular $d(c(a+t),c'(t)) = d(c(a),c'(0)) \geq d(c(0),c'(0))$, and hence $\pi(c'(t)) = c(t)$ for all t.

Given $t < t'$, we consider the quadrilateral in X which is the union of the geodesic segments $[c(t),c(t')]$, $[c(t'),c'(t')]$, $[c'(t'),c'(t)]$ and $[c'(t),c(t)]$. We divide this into two geodesic triangles by introducing the diagonal $[c(t),c'(t')]$. Let $t_m = (t+t')/2$. By the triangle inequality, $d(c(t_m),c'(t_m)) \leq d(c(t_m),m) + d(m,c'(t_m))$, where m is the midpoint of $[c(t),c'(t')]$. And by hypothesis, $d(c(t_m),c'(t_m)) = D$.

Consider the parallelogram in the Euclidean plane obtained by joining comparison triangles $\overline{\Delta}(c(t),c'(t'),c'(t))$ and $\overline{\Delta}(c(t),c'(t'),c(t'))$ along the side $[\overline{c(t)},\overline{c'(t')}]$. We have $d(\overline{c(t_m)},\overline{c'(t'_m)}) = d(\overline{c(t_m)},\overline{m}) + d(\overline{m},\overline{c'(t'_m)}) = D$. But by the CAT(0) inequality, $d(\overline{c(t_m)},\overline{m}) \geq d(c(t_m),m)$ and $d(\overline{m},\overline{c'(t'_m)}) \geq d(m,c'(t'_m))$. Thus we must have equality throughout, and by the flat triangle lemma, the convex hull of $\{c(t),c'(t),c(t'),c'(t')\}$ is isometric to a Euclidean rectangle.

Sending $t \to -\infty$ and $t' \to \infty$, it follows that the map $\mathbb{R} \times [0,D] \to X$ which sends (t,s) to the point a distance s from $c(t)$ on the geodesic segment $[c(t),c'(t)]$ is an isometry onto the convex hull of $c(\mathbb{R}) \cup c'(\mathbb{R})$. □

By means of a (somewhat subtle) limiting and compactness argument based on the Flat triangle lemma, one can prove the following result. This result was proved in the setting of smooth manifolds by Eberlein [57]. Gromov pointed out that it should work in arbitrary CAT(0) spaces and I provided a proof of Gromov's suggestion [27]. Bowditch [24] provided an alternative proof that works in even greater generality.

Theorem 2.10 (Flat plane theorem) *Let X be a proper CAT(0) space that admits a cocompact action by a group of isometries. Then X is δ-hyperbolic for some $\delta > 0$ if and only if it does not contain a subspace isometric to the Euclidean plane.*

Product decompositions and isometries

In view of the Flat strip theorem, the terms *parallel* and *asymptotic* are synonymous when applied to geodesic lines in a CAT(0) space. From the Flat strip theorem we deduce:

Theorem 2.11 (A product decomposition theorem) *Let X be a CAT(0) space and let $c : \mathbb{R} \to X$ be a geodesic line.*

1. *The union of the images of all geodesic lines $c' : \mathbb{R} \to X$ parallel to c is a convex subspace X_c of X.*
2. *Let p be the restriction to X_c of the projection on the complete convex subspace $c(\mathbb{R})$. Let $X_c^0 = p^{-1}(c(0))$. Then, X_c^0 is convex (in particular it is a CAT(0) space) and X_c is canonically isometric to the product $X_c^0 \times \mathbb{R}$.*

Proof [33] Given two points $x_1, x_2 \in X_c$ we fix geodesic lines c_1 and c_2 parallel to c, such that x_1 lies in the image of c_1 and x_2 lies in the image of c_2. Because c_1 is parallel to c_2, we can apply the Flat strip theorem and deduce that the convex hull of $c_1(\mathbb{R}) \cup c_2(\mathbb{R})$ is isometric to a flat strip; in particular it is the union of images of geodesic lines parallel to c. This proves that X_c is convex.

Given $x \in X_c$, let c_x be the unique geodesic line in X_c with $p(c_x(0)) = c(0)$.

Let $j : X_c^0 \times \mathbb{R} \to X_c$ be the bijection defined by $j(x,t) = c_x(t)$. Using the Flat strip theorem, it is clear that this map is an isometry provided $d(x,x') = d(c_x(\mathbb{R}), c_{x'}(\mathbb{R}))$ for all $x, x' \in X_c^0$. To complete the proof one has to check that, given three geodesic lines $c_i : \mathbb{R} \to X$, $i = 1,2,3$, if the union of each pair of these lines is isometric to the union of two parallel lines in $\mathbb{E}^2$, and if $p_{i,j}$ is the map that assigns to each point of $c_j(\mathbb{R})$ the unique closest point on $c_i(\mathbb{R})$, then $p_{1,3} \circ p_{3,2} \circ p_{2,1} = p_{1,1}$, the identity of $c_1(\mathbb{R})$. (See [33, II.2.15].) □

The above splitting forms the basis of a number of important results concerning the structure of groups which act by isometries on CAT(0) spaces.

Definition 2.12 Given an isometry γ of a metric space X, the *translation length* of γ is defined to be $|\gamma| := \inf\{d(x, \gamma.x) \mid x \in X\}$. We also define $Min(\gamma) := \{x \in X \mid d(x, \gamma.x) = |\gamma|\}$. If $Min(\gamma) \neq \emptyset$, then γ is called *semisimple*. If $|\gamma| > 0$ and $Min(\gamma) \neq \emptyset$, then γ is called a *hyperbolic isometry*.

Proposition 2.13 *If a group Γ acts properly and cocompactly by isometries on a CAT(0) space X, then every element $\gamma \in \Gamma$ of infinite order acts as a hyperbolic isometry. If γ is a hyperbolic isometry of X then:*

1. *$Min(\gamma)$ is closed and convex.*
2. *$Min(\gamma)$ is a union of parallel geodesic lines each of which is γ-invariant. (These lines are called axes for γ.)*
3. *$Min(\gamma)$ is isometric to a product $Y \times \mathbb{R}$, and the restriction of γ to $Min(\gamma)$ is of the form $(y,t) \mapsto (y, t + |\gamma|)$, where $y \in Y, t \in \mathbb{R}$.*
4. *Every isometry α which commutes with γ leaves $Min(\gamma) = Y \times \mathbb{R}$ invariant, and its restriction to $Y \times \mathbb{R}$ is of the form (α', α''), where α' is an isometry of Y and α'' is a translation of $\mathbb{R}$.*

Exercise 2.14 Prove Theorem 2.1(8).

The following result places severe restrictions on the way in which central extensions can act by isometries on CAT(0) spaces. Notice that the conditions on the action are very weak here — in particular no discreteness is assumed.

Theorem 2.15 *Let X be a CAT(0) space and let Γ be a finitely generated group acting by isometries on X. If Γ contains a central subgroup $A \cong \mathbb{Z}^n$ that acts faithfully by hyperbolic isometries (apart from the identity element), then there exists a subgroup of finite index $H \subset \Gamma$ which contains A as a direct factor.*

Proof [33] Fix $\alpha \in A$. According to 2.13(4), the action of Γ leaves $Min(\alpha) = Y \times \mathbb{R}$ invariant, and the restriction of each $\gamma \in \Gamma$ to $Y \times \mathbb{R}$ is of the form (γ', γ''), where γ' is an isometry of Y and γ'' is a translation of $\mathbb{R}$. The map $\gamma \mapsto \gamma''$ defines a homomorphism from Γ to a finitely generated group of translations of $\mathbb{R}$. Such a group of translations is isomorphic to $\mathbb{Z}^m$, for some m, so we have a surjective homomorphism $\psi : \Gamma \to \mathbb{Z}^m$. The image under ψ of A is non-trivial, because it contains α.

We compose ψ with the projection of $\mathbb{Z}^m$ onto a suitable direct summand so as to obtain a homomorphism $\phi : \Gamma \to \mathbb{Z}$ such that $\phi(A)$ is non-trivial. We then choose $a \in A$ so that $\phi(a)$ generates $\phi(A)$. Let $H_0 = \phi^{-1}(\phi(A))$ and note that H_0 has finite index in Γ. The map $\phi(a) \mapsto a$ splits $\phi|_{H_0}$, giving $H_0 = \ker\phi \times \langle a \rangle$ (since a is central) and $A = A' \times \langle a \rangle$, where $A' = A \cap \ker\phi$. By induction on m (the rank of A) we may assume that A' is a direct factor of a subgroup of finite index $H' \subset \ker\phi$. Let $H = H' \times \langle a \rangle$. □

To give a taste of how one might use a result such as this, let me mention two applications. The first shows that (2.15) obstructs a suggested approach to the construction of faithful linear representations of mapping class groups. An unpublished result of Geoff Mess (see [33, II.7]) shows that centralizers in the mapping class group $\mathcal{M}$ of closed surfaces of genus at least three are not virtually direct products. Thus (2.15) restricts the ways in which $\mathcal{M}$ can act on CAT(0) spaces (cf. [78]).

Recall that a Coxeter group ("generalized reflection group") is a group of the form $\langle s_1, \ldots, s_n \mid s_i^2 = (s_i s_j)^{m_{ij}} = 1, i, j = 1, \ldots, n\rangle$. Gabor Moussong [89] shows that every Coxeter group acts properly and cocompactly by isometries on a CAT(0) space, thus:

Corollary 2.16 *The mapping class groups of closed surfaces of genus at least three do not admit faithful representations into Coxeter groups.*

Combining Corollary 2.16 with some standard (but non-trivial) results in 3-manifold topology, in [33, II.7] we prove:

Theorem 2.17 *If a closed 3-manifold M admits a metric that has non-positive curvature (in the sense of Alexandrov), and if the centre of $\pi_1 M$ is non-trivial, then M has a finite-sheeted covering $\widehat{M} = \Sigma \times S^1$, where Σ is a closed surface.*

Flat torus theorem

We have seen (Flat plane theorem) that the existence of flat planes is the only obstruction that can prevent a cocompact CAT(0) space from being hyperbolic. But how might such planes arise? Here we find another beautiful connection between geometry and group theory, first discovered in the setting of smooth manifolds by Gromoll-Wolf [66] and Lawson-Yau [79].

Theorem 2.18 (Flat torus theorem) *Let A be a free abelian group of rank n acting properly by semi-simple isometries on a CAT(0) space X. Then:*

1. *$Min(A) = \bigcap_{\alpha \in A} Min(\alpha)$ is non-empty and splits as a product $Y \times \mathbb{E}^n$.*
2. *Every element $\alpha \in A$ leaves $Min(A)$ invariant and respects the product decomposition; α acts as the identity on the first factor Y and as a translation on the second factor $\mathbb{E}^n$.*

3. *The quotient of each n-flat* $\{y\} \times \mathbb{E}^n$ *by the action of* A *is an n-torus.*
4. *If an isometry of* X *normalizes* A*, then it leaves* $Min(A)$ *invariant and preserves the product decomposition.*
5. *If a subgroup* $\Gamma \subset Isom(X)$ *normalizes* A*, then a subgroup of finite index in* Γ *centralizes* A*. Moreover, if* Γ *is finitely generated, then* Γ *has a subgroup of finite index that contains* A *as a direct factor.*

The proof proceeds by induction on the rank of A; the base step is contain in (2.13). In the inductive step one looks at the action of A on $Min(a_1)$ where a_1 is a primitive element of A. See [33, II.7] for a complete proof, generalizations, and proofs of the following consequences.

Corollary 2.19 *If* Γ *is the fundamental group of a compact geodesic space of non-positive curvature, then*

1. *every abelian subgroup of* Γ *is finitely generated,*
2. *virtually abelian subgroups of* Γ *satisfy the ascending chain condition,*
3. *every solvable subgroup of* Γ *is virtually abelian.*

The ideas used to prove the Product Decomposition Theorem and the Flat torus theorem can also be used to prove the following theorem, which was first proved for smooth manifolds in [66] and [79]. In the setting of CAT(0) spaces this result is due to Claire Baribaud [11]. A more general result is proved in [33, II.6.22].

Theorem 2.20 (Splitting theorem) *Let* Y *be a compact geodesic space of non-positive curvature in which every local geodesic can be extended to a locally isometric embedding of the real line. If the fundamental group of* Y *splits as a product* $\Gamma = \Gamma_1 \times \Gamma_2$ *and* Γ *has trivial centre, then* Y *splits as a product* $Y = Y_1 \times Y_2$*, isometrically, where* $\pi_1 Y_i = \Gamma_i$ *for* $i = 1, 2$*.*

A fundamental dichotomy in group theory?

Let me end this lecture by returning, in the light of the Flat torus theorem, to the dichotomy exposed by the Flat plane theorem.

Question 2.21 Suppose that Γ is the fundamental group of a compact non-positively curved space. If Γ is *not* hyperbolic, must it contain a free abelian subgroup of rank two ?

I suspect that the answer to this question is no, but in restricted settings the answer is yes: Bangert and Schroeder [10] proved that the answer is yes for real-analytic manifolds; Lee Mosher and I [34] proved the same for closed topological 3-manifolds (following earlier work of Mosher [88], Schroeder [101] and Buyalo [41]); and the answer is also yes in the case where Γ is a cocompact lattice in a semi-simple Lie group [83].

An intriguing possibility is that the dichotomy suggested in (2.21) may have little to do with non-positive curvature, *per se*, but instead it might be a special case of a much more general phenomenon.

Question 2.22 Let Γ be a finitely presented group. Suppose that Γ admits an Eilenberg-Maclane space $K(\Gamma, 1)$ that has only finitely many cells in each dimension (equivalently, suppose that the cohomology of Γ is finitely generated in each dimension). If Γ is *not* hyperbolic, must it contain a subgroup of the form $\langle x, y \mid y^{-1}x^p y = y^q \rangle$?

Question 2.21 is a special case of Question 2.22. Another interesting special case of Question 2.22 is where Γ is a subgroup of a hyperbolic group.

The dichotomy suggested in Theorem 2.2 does not hold for finitely presented groups in general. Thus far two types of counterexamples are known: Noel Brady [25] showed that there exist hyperbolic groups G and finitely presented subgroups $H \subset G$ that are not hyperbolic (cf. (1.36)), and Slava Grigorchuk [65] has constructed finitely presented HNN extensions of the form $H = B*_\phi$ where B is an infinite, finitely generated, torsion group and $\phi : B \to B$ is an epimorphism.

In each of these examples, the third homology group of H is not finitely generated and therefore H does not have an Eilenberg-Maclane complex $K(H, 1)$ with finitely many 3-cells.

Lecture 3: Building CAT(0) spaces of interest in group theory

In Lecture 1 we saw how the basic decision problems of combinatorial group theory lead to the study of non-positive curvature. Lecture 2 was about the basic structure of CAT(0) spaces with emphasis on the close connections between geometry, topology and group theory in the presence of non-positive curvature. The purpose of Lecture 3 is to put some meat on the attractive theoretical skeleton that has emerged: I shall explain various examples and general methods for constructing CAT(0) spaces, and I shall try to illustrate how such constructions can give rise to interesting new groups and new insights into well-known groups. Let me begin by mentioning one result of the latter kind.

Theorem 3.1 *Every Coxeter group has a solvable conjugacy problem.*

Surprisingly, no algebraic proof of this result is known. The only known proof relies on the theorem of Moussong that I mentioned earlier: every finitely generated Coxeter group W acts properly and cocompactly by isometries on a piecewise-Euclidean complex. In (4.5) I shall explain why groups which act in this way have a solvable conjugacy problem.

Example 3.2 (1) A simply-connected, complete, Riemannian manifold is a CAT(0) space if and only if all of its sectional curvatures are non-positive (see [48], [33, II.1A]). Examples include $\mathbb{E}^n$, $\mathbb{H}^n$, and any symmetric space of non-compact type.

(2) *Metric graphs.* A metric graph (1-complex) X consists of a set of vertices and a set of edges joining them; each edge is given a metric that makes it locally isometric to a closed interval of the real line; a (pseudo)metric is defined on X by setting the distance between two points equal to the length of the shortest path between them, where length is measured in the given (local) metrics on the edges.

I encourage you to consider the many pathologies that can arise if one does not impose any restrictions on the lengths of the edges (particularly if there are vertices of infinite valence) cf. [33, I.1.9]. I also encourage you to check that if there are only finitely many different edge lengths then the metric defined above makes X a complete geodesic space of non-positive curvature.

Truncated hyperbolic space. Given a geodesic space X with metric d and an arc-connected subspace $Y \subset X$, there are two metrics with which one might endow Y: one might simply define the distance between $y, y' \in Y$ to be $d(y, y')$, or one might define the distance to be the infimum of the lengths of paths that join y to y' in Y (where length is measured using d). The second construction is called the *induced path metric on* Y. In order to see the difference between these metrics, consider the case where Y is the unit circle in the Euclidean plane.

If one removes a disjoint collection of open horoballs from real hyperbolic space $\mathbb{H}^n$ then the induced path metric is a CAT(0) space. This is proved in [33, II.11]. It follows that non-uniform lattices in $SO(n, 1)$ act properly and cocompactly by isometries on CAT(0) spaces (namely these truncated hyperbolic spaces). The presence of nilpotent subgroups that are not virtually abelian show that non-uniform lattices in other rank 1 Lie groups do not admit such actions (by 2.1(5)) and hence the corresponding truncated symmetric spaces are not non-positively curved.

Polyhedral complexes

The simplest examples of geodesic metric spaces that are not manifolds are provided by metric graphs. A wider and much more interesting class of spaces is provided by the higher dimensional analogues of graphs, *metric polyhedral complexes*. Roughly speaking, to construct a piecewise-Euclidean complex K one takes the disjoint union of a family of convex polyhedra from Euclidean space $\mathbb{E}^n$ and one identifies them along isometric faces. As with graphs, a (pseudo)metric on K is defined by setting the distance between each pair of points equal to the length of paths joining them, where length is measured in the given metrics on the individual cells. In my thesis [26] I proved that if the set of isometry types of the cells, denoted $Shapes(K)$, is finite, then every pair of points in K can be joined by a shortest path, in other words K is a geodesic space.

An entirely analogous construction gives piecewise hyperbolic complexes (obtained by identifying faces among a disjoint union of convex polyhedra from hyperbolic space $\mathbb{H}^n$), or piecewise spherical complexes (obtained by identifying faces among a disjoint union of convex polyhedra from the n-sphere $\mathbb{S}^n$). More generally, one can take convex polyhedra from any simply-connected manifold of constant sectional curvature, and this gives the general notion of a metric polyhedral complex (see [33, I.7]).

There are various ways of characterizing complexes whose curvature is bounded above (see [33, II.5]); we mention two of them.

Theorem 3.3 *Let K be a piecewise Euclidean or piecewise hyperbolic complex with $Shapes(K)$ finite. The following conditions are equivalent.*

1. *K is non-positively curved.*

2. *Every $x \in K$ has a neighbourhood in which there is a unique geodesic joining each pair of points.*
3. *K satisfies the link condition.*

The *link condition* can be described as follows. First think of a vertex v of a convex polyhedron P (e.g. a solid cube) in Euclidean space of some dimension and imagine how a small (radius ε) sphere centred at v intersects P. The intersection is a small spherical polyhedron (a spherical triangle if P is a cube). Now rescale your mental picture so that this spherical polyhedron is based on a sphere of radius one rather than radius ε. It is then called the link of v in P.

If K is as in Theorem 3.3, then the *link* of a vertex $v \in K$, denoted $Lk(v)$, is obtained by taking the link of v in each of the individual cells incident at v and assembling them according to the way in which the cells intersect. Thus $Lk(v)$ is a piecewise spherical complex. It is the analogue in metric complexes of the unit tangent space in differential geometry: the points of $Lk(v)$ are the directions of geodesics issuing from v, and the distance between each pairs of points should be thought of as the angle at v between the geodesics issuing in these directions. One says that K *satisfies the link condition* if for all vertices $v \in K$ and all $x, y \in Lk(v)$, if $d(x,y) < \pi$ then there is a unique geodesic joining x to y in $Lk(v)$.

2-dimensional complexes

I shall describe a number of constructions of non-positively curved 2-dimensional complexes; further constructions are given in [33, II.5]. It is more difficult to construct examples in higher dimensions, but examples do exist, for instance Bruhat-Tits buildings of affine type (see [37] and [36]). One of the advantages of working with 2-dimensional complexes[8] is that it is easy to check the link condition.

Example 3.4 (The link condition for 2-complexes) Let K be a piecewise Euclidean or piecewise hyperbolic 2-complex and let $v \in K$ be a vertex. Then $Lk(v)$, as defined above, is a metric graph and K is non-positively curved if and only if there does not exist a vertex such that $Lk(v)$ contains an embedded circuit of length strictly less than 2π.

Figure 3.1 should be compared with Figure 1.1.

Squared 2-complexes

Squared 2-complexes are, by definition, piecewise Euclidean 2-complexes in which all of the 2-cells are squares of side length 1. In this simple setting the link condition reduces to the statement that the link of each vertex contains no embedded circuits of combinatorial length less than four.

Exercise Check that you thoroughly understand the following assertion: the product of any two metric graphs with edge lengths 1 is a squared 2-complex of non-positive curvature. (The links of vertices in this case are complete bipartite graphs.)

[8] hereafter abbreviated to "2-complexes"

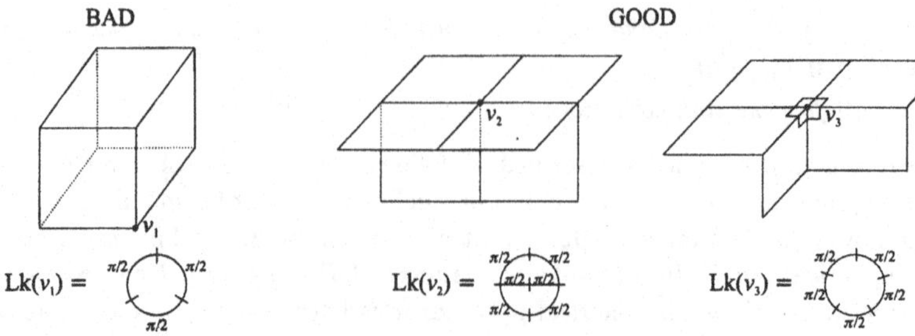

Figure 3.1. $Lk(v)$

In what follows we shall use the term *tree* to mean a simply connected metric graph all of whose edges have length 1.

Theorem 3.5 *Let K be a compact, connected, non-positively curved squared complex. If the link of each vertex in K is isomorphic to a complete bi-partite graph, then the universal cover of K is the product of two trees.*

This result seems to have been discovered independently by a number of authors [115], [7], [39]. A detailed proof can be found in [35].

This theorem relates the study of squared 2-complexes to the study of the automorphism groups of products of trees. This brings us into a very rich world, because trees are combinatorial analogues of symmetric spaces of non-compact type, and in keeping with this analogy one finds that many of the groups which act on products of trees enjoy remarkable geometric and arithmetic properties. The most striking result in this direction is the following construction due to Marc Burger and Shahar Mozes [39].

Theorem 3.6 (Burger-Mozes) *There exist finitely presented simple groups that act freely and cocompactly by isometries on a product of two trees. Moreover, one can construct such groups by amalgamating two finitely generated free groups along subgroups of finite index.*

But the number of generators and relations is rather large. Prior to the work of Burger and Mozes, my student Dani Wise had (by hand) produced many interesting examples of non-positively curved squared complexes including some whose fundamental groups have no non-trivial finite quotients [115].

Notice that by (1.13) any group Γ which acts properly and cocompactly on a product of trees in quasi-isometric to a product of finitely generated free groups, but (3.6) shows that Γ may have utterly different residual properties to products of free groups.

Ian Acheson and Dani Wise [115] (independently) noticed that work of Weinbaum [114], originally phrased in the language of small cancellation theory, essentially proves the following result.

Theorem 3.7 *The fundamental group of the complement of every tame, prime, alternating link in $\mathbb{S}^3$ is the fundamental group of a compact non-positively curved squared complex.*

Remark It follows from (3.12) that one can remove the hypothesis that the link is prime at the expense of replacing the conclusion "squared" by "polyhedral". The complex in (3.7) has two vertices, one edge for each region in the plane of projection and one square for each crossing. The idea of the construction is this: one regards the alternating projection as lying on a plane in $\mathbb{R}^3$ and one places one vertex a above the plane and one b below it. For each complementary region x_i of the projection one introduces an edge labelled x_i oriented from a to b. Then for each crossing in the projection of the link one attaches a square (2-cell) by gluing the boundary circuit to the path that traces out the edges $x_{i_1}, x_{i_2}^*, x_{i_3}, x_{i_4}^*$ in order, where the asterisk denotes reversed orientation and the regions $x_{i_1}, x_{i_2}, x_{i_3}, x_{i_4}$ are those that provide the corners at the crossing; they are read from above the plane in clockwise order and the strand of the link that is between x_{i_1} and x_{i_2} passes over the other strand. The natural presentation of the fundamental group of this complex is essentially Dehn's presentation of the link group [81]. (See [115] or [33, II.7] for details.)

Exercise The Hopf link consists of two circles that cannot be separated in 3-dimensional space and are such that the link has a projection with only two crossings. Show that the complex that the above construction ascribes to this link is a torus, but that this torus is not a metric product.

In [35] the geometry of non-positively curved squared complexes is used to investigate subgroups of the direct product of two free groups. In particular we give a geometric explanation of the following theorem of Baumslag and Roseblade [15].

Theorem 3.8 ([15]) *If F_1 and F_2 are free groups and H is a finitely presented subgroup of $F_1 \times F_2$, then either H is free or else it contains a subgroup of finite index $H_1 \times H_2$, where $H_1 \subset F_1$ and $H_2 \subset F_2$.*

The modified Rips construction

The following algorithmic construction gives rise to a host of interesting examples of non-positively curved complexes and allows one to translate the pathologies of arbitrary group presentations into pairs of groups $N \subset \Gamma$, where Γ is hyperbolic and N is finitely generated. In Lecture 4 I shall explain how one can exploit this transfer of pathology to expose the diverse nature of decision problems for subgroups of groups which act properly and cocompactly on CAT(0) spaces.

Theorem 3.9 *There is an algorithm that associates to every finite presentation a short exact sequence*

$$1 \to N \to \Gamma \to G \to 1,$$

where G is the group given by the presentation, N is a finitely generated group and Γ is the fundamental group of a compact, negatively curved, piecewise hyperbolic

2-complex K. Moreover one can remetrize K as a piecewise Euclidean squared complex of non-positive curvature.

This construction is based closely on an idea of E. Rips. In [99] he proved a version of this theorem in which instead of forcing Γ to be a fundamental group of the above type, he arranged for it to satisfy an arbitrarily strict small cancellation condition. The details of the adaptation given above were worked out by Wise in [116] and a slightly different account is given in [33, II.5].

In outline, Rips's idea is that given $G = \langle \mathcal{X} \mid \mathcal{R} \rangle$, one should introduce new generators $\underline{a} = \{a_1, \ldots, a_n\}$, "unwrap" the old relations, and add a new set of relations U to make the subgroup generated by $\underline{a}$ normal. To "unwrap" $r \in \mathcal{R}$ we replace $r = 1$ by a relation of the form $r = w_r(\underline{a})$; let $\tilde{\mathcal{R}}$ denote the set of these new relations. To construct U, for each $x \in \mathcal{X}$ and $a_i \in \underline{a}$ we choose words $u_{i,x}$ and $u'_{i,x}$ and define $x^{-1}a_ix = u_{i,x}(\underline{a})$ and $xa_ix^{-1} = u'_{i,x}(\underline{a})$. The key to the construction is that the words $w_r, u_{i,x}$ and $u'_{i,x}$ must be long enough and sufficiently independent to ensure that

$$\Gamma := \langle \mathcal{X}, \underline{a} \mid \tilde{\mathcal{R}}, U \rangle$$

satisfies a small cancellation condition. (See [81] for background on small cancellation theory.)

In Rips's original construction one only needs two new generators, but in (3.9) one needs more in order to promote the small cancellation condition to the (more restrictive) link condition for an associated polyhedral complex.

Example 3.10 (Non-hopfian groups) Following earlier work of Baumslag and Solitar [16] and Meier [85], in [116] Dani Wise considered the following groups:

$$T(n) = \langle a, b, t_a, t_b \mid [a,b] = 1,\ t_a^{-1}at_a = (ab)^n,\ t_b^{-1}bt_b = (ab)^n \rangle.$$

If $n \geq 2$, certain non-trivial commutators, for example $g_0 = [t_a(ab)t_a^{-1}, b]$, lie in the kernel of the epimorphism $T(n) \twoheadrightarrow T(n)$ given by $a \mapsto a^n, b \mapsto b^n, t_a \mapsto t_a, t_b \mapsto t_b$. Groups which admit such self-surjections with kernel are called *non-Hopfian.*

$T(n)$ is the fundamental group of the non-positively curved 2-complex $X(n)$ that one constructs as follows: take the (skew) torus formed by identifying opposite sides of a rhombus with sides of length n and small diagonal of length 1; the loops formed by the images of the sides of the rhombus are labelled a and b respectively; to this torus attach two tubes $S \times [0,1]$, where S is a circle of length n; one end of the first tube is attached to the loop labelled a and one end of the second tube is attached to the loop labelled b; in each case the other end of the tube wraps n times around the image of the small diagonal of the rhombus. One checks easily that this complex satisfies the link condition.

New spaces from old

A natural way to construct interesting new metric spaces is to take a disjoint collection of known metric spaces and glue them together along subspaces. Intuitively

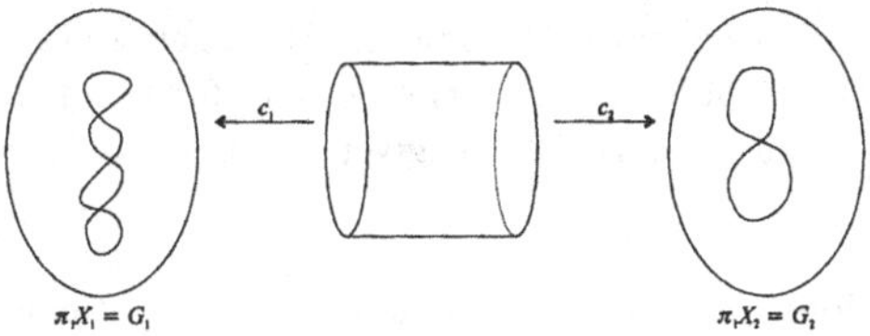

Figure 3.2. Proof of Corollary 3.12

speaking, this means that we glue the spaces by means of an equivalence relation and then define the distance between points in the resulting space by taking the infimum of the lengths (measured in the original metrics) of paths joining the points, as we did in the construction of polyhedral spaces. In order to make this description precise we need two technical devices. First, given two metric spaces X_1, X_2, we extend the given metrics to a metric on their disjoint union $X_1 \coprod X_2$ by defining $d(x, y) = \infty$ if $x \in X_1$ and $y \in X_2$. Secondly, given a metric space X and an equivalence relation $\sim$ on X, *the quotient pseudometric* $\overline{d}$ on the set of equivalence classes $X/\sim$ is given by the formula:

$$\overline{d}(\overline{x}, \overline{y}) = \inf \sum_{i=1}^{n} d(x_i, y_i),$$

where the infimum is taken over all sequences $C = (x_1, y_1, x_2, y_2, \ldots, x_n, y_n)$ of points of X such that $x_1 \in \overline{x}$, $y_n \in \overline{y}$, and $y_i \sim x_{i+1}$ for $i = 1, \ldots, n-1$. ($\overline{d}$ may fail to be a metric because the distance between distinct classes $\overline{x}$ and $\overline{y}$ is zero.)

Theorem 3.11 **([33, II.11])** *Let X and A be metric spaces of non-positive curvature and suppose that A is compact. Let $\phi_1, \phi_2 : A \to X$ be local isometries and let $\overline{X} = X \coprod (A \times [0,1])/\sim$, where $\sim$ is generated by $(a, 0) \sim \phi_1(a)$ and $(a, 1) \sim \phi_2(a)$. Then the quotient pseudometric on $\overline{X}$ is actually a metric and $\overline{X}$ is non-positively curved.*

We give one application to indicate how this result can be used to mimic group theoretic constructions in the world of non-positive curvature.

Corollary 3.12 *If G_1 and G_2 act properly and cocompactly by isometries on some CAT(0) space, then so too does any amalgamated free product of the form $G_1 *_C G_2$, where C contains a cyclic subgroup of finite index.*

Proof We sketch the proof in the torsion-free case with $C = \mathbb{Z}$; see [33, II.11] for the general case. For $i = 1, 2$ let X_i be a compact non-positively curved space with $\pi_1 X_i = G_i$. The idea of the proof is shown in Figure 3.2.

The generators of the cyclic subgroups identified by the amalgamated free product can be represented by closed geodesics (i.e. locally isometric embeddings of circles), $c_1 : S_1 \to X_1$ and $c_2 : S_2 \to X_2$. By scaling the metric on X_1, we may assume that these geodesics have the same length. Then we can apply the theorem with $X = X_1 \coprod X_2$ and $A = S_1 = S_2$ and $\phi_i = c_i$ for $i = 1, 2$. The Seifert - van Kampen theorem implies that $\pi_1\overline{X}$ is the given amalgamated free product $G_1 *_{\mathbb{Z}} G_2$. □

Corollary 3.12 does not extend (without further hypotheses) to amalgamated free products along finitely generated free groups (or free abelian) groups of higher rank. The question of when these more general amalgamations do act nicely on CAT(0) spaces is discussed at some length in [33, III]. Note too that in the light of 2.1(8), we know that $\mathbb{Z}*_{\mathbb{Z}} = \langle x, t \mid t^{-1}xt = x^2\rangle$ provides a counterexample to the most naive HNN analogue of (3.12).

An embedding theorem

The following is one a number of embedding theorems of a similar ilk proved in [31].

Theorem 3.13 *Every compact, connected, non-positively curved space X admits an isometric embedding into a compact, connected, non-positively curved space $\overline{X}$ such that $\overline{X}$ has no non-trivial finite-sheeted coverings (equivalently $\pi_1\overline{X}$ has no non-trivial finite quotients). If X is a polyhedral complex of dimension $n \geq 2$, then one can arrange for $\overline{X}$ to be a complex of the same dimension.*

Given an arbitrary group Γ_0 with finite generating set $\{a_1, \dots, a_n\}$, in order to embed Γ_0 in a group with no non-trivial finite quotients one can proceed as follows. First, to reduce to the case where the a_i have infinite order, replace Γ_0 by $\Gamma = \Gamma_0 * \langle t\rangle$ and $\{a_1, \dots, a_n\}$ by $\{t, ta_1, \dots, ta_n\}$. Let G be a torsion-free group (such as that in (3.6)) that has no finite quotients. Form $\Gamma *_{\mathbb{Z}} G$ by identifying a_1 with an element of infinite order in G, then identify the resulting group with another copy of G, this time identifying a_2 with an element of infinite order in this second copy of G. Repeat a total of n times. In the group that one obtains, Γ amalgamated with n copies of G, each along a copy of $\mathbb{Z}$, all of the obvious generators have been identified with something that must have trivial image in every finite quotient.

In the light of (3.6) and (3.12), the above operations can be carried out within the class of fundamental groups of compact non-positively curved spaces. Moreover, by making the gluing tubes sufficiently long one can ensure that the natural embedding of X into the final space is isometric. This proves (3.13). An alternative proof, which uses the elementary construction from (3.10) instead of (3.6), is given in [31].

3.14 Complexes of groups

I want to mention one other general technique for constructing group actions on CAT(0) spaces: complexes of groups in the sense of Haefliger. The foundations of the subject are rather technical, so I am afraid that I can only give a rather vague outline of the theory — see [71, 72] and [33] (II.12) and (III) for details.

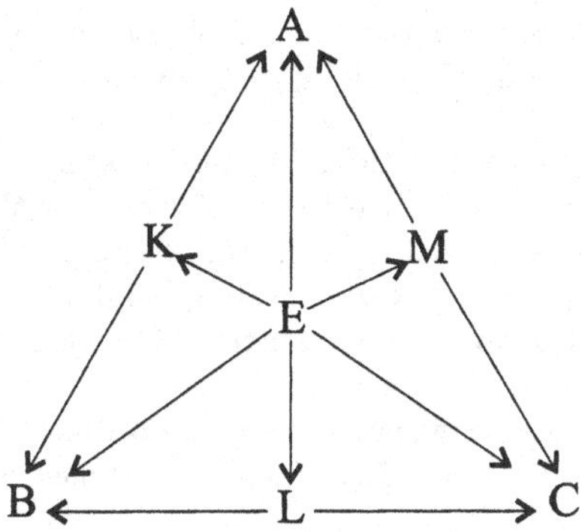

Figure 3.3. A triangle of groups

To illustrate the construction, let's consider *triangles of groups*. A triangle of groups is a diagram of groups and monomorphisms as shown is Figure 3.3.

One way in which such diagrams are obtained is by recording the stabilizers of vertices, edges and the 2-cell in a strict fundamental domain for the simplicial action of a group Γ on a simply connected simplicial 2-complex X; these stabilizers are called the *local groups* of the complex. If the triangle of groups arises in this way, then by taking the pushout of the diagram in the category of groups one recovers Γ, and hence it is reasonable to call Γ the *fundamental group* of the triangle of groups. Moreover, by taking the geometric realization of the poset of the cosets of the local groups in Γ, one can recover the complex X and the action of Γ on it (see [71], [108], [26]).

A similar process of recovery works for more general group actions on complexes where there is a strict fundamental domain, and is explained in [72] and [33, II.12]. In this more general setting, the stabilizers are arranged not into a triangle of groups but into a complex of groups in the sense of Haefliger. A more sophisticated notion is needed in order to handle actions which do not have a strict fundamental domain (see [72] or [33, III.C]).

The theory of complexes of group generalizes the Bass-Serre theory of graphs of groups [105]. In contrast to the 1-dimensional case (graphs), complexes of groups do not in general arise as the pattern of stabilizers of a group action on a simply connected space, i.e. they may not be "developable". However, if the complex of groups is non-positively curved in a suitable sense (see [71, 72], [33, III.C], [108]), then it is indeed developable, and its fundamental group will act on a CAT(0) space in a manner dictated by the complex of groups.

3-manifolds of non-positive curvature

Let me finish this lecture by turning from complexes to manifolds. In this section I shall assume more knowledge of topology than is required elsewhere in this series.

Theorem 3.15 *Let M be a compact 3-manifold with non-empty boundary. If every 2-sphere in M bounds a ball, then M supports a metric of non-positive curvature.*

Remarks (1) Throughout this article, we use the term *non-positive curvature* in the sense of A.D. Alexandrov. In the setting of the above theorem one can always arrange for the metrics considered to be smooth Riemannian metrics on the interior of the manifold, but the boundary will not in general be locally convex — indeed if it were then it would perforce be π_1-injective (2.3) which cannot happen if $\pi_1 M$ is free.

(2) The case where the boundary of M has zero Euler characteristic was dealt with by Leeb in his thesis [80]. The argument sketched below is worked out in detail in [29].

The first step in the proof involves cutting the given manifold up according to the JSJ decomposition [75], [76], [112]. This means that one cuts along a maximal non-parallel family of π_1-embedded discs, tori and Klein bottles. Thurston's Geometrization Theorem for Haken manifolds states that the pieces each have a geometric structure. It is easy to show that each piece supports a metric of non-positive curvature such that the boundary tori and Klein bottles are convex. In order to reassemble the original manifold we must perturb the metrics on the pieces so that the pairs of tori and Klein bottles to be glued are isometric.

On the geometrically finite hyperbolic pieces, the tori and Klein bottles correspond to cusps truncated by horoballs (cf. 3.2(3)), and by truncating further down the cusp one can shrink the area of each such boundary component without changing the metric away from the cusp. One can also adjust the conformal type of each such boundary component using a standard warping technique, which is explained nicely by Viktor Schroeder in [102]. (Schroeder also tells me that there is a more detailed account by Buyalo in an appendix to the Russian edition of [118].)

Having organized the tori and Klein bottle boundary components on the geometric pieces appropriately so that they can be glued (the Seifert fibred pieces do not cause any serious problems and are dealt with in [80]), the only problem that we face in trying to reassemble the original manifold is that we must be able to perform the boundary sums (the inverse operation of cutting along discs) in a manner that preserves non-positive curvature. For this one uses gluing tubes of different types:

3.16 Gluing tubes

Suppose that we have two 3-manifolds with flat boundary components and we wish to glue them along isometric discs in these components. To achieve this we take a tube of *Euclidean type* (the product of an appropriately metrized topological disc D with $[0,1]$) and we attach the ends $D \times \{0\}$ and $D \times \{1\}$ to the discs on the boundaries of the manifolds. The resulting manifold is the boundary sum of the pieces, and if the pieces were non-positively curved, so is the result, by (3.11).

If the boundary of the pieces were not flat but instead contained an isometrically embedded hyperbolic disc (of small radius)[9], then instead of using a product cylinder one would use a tube of *hyperbolic type* (the intersection of a tubular neighbourhood of a geodesic c in $\mathbb{H}^3$ with the complement of two disjoint open half-spaces whose bounding hyperplanes are orthogonal to c).

The use of suitable tubes also allows one to combine pieces with hyperbolic patches on their boundary with pieces that have Euclidean boundary. For this one uses a tube of *mixed type*: such a tube is a compact subset of $\mathbb{H}^3$ obtained by intersecting a tubular neighbourhood of a geodesic c, with the complement of an open horoball about one endpoint of c, *and* a closed half-space (containing the open horoball) that is bounded by a hyperplane orthogonal to c. In the induced path metric, this tube is non-positively curved; both ends of the tube are convex in this metric; the end lying on the bounding hyperplane is isometric to a disc in the hyperbolic plane, and the end on the horosphere is isometric to a Euclidean disc.

The full process of reassembling the 3-manifold also requires that one combine certain scaled copies of the tubes of mixed type. See [29] for details.

Lecture 4: Decision problems

In the first half of this lecture we return to our point of departure in Lecture 1: the basic decision problems of combinatorial group theory as framed as Dehn. I shall describe how one solves the word and conjugacy problems for groups which act properly and cocompactly by isometries on CAT(0) spaces and I shall describe a solution to the conjugacy problem for hyperbolic groups. I shall also discuss the isomorphism problem briefly (without proofs) and explain how it is connected to classification problems in geometry and topology, following [33, III].

In the second half of this lecture I shall present a number of recent results that relate to the following question: *Let X be a compact non-positively curved space and let $\Gamma \leq \pi_1 X$ be a subgroup. Under what circumstances can one deduce that Γ is the fundamental group of a compact non-positively curved space?*

Non-positive curvature and the word and conjugacy problems

We discussed the word problem for hyperbolic groups in Lecture 1, indeed it was the bridge that led us to into our discussion of negative curvature. Recall:

Theorem 4.1 *The following statements are equivalent for f.p. groups:*

1. *Γ is a hyperbolic group.*
2. *Γ has a finite Dehn presentation.*
3. *Γ satisfies a linear isoperimetric inequality.*
4. *Γ satisfies a sub-quadratic isoperimetric inequality.*

[9] This is what happens on boundary components of higher genus associated to geometrically finite hyperbolic pieces.

There exist non-hyperbolic groups that act properly and cocompactly by isometries on CAT(0) spaces, e.g. $\mathbb{Z}^2$, so point (3) of the above theorem shows the quadratic bound in the following result is sharp.

Theorem 4.2 *If Γ acts properly and cocompactly by isometries on a CAT(0) space X, then it is finitely presented and satisfies a quadratic isoperimetric inequality.*

Proof I'll sketch the proof, following [33, III]. Fix $x_0 \in X$ and $D > 0$ such that X is the union of the balls $\gamma.B(x_0, D/3)$. General considerations of group actions (cf. 1.13) yield:

$\mathcal{A} = \{a \in \Gamma \mid d(a.x_0, x_0) \leq 2D+1\}$ generates Γ. Given $\gamma \in \Gamma$, if $d(x_0, \gamma.x_0) \leq 4D+1$ then $\gamma = a_1 \dots a_l$ for some $a_i \in \mathcal{A}$ and $l \leq 4$. Let $\mathcal{R}$ be the set of words in the free group $F(\mathcal{A})$ that have length at most 10 and represent the identity in Γ. Then $\langle \mathcal{A} \mid \mathcal{R} \rangle$ is a finite presentation for Γ.

We shall see that if a word $w = b_1 \dots b_m$, with $b_j \in \mathcal{A}^{\pm 1}$, represents $1 \in \Gamma$ then w can be expressed in the free group $F(\mathcal{A})$ as a product of at most $(D+1)m^2$ conjugates of relators $r \in \mathcal{R}$. The proof will also show that the length of the conjugating elements is bounded by a linear function of m.

To each $\gamma \in \Gamma$ we associate a word σ_γ in the generators $\mathcal{A}$ as follows. Let c_γ be the unique geodesic joining x_0 to $\gamma.x_0$ in X. For each integer $i < d(x_0, \gamma.x_0) + 1$ let $\sigma_\gamma(i) \in \Gamma$ be such that $d(c_\gamma(i), \sigma_\gamma(i).x_0) \leq D$, with $\sigma_\gamma(0) = 1$ and $\sigma_\gamma(i) = \gamma$ when i is large enough. Note that $a_i = \sigma_\gamma(i)^{-1}\sigma_\gamma(i+1) \in \mathcal{A}$. Define σ_γ to be $a_0 \dots a_n$, where n is the least integer greater than $d(x_0, \gamma.x_0)$. Note that $\gamma = \sigma_\gamma$ in Γ. We choose σ_1 to be the empty word.

We fix $\gamma \in \Gamma$ and $b \in \mathcal{A}$ and compare σ_γ with $\sigma_{\gamma'}$ where $\gamma' = \gamma b$. By appending letters a_i that represent the identity if necessary, we may write $\sigma_\gamma = a_0 \dots a_n$ and $\sigma_{\gamma'} = a_0' \dots a_n'$, where $n = n(\gamma, \gamma') = \max\{|\sigma_\gamma|, |\sigma_{\gamma'}|\}$. Now $d(\gamma.x_0, \gamma'.x_0) \leq 2D + 1$ (because $b \in \mathcal{A}$), so from the convexity of the metric on X (2.4) we have $d(c_\gamma(i), c_{\gamma'}(i)) < 2D + 1$ for all i and hence $d(\sigma_\gamma(i).x_0, \sigma_{\gamma'}(i).x_0) < 4D + 1$. If for each i we choose a word $\alpha(i)$ of length at most 4 that is equal to $\sigma_\gamma(i)^{-1}\sigma_{\gamma'}(i)$, then $\alpha(i)^{-1}a_{i+1}\alpha(i+1){a'_{i+1}}^{-1} \in \mathcal{R}$. We choose $\alpha(n)$ to be b (the difference between γ and γ'), and α_0 to be the empty word.

If we write $p_\gamma(i)$ for the word $a_0 \dots a_i$ (the i-th prefix of σ_γ), and define $p_{\gamma'}(i)$ similarly, then we have the following equality in the free group on $\mathcal{A}$:

$$\sigma_\gamma b \sigma_{\gamma'}^{-1} = \prod_{i=1}^{n(\gamma,\gamma')} p_{\gamma'}(i-1)\left[\alpha(i-1)^{-1}a_i\alpha(i){a'_i}^{-1}\right] p_{\gamma'}(i-1)^{-1}.$$

Each of the words in square brackets belongs to $\mathcal{R}$.

Finally, we consider the given word $w = b_1 \dots b_m$ that represents the identity in Γ. Let $\gamma_0 = 1$ and let $\gamma_j \in \Gamma$ be the element represented by $b_1 \dots b_j$. Note $\gamma_m = 1$ and $\sigma_{\gamma_m} = \sigma_{\gamma_0}$ is the empty word. By definition, in $F(\mathcal{A})$ we have the equality

$$w = \prod_{j=1}^{m} \sigma_{\gamma_{j-1}} b_j \sigma_{\gamma_j}^{-1}.$$

By replacing each factor on the right hand side of this equality with the right hand side of the previous equation (with $\gamma = \gamma_{j-1}$, $b = b_j$, and $\gamma' = \gamma_j$), we obtain the desired quadratic inequality. (To check the bounds on the number of factors note that $d(\gamma_j.x_0, x_0) \leq (2D+1)|w|/2$.) □ □

Remarks (1) In the course of the preceding proof we proved that every relation $w = 1$ is a consequence of the given relations $\mathcal{R}$.

(2) The idea behind the preceding proof becomes more transparent if one translates it into the language of van Kampen diagrams (cf. [28]).

The conjugacy problem

The discussion in this section follows [33, III]. Many of the basic ideas can be found in [68, 69].

The way in which we shall solve the conjugacy problem for groups that act properly and cocompactly by isometries on CAT(0) spaces is motivated by the following geometric observation: given $l > 0$, if two loops of length $\leq l$ are freely homotopic in a complete non-positively curved space, then one can continuously deform one to the other through intermediate loops of length $\leq l$. Free homotopy classes of loops in a space X correspond to conjugacy classes in $\pi_1 X$, and quasifying the above observation we are led to the following definition in which the elements $w_i^{-1} u w_i \in \Gamma$ are the analogues of the intermediate loops in the free homotopy.

Definition 4.3 (q.m.c.) A group Γ with finite generating set $\mathcal{A}$ is said to have the *quasi-monotone conjugacy property (q.m.c)* if there is a constant $K > 0$ such that if two words $u, v \in F(\mathcal{A})$ are conjugate in Γ, then there is a word $w = a_1 \ldots a_n$ with $a_i \in \mathcal{A}^{\pm 1}$, such that $w^{-1}uw = v$ and $d(1, w_i^{-1}uw_i) \leq K \max\{|u|, |v|\}$ for $i = 1, \ldots, n$, where $w_i = a_1 \ldots a_i$.

The existence of K does not depend on the choice of generating set $\mathcal{A}$ but its value does.

Proposition 4.4 *If a group Γ acts properly and cocompactly by isometries on a CAT(0) space, then it has q.m.c.*

Proof[Sketch] Suppose that Γ acts on X. Fix $x_0 \in X$ and let $\mathcal{A}$ and σ_γ be as in the proof of (4.2). Given words u and v such that $\gamma^{-1}u\gamma = v$ in Γ, one verifies that $w := \sigma_\gamma$ satisfies the requirements of (4.3) (with respect to a constant that depends only on the parameters of the quasi-isometry $\gamma \mapsto \gamma.x_0$ described in (1.13)). To perform this verification, one uses the convexity of the metric on X (2.4) to compare the geodesic quadrilateral in X with vertices $\{x_0, \gamma.x_0, u.x_0, \gamma u.x_0 = v\gamma.x_0\}$ to a quadrilateral Q in the Cayley graph $\mathcal{C}_\mathcal{A}(\Gamma)$. The vertices of Q are $\{1, \gamma, u, \gamma u = v\gamma\}$, two of its sides are labelled σ_γ, and the other two sides are labelled u and v. □

4.5 An algorithm to determine conjugacy

Let Γ be a group with finite generating set $\mathcal{A}$. Suppose that Γ has a solvable word problem and also has the q.m.c. property. Let the constant K be as in (4.3).

For each positive integer n, we consider the set $B(n)$ of words in $F(\mathcal{A})$ that have length at most n. Because Γ has a solvable word problem, given a pair of words $v_1, v_2 \in B(n)$ one can decide if there exists $a \in \mathcal{A}^{\pm 1}$ such that $a^{-1}v_1a = v_2$ in Γ; if such an a exists we write $v_1 \sim v_2$.

Consider the (algorithmically constructed) finite graph $\mathcal{G}(n)$ with vertex set $B(n)$ that has an edge joining v_1 to v_2 if and only if $v_1 \sim v_2$. The q.m.c. property says precisely that two words u and v are conjugate in Γ if and only if u and v lie in the same path connected component of $\mathcal{G}(n)$, where $n = K \max\{|u|, |v|\}$. Thus we may decide if u and v represent conjugate elements of Γ.

The conjugacy problem for hyperbolic groups

Hyperbolic groups also satisfy that q.m.c. condition and hence have a solvable conjugacy problem. I shall describe a more efficient way to solve the conjugacy problem in this case. (David Epstein has recently discovered an even quicker algorithm.)

In a *free group* it is easy to solve the conjugacy problem. By definition, a word $w = a_0 \ldots a_n$ in the free group on $\mathcal{A}$ is *cyclically reduced* if $a_i^{-1} \neq a_{i+1}$ for $1 = 0, \ldots, n-1$, and $a_0^{-1} \neq a_n$. Given two words u and v, in order to check if they define conjugate elements of $F(\mathcal{A})$ one cyclically reduces both words and then looks to see if one of the resulting words is a cyclic permutation of the other.

We wish to quasify this simple solution to the conjugacy problem in order to solve the conjugacy problem for hyperbolic groups. With this is mind, note that a word in $F(\mathcal{A})$ is cyclically reduced if and only if it and all of its cyclic permutations are geodesic words. And recall (1.26) that geodesics in hyperbolic groups are well-approximated by local-geodesics. These facts motivate:

Proposition 4.6 *Let Γ be a group that is δ-hyperbolic with respect to the finite generating set $\mathcal{A}$. There exist constants L and K, depending only on δ, such that if $u, v \in F(\mathcal{A})$ represent conjugate elements of Γ, and if u, v and all of their cyclic permutations are L-local geodesics, then*

1. $\max\{|u|, |v|\} \leq K$, *or else*
2. *there exists a word $w \in F(\mathcal{A})$ of length at most K such that $w^{-1}u'w = v'$ in Γ, where u' and v' are cyclic permutations of u and v.*

See [33, III] for a proof.

4.7 An algorithm to decide conjugacy in hyperbolic groups

Let Γ be a group that is δ-hyperbolic with respect to the finite generating set $\mathcal{A}$. Given two words u and v over the alphabet $\mathcal{A}^{\pm 1}$, one looks for subwords of length $\leq L$ in u, v and their cyclic permutations that are not geodesic, where L is the constant of (4.6). If such a subword is found, one replaces it by a geodesic word representing the same group element. One continues this process until u and v have been replaced by (conjugate) words u' and v' all of whose cyclic permutations are L-local geodesics. Let u' and v' be the resulting words. The preceding proposition provides a *finite* set of words Σ such that u is conjugate to v in Γ if and only if $w^{-1}u'w = v'$ in Γ for some $w \in \Sigma$. And using Dehn's algorithm one can decide whether any of the putative relations $w^{-1}u'w = v'$ is actually valid in Γ.

The conjugacy problem for subgroups

Suppose that Γ acts properly and cocompactly by isometries on a CAT(0) space. We shall see a little later that Γ may contain finitely presented subgroups with an unsolvable conjugacy problem. On the other hand:

Theorem 4.8 ([29]) *If Γ acts properly and cocompactly by isometries on a CAT(0) space and if Q is a group in which the generalized word problem is solvable, then the kernel of any homomorphism $\Gamma \to Q$ has a solvable conjugacy problem.*

A key point in the proof of this result is that there is an algorithm which given $\gamma \in \Gamma$ will compute a finite set of generators for $C_\gamma(\Gamma)$; this uses ideas of Hamish Short [107].

Isomorphism problems

The ideas used in the proof of the following beautiful theorem of Zlil Sela [104] have not been discussed in these notes.

Theorem 4.9 *The isomorphism problem is solvable among torsion-free hyperbolic groups.*

Distinguishing among non-positively curved manifolds

Closed 2-manifolds were classified in the nineteenth century — they are determined up to homeomorphism by orientability and Euler characteristic. If Thurston's Geometrization Conjecture [112] is true then the homeomorphism problem for compact 3-manifolds is also solvable. In other words, there is an algorithm which takes as input pairs of compact 3-manifolds and answers YES or NO, after a finite amount of time, according to whether or not the manifolds are homeomorphic. (We are implicitly assuming that the manifolds under consideration are described as finite objects. For the sake of argument, let us suppose that they are specified as finite simplicial complexes.)

For each $n \geq 4$, there is an algorithm that associates to any finite presentation $\mathcal{P}$ a closed n-manifold with fundamental group $|\mathcal{P}|$. Using this fact, Markov [84] proved that there does not exist an algorithm to decide homeomorphism among (smooth, PL or topological) manifolds in general.

On the other hand, if one restricts one's attention to manifolds of negative curvature then the homeomorphism problem is solvable. To see this one needs the following deep topological rigidity theorem of Farrell and Jones [61].

Theorem 4.10 *Let $n \geq 5$ and let M and N be closed non-positively curved n-manifolds. If $\pi_1 M \cong \pi_1 N$, then M and N are homeomorphic.*

As Sela pointed out in [104], combining (4.9) and (4.10) one gets:

Theorem 4.11 *Let $n \geq 5$ be an integer. There exists an algorithm which takes as input two compact n-manifolds that support metrics of negative curvature, and which (after a finite amount of time) will stop and answer YES or NO according to whether or not the manifolds are homeomorphic.*

Cautious readers should interpret (4.12) as a statement regarding the homeomorphism problem for recursive classes of negatively curved manifolds.

Question 4.12 Is there an algorithm to decide homeomorphism among closed non-positively curved manifolds in each dimension $n \geq 4$?

I strongly suspect that the answer to this question is no. In the light of (4.11), in order to provide a negative answer one should show that the isomorphism problem for the fundamental class of such manifolds is unsolvable. For the moment I cannot even prove this with manifolds replaced by complexes. The only result in this direction is:

Theorem 4.13 ([13]) *There is a closed polyhedral manifold M of non-positive curvature and a recursive sequence of covering spaces $\widehat{M_i} \to M$, $i \in \mathbb{N}$, with each $\pi_1 \widehat{M_i}$ finitely presented, such that there does not exist an algorithm to determine whether or not M_i is homotopy equivalent to M_0.*

The subgroup question

Let me now turn to the basic question that I articulated at the beginning of this lecture:

Question 4.14 (The subgroup question[10]) Let X be a compact non-positively curved space and let $\Gamma \leq \pi_1 X$ be a subgroup. Under what circumstances can one deduce that Γ is the fundamental group of a *compact* non-positively curved space?

One might regard this question as a special case of the broader question: might there exist an algebraic characterization of the groups Γ that act properly and cocompactly by isometries on CAT(0) spaces? However (4.14) has other facets. For example, since any compact non-positively space X has the homotopy type of $K(\pi_1 X, 1)$, Question (4.14) can be rephrased as: under what circumstances does a finitely presented covering space of a non-positively curved space have the homotopy type of a compact non-positively curved space? In turn, this question is related to the existence of compact cores, in the sense of [103].

[10]In order to allow torsion, one ought to restate this question in terms of proper cocompact actions by groups of isometries. I have chosen the simpler form for expository reasons.

Subgroups of hyperbolic groups

Recall the modified Rips construction (3.9). *In the following discussion we shall maintain the notation established in (3.9).*

We first apply (3.9) in tandem with a theorem of Robert Bieri [19]:

Theorem 4.15 *If a normal subgroup of a group of cohomological dimension* 2 *is finitely presented, then that subgroup is either free or of finite index.*

It is not hard to show that the group N in the Rips construction (3.9) is not free if G is infinite and therefore:

Corollary 4.16 *In the notation of (3.9), the subgroup N is finitely presentable if and only if G is finite.*

Remarks (1) Baumslag, Miller and Short [14] observed that since there is no algorithm to decide if a group given by an arbitrary finite presentation is finite, (4.16) can be used to show that there is no algorithm to decide if a finitely generated subgroup of a hyperbolic group is finitely presented. See [14] for further results in this vein.

(2) We have just seen that one can produce finitely generated subgroups of hyperbolic groups that are not finitely presented and hence not hyperbolic. It is much more difficult to construct finitely presented subgroups of hyperbolic groups which are not themselves hyperbolic. At present the only construction of such subgroups is that given by Noel Brady [25]. In [62] Steve Gersten shows that if G is a hyperbolic group of cohomological dimension two, then every finitely presented subgroup of G is also hyperbolic.

Positive answers to the subgroup question in low dimensions

In low dimensions there are clear positive answers to the subgroup closure question (4.14). If X is 1-dimensional then in order to get the desired action it is sufficient to assume that $\Gamma \subset \pi_1 X$ is finitely generated — subgroups of free groups are free! In (4.16) we saw that in dimension 2 it is not sufficient to assume that Γ is finitely generated — if Γ is to be the fundamental group of a compact non-positively curved space then it must be finitely presented. However the absence of finite presentability turns out to be the only obstruction in dimension 2. This follows from a very general result that I learnt from Peter Shalen:

Theorem 4.17 (Subgroup closure theorem) *Let $\mathbb{K}$ be a class of combinatorial CW complexes that is closed under passage to connected subcomplexes and covering spaces. If $K \in \mathbb{K}$ and H is a finitely presented subgroup of $\pi_1 K$, then there is a compact $X \in \mathbb{K}$ of dimension at most 2 such that $\pi_1 X \cong H$.*

By definition, a combinatorial map between CW complexes is one that takes open cells homeomorphically onto cells. A complex is called combinatorial if all of its attaching maps are combinatorial.

When endowed with the unique path metric that makes the covering map a local isometry, a covering of any non-positively curve space is obviously non-positively curved. Thus one can apply the above theorem to any class of non-positively curved complexes that is closed under passage to subcomplexes. It is clear from the link condition (3.4) that 2-dimensional complexes satisfy this condition, thus:

Corollary 4.18 *If X is a compact non-positively curved 2-complex, then every finitely presented subgroup of $\pi_1 X$ is also the fundamental group of such a complex.*

One proves the Subgroup closure theorem by means of a tower argument. Towers were developed in the context of 3-manifolds by C.D. Papakyriakopoulos [94] and adapted to the setting of combinatorial complexes by Jim Howie [74].

Definition 4.19 (of a tower) A *tower map* (of height h) is a map $g : K_h \to L_0$ between connected CW complexes that admits a decomposition

$$g = i_0 \circ p_1 \circ \cdots \circ p_h \circ i_h,$$

where each $i_r : K_r \to L_r$ is the inclusion of a connected subcomplex, and each $p_r : L_r \to K_{r-1}$ is a connected covering. A *tower lifting* of a continuous map $f : Y \to L$ of CW-complexes is a factorization $f : Y \xrightarrow{f'} K \xrightarrow{g} L$ where g is a tower map.

Howie [74, 3.1] gives an integer valued measure of complexity for combinatorial maps that is reduced by passage to proper coverings. Thus he proves:

Lemma 4.20 *If Y and L are connected CW complexes and Y is compact, then every combinatorial map $f : Y \to L$ has a (maximal) tower lift $f : Y \xrightarrow{f'} K \xrightarrow{g} L$ with $f'_* : \pi_1 Y \to \pi_1 K$ surjective.*

One uses (4.20) in tandem with the following enhanced version of van Kampen's Lemma [77], a detailed proof of which can be found in [33]; see also [62].

Lemma 4.21 *If X is a combinatorial complex, G is a finitely presented group and $\psi : G \to \pi_1(X, x_o)$ is a homomorphism, then there exists a compact 2-complex Y with $\pi_1(Y, y_o) \cong G$ and a combinatorial map $f : Y \to X$ such that $f(y_o) = x_o$ and the following diagram commutes:*

$$\begin{array}{ccc} G & \xrightarrow{\ \psi\ } & \pi_1(X, x_0) \\ \downarrow \cong & & \| \\ \pi_1(Y, y_0) & \xrightarrow{\ f_*\ } & \pi_1(X, x_0). \end{array}$$

When taken together, these lemmas prove the Subgroup closure theorem.

3-manifolds

The examples described in (4.23) show that (4.18) cannot be extended to complexes of any dimension greater than 2. But in the setting of manifolds there is a subgroup closure result in dimension 3.

Theorem 4.22 *If M is a compact non-positively curved 3-manifold, then every finitely generated subgroup of $\pi_1 M$ is the fundamental group of a compact non-positively curved 3-manifold.*

This theorem is proved in [29]; it is a consequence of Scott's Compact Core Theorem [103] and (3.15). The manifolds considered may have boundary.

Cohomological obstructions

Example 4.23 Let F be a free group of rank two, let P_n be the direct product of n copies of F and let $h_n : P_n \to \mathbb{Z}$ be a map whose restriction to each factor is a surjection. Let $SB_n = \ker h_n$.

Following earlier work of Stallings [109] on the case $n = 3$, in [20] Bieri showed that if $n \geq 3$ then SB_n is finitely presented and that there is an Eilenberg-Maclane space $K(SB_n, 1)$ with a finite $(n-1)$-skeleton, but $H^n(SB_n, \mathbb{Z})$ is not finitely generated and hence there is no $K(SB_n, 1)$ with a finite n-skeleton. It follows from 2.1(7) that SB_n is not the fundamental group of any compact non-positively curved space. (It does however satisfy the other properties listed in (2.1) — see [30].)

Bad subgroups via fibre products

In this section I shall explain how Baumslag, Miller, Short and I exploited the modified Rips construction (3.9) in [13].

The Rips construction encodes the pathologies of arbitrary finite group presentations into pairs of groups $N \subset \Gamma$, where Γ is the fundamental group of a compact negatively curved complex and N is a finitely generated subgroup. Here is an example of how this pathology gets transmitted:

Proposition 4.24 *In the notation of (3.9): if $G = \Gamma/N$ has an unsolvable word problem, then $N \subset \Gamma$ (which is finitely generated) has an unsolvable conjugacy problem.*

Proof We continue with the notation of (3.9). Fix $a \in \underline{a}$. Since Γ is torsion-free and hyperbolic, the centralizer of a is cyclic (1.36). Moreover, one can arrange for a to be the generator of its centralizer (see [33, II.5] or [13]).

Given a word w in the generators $\mathcal{X}$ one can use the relations U to rewrite $w^{-1}aw$ as a word $\hat{w}$ in the generators $\underline{a}$. If $w \in N$ then obviously $\hat{w}$ is conjugate to a in N. Conversely, if $n^{-1}an = \hat{w}$, then $n^{-1}w$ centralizes a and hence $w = na^r$ for some r. It follows that $w \in N$ if and only if $\hat{w}$ is conjugate to a in N. But w is in N if and only if $w = 1$ in $G = \Gamma/N$, and we are assuming that there is no algorithm to decide if $w = 1$ in G. □

Constructing finitely *presented* subgroups which display the sort of pathologies described in (4.24) is much harder. The key to doing so is the following technical theorem. A group G is said to be of type F_3 if there is a $K(G,1)$ with a finite 3-skeleton.

Theorem 4.25 (The 1-2-3 theorem) *Suppose that* $1 \to N \to \Gamma \xrightarrow{p} G \to 1$ *is exact, and consider the fibre product:*

$$P := \{(\gamma_1, \gamma_2) \mid p(\gamma_1) = p(\gamma_2)\} \subseteq \Gamma \times \Gamma.$$

If N is finitely generated, Γ is finitely presented and G is of type F_3, then P is finitely presented.

The name of this theorem comes from the fact that the groups N, Γ and G are assumed to be of type F_1, F_2 and F_3 respectively. (Results similar to this, cast in the language of pictures, can be found in [5].)

The product of non-positively curved complexes is non-positively curved, so we may use the 1-2-3 theorem in tandem with the modified Rips construction in order to construct non-positively curved 4-dimensional complexes whose finitely presented subgroups display various pathologies.

Theorem 4.26 *There exist hyperbolic groups Γ and finitely presented subgroups $P \subset \Gamma \times \Gamma$ such that there is no algorithm to decide membership of P, and the conjugacy problem for P is unsolvable. (And Γ is the fundamental group of a compact negatively curved 2-complex.)*

A subtle argument involving (4.26), a judicious choice of HNN extension and a careful analysis of centralizers, yields:

Theorem 4.27 *There exists a compact non-positively curved 4-complex X and a recursive class of finitely presented subgroups $H_n \subset \pi_1 X$ ($n \in \mathbb{N}$) such that there is no algorithm to determine if H_n is (abstractly) isomorphic to H_0.*

By a process of relative hyperbolization one can isometrically embed X in a polyhedral manifold, hence (4.13).

Let me finish be stating a result intended to illustrate how subtle the Subgroup question (4.14) can be:

Theorem 4.28 ([29]) *There exists a closed non-positively curved manifold of dimension nine whose fundamental group has a subgroup $H \subset \Gamma$ such that:*

1. *H has a finite $K(H,1)$,*
2. *H satisfies a quadratic isoperimetric inequality,*
3. *H has a solvable conjugacy problem,*
4. *the centralizer $C_H(h)$ of every $h \in H$ is finitely presented, but*
5. *there exist $h \in H$ such that $C_H(h)$ does not satisfy a quadratic isoperimetric inequality, and therefore H does not act properly and cocompactly by isometries on any CAT(0) space.*

References

[1] S. B. Alexander and R. L. Bishop, "The Hadamard–Cartan theorem in locally convex spaces", *Enseign. Math.* **36** (1990) 309–320.

[2] A. D. Alexandrov, "A theorem on triangles in a metric space and some of its applications", *Trudy Mat. Inst. Steklov* **38** (1951) 5-23.

[3] J. M. Alonso, "Inégalités isopérimétriques et quasi-isométries", *C.R. Acad. Sci. Paris Sér. 1 Math.* **311** (1990) 761–764.

[4] J. M. Alonso and M. R. Bridson, "Semihyperbolic groups", *Proc. London Math. Soc.* (3) **70** (1995) 56–114.

[5] Y.G. Baik, J. Harlander and S.J. Pride, *The Geometry of Group Extensions*, Preprint, University of Glasgow, 1997.

[6] W. Ballmann, *Lectures on Spaces of Nonpositive Curvature*, DMV Seminar, Band 25, Birkäuser, Basel, Boston, Berlin 1995.

[7] W. Ballmann and M. Brin, "Polygonal complexes and combinatorial group theory", *Geom. Dedicata* **50** (1994) 164-191.

[8] W. Ballmann and S. Buyalo, "Non-positively curved metrics on 2-polyhedra", *Math. Z.* **222** (1996) 97–134.

[9] W. Ballmann, M. Gromov, V. Schroeder, "Manifolds of nonpositive curvature", *Prog. Math.* **61** Birkäuser, Boston, 1985

[10] V. Bangert and V. Schroeder "Existence of flat tori in analytic manifolds of nonpositive curvature *Ann. Sci. Ecole Num. Sup.* **24** (1991) 605–634.

[11] C. Baribaud, *Travail de diplôme*, Univ. de Genève, 1993.

[12] G. Baumslag, M. R. Bridson, C. F. Miller III and H. B. Short, "Subgroups of automatic groups and their isoperimetric inequalities", *J. London Math. Soc.(2)* **56** (1997) 292–304.

[13] G. Baumslag, M. R. Bridson, C. F. Miller III and H. B. Short, *Fibre products, non-positive curvature and decision problems*, Preprint, Oxford 1998.

[14] G. Baumslag, C. F. Miller III and H. Short, "Unsolvable problems about small cancellation groups and word hyperbolic groups", *Bull. London Math. Soc.* **23** (1993) 97–101.

[15] G. Baumslag and J.E. Roseblade, "Subgroups of Direct Products of Two Free Groups" J. London Math. Soc. (2) **30** (1984) 44–52.

[16] G. Baumslag and D. Solitar "Some two-generator one-relator non-Hopfian groups", *Bull. Amer. Math. Soc.* **68** (1962) 199–201.

[17] Nadia Benakli, "Polyèdres à géométrie locale donnée", *C. R. Acad. Sci. Paris* **313** (1991) 561–564.

[18] M. Bestvina and N. Brady, "Morse theory and finitenesss properties of groups", *Invent. Math* **129** (1997) 445–470.

[19] R. Bieri, "Normal subgroups in duality groups and in groups of cohomological dimension 2", *J. Pure Appl. Algebra* **7** (1976) 32–52.

[20] R. Bieri, "Homological dimension of discrete groups", Queen Mary Lecture Notes.

[21] O. V. Bogopolskii and V. N. Gerasimov, "Finite subgroups of hyperbolic groups", *Algebra and Logic* **34** (1995) 343–345.

[22] B. Bowditch, "Notes on Gromov's hyperbolicity criterion for path-metric spaces in Group theory from a geometrical viewpoint", eds. E. Ghys, A. Haefliger, A. Verjovsky, Proc. ICTP Trieste, World Scientific, Singapore (1991) 373–464.

[23] B. Bowditch "A short proof that a subquadratic isoperimetric inequality implies a linear one", *Michigan J. Math.* **42** (1995) 103–107.

[24] B. Bowditch "Minkowskian subspaces of non-positively curved metric spaces", *Bull. London Math. Soc.* **27**(1995) 575–584.

[25] N. Brady, "A non-hyperbolic finitely presented subgroup of a hyperbolic group",

J. London Math. Soc., to appear.
[26] M. R. Bridson, "Geodesics and curvature in metric simplicial complexes", in *Group theory from a geometrical viewpoint*, eds. E. Ghys, A. Haefliger, A. Verjovsky *Proc. ICTP Trieste*, World Scientific, Singapore (1991) 373–464.
[27] M. R. Bridson, "On the existence of flat planes in spaces of nonpositive curvature", *Proc. Amer. Math. Soc.* **123** (1995) 223–235.
[28] M. R. Bridson, "On the geometry of normal forms in discrete groups", *Proc. London Math. Soc.(3)* **67** (1993) 596–616.
[29] M. R. Bridson, "On the subgroups of semihyperbolic groups", Preprint, Oxford 1998.
[30] M. R. Bridson "Doubles, finiteness properties of groups and the quadratic isoperimetric inequality *J. Algebra*, to appear.
[31] M. R. Bridson, "Controlled embeddings into groups that have no non-trivial finite quotients", *J. Geom. Topol.*, to appear.
[32] M.R. Bridson and S.M. Gersten "The optimal isoperimetric inequality for torus bundles over the circle", *Quart. J. Math.* **47** (1996), 1–23.
[33] M.R. Bridson and A. Haefliger, *Metric spaces of non-positive curvature*, Springer-Verlag, to appear.
[34] M. R. Bridson and L. Mosher, "Metrics of non-positive curvature on 3-manifolds", in preparation.
[35] M. R. Bridson and D. Wise, "VH-complexes, towers, and subgroups of $F \times F$", *Math. Proc. Cambridge Philos. Soc.*, to appear.
[36] K. S. Brown *Buildings*, Springer Verlag, Berlin, Heidelberg, New York. 1988.
[37] F. Bruhat et J. Tits, "Groupes réductifs sur un corps local. I. Données radiciells valuées", *I.H.E.S Publ. Math.* **41** (1972) 5–251.
[38] Y. Burago, M. Gromov and G. Perel'man, "A. D. Alexandrov spaces with curvature bounded below", *Russian Math. Surveys* **47**:2 (1992) 3–51.
[39] M. Burger and S. Mozes, "Finitely presented simple groups and products of trees", *C.R. Acad. Sci. Paris Sér 1 Math* **324.I** (1997) 747–752.
[40] H. Busemann, "Spaces with non-positive curvature", *Acta Math.* **48** (1947) 234–267.
[41] S. Buyalo, "Euclidean planes in open 3-dimensional manifolds of non-positive curvature", *St. Petersburg Math J.* **3** (1992) 83–96.
[42] J. W. Cannon, " The combinatorial structure of cocompact discrete hyperbolic groups" *Geom. Dedicata* **16** (1984) 123–148.
[43] J. W. Cannon, "The theory of negatively curved spaces and groups" in *Ergodic theory, symbolic dynamics and hyperbolic spaces*, eds. T. Bedford, M. Keane, C. Series. Oxford Univ. Press, New York (1991) 315–369.
[44] E. Cartan, *Leçons sur la géométrie des espaces de Riemann*, Gauthier-Villars, Paris 1928.
[45] A. Cayley "On the theory of groups", *Proc. London Math. Soc.*, **9** (1878) 126–133.
[46] B. Chandler and W. Magnus, "The history of combinatorial group theory", Springer Verlag, New York 1982.
[47] R. Charney and M. Davis, "Strict hyperbolization", *Topology* **34** (1995) 329–350.
[48] J. Cheeger and B. Ebin, *Comparison Theorems in Riemannian Geometry*, North Holland, Amsterdam 1975.
[49] M. Coonaert, T. Delzant and A. Papadopoloulos, *Notes sur les groupes hyperboliques de Gromov*, Springer Verlag, LN **1441** Berlin, Heidelberg, New York.
[50] M.W. Davis, "Groups generated by reflections and aspherical manifolds not covered Euclidean space", *Ann. Math. (2)* **117** (1983) 293–324.
[51] M. Davis and T. Januszkiewicz, "Hyperbolization of polyhedra", *J. Differential Geom.* **34** (1991) 347–388.
[52] M. Dehn, *Papers on group theory and topology*, translated and introduced by John

Stillwell, Springer Verlag Berlin, Heidelberg, New York 1987.

[53] M. Dehn, "Transformationen der Kurven auf zweiseitigen Flächen", *Math. Ann.* **72** (1912) 413–421.

[54] M. Dehn, "Uber unendliche diskontinuierliche Gruppen", *Math. Ann.* **71** (1912) 116–144.

[55] T. Delzant, "Sous-groupes distingués et quotients des groupes hyperboliques", *Duke Math. J.* **83** (1996) 661–682.

[56] W. von Dyck, "Gruppentheoretische Studien", *Math. Ann.* **20** (1882) 1–45.

[57] P. B. Eberlein, "Geodesic flow in certain manifolds without conjugate points", *Trans. Amer. Math. Soc.* **167** (1972) 151–170.

[58] V.A. Efromovich "The proximity geometry of Riemannian manifolds", *Uspekhi Math. Nauk* **8** (1953) 189–195,

[59] D.B.A. Epstein, J.W. Cannon, D.F. Holt, S.V.F. Levy, M.S. Paterson, and W.P. Thurston, *Word Processing in Groups*, Jones and Bartlett, Boston MA, 1992.

[60] B. Farb "The quasi-isometry classification of lattices in semisimple Lie groups", *Math. Res. Letters* **4** (1997) 705–718.

[61] T. Farrell and L. Jones, "Topological rigidity for compact non-positively curved manifolds", in "Differential geometry, Los Angeles 1990", *Proc. Sympos. Pure Math.* **54**, 1993, 224–274

[62] S.M. Gersten "Subgroups of word hyperbolic groups in dimension 2", *J. London Math. Soc.* **54** (1996) 261–283.

[63] S.M. Gersten and H. Short "Rational subgroups of biautomatic groups", *Ann. Math.* **134** (1991) 125–158.

[64] E. Ghys and P. de la Harpe (editors) "Sur les groupes hyperboliques d'après Mikhael Gromov", *Prog. in Math. 83* Birkäuser, Boston 1990.

[65] R. Grigorchuik "On the system of defining relations and the Schur multiplier of periodic groups generated by finite automata", in *Proceedings of Groups St Andrews 1997 in Bath*, eds. C.M. Campbell, E.F. Robertson, N. Ruskuc & G.C. Smith, Cambridge University Press, Cambridge 1998.

[66] D. Gromoll and J. Wolf, "Some relations between the metric structure and the algebraic structure of the fundamental group in manifolds of non-positive curvature", *Bull. Amer. Math. Soc.* **77** (1971) 545–552.

[67] M. Gromov, "Infinite groups as geometric objects", *Proc. ICM*, Warsaw 1983 PWN, Warsaw 1984, 385–392

[68] M. Gromov, "Hyperbolic groups", in *Essays in group theory*, ed. S. M. Gersten, Springer Verlag, MSRI **8** (1987) 75–263.

[69] M. Gromov, "Asymptotic invariants of infinite groups", in *Geometric Group Theory* LMS Lecture Notes, **182** eds. G. A. Niblo and M. A. Roller, Cambridge Univ. Press 1993.

[70] M. Gromov, "Groups of polynomial growth and expanding maps", *Publ. Math. I.H.E.S.* **53** (1981) 53-78.

[71] A. Haefliger "Complexes of groups and orbihedra" in "Group theory from a geometric viewpoint" eds. E. Ghys, A. Haefliger, A. Verjovsy *Proc. ICTP Trieste*, World Scientific, Singapore (1991) 504–540.

[72] A. Haefliger, "Extension of complexes of groups", *Ann. Inst. Fourier* **42** (1992) 275–311.

[73] F. Haglund "Les polyèdres de Gromov", *C. R. Acad. Sci. Paris Sér. 1 Math* **313** (1991) 603–606.

[74] J. Howie, "On pairs of 2-complexes and systems of equations over groups", *J. Reine Angew. Math.* **324** (1981) 165–174.

[75] W. Jaco and P. Shalen, "Seifert fibered spaces in 3-manifolds", *Memoirs Amer.*

Math. Soc. **220** (1979).

[76] K. Johannson, *Homotopy equivalence of 3-manifolds with boundaries*, Springer Verlag, LN **761**, Berlin-Heidelberg- New York.

[77] E. R. van Kampen, "On some lemmas in the theory of groups", *Amer. J. Math.* **55** (1933) 268–273.

[78] M. Kapovich and B. Leeb, "Actions of discrete groups on non-positively curved spaces", *Math. Ann.* **306** (1996) 341–352.

[79] H. B. Lawson and S. T. Yau, "Compact manifolds of nonpositive curvature", *J. Differential Geom.* **7** (1972) 211–228.

[80] B. Leeb "3-manifolds with(out) metrics of non-positive curvature", Ph.D. thesis, University of Maryland, 1993.

[81] R. C. Lyndon and P. E. Schupp, "Combinatorial group theory", Springer Verlag, Berlin, Heidelberg, New York 1977.

[82] I. G. Lysenok, "On some algorithmic properties of hyperbolic groups", *Math. USSR Izv.* **35** (1990) 145–163.

[83] G. A. Margulis, *Discrete subgroups of semisimple Lie groups*, Ergeb. Math. **17** Springer Verlag, Berlin, Heidelberg, New York 1991.

[84] A. A. Markov, "Insolubility of the problem of homeomorphy", in *Proc. ICM Cambridge* (1958) 300-306. Cambridge University Press, 1960.

[85] D. Meier, "Non-Hopfian groups" *J. London Math. Soc.*(2) **26** (1982) 265–270.

[86] C.F. Miller III, "Decision problems for groups: survey and reflections" in *Algorithms and Classification in Combinatorial Group Theory*, eds. G Baumslag and C. F. Miller III, MSRI Publ. **23**, Springer-Verlag, Berlin, New York, Heidelberg 1992, 1–59.

[87] J. Milnor "A note on curvature and the fundamental group", *J. Differential Geom.* **2** (1968) 1-7.

[88] L. Mosher "Geometry of cubulated 3-manifolds", *Topology* **34** (1995) 789–814.

[89] G. Moussong " Hyperbolic Coxeter groups", Ph.D Thesis, The Ohio State University, 1988.

[90] A. Yu. Ol'shanskii "Hyperbolicity of groups with subquadratic isoperimetric inequalities", *Internat. J. Algebra Comput.* **1** (1991) 281–289.

[91] A. Yu. Ol'shanskii "Almost every group is hyperbolic", *Intl. J. Alg. Comp.* **2** (1992) 1–17.

[92] A. Yu. Ol'shanskii "SQ-universality of hyperbolic groups", *Mat. Sb.* **186** (1995) 119-132.

[93] P. Pansu "Croissance des boules et des géodesiques fermées dans les nilvariétés", *Ergodic Theory Dynamical Systems* **3** (1983) 415-445.

[94] C. D. Papakiakopolous, "On Dehn's lemma and the asphericity of knots" *Ann. of Math.* **66** (1957) 1–26.

[95] P. Papasoglou "On the sub-quadratic isoperimetric inequality" in *Geometric group theory*, Walter de Gruyter, Berlin-New York, 1995, 149–158.

[96] F. Paulin, "Construction of hyperbolic groups via hyperbolization of polyhedra" in *Group theory from a geometric viewpoint*, eds. E. Ghys, A. Haefliger, A. Verjovsy *Proc. ICTP Trieste*, World Scientific, Singapore 1991, 313–373.

[97] H. Poincaré, "Analysis Situs", *J. d'Ecole poly. Norm.* **1** (1895) 1–121.

[98] H. Poincaré, "Cinquième complement à l'analysis situs", *Rend. Circ. Mat. Palermo* **18** (1904) 45–110.

[99] E. Rips, "Subgroups of small cancellation groups *Bull. London Math. Soc.* **14** (1982) 45–47.

[100] M. Sageev, "Ends of group pairs and non-positively curved cube complexes", *Proc. London Math. Soc.* **71** (1995) 585-617.

[101] V. Schroeder, "Codimension one tori in manifolds of non-positive curvature", *Geom.*

Dedicata **33** (1990) 251–263.

[102] V. Schroeder, "A cusp closing theorem", *Proc. Amer. Math. Soc.* **106** (1989) 797–802.

[103] G. P. Scott, "Compact submanifolds of 3-manifolds", *J. London Math. Soc.* **7** (1973) 246–250.

[104] Z. Sela "The isomorphism problem for hyperbolic groups I", *Ann. of Math.*(2) **141** (1995) 217–283.

[105] J-P. Serre, *Trees*, Springer Verlag, Berlin, Heidelberg, New York 1980. Translation by J. Stillwell of "Arbres, Amalgames, SL_2", *Astérisque* **46** (1977).

[106] H. B. Short (editor) "Notes on negatively curved groups Group theory from a geometrical viewpoint" Eds. E. Ghys, A. Haefliger, A. Verjovsky, *Proc. ICTP Trieste*, World Scientific, Singapore (1991) 3–64.

[107] H. B. Short, "Groups and combings", Preprint, ENS Lyon, 1990.

[108] J.R. Stallings, "Triangles of groups" in *Group theory from a geometrical viewpoint*, eds E. Ghys, A. Haefliger, A. Verjovsky *Proc. ICTP Trieste*, World Scientific, Singapore (1991) 491–503.

[109] J.R. Stallings, "A finitely presented group whose 3-dimensional integral homology is not finitely generated", *Amer. J. Math.* **85** (1963) 541–543.

[110] A.S. Švarc, "Volume invariants of coverings", *Dokl. Ak. Nauk.* **105** (1955) 32-34.

[111] W. P. Thurston, "The geometry and topology of 3-manifolds", Lecture notes, Princeton University, 1980.

[112] W. P. Thurston, "Three dimensional manifolds, Kleinian groups and hyperbolic geometry", *Bull. Amer. Math. Soc.* **6** (1982) 357-381.

[113] J. Tits, "On buildings and their Applications", in *Proc. ICM, Vancouver* 1974 **1** Canad. Math. Congress, Montreal 1975 209–220.

[114] C.M. Weinbaum "The word and conjugacy problem for the knot group of any tame prime alternating knot", *Proc. Amer. Math. Soc.* **22** (1971) 22–26.

[115] D.T. Wise *Nonpositively curved squared complexes, aperiodic tilings, and non-residually finite groups*, Ph.D Thesis, Princeton Univ., 1996.

[116] D.T. Wise, "Incoherent negatively curved groups", *Proc. Amer. Math. Soc.* **126** (1998) 957–964.

[117] D.T. Wise, "An automatic group that is not Hopfian", *J. Algebra* **180** (1996) 845–847.

[118] J.A. Wolf, *Spaces of constant curvature*, McGraw-Hill, New York, 1967.

GROUP-THEORETIC APPLICATIONS OF NON-COMMUTATIVE TORIC GEOMETRY

CHRISTOPHER J. B. BROOKES

Corpus Christi College, Cambridge CB2 1RH, England

1 Introduction

These lectures consider the way some geometric techniques can be used to approach questions about 'small' groups, in particular, their finite presentability and the structure of their automorphism groups. By a 'small' group I mean one containing no non-Abelian free subgroups. Most of the time this will be in the context of discrete groups but towards the end I shall observe that one can glean related information using counting methods in the world of pro-p groups. The common theme is that a small group is in some sense at most half the size of a free group. The geometric and pro-p approaches give different ways of making this latter statement more precise. Both techniques involve looking only at the residually finite images of the group; there is no information to be had in this way about infinite simple groups for example. At the other extreme, finite groups also do not register on the scale.

In fact we shall be looking at various aspects of the representation theory of the group and consequently group rings come into play. However, rather than concentrating on group rings, I shall try and convince you that some non-commutative relatives of familiar commutative Noetherian rings are worth studying and that their representation theory has applications in group theory.

When looking at co-ordinate rings in algebraic geometry, one way of defining dimension is via the transcendence degree of the fraction field. For example the complex group algebra of a free Abelian group of rank n has dimension n while that of a finite Abelian group has dimension 0. Our geometric approach uses an analogue of this version of dimension, while the pro-p method is based on one more akin to that defined via the growth rate of images of the co-ordinate ring in characteristic p.

The analogy with algebraic geometry is a close one. The complex torus $(\mathbb{C}^\times)^n$ has co-ordinate ring $\mathbb{C}H$, the group ring of a free Abelian group H of rank n, or equivalently the complex Laurent polynomial ring $\mathbb{C}[X_1, X_1^{-1}, \ldots, X_n, X_n^{-1}]$. The prime spectrum Spec $\mathbb{C}H$ is the set of prime ideals of $\mathbb{C}H$ and, by the Nullstellensatz, the maximal ideals of $\mathbb{C}H$ are the kernels of the ring maps $\mathbb{C}H \longrightarrow \mathbb{C}$ and are therefore in 1–1 correspondence with the points of $(\mathbb{C}^\times)^n$. For any ideal I of $\mathbb{C}H$ one defines $V(I)$ to be the closed subset consisting of the points $(x_1, \ldots, x_n) \in (\mathbb{C}^\times)^n$ at which each Laurent polynomial in I vanishes. This corresponds to the set of maximal ideals of $\mathbb{C}H/I$. More generally, for a finitely generated $\mathbb{C}H$-module A, one is interested in $\{P \in \text{Spec } \mathbb{C}H : A > AP\}$. In fact, if $A > AP$ there usually is a proper descending filtration $A > AP > AP^2 > \ldots$. What I want to present is an approach to groups G which involves keeping a record of the existence or otherwise of certain well-chosen proper filtrations of the integral group ring $\mathbb{Z}G$. Incidentally, the pro-p group approach focuses on just one filtration of $\mathbb{F}_pG$, namely that given by powers of the augmentation ideal $\mathfrak{g}$, the kernel of the map $\mathbb{F}_pG \longrightarrow \mathbb{F}_p$ with each $g \longrightarrow 1$. This is analogous to considering just local information in the geometric set up.

Of course, considering filtrations of group rings is not new. These lectures are descendants of those given by Jim Roseblade [45] at Groups-St Andrews 1985, where he talked about applications of commutative algebra and valuation theory to the study of group rings. My aim is to report on some developments since then, and in particular on my recent work with John Groves [14, 15, 16, 17], which extends Jim's discussion to a non-commutative setting. This involves looking at the representation theory of torsion-free nilpotent groups rather than that of free Abelian ones. I shall concentrate on groups of nilpotency class 2 in order to explain how the non-commutativity crucially influences the representation theory and thereby imposes constraints on small finitely presented groups.

There is also a connection with the large body of work in geometric group theory inspired by the Nielsen-Thurston theory of surface automorphisms [19]. There an automorphism of a closed surface is studied via its induced action on the boundary of the compactification of the Teichmüller space of the surface. Each boundary point may be represented by an action of the fundamental group on an $\mathbb{R}$-tree obtained from the leaf space of a measured codimension 1 lamination of the surface. Culler, Morgan and Shalen [21, 33, 34, 35, 36] generalised this theory to higher dimensional manifolds, defining a natural compactification of the moduli space of hyperbolic structures on the manifold using methods from algebraic geometry. Again the boundary points are represented by certain $\mathbb{R}$-tree actions of the fundamental group. Shortly afterwards Bestvina [4] and Paulin [38] provided an alternative approach to these degenerations of hyperbolic structure. Along the same lines Culler and Vogtmann [22] considered the space of all free properly discontinuous $\mathbb{R}$-tree actions of a free group F. This is contractible and has a compactification whose boundary points are also represented by $\mathbb{R}$-tree actions of F, this time with cyclic arc stabilisers. This compact space, which has become known as 'Outer Space', has been fundamental in the recent advances in the study of automorphisms of free groups. Moreover a similar approach has proved fruitful in studying the structure and automorphisms of word hyperbolic groups (see for example [41]).

The filtrations in these lectures also arise in connection with a particular type of tree action and we shall study group automorphisms via a construction related to the compactification of complex varieties.

2 Introducing some crossed products

Let me start by fixing some notation which I shall keep standard throughout. We shall be dealing with a finitely generated group G, which will often be finitely presented but need not be at this stage. Suppose it has a torsion-free nilpotent quotient $G/K = H$. The lack of torsion is not important, but it does help make things clearer. Think of H as a group of unitriangular integral matrices if you wish.

We want to make use of the representation theory of H and so we look at Abelian sections of G on which K acts trivially via conjugation. I shall restrict attention to a section $K/L = A$, where L is normal in G and $L \geq K'$. This has the attraction that A, when viewed as a $\mathbb{Z}H$-module via conjugation, is finitely generated. This is because the finitely generated nilpotent group H is necessarily finitely presented.

The main problem is that even were A cyclic as a $\mathbb{Z}H$-module one would have $A \cong \mathbb{Z}H/I$, where I is a 1-sided ideal rather than 2-sided. The ring $\mathbb{Z}H$ is non-commutative whenever H is, and there is no immediate relationship between 1-sided and 2-sided ideals. The obvious analogue of $V(I)$ as defined for the commutative ring $\mathbb{C}H$, is not going to yield a rich enough theory. On the other hand $\mathbb{Z}H$ is almost as well behaved as a non-commutative ring can be. It is Noetherian and its prime ideals are well understood thanks to Zaleskii [52] and Roseblade [44]. Moreover it has no non-zero zero divisors and satisfies both the right and left Ore conditions; this ensures it has a unique full division ring of quotients. If instead we only invert some of the non-zero elements we get other interesting localisations of the group ring.

For example, assuming that $H/H' = Q$ is free Abelian, again just for convenience, we can recover a non-commutative version S of a group ring of Q by inverting all the non-zero elements of the subring $\mathbb{Z}H'$ of $\mathbb{Z}H$. Thus $S = \mathbb{Z}H(\mathbb{Z}H'\backslash\{0\})^{-1}$ and contains the division ring $D = \mathbb{Z}H'(\mathbb{Z}H'\backslash\{0\})^{-1}$. Given a basis $x_1, \ldots, x_n$ of the free Abelian group Q we can pick $h_1, \ldots, h_n$ in H with $x_i = H'h_i$, and $h_1, \ldots, h_n$ will generate H. Each element of S is uniquely expressible in the form $\sum d_{(r_1,\ldots,r_n)} h_1^{r_1} \ldots h_n^{r_n}$, where $(r_1, \ldots, r_n) \in \mathbb{Z}^n$, $d_{(r_1,\ldots,r_n)} \in D$, and only finitely many $d_{(r_1,\ldots,r_n)}$ are non-zero. For example, when H is Abelian and so equal to Q, we get $S = \mathbb{Q}H$ the rational group ring. In general the partial localisation S of $\mathbb{Z}H$ is an example of a crossed product of a division ring D by the free Abelian group Q. Let me remind you of the general definition of such a crossed product. One has a Q-graded D-vector space; there is a basis $\{e_q; q \in Q\}$ indexed by the group and every element is uniquely expressible in the form $\sum d_q e_q$ with each $d_q \in D$ and only finitely many d_q non-zero. Thus $S = \bigoplus_{q \in Q} De_q$ and each De_q is the q-component. An associative multiplication respecting this grading is defined using maps $\tau : Q \longrightarrow \text{Aut } D$ and $\theta : Q \times Q \longrightarrow D\backslash\{0\}$ such that

$$d_{q_1} e_{q_1} . d_{q_2} e_{q_2} = d_{q_1} d_{q_2}^{\tau(q_1)} \theta(q_1, q_2) e_{q_1 q_2}.$$

Thus $d_{q_1} e_{q_1} . d_{q_2} e_{q_2}$ belongs to $De_{q_1 q_2}$ and clearly the multiplicative identity will lie in De_1; we may as well assume that it is e_1. Note that each e_q is a unit and associativity ensures that θ satisfies some cocycle condition. τ itself is not necessarily a homomorphism but it does induce a homomorphism from Q to the outer automorphism group of D. In our example we can set $q = x_1^{r_1} \ldots x_n^{r_n} \in Q$, $e_q = h_1^{r_1} \ldots h_n^{r_n}$ and $d_q = d_{(r_1,\ldots,r_n)}$.

At first sight we might expect these crossed products to be rather uninteresting generalisations of group rings and expect $\mathbb{C}Q$ and $\mathbb{Q}Q$ to be typical examples. However, even when τ is trivial, D is a (commutative) field but the cocycle θ is non-trivial, such crossed products need not behave like commutative algebras. Moreover they are of independent interest, surfacing as the 'quantum torus' in various guises in the theory of quantum groups and C^*-algebras, usually as a completed version of a crossed product of $\mathbb{C}$ by a free Abelian group Q of rank 2. (See for example the study by Rieffel [39, 40] of the K-theory for the C^*-algebra version.) The representation theory is of a different flavour to that of $\mathbb{C}Q$ when the image of the cocycle θ does not consist simply of roots of unity. For example, if H is free nilpotent of class 2 on generators $h_1, \ldots, h_n$ then the crossed product S obtained by inverting the non-zero central elements of $\mathbb{Z}H$ is a simple ring and so Spec S is of little use.

However we should observe that there are good subrings of S with non-zero prime ideals. For example if $Q = \langle q \rangle$, an infinite cyclic group, then there is $S_+ = \bigoplus_{i\geq 0} De_{q^i}$ containing $P_+ = \bigoplus_{i>0} De_{q^i}$, and $S_- = \bigoplus_{i\leq 0} De_{q^i}$ containing $P_- = \bigoplus_{i<0} De_{q^i}$. These exist no matter what τ and θ are.

It is time to think about a standard construction in algebraic geometry.

3 Toric varieties

Let me start with an example and consider the complex group ring of an infinite cyclic group. This is the Laurent polynomial ring $S = \mathbb{C}[X, X^{-1}]$ and the prime ideals of S are the zero ideal and the maximal ideals; the latter, being kernels of ring homomorphism $\mathbb{C}[X, X^{-1}] \longrightarrow \mathbb{C}$, are determined by the image of X and are in 1–1 correspondence with $\mathbb{C}\backslash\{0\}$. The subring $S_+ = \mathbb{C}[X]$ has prime ideal $P_+ = X\mathbb{C}[X]$. The rest of Spec S_+ consists of the image of Spec S under the restriction map

$$\text{res} : \text{Spec } S \longrightarrow \text{Spec } S_+ \quad P \longrightarrow P \cap S_+.$$

Considering the maximal ideals, the restriction map induces an embedding of $\mathbb{C}^\times$ into $\mathbb{C}$ with the zero of $\mathbb{C}$ corresponding to P_+, the kernel of the map $\mathbb{C}[X] \longrightarrow \mathbb{C}$ with $X \longrightarrow 0$.

Similarly, letting $S_- = \mathbb{C}[X^{-1}]$, we get a restriction map inducing embeddings Spec $S \longrightarrow$ Spec S_- and $\mathbb{C}^\times \longrightarrow \mathbb{C}$ with the extra point of $\mathbb{C}$ corresponding to $P_- = X^{-1}\mathbb{C}[X^{-1}]$, the kernel of the map $S_- \longrightarrow \mathbb{C}$ with $X_- \longrightarrow 0$. Glueing Spec S_- and Spec S_+ together along Spec S corresponds to forming the complex projective line $\mathbb{P}^1(\mathbb{C})$.

Complex toric varieties are complex algebraic varieties that contain a torus $\mathbb{T} = (\mathbb{C}^\times)^n$ embedded as an open subvariety so that the translation action of $\mathbb{T}$ on itself extends to an action on the whole variety. Their special attraction is that their geometric properties are bound up with combinatorics of a 'fan' of cones in a n-dimensional real vector space. Regarding $(\mathbb{C}^\times)^n$ as the set of maximal ideals of $\mathbb{C}Q$ for a free Abelian group Q of rank n, the real vector space in question is $\text{Hom}(Q, \mathbb{R})$, the dual consisting of group homomorphisms χ of Q to the additive group of $\mathbb{R}$. We are interested in polyhedral convex cones of the form

$$\sigma = \{\lambda_1\chi_1 + \ldots + \lambda_s\chi_s \in \text{Hom}(Q, \mathbb{R}) : \lambda_i \geq 0\}$$

for some $\chi_1, \ldots, \chi_s$; in fact only rational ones are important, those which are the intersection of finitely many half-spaces, each of the form $\{\chi \in \text{Hom}(Q, \mathbb{R}) : \chi(q) \geq$

$0\}$ for some $q \in Q$. Associated with each cone σ we can define a semigroup $Q_\sigma = \{q \in Q : \chi(q) \geq 0 \text{ for all } \chi \in \sigma\}$ and form the subring $\mathbb{C}Q_\sigma$ of S. In our 1-dimensional example for the infinite cyclic group Q, both S_+ and S_- are such subrings, corresponding to the cones $\mathbb{R}_+ = \{\lambda \in \mathbb{R} : \lambda \geq 0\}$ and $\mathbb{R}_- = \{\lambda \in \mathbb{R} : \lambda \leq 0\}$. In fact S itself corresponds to a cone, namely $\{0\} \subseteq \mathbb{R}$. The reason for concentrating on cones in $\mathrm{Hom}(Q, \mathbb{R})$ rather than in the real vector space $Q \otimes_{\mathbb{Z}} \mathbb{R}$ is for convenience, since if $\sigma_1 \subseteq \sigma_2$ then there is a restriction map $\mathrm{Spec}\ \mathbb{C}Q_{\sigma_1} \longrightarrow \mathrm{Spec}\ \mathbb{C}Q_{\sigma_2}$. To ensure these maps are well behaved we want all our semigroups Q_σ to generate the whole of the group Q; equivalently we want σ to be *strongly convex*, that is to contain no non-zero linear subspace. If σ_1 is a face of σ_2 and they are both strongly convex, the restriction map is an embedding. We consider a *fan* $\mathcal{F}$, a set of cones satisfying

(i) each face of a cone in $\mathcal{F}$ is also a cone in $\mathcal{F}$, and

(ii) the intersection of two cones in $\mathcal{F}$ is a face of each.

Using the restriction maps between the spectra of the subrings $\mathbb{C}Q_\sigma$ for $\sigma \in \mathcal{F}$, we can glue them together, just as we did in the example, to form a variety, a *toric variety*. This method of associating a variety with a fan was first introduced by Demazure [23] in 1970, and extended to non-smooth varieties by Kempf, Knudsen, Mumford and Saint-Donat in [28]. I recommend Fulton's book [25] as a very readable introduction to the subject.

Much of the motivation for the development of the theory of toric varieties was to provide various examples of compactifications. A complex variety is compact in classical topology precisely when it is complete (proper) as an algebraic variety. For a toric variety this happens precisely when the union of the cones in the fan $\mathcal{F}$ gives the whole of $\mathrm{Hom}(Q, \mathbb{R})$. Moreover, given the variety associated with a fan $\mathcal{F}$ we can extend $\mathcal{F}$ to a fan $\mathcal{F}_1$, the union of whose cones is the whole of $\mathrm{Hom}(Q, \mathbb{R})$, and so we can compactify the original variety. If instead we start with an ideal I of $\mathbb{C}Q$ and look at the closed subset $V(I) \in (\mathbb{C}^\times)^n$, then certainly the closure of $V(I)$ in a compact complex variety containing our torus will also be compact. However compactification of $V(I)$ may be possible even if we embed the torus in a non-compact variety. In fact, for any such ideal I, there is a polyhedral subset of $\mathrm{Hom}(Q, \mathbb{R})$, minimal in the sense that it is itself a union of a fan associated with a toric variety in which the closure of $V(I)$ is compact, and, furthermore, lies in the union of any such fan. This follows from the work of Bieri and Groves [5], who confirmed a polyhedrality conjecture of Bergman [2] made at the time of the birth of toric varieties. When I is a principal ideal the construction is quite straightforward, being closely related to the Newton polygon.

It should perhaps not be a surprise that analogous sets arise in the case of a non-commutative crossed product DQ, since often in the theory of toric varieties it is the *support* $\{q \in Q : \lambda_q \neq 0\}$ of an element $\sum \lambda_q q$ of $\mathbb{C}Q$ which is important rather than the element itself. It is the Q-grading and not the commutativity of the ring which is crucial.

However, before contemplating the definitions appropriate in the general case, I promised you some trees.

4 Filtrations and trees

I think the best thing for me to do is to give just one well-known example.

Let G be the wreath product $C_\infty \mathrm{wr}\, C_\infty$ of two infinite cyclic groups. Thus G is metabelian , containing an Abelian normal subgroup A with $Q = G/A$ infinite cyclic. There is a 2×2 matrix representation with

$$G = \{\begin{pmatrix} X^i & \xi \\ 0 & 1 \end{pmatrix} : i \in \mathbb{Z}, \xi \in \mathbb{Z}[X, X^{-1}]\}$$

and

$$A = \{\begin{pmatrix} 1 & \xi \\ 0 & 1 \end{pmatrix} : \xi \in \mathbb{Z}[X, X^{-1}]\}.$$

Form $S = \mathbb{Q}[X, X^{-1}]$ and $M = A \otimes_{\mathbb{Z}[X,X^{-1}]} S$. As usual we study the subring $S_+ = \mathbb{Q}[X]$ with prime ideal $P_+ = X\mathbb{Q}[X]$. There is a filtration of S

$$\ldots > X^{-1}\mathbb{Q}[X] > \mathbb{Q}[X] > X\mathbb{Q}[X] > \ldots$$

corresponding to filtration

$$\ldots > X^{-1}\mathbb{Z}[X] > \mathbb{Z}[X] > X\mathbb{Z}[X] > \ldots$$

of $\mathbb{Z}[X, X^{-1}] \cong A$. Within the group there is a monoid

$$G_+ = \{\begin{pmatrix} X^i & \xi \\ 0 & 1 \end{pmatrix} : i \geq 0, \xi \in \mathbb{Z}[X]\}$$

which gives rise to a partial order relation on G:

$$g_1 \geq g_2 \iff g_1 G_+ \subseteq g_2 G_+.$$

This determines a tree upon which G acts. The vertices are indexed by the cosets gG_+. The height of vertex gG_+ is $i \in \mathbb{Z}$ when

$$g = \begin{pmatrix} X^i & \xi \\ 0 & 1 \end{pmatrix}.$$

An (undirected) edge joins vertex g_1G_+ to g_2G_+ precisely when $g_1 \geq g_2$ and the heights of the two vertices differ by 1. The action of G corresponds to left multiplication on the cosets. The tree action is in fact *minimal*; there are no proper G-invariant subtrees. Moreover the tree has infinitely many positive ends at infinity, reached via rays passing through vertices of increasing height, but only a single negative end at infinity; any two rays passing through vertices of decreasing height eventually have coincident tracks. Such a tree action is of the type associated by Bass-Serre theory [46] with ascending HNN-extensions; in this case we can express G as an HNN extension with base group

$$A_0 = \{\begin{pmatrix} 1 & \xi \\ 0 & 1 \end{pmatrix} : \xi \in \mathbb{Z}[X]\}.$$

The height function on the vertices can be extended to give a G-equivariant map from the tree to the real line with G acting via translations, with

$$g = \begin{pmatrix} X^i & \xi \\ 0 & 1 \end{pmatrix}$$

shifting the line by i. This translation corresponds to a group homomorphism $G \longrightarrow \mathbb{Z}$ inducing a character $\chi : Q \longrightarrow \mathbb{Z}$.

Of course there are similar constructions associated with S_- and P_-, and for the group $G = C_p \mathrm{wr}\, C_\infty$ in which case $S = \mathbb{F}_p[X, X^{-1}]$.

In general the idea is to study those filtrations of A arising from prime ideals of subrings analogous to P_+ in S_+, or alternatively view certain families of partial orders and tree actions of G. The motivation is a theorem of Bieri, Neumann and Strebel [7]. They defined a subset Σ^c of $\mathrm{Hom}(G, \mathbb{R})$ and showed that for a small finitely presented group G then Σ^c contains no non-zero linear subspaces. (For brevity we shall say that Σ^c contains no lines.) Σ^c was originally defined and investigated by Bieri and Strebel in the context of metabelian groups [8, 9, 48]. Neumann [37] approached the subject by considering those non-zero $\chi \in \mathrm{Hom}(G, \mathbb{Z})$ for which $\ker \chi$ is a finitely generated group; this happens precisely when neither χ nor $-\chi$ lie in Σ^c. Brown [18] reinterpreted the definition of Σ^c in terms of tree actions of the group; roughly speaking non-zero χ lies in Σ^c when there is a minimal tree action of G with associated height function χ, so that there is a single negative end and infinitely many positive ones, just as in our example. However we need to allow ourselves to use $\mathbb{R}$-trees, rather than simplicial trees when χ is not a discrete homomorphism. In the proof of the Bieri-Neumann-Strebel theorem one supposes χ and $-\chi$ both lie in Σ^c. The idea is to fuse the minimal $\mathbb{R}$-tree actions associated with χ and $-\chi$ together to give a new $\mathbb{R}$-tree action with infinitely many positive and infinitely many negative ends at infinity relative to a height function associated with χ. From the general theory about actions on $\mathbb{R}$-trees one deduces that G contains a non-Abelian free subgroup.

Σ^c is a closed subset of the real vector space $\mathrm{Hom}(G, \mathbb{R})$ in the usual Euclidean topology, and is closed under homothety; if $\chi \in \Sigma^c$ then $\lambda\chi \in \Sigma^c$ when $\lambda > 0$. To fit in with the language of fans it is geometrically convenient to include the zero map in Σ^c, but the original authors preferred to consider homothety classes in $\mathrm{Hom}(G, \mathbb{R}) \backslash \{0\}$ and talk of subsets of the sphere S^{n-1}. The work of Bieri and Groves [5] I mentioned earlier establishes that for a finitely generated metabelian group, where there is an Abelian normal subgroup A with $Q = G/A$ Abelian, Σ^c is a finite union of closed convex rational polyhedra. The point is that in this case the filtrations of the finitely generated $\mathbb{Z}Q$-module A give the full story. While the integral coefficents require more careful handling the theory is not vastly different from the geometric situation for $\mathbb{C}Q$. In spirit Σ^c is analogous to the minimal union of cones in a fan required for the compactification of $V(I)$ for an ideal I.

In general Σ^c need not be a rational polyhedron. For example there is a plentiful supply of finitely presented groups of orientation-preserving piecewise linear homeomorphisms of the unit interval for which Σ^c consists of a pair of rays emanating from the origin, and neither ray need be rational. However we do know that the rational polyhedrality of Σ^c holds for fundamental group of a compact connected smooth 3-manifold as long as we avoid counterexamples to the Poincaré conjecture. When G is the fundamental group of a compact connected smooth n-manifold Y one can identify $\mathrm{Hom}(G,\mathbb{R})$ with the de Rham cohomology group $H^1(Y;\mathbb{R})$ and relate the points of Σ^c to the existence or otherwise of certain 1-forms on Y. For example Levitt [29] characterised Σ^c in this way for closed Y, and in general one can quite easily see that $\Sigma^c \subseteq \Sigma^c(Y)$ where the latter consists of those cohomology classes that cannot be represented by a 1-form non-vanishing on Y (and ∂Y). $\Sigma^c(Y)$ is equal to $-\Sigma^c(Y)$ and is significant since Tischler [49] observed that non-zero $\chi : G \longrightarrow \mathbb{Z}$ is associated with a smooth fibration of Y over a circle precisely when $\chi \notin \Sigma^c(Y)$. For a compact 3-manifold Y, Thurston [50] used a norm on the second homology groups $H_2(Y;\mathbb{R})$ and $H_2(Y,\partial Y;\mathbb{R})$ to show that $\Sigma^c(Y)$ is a rational polyhedron and Bieri, Neumann and Strebel [7] established that $\Sigma^c = \Sigma^c(Y)$ as long as Y contains no fake cells and an exotic homotopy $\mathbb{R}P^2 \times S^1$ is not involved.

5 A geometric set Δ

We return to our general crossed product $S = DQ$. For each $\chi \in \mathrm{Hom}(Q,\mathbb{R})$ define

$$S = \bigoplus_{q\in Q} De_q, \quad S_\chi = \bigoplus_{\chi(q)\geq 0} De_q, \quad P_\chi = \bigoplus_{\chi(q)>0} De_q,$$

taking P_χ to be the zero ideal when χ is zero. Thus P_χ is a prime ideal of the subring S_χ of S. For a finitely generated S-module M we record those χ giving rise to proper filtrations.

$$\Delta(M) = \{\chi : YS_\chi > YP_\chi \text{ for some finite generating set } Y \text{ of } M \text{ as } S-\text{module}\}.$$

In fact if $YS_\chi > YP_\chi$ holds for some generating set Y then it holds for any other generating set. Note that 0 lies in $\Delta(M)$ when M is non-zero.

Recall our standard notation that G is a finitely generated group with normal subgroups K and L with $G/K = H$ torsion-free nilpotent and $K/L = A$ Abelian. We can form various crossed products S by inverting non-zero elements of a subgroup ring; last time I chose $\mathbb{Z}H'$ but in fact $\mathbb{Z}H_1$ for any $H_1 \geq H'$ with H/H_1 torsion-free will do. The key point is that, setting $M = A\otimes_{\mathbb{Z}H} S$, we can see with a little effort that $\Delta(M) \subseteq \Sigma^c$ and so the Bieri-Neumann-Strebel result implies that $\Delta(M)$ contains no lines when G is a small finitely presented group.

$\Delta(M)$ is closed under homothety. When Q is infinite cyclic then there are four possibilities for non-zero M; it is either just the origin of the vector space $\mathbb{R}$, the non-negative half-line, the non-positive half-line or the whole of $\mathbb{R}$. We have $\Delta(\mathbb{C}Q) = \mathbb{R}$ since $S_+ > P_+$ and $S_- > P_-$. In the general case a weak polyhedrality result holds.

Theorem 1 ([15]) *Let S be the crossed product of a division ring be a free Abelian group of finite rank, and let M be a non-zero finitely generated S-module. Then*

$$\Delta(M) = \Delta_1(M) \cup \Delta_2(M)$$

where $\Delta_1(M)$ is the non-empty finite union of closed convex rational polyhedra of dimension m, and $\Delta_2(M)$ lies in a finite union of closed convex rational polyhedra of dimension less than m.

As yet we have been unable to come up with an example of $\Delta(M)$ which is not polyhedral. In the group ring case Bieri and Groves [5] proved for a prime ideal I of $\mathbb{C}Q$ that $\Delta(\mathbb{C}Q/I)$ is a homogeneous polyhedron; in other words $\Delta_2(\mathbb{C}Q/I)$ can be taken to be empty and $\Delta(\mathbb{C}Q/I)$ is the union of the cones in a fan. It may be just a weakness of our method of proof but the lower dimensional inexactitude in the general case seems to be tied up with questions about module extensions, standard material when S is commutative but not so straightforward in general. However note that $\Delta_1(M)$ is uniquely determined; it is the closure of the set of *regular* points, those that lie at the centre of an m-dimensional ball contained in $\Delta(M)$. The polyhedrality implies that there are only finitely many vector spaces, the *carrier spaces* for $\Delta(M)$, spanned by such m-dimensional balls. It is these carrier spaces we shall be concerned with when studying automorphisms.

We can now define the dimension of a non-zero finitely generated S-module M to be the m appearing in Theorem 1. In Noetherian ring theory there are several possible concepts of dimension that coincide in the commutative case but differ in general. The dimension just defined coincides with one of these, namely Gelfand-Kirillov dimension, defined in terms of growth rates (see for example [32], Chapter 8). A non-zero S-module has dimension 0 if and only if it is finite dimensional as a D-vector space. In due course we shall see that the cocycle used in the definition of the crossed product S influences the dimensions possible for finitely generated S-modules; roughly speaking, the larger the image of the map $Q \times Q \longrightarrow D^{\times}$, the larger the dimension of a finitely generated S-module has to be. On the other hand, motivated by the Bieri-Neumann-Strebel theorem, we shall be interested in modules M for which $\Delta(M)$ contains no lines; this turns out to imply $\dim(M) \leq n/2$, where n is the rank of Q. These upper and lower bounds can be incompatible, and balancing them against each other will tell us about the constraints on the images of small finitely presented groups. These are to be seen even in the 'quantum' case.

6 The quantum case

Let $S = \mathbb{Z}H(\mathbb{Z}Z\backslash\{0\})^{-1}$ with H finitely generated torsion-free nilpotent of class 2 with centre Z. Here D is the fraction field of $\mathbb{Z}Z$ and is central in S, and $Q = H/Z$ is Abelian. Let us start by supposing we have a 0-dimensional finitely generated S-module, in other words one which is finite dimensional as a D-vector space. Then H is acting as vector space automorphisms and so every element of H' acts with determinant 1. But $H' \leq Z$ and so such an element also acts as a scalar multiple of the identity. These scalars must therefore be roots of unity, and, since Z is torsion-free, they are actually 1. But H' is non-trivial and we have a contradiction; such modules cannot exist.

The next step is to use this simple argument to link the dimension of a module to the ranks of Abelian subgroups of H. For an m-dimensional S-module M, with $m > 0$, the polyhedron $\Delta_1(M)$ is m-dimensional an so there is $\chi \in \Delta_1(M)$ with

$\mathrm{rk}(Q/\ker\chi) = m$. Such a χ has the smallest kernel possible among elements of $\Delta(M)$; if $\mathrm{rk}(Q/\ker\chi) > m$ then χ does not lie in an m-dimensional vector subspace of $\mathrm{Hom}(Q, \mathbb{R})$ and so cannot lie in $\Delta(M)$. It is therefore not a surprise to find that for such a χ and any generating set Y of M as an S-module we have $YS_\chi > YP_\chi$, since $\chi \in \Delta(M)$, with YS_χ/YP_χ a finite dimensional D-vector space. Noticing that S_χ/P_χ is again a crossed product of the form $D(\ker\chi) \cong \mathbb{Z}B(\mathbb{Z}Z\backslash\{0\})^{-1}$ for some B with $H > B > Z$ and both H/B and B/Z torsion-free. The same argument as used in the 0-dimensional case now gives that B is Abelian. So

$$\dim(M) = \mathrm{rk}(H/B) \textit{ for some Abelian } B. \tag{1}$$

This confirms the crucial role in the representation theory played by the Abelian subgroups of H, or equivalently the commutative subalgebras of S. Some time ago Ken Brown and I [13] showed that even when H is not finitely generated, but still torsion-free nilpotent of class 2, that $\mathbb{Z}H(\mathbb{Z}Z\backslash\{0\})^{-1}$ can be Noetherian. This happens if H satisfies the maximum condition on Abelian subgroups B containing Z, for example when H is free nilpotent class 2 on countably many generators. Building on an investigation by McConnell and Pettit [31] of the homological properties of S, I have shown [12] that, in the finitely generated case, (1) implies that the global dimension of S is equal to the maximal rank possible for B/Z for Abelian subgroups $B \geq Z$.

The lower bound on the dimension of non-zero finitely generated S-modules given by (1) has immediate consequences. If H is free nilpotent of class 2 on n generators then, for any Abelian subgroup $B \geq Z$ we have $\mathrm{rk}(B/Z) \leq 1$ and so $\mathrm{rk}(H/B) \geq n-1$. Thus $\dim(M) \geq n-1$. However we saw that for $\Delta(M)$ to contain no lines we need $\dim(M) \leq (n/2)$, and so this is not possible if $n \geq 3$. Applying this to our group G with normal subgroups K and L with $G/K = H$, $K/L = A$ and $M = A \otimes_{\mathbb{Z}H} S$, we deduce that M must be zero, and so A must be a torsion $\mathbb{Z}Z$-module, if $\Delta(M)$ contains no lines and H is free nilpotent of class 2 on more than 2 generators.

Another popular family of torsion-free nilpotent groups is that of the discrete Heisenberg groups,

$$\langle h_1, h_2, \ldots, h_{2r-1}, h_{2r}, z : [h_{2i-1}, h_{2i}] = z \ (1 \leq i \leq r), [h_j, h_k] = 1 \text{ otherwise}, \\ [h_j, z] = 1 \ (j \leq 2r)\rangle.$$

Here Z is infinite cyclic, generated by z, $n = \mathrm{rk}(H/Z) = 2r$ and $\mathrm{rk}(B/Z) \leq r = (n/2)$ for any Abelian subgroup B containing Z. So the dimension of a non-zero finitely generated S-module is at least $n/2$, with equality if $\Delta(M)$ contains no lines. It is tempting to call the modules of dimension $n/2$ *holonomic*, by analogy with the terminology in the theory of algebraic D-modules, and expect an extensive theory featuring good duality properties.

The easiest examples of differential operator algebras are the Weyl algebras, generated by the operators $Y_1, \partial/\partial Y_1, \ldots, Y_r, \partial/\partial Y_r$ on the polynomial algebra $\mathbb{C}[Y_1, \ldots, Y_r]$. This is isomorphic to the algebra

$$A_r = \mathbb{C}\langle X_1, X_2, \ldots, X_{2r-1}, X_{2r} : X_{2i-1}X_{2i} - X_{2i}X_{2i-1} = 1 \ (1 \leq i \leq r), \\ X_jX_k = X_kX_j \text{ otherwise}\rangle.$$

The crossed products $S = \mathbb{C}H(\mathbb{C}Z\backslash\{0\})^{-1}$ for the discrete Heisenberg groups H can be regarded as their multiplicative analogues. The Weyl algebras are fundamental to the representation theory of nilpotent Lie algebras since they appear as images of their enveloping algebras. The Bernstein inequality [3] gives a lower bound on the (GK-)dimension of a non-zero module M for the Weyl algebra A_r, namely $\dim(M) \geq r = \dim(A_r)/2$. The holonomic modules are those for which the bound is attained. Their importance is that they arise naturally in symplectic geometry. A_r-modules M are studied by considering graded versions $\mathrm{gr}(M)$ which are modules for the graded polynomial algebra $\mathbb{C}[X_1, X_2, \ldots, X_{2r}]$, the co-ordinate ring of $\mathbb{C}^{2r}$. One defines a variety, the characteristic variety, for such a graded module in the usual way. It may be regarded as a subvariety of the cotangent bundle. It is involutive, which is equivalent to saying that the radical of the annihilator of $\mathrm{gr}(M)$ is closed under the Poisson bracket on $\mathbb{C}[X_1, X_2, \ldots, X_{2r}]$ associated with the standard symplectic structure on $\mathbb{C}^{2r}$. This was first proved in an analytic context but Gabber [26] has since given an algebraic proof. Bernstein's inequality follows from a general inequality for the dimension of an involutive variety. The holonomic modules are those that yield Lagrangian characteristic varieties. A very readable introduction to Weyl algebras, and algebraic D-modules in general, is Coutinho's book [20]. However he suppresses all homological algebra and does not prove Gabber's theorem. An alternative is [11] where the homological theory is developed using derived categories. A proof of Gabber's theorem can be found in [10].

7 The structure of Δ

Before turning to study automorphisms where we shall want to know about the rigidity of $\Delta(M)$ I had better describe some of its properties and how they are linked to the algebraic structure of M.

First, if M is an extension of M_1 by M_2 then $\Delta(M) = \Delta(M_1) \cup \Delta(M_2)$. Thus if M can be broken up into finitely many 'building blocks' we can restrict our attention to them. When kQ is a group algebra for a field k, we concentrate in this way on cyclic modules of the form kQ/P with P a prime ideal of kQ. These are examples of *critical* modules where the dimension of any proper quotient is strictly smaller that that of the module itself. For a critical module M, $\Delta_1(M_1) = \Delta_1(M)$ for any non-zero submodule M_1. In general S, viewed as a module for itself, is always of this type. Moreover any induced module $M_0 \otimes_{DQ_0} S$, where Q/Q_0 is torsion-free and M_0 is a critical DQ_0-module, is also critical. The crucial point is that any non-zero S-module M contains a critical one M_1, though this requires some work to prove. Applying this to the quotient M/M_1 and repeating yields a chain of submodules of M with critical factors. For a finitely generated, and hence Noetherian, S-module this process terminates and so critical modules do fulfill the role of building blocks.

There are two canonical subgroups of Q associated with $\Delta(M)$, or more precisely with the top dimensional polyhedral part $\Delta_1(M)$. For any subgroup Q_0 of Q there is a restriction map

$$\mathrm{res}^Q_{Q_0} : \mathrm{Hom}(Q, \mathbb{R}) \longrightarrow \mathrm{Hom}(Q_0, \mathbb{R}).$$

The image of $\Delta_1(M)$ under this map is $\Delta_1(M_0)$, where M_0 is a finitely generated DQ_0-submodule of M with $M = M_0S$.

We define Q_1 to be the maximal subgroup with respect to the condition that $\mathrm{res}^Q_{Q_1}(\Delta_1(M)) = \{0\}$. For example $Q_1 = Q$ precisely when M has finite dimension as a D-vector space; in general Q_1 is maximal subject to every finitely generated DQ_1-submodule of M being a finite dimensional D-vector space.

The other canonical subgroup Q_2 contains Q_1 and is the subgroup minimal with respect to Q/Q_2 being torsion-free and $\Delta_1(M) = (\mathrm{res}^Q_{Q_2})^{-1}(\mathrm{res}^Q_{Q_2})(\Delta_1(M)))$. This is equivalent to identifying the maximal Cartesian factor of $\Delta_1(M)$; thus $\Delta_1(M) = \Delta' \times \mathbb{R}^l$ where $l = \mathrm{rk}(Q/Q_2)$, $\mathbb{R}^l = \ker \mathrm{res}^Q_{Q_2}$ and $\Delta' = \Delta_1(M_2)$ for some finitely generated DQ_2-module M_2 of M. Algebraically, for critical modules *Cartesian factors correspond to induction*; Q_2 is the subgroup minimal with respect to Q/Q_2 being torsion-free and M containing a non-zero submodule of the form $M_2 \otimes_{DQ_2} DQ$. For example for S itself, Q_2 is trivial since $\Delta_1(S) = \mathrm{Hom}(Q, \mathbb{R})$.

8 Automorphisms

Consider a ring automorphism γ of S preserving the Q-grading. Any automorphism of H will induce such an automorphism when $S = \mathbb{Z}H(\mathbb{Z}Z\backslash\{0\})^{-1}$. In general the unit group of $S = \bigoplus De_q$ consists of non-zero homogeneous elements of the form λe_q with λ non-zero. So γ will leave $D = De_1$ invariant and permute the homogeneous components De_q. Thus γ induces an automorphism of Q, and hence of $\mathrm{Hom}(Q, \mathbb{R})$, but one may not get every automorphism of Q this way. For an S-module M one can define $M\gamma$ via $(m\gamma)s = (ms^{\gamma^{-1}})\gamma$. If S is the group ring kQ and $M = kQ/I$ then $M\gamma \cong kQ/I^\gamma$. In general $\Delta(M\gamma)$ is the image of $\Delta(M)$ under the automorphism induced by γ on $\mathrm{Hom}(Q, \mathbb{R})$. So if $M\gamma \cong M$, and so $\Delta_1(M\gamma) = \Delta_1(M)$, then the carrier spaces of $\Delta_1(M\gamma)$ and $\Delta_1(M)$ are the same; they have just been permuted by the action of γ.

Jim Roseblade in his lectures [45] in 1985 described a theorem, generalising one due to Bergman [2], about those γ that stabilise prime ideals P of a group ring kQ. He and Brewster [43] showed that if $P^\gamma = P$ then γ acts finitely on Q_2/Q_1 where Q_1 and Q_2 are the canonical subgroups defined for the module kQ/P. In fact they proved a slightly stronger result showing the finiteness of the action of γ on Q_2/Q_0 where $Q_0 = Q \cap (1 + P)$. Their proof was by induction using valuations on the fraction field of kQ/P, and a little Galois theory to make the final step from Q_2/Q_1 to Q_2/Q_0. A little later Bieri and Groves [6] provided a more geometric approach using the rigidity of $\Delta(kQ/P)$. In the non-degenerate case when $Q_1 = 1$ and $Q_2 = Q$ they showed that the 1-dimensional subspaces of $\mathrm{Hom}(Q, \mathbb{R})$ obtained by intersecting a number of carrier spaces actually span $\mathrm{Hom}(Q, \mathbb{R})$. These finitely many 1-dimensional spaces are permuted under the action of γ when $P^\gamma = P$ and so some positive power of γ fixes them and hence fixes the whole of $\mathrm{Hom}(Q, \mathbb{R})$.

Farkas and Snider [24] showed that the stabilizer of a semiprime ideal I is commensurable with a (Zariski-) closed subgroup of $\mathrm{Aut}(Q) \cong GL(n, \mathbb{Z})$, and thus arithmetic. Arithmetic groups are finitely presented and Bachmuth, Baumslag, Dyer and Mochizuki [1] used this approach to show the automorphism group of a 2-generated metabelian group G induces a finitely presented group of automorphisms of G/G'.

In the non-commutative case critical modules M take the role of kQ/P. Since $\Delta_1(M) = \Delta_1(M_1)$ for any non-zero $M_1 \leq M$, we cannot hope to distinguish between M and its non-zero submodules. So we define an equivalence relation, usually known as *similarity*, on critical modules: $M \sim N$ if M and N have isomorphic (essential) non-zero submodules. (The 'essential' is tautologous in the critical case, but is necessary when extending the definition to all S-modules.) Another way of phrasing this is to say that M and N are similar if they have isomorphic injective hulls. In fact this latter version of the definition is used for any finitely generated module over a Noetherian ring. We denote the similarity class of M by $[M]$.

Theorem 2 ([16]) *Let M be a critical S-module. Then*

$$\mathrm{Stab}[M] = \{\gamma \in \mathrm{Aut}(S) : M\gamma \sim M\}$$

acts finitely on Q_2/Q_1, where Q_1 and Q_2 are the canonical subgroups of Q associated with M.

Unfortunately we have been unable to prove enough about the 1-dimensional intersections of carrier spaces to follow the same geometric argument as Bieri and Groves. Instead the proof is an inductive one which appears rather opaque.

When $S = \mathbb{Z}H(\mathbb{Z}Z\backslash\{0\})^{-1}$ for a torsion-free finitely generated nilpotent group of class 2 we can make use of our knowledge of $\mathrm{Aut}(H)$. Building on work of Auslander and Baumslag, Segal showed in his book [46] that $\mathrm{Aut}(H)$ is arithmetic; it is isomorphic via a map ϕ to a subgroup of $GL(t, \mathbb{Q})$ commensurable to a closed subgroup of $GL(t, \mathbb{Z})$. Indeed from H one can form a canonical rational Lie algebra Λ and $\mathrm{Aut}(\Lambda)$ can be identified with a subgroup of $GL(t, \mathbb{Q})$ where $t = \dim_{\mathbb{Q}}\Lambda$. Then $\phi(\mathrm{Aut}(H))$ is commensurable with the stabilizer $N_{\mathrm{Aut}}(\Lambda)(L)$ of some lattice L in Λ.

Theorem 3 ([17]) *Let H be a torsion-free finitely generated nilpotent group. Let R be a commutative Noetherian ring and let M be a finitely generated RH-module. Identify $\mathrm{Aut}(H)$ with a subgroup of $GL(t, \mathbb{Q})$ as above. Then $\mathrm{Stab}_{\mathrm{Aut}(H)}[M]$ is commensurable with a closed subgroup of $\mathrm{Aut}(H)$ and thus is arithmetic.*

To deduce Theorem 3 requires a few reductions to translate from a general module over a group ring RH to a critical S-module for some crossed product S, but before saying a few words about that let me state a result about automorphisms of groups also following from Theorem 2.

Theorem 4 ([17]) *Let G be a finitely generated metanilpotent group with Fitting subgroup F. Suppose no subgroup of finite index in G has a quotient isomorphic* *to a wreath product $C_p \mathrm{wr}\, C_\infty$ for any prime p. Then the group of automorphisms induced on G/F by the automorphisms of G has a subgroup of finite index acting nilpotently on G/F.*

We saw that the existence of a $C_p \mathrm{wr}\, C_\infty$ quotient ensures a line in Σ^c, and that this cannot happen for a small finitely presented group. Since subgroups of small finitely presented groups inherit the property we deduce that any of their metanilpotent images satisfy the condition of the theorem.

The module interpretation of the wreath product condition is that, if we set $G/F = H$ and $A = F/F'$, it is equivalent to saying that no $\mathbb{Z}H$-module section of A is of the form $A_0 \otimes_{\mathbb{Z}H_0} \mathbb{Z}H$ with $|H : H_0| = \infty$ and A_0 a non-zero $\mathbb{Z}H_0$-module. In other words no section involves 'serious' induction.

9 Representation theory of nilpotent groups of class 2

Up to now, when faced with a $\mathbb{Z}H$-module A for a torsion-free nilpotent group H with centre Z, we have formed $M = A \otimes_{\mathbb{Z}H} S$ where $S = \mathbb{Z}H(\mathbb{Z}Z\backslash\{0\})^{-1}$. The problem is that if every element of M has non-zero annihilator in $\mathbb{Z}Z$ then M is zero, and we learn nothing. However there are crossed products better suited to the module in question.

Pick $a \in A$ with annihilator I in $\mathbb{Z}Z$ maximal among such annihilators. The maximality ensures that I is a prime ideal and that $a\mathbb{Z}H$ is $\mathbb{Z}Z/I$-torsion-free. We may as well assume that $a\mathbb{Z}H$ is *faithful for* Z, in other words $(I+1) \cap Z = \{1\}$; otherwise we could study it as a $\mathbb{Z}(H/Z_1)$-module for some $Z_1 \neq \{1\}$. Now set D to be the field of fractions of $\mathbb{Z}Z/I$ and $S = (\mathbb{Z}H/I\mathbb{Z}H) \otimes_{(\mathbb{Z}Z/I)} D$, of the form DQ with $Q = H/Z$. Then $M = a\mathbb{Z}H \otimes_{\mathbb{Z}H} S$ is an S-module which can be studied as before; we are in the quantum case, and D is the centre of S since Z is the centre of H and $a\mathbb{Z}H$ is faithful for Z. By passing to a submodule if necessary, M can be taken to be critical, and if *no non-zero $\mathbb{Z}H$-submodule of A is of the form $A_0 \otimes_{\mathbb{Z}H_0} \mathbb{Z}H$ with $|H : H_0| = \infty$* then the canonical subgroup Q_2 for M is equal to Q.

I shall now try to convince you that such a lack of serious induction would impose a lower bound on $\dim(M)$. Again the Abelian subgroups $B \geq Z$ are involved. Pick one such B. We repeat the maximal annihilator argument as before but this time make sure that $a \in A$ has annihilator P in $\mathbb{Z}B$ maximal among all such annihilators. This again is a prime ideal and there is a useful Clifford theory lemma of Roseblade [43] which says that $a\mathbb{Z}H \cong a\mathbb{Z}N \otimes_{\mathbb{Z}N} \mathbb{Z}H$ where N is the stabiliser of P in H. The lack of serious induction would force N to be of finite index in H. However we already know from Roseblade and Brewster [44] how the action of N on B is linked with the structure of $\mathbb{Z}B/P$. Since our module is faithful for Z, we deduce $P = I\mathbb{Z}B$ and so $(\mathbb{Z}B/P) \otimes_{(\mathbb{Z}Z/I)} D = DQ_0$ where $Q_0 = B/Z$. Thus P, despite its maximality, is as small as it can be while containing I. Thus M is DQ_0-torsion free and so $\Delta_1(M)$ maps to the whole of $\mathrm{Hom}(Q_0, \mathbb{R})$ under restriction. So $\dim(M) \geq \mathrm{rk}(B/Z)$ for an Abelian subgroup $B \geq Z$, given the assumption about the lack of induced modules.

If $\Delta(M)$ contains no lines it certainly has no Cartesian factors. Recall that this implies there is no serious induction. We know several things about $\dim(M)$ in this context; it is equal to $\mathrm{rk}(H/B)$ for some Abelian $B \geq Z$ and also it is at least $\mathrm{rk}(B/Z)$ for any such B. From this it is easy to see that $\dim(M) \geq \frac{1}{2}\mathrm{rk}(Q)$. But if $\Delta(M)$ contains no lines we saw that $\dim(M) \leq \frac{1}{2}\mathrm{rk}(Q)$. So we have equality and M has to be holonomic if it is faithful for Z.

This applies for the modules arising from groups G for which Σ^c contains no lines, for example small finitely presented groups. In particular this is true for the examples of Robinson and Strebel [42] of finitely presented Abelian-by-nilpotent groups.

Similar methods give structure theorems for $\mathbb{Z}H$-modules A for torsion-free nilpotent groups of higher class [17]. For example, if A is $\mathbb{Z}Z/I$-torsion-free for some ideal I of $\mathbb{Z}Z$, faithful for Z, and no submodule has serious induction, set D to be the field of fractions of $\mathbb{Z}Z/I$ and form the crossed product $D(H/Z) = \mathbb{Z}H \otimes_{\mathbb{Z}Z/I} D$. Then $M = A \otimes_{\mathbb{Z}H} D(H/Z)$ is $D(H'/Z)$-torsion-free. Note that H/Z is no longer Abelian, but it is still possible to define a suitable dimension for such a module M. This time however we deduce the strict inequality $\dim(M) > \frac{1}{2}h(H/Z)$ when H is of class at least 3; here $h(\ )$ denotes Hirsch length, the appropriate rank for non-Abelian nilpotent groups. In the class 2 case we balanced a lower bound of this type against an upper bound. It is tempting to conjecture that such an upper bound, namely $\dim(M) \leq \frac{1}{2}h(H/Z)$, would again apply for modules M arising from small finitely presented groups, and so it would follow that such an M could not be faithful for Z. Indeed we would deduce, roughly speaking, that H only acts at worst as a class 2 group on the building blocks of A. At various points in the argument we would have to pass to a subgroup of finite index to regain a torsion-free quotient and so the final statement would have to take this into account. The extra difficulty in proving this conjecture is that H/Z is non-Abelian for groups H of class greater than 2, while the approach relying upon Σ^c only records information about characters factoring through the Abelianisation H/H'. However pro-p group methods, considering the constraints on the virtually Abelian images of G, do deliver the desired upper bound, but for a slightly different class of groups.

10 Pro-p methods

In the second half of his lectures in Galway Efim Zelmanov talked about Golod-Shafarevich inequalities in the context of pro-p groups, with applications to discrete residually-p groups G. I refer you to [53] for the details but let me remind you of a few things.

The idea is to consider presentations of pro-p groups $\hat{G} = \hat{F}/\hat{R}$ where $\hat{F}$ is a free pro-p group of finite rank, d say, and $\hat{R}$ is the closed normal subgroup generated by a set of relators. In fact $\hat{F}$ is the (closed) multiplicative group generated by the units $1 + X_1, \ldots, 1 + X_d$ of the formal non-commuting power series algebra $\mathbb{F}_p[[X_1, \ldots, X_d]]$, which is the completion on the free associative algebra on $X_1, \ldots X_d$. Each relator can therefore be thought of as being of the form $1 + f_\alpha(X_1, \ldots, X_d)$ with $f_\alpha(X_1, \ldots, X_d)$ of degree at least 1. (Here the degree is the least degree of a monomial appearing in f_α.) Suppose that each f_α is of degree at least 2 and that there are exactly r_i relators with f_α of degree i. Let $H_R(t) = r_2t^2 + r_3t^3 + \ldots$. The set of relators is *small in the sense of Golod-Shafarevich* if the degrees of all the f_α are at least 2 and there is a real number t_0 with $0 < t_0 < 1$ such that $1 - dt_0 + H_R(t_0) < 0$. The key point is that if $\hat{G}$ has a presentation with a small set of relators then the group is infinite in a strong way. To make this more precise observe that the completed group algebra $\widehat{\mathbb{F}_pG}$ obtained by setting all the f_α from the relators to be zero, is an image of $\mathbb{F}_p[[X_1, \ldots, X_d]]$. Let $\mathfrak{g}$ be the image of the prime ideal of $\mathbb{F}_p[[X_1, \ldots, X_d]]$ consisting of elements of degree at least 1. Let

$$H_B(t) = 1 + \sum_{i=1}^{\infty} \dim(\mathfrak{g}^i/\mathfrak{g}^{i+1})t^i.$$

This Poincaré series records the growth of the finite dimensional unipotent representations of $\hat{G}$ in characteristic p, using the filtration $\{\mathfrak{g}^i\}$ of $\widehat{\mathbb{F}_pG}$. The two Poincaré series are related by

$$H_B(t)(1 - dt + H_R(t))/(1 - t) \geq 1/(1 - t).$$

Wilson [51] generalized Golod's original idea by observing that if one has a small set of generators one can add in some more, retaining the smallness but also ensuring that in the resulting pro-p group quotient the image of the discrete group generated by $1+X_1, \ldots, 1+X_d$ is periodic with all elements of p-power order. Thus every pro-p group with a presentation with a small set of relators in the Golod-Shafarevich sense has a pro-p image, again with a presentation with a small set of relators, but also containing a dense discrete infinite periodic residually-p group. Zelmanov [53] has established that such a pro-p group has a non-Abelian free pro-p subgroup. Thus to avoid non-Abelian free pro-p groups one has to study pro-p groups without a presentation with a small set of relators. For these groups one can deduce a Golod-Shafarevich inequality between the minimal numbers of generators and relators. Building on work of Lubotsky [30] linking presentations of discrete groups with those of their pro-p completions, Wilson [51] showed that for a discrete group G, if, for all primes p, the pro-p completion of all subgroups G_0 of finite index in G cannot be presented using a small set of relators then there is a constant C such that if $|G : G_0| = r$ then one needs at most $Cr^{\frac{1}{2}}$ generators for G_0/G_0'. Note that in a free group G of rank d one requires $dr - (r - 1)$ generators for the Abelianisation of such a subgroup G_0, and so an upper bound can be no better than linear in r rather than of the form $Cr^{\frac{1}{2}}$. Thus in this sense a finitely presented group not giving rise to non-Abelian free pro-p groups is at most half the size of a free group. Groves and Wilson [27] investigated the effect on the soluble images of G of the bound $Cr^{\frac{1}{2}}$ and showed that, in the context of the last paragraph of Section 9,

$$\dim(M) \leq \frac{1}{2}h(H/Z).$$

Thus, for such a G, we are again in a position to play upper and lower bounds off against each other.

Theorem 5 ([12]) *Let G be a finitely presented group and suppose that for every prime p no subgroup of finite index has pro-p completion containing a non-Abelian free pro-p group. Then all metanilpotent images $\bar{G}$ of G have a subgroup $\bar{G}_0$ of finite index with $\bar{G}_0/\mathrm{Fitt}(\bar{G}_0)$ nilpotent of class at most 2.*

This was already known for 2-generator groups as a result of combining Theorem A of Groves and Wilson [27] with Zelmanov's work.

I believe that the conclusions should still hold simply if G contains no non-Abelian discrete free subgroup.

References

[1] S. Bachmuth, G. Baumslag, J. Dyer and H.Y. Mochizuki, Automorphism groups of 2-generator metabelian groups, J. London Math. Soc. (2) 36 (1987), 393–406.

[2] G. M. Bergman, The logarithmic limit-set of an algebraic variety, Trans. Amer. Math. Soc. 157 (1971), 459–469.

[3] I.N. Bernstein, The analytic continuation of generalized functions with respect to a parameter, Funct. Anal. Appl. 6 (1972), 273–285.

[4] M. Bestvina, Degenerations of the hyperbolic space, Duke Math. J. 56 (1988), 143–161.

[5] R. Bieri and J.R.J. Groves, The geometry of the set of characters induced by valuations, J. Reine Angew. Math. 347 (1984), 168–195.

[6] R. Bieri and J.R.J. Groves, A rigidity property for the set of all characters induced by valuations, Trans. Amer. Math. Soc. 294 (1986), 425–434.

[7] R. Bieri, W.D. Neumann and R. Strebel, A geometric invariant of discrete groups, Invent. Math. 90 (1987), 451–477.

[8] R. Bieri and R. Strebel, Valuations and finitely presented metabelian groups, Proc. London Math. Soc. (3) 41 (1980), 439–464.

[9] R. Bieri and R. Strebel, A geometric invariant for modules over an abelian group, J. Reine Angew. Math. 322 (1981), 170–189.

[10] J-E. Björk, Analytic D-modules and applications (Mathematics and its applications 247, Kluwer, 1993).

[11] A. Borel *et al*, Algebraic D-modules (Academic Press, 1987).

[12] C.J.B. Brookes, Crossed products and finitely presented groups, preprint.

[13] C.J.B. Brookes and K.A. Brown, Primitive group rings and Noetherian rings of quotients, Trans. Amer. Math. Soc. 288 (1985), 605–623.

[14] C.J.B. Brookes and J.R.J. Groves, Modules over nilpotent group rings, J. London Math. Soc. (2) 52 (1995), 467–481.

[15] C.J.B. Brookes and J.R.J. Groves, Modules over crossed products of a division ring by a free Abelian group I, preprint.

[16] C.J.B. Brookes and J.R.J. Groves, Modules over crossed products of a division ring by a free Abelian group II, preprint.

[17] C.J.B. Brookes and J.R.J. Groves, Some infinite soluble groups, their modules and the arithmeticity of associated automorphism groups, preprint.

[18] K.S. Brown, Trees, valuations and the Bieri-Neumann-Strebel invariant, Invent. Math. 90 (1987), 479–504.

[19] A.J. Casson and S.A. Bleiler, Automorphisms of surfaces after Nielsen and Thurston (London Math. Soc. Student Texts 9, Cambridge University Press, 1988).

[20] S.C. Coutinho, A primer of algebraic D-modules (London Math. Soc. Student Texts 33, Cambridge University Press, 1995).

[21] M. Culler and P.B. Shalen, Varieties of group representations and splittings of 3-manifolds, Ann. of Math. 117 (1983), 109–146.
[22] M. Culler and K. Vogtmann, Moduli of graphs and automorphisms of free groups, Invent. Math. 84 (1986). 91–119.
[23] M. Demazure, Sous-groupes algèbriques de rang maximum du groupe de Cremona, Ann. Sci. Ecole Norm. Sup. 3 (1970), 507–588.
[24] D.R. Farkas and R.L. Snider, Arithmeticity of stabilizers of ideals in group rings, Invent. Math. 75 (1984), 75–84.
[25] W. Fulton, Introduction to toric varieties (Ann. of Math. Studies 131, Princeton University Press, 1993).
[26] O. Gabber, The integrability of the characteristic variety, Amer. J. Math. 103 (1981), 445–468.
[27] J.R.J. Groves and J.S. Wilson, Finitely presented metanilpotent groups, J. London Math. Soc. (2) 50 (1994), 87–104.
[28] G. Kempf, F. Knudsen, D. Mumford and B. Saint-Donat, Toroidal embeddings I (Lecture Notes in Mathematics 339, Springer, 1973).
[29] G. Levitt, 1-formes fermées singulières et groupe fondamental, Invent. Math. 88 (1987), 635–667.
[30] A. Lubotsky, Group presentation, p-adic analytic groups and lattices in $SL_n(\mathbb{C})$, Ann. of Math. (2) 118 (1983), 115–130.
[31] J.C. McConnell and J.J. Pettit, Crossed products and multiplicative analogues of Weyl algebras, J. London Math. Soc. (2) 38 (1988), 47–55.
[32] J.C. McConnell and J.C.Robson, Noncommutative noetherian rings (Wiley, 1987).
[33] J.W. Morgan, Group actions on trees and the compactification of the spaces of classes of $SO(n,1)$-representations, Topology 25 (1986),1–33.
[34] J.W. Morgan and P.B. Shalen, Valuations, trees and degenerations of hyperbolic structures I, Ann. of Math. 120 (1984), 401–476.
[35] J.W. Morgan and P.B. Shalen, Degenerations of hyperbolic structures II: measured laminations in 3-manifolds, Ann. of Math. 127 (1988), 403–456.
[36] J.W. Morgan and P.B. Shalen, Degenerations of hyperbolic structures III: actions of 3-manifold groups on trees and Thurston's compactness theorem, Ann. of Math. 127 (1988), 457–519.
[37] W.D. Neumann, Normal subgroups with infinite cyclic quotient, Math. Sci. 4 (1979), 143–148.
[38] F. Paulin, Topologie de Gromov équivariante, structures hyperboliques et arbres réels, Invent. Math. 94 (1988), 53–80.
[39] M.A. Rieffel, C^*-algebras associated with irrational rotations, Pacific J. Math. 95 (1981), 415–429.
[40] M. A. Rieffel, K-theory of crossed products of C^*-algebras by discrete groups, in Group actions on rings (Contemp. Math. 43, Amer. Math. Soc.,1985), 253–265.
[41] E. Rips and Z. Sela, Structure and rigidity in hyperbolic groups I, Geom. Funct. Anal. 4 (1994), 337–371.
[42] D.J.S. Robinson and R. Strebel, Some finitely presented soluble groups which are not nilpotent by abelian by finite, J. London Math. Soc. (2) 26 (1982), 435–440.
[43] J.E. Roseblade, Group rings of polycyclic groups, J. Pure Appl. Algebra 3 (1973), 307–328.
[44] J.E. Roseblade, Prime ideals in group rings of polycyclic groups, Proc. London Math. Soc. (3) 36 (1978), 385–447.
[45] J.E. Roseblade, Five lectures on group rings, in Proceedings of Groups-St Andrews 1985 (London Math. Soc. Lecture Notes 211, Cambridge University Press, 1986), 93–109.

[46] D. Segal, Polycyclic groups (Cambridge University Press, 1983).
[47] J-P. Serre, Trees (Springer, 1980).
[48] R. Strebel, Finitely presented soluble groups, in Group theory (ed. K.W. Gruenberg and J.E. Roseblade, Academic Press, 1984), 257–314.
[49] D. Tischler, On fibering certain foliated manifolds over S^1, Topology 9 (1970), 153–154.
[50] W.P. Thurston, A norm on the homology of 3-manifolds, Memoirs Amer. Math. Soc. 339 (1986), 100-130.
[51] J.S. Wilson, Finite presentations of pro-p groups and discrete groups, Invent. Math. 105 (1991), 177–183.
[52] A.E. Zaleskii, Irreducible representations of finitely generated nilpotent torsion-free groups, Mat. Zametki 9 (1971) 199–210; Math. Notes 9 (1971), 117–123.
[53] E.I. Zelmanov, Lie ring methods in the theory of nilpotent groups, in Proceedings of Groups '93 Galway/ St Andrews (London Math. Soc. Lecture Notes 212, Cambridge University Press, 1995), 567–585.

THEOREMS OF KEGEL-WIELANDT TYPE

ANGEL CAROCCA* and RUDOLF MAIER†

*Facultad de Matemáticas, Pontifícia Universidad Católica de Chile, Casilla 306, Santiago 22, Chile

†Universidade de Brasília, Departamento de Matemática IE , BR-70910-900 Brasília - DF, Brazil

1991 Mathematics Subject Classification: Primary 20F17, 20D40; Secondary 20D25, 20E28.

For not explicitly explained concepts used in this report, the cited literature or usual books in finite group theory may be consulted.

Based on W. Burnside's famous theorem, published in 1904 [5], which states that a finite group is soluble, if its order is divisible by at most two distinct prime numbers, P. Hall characterizes the soluble groups in the decade 1928 - 1937, by means of the existence of Sylow complements and Sylow systems [15, 16, 17]. In particular:

> *A finite group is soluble if and only if it is the product of pairwise permutable Sylow subgroups.*

With this characterization arises the question:

> *If a finite group is the product of pairwise permutable nilpotent subgroups, the group is soluble?*

In 1951, H. Wielandt [28] begins with a series of theorems which he finishes in 1958 [29] and which together with a result of O. H. Kegel 1962 [21] answers affirmatively to this question.

Meanwhile, in 1953, B. Huppert [19] solves a particular case of the problem, which merits to be mentioned:

> *A finite group which is the product of pairwise permutable cyclic subgroups is supersoluble.*

The general background of these theorems is: Given is a group G in a factorized form $G = G_1G_2...G_r$ with subgroups $G_1, ..., G_r$ of G such that $G_iG_j = G_jG_i$ for all indices $i, j \in \{1, 2, ..., r\}$ where $r \geq 2$. What can we say about G if special properties of the factors $G_1, ..., G_r$ are known? In other words: If the G_i belong to a certain class $\mathcal{K}$ of groups, can we describe in some satisfactory manner the class $\mathcal{L}$ to which G belongs?

Sometimes the case $r = 2$, when the group $G = G_1G_2$ is just the product of two factors, is of interest, since the problem often reduces from an arbitrary number r of factors to $r = 2$.

In 1955, N. Itô [20] shows that *any (not necessarily finite) group is metabelian, whenever it is the product of two abelian subgroups.*

In 1956, P. Hall and G. Higman [18] introduce the concept of p-solubility of a finite group (the localized form of solubility for an individual prime number p) and its p-length which measures the embedding complexity of the Sylow-p-subgroups. They prove Theorems of the type: An uncomplicated inner structure of the Sylow-p-subgroups of a p-soluble group guarantees their uncomplicated embedding, that is, the whole group has a controlled p-length.

Keeping in mind that a group which is the product of two abelian subgroups is always metabelian, whereas a finite group $G = HK$ which is the product of two nilpotent subgroups H and K is only soluble but in general not metanilpotent (consider the symmetric group S_4), R. Maier [22] investigates in 1972 the question which further property would characterize the metanilpotency of G. The result is:

> *The product $G = HK$ of two nilpotent subgroups H and K is metanilpotent, if and only if G is of p-length at most one for all primes p.*

The proof of this Theorem is a not quite evident reduction of a minimal counterexample to abelian factors and so Itô's theorem can be applied.

In the same paper, an interesting observation is made: The intersection $H \cap K$ of a product $G = HK$ of two nilpotent subgroups is always a subnormal subgroup of G. This fact leads to the question:

> *If a finite group $G = HK$ is the product of two (not necessarily nilpotent) subgroups H and K such that a subgroup $X \leq H \cap K$ is subnormal in both H and K, is it true that X is also subnormal in G?*

The positive answer to this problem is conjectured and solved in the case where the subgroup X is soluble in 1977 by Maier [25]. The general affirmative answer Wielandt [31] obtains in 1982 (see also [26]). For infinite groups (except for special classes of infinite groups for which the problem reduces to the finite case) this remains an open problem, even when the factors are nilpotent.

In 1976, Maier [23] tries to localize Huppert's theorem on the supersolubility of the product of pairwise permutable cyclic groups for p-soluble groups. He introduces the *polycyclic length* of a p-group (= the smallest length of a subnormal chain with cyclic factors) and proves:

> *Let p be an odd prime number, $G = G_1G_2\cdots G_r$ a p-soluble group which is the product of the pairwise permutable subgroups $G_1, \cdots, G_r$. If the polycyclic length of the Sylow-p-subgroups of all factors G_i is limited by $\frac{p-1}{2}$, then G is of p-length at most one.*

As a consequence one can state:

> *Let the G_i have cyclic Sylow-p-subgroups. If G has odd order, or if the G_i are p-nilpotent, then G is p-supersoluble.*

For $p = 2$, an interesting 2-nilpotency criterion emerges (see Maier [24]), which contains the classical result of the 2-nilpotency of a group with cyclic Sylow-2-subgroups:

> *A group is 2-nilpotent, whenever it can be written as the product of pairwise permutable subgroups, each of which is the direct product of a cyclic 2-group by a group of odd order.*

More recent results

A very special case of a product $G = G_1G_2\cdots G_r$ with pairwise permutable subgroups $G_1, \cdots, G_r$ we have when all the factors G_i are normal subgroups of G; in particular, when $G = G_1 \times G_2 \times \cdots \times G_r$ is a direct product. The gap between a general product of pairwise permutable subgroups and a direct product is large: If we take two elements x and y (or two subgroups X and Y) of distinct factors of a subgroup product, there will be in general no immediate relation between them. In a direct product, however, they commute. To create intermediate situations, it seems therefore reasonable to consider products of subgroups in which the elements (or subgroups) of distinct factors are linked by certain relations.

Some interesting considerations in this direction are made in 1989 by M. Asaad and A. Shaalan in [1]: They study finite factorized groups $G = HK$ such that H permutes with all subgroups of K and vice versa. This includes of course the possibility of H and K being normal in G. For such products they prove some theorems on supersoluble groups which were known to be true in the case of normal factors. In [7], A. Carocca localizes Asaad and Shaalan's results to the p-supersoluble case.

Whereas the product $G = HK$ of supersoluble normal subgroups H and K is known to be in general not supersoluble, Asaad and Shaalan prove that the supersolubility of G is secured if one knows the permutability of every subgroup of H with every subgroup of K. This includes of course the case of a direct product.

These ideas have much wider application, however: In 1992, Maier observes in [27] that for subgroup products $G = HK$ such that every subgroup of H is permutable with every subgroup of K, infinitely many analogues of Asaad and Shaalan's result are true:

> *If one knows that H and K both belong to a saturated formation $\mathcal{F}$ which contains the class of all supersoluble groups, G also belongs to $\mathcal{F}$.*

In 1996, A. Ballester-Bolinches and M. D. Pérez-Ramos [3] prove this result without the hypothesis of $\mathcal{F}$ being saturated.

In 1992, Carocca introduces in his doctoral thesis [8], the concept of mutually and totally f-permutable subgroups: Let f be a functor which to every finite group X associates a family of subgroups fX of X, such that $f\mu X = \mu fX$ for every X and every isomorphism μ of X. Two subgroups H and K of a group G then are called *mutually f-permutable*, if H permutes with every subgroup in $fK \cup \{K\}$ and K permutes with every subgroup in $fH \cup \{H\}$. They are called *totally f-permutable*, if every subgroup in $fH \cup \{H\}$ permutes with every subgroup in $fK \cup \{K\}$. He then considers groups $G = G_1G_2\cdots G_r$ such that G_i and G_j are mutually (or totally) f-permutable for $1 \leq i \neq j \leq r$ for some special choices of the functor f. In particular, he studies $f = syl, f = m, f = sn, f = sf$ and $f = s$, that is, when fX is the set of all Sylow, all maximal, all subnormal, all self-normalizing and all subgroups of X, respectively. Some of his results are:

1) *If the G_i are p-soluble for some prime p and mutually syl-permutable, then G is p-soluble.*
2) *If the G_i are > 1 and mutually m- or syl-permutable and if G is simple, then* $G_1 = G_2 = \cdots = G_r = G$.
3) *If the G_i are p-supersoluble for some prime p and mutually s-permutable and if G' is p-nilpotent, then G is p-supersoluble.*
4) *If the G_i are pairwise totally s-permutable and belong to a saturated formation $\mathcal{F}$ which contains the supersoluble groups, then G belongs to $\mathcal{F}$.*
5) *If the G_i are soluble and mutually sn-permutable, then G is soluble.*
6) *If the G_i are soluble and totally sf-permutable, then G is soluble.*

Remarks

The proofs of 1) and 2) use Hall [17] and Wielandt [30]. Applying 3) for $r = 2$ and all p, a result of Asaad and Shaalan [1] is obtained, which for $G_1, G_2 \trianglelefteq G$ passes into the classical result due to Baer [2]. 4) is the generalization of Maier [27] to an arbitrary number r of factors (see also Carocca [9]). In view of the recent result of Ballester-Bolinches/Pérez-Ramos [3], it is natural to conjecture that the following should be true:

> **Conjecture 4′)** *Let $G = G_1G_2\cdots G_r$ be a finite group which is the product of the pairwise totally s-permutable subgroups $G_1, \cdots, G_r$ and let $\mathcal{F}$ be a formation which contains all supersoluble groups. If the G_i belong to $\mathcal{F}$, then also G is an $\mathcal{F}$-group.*

(While writing this report we are informed that this conjecture has really been proved by Ballester-Bolinches, Pedraza-Aguilera and Perez-Ramos; see [4]).

5) is of some own interest and is to be published in [12]. Since the only selfnormalizing subgroup of a nilpotent group is the group itself, result 6) is an interesting generalization of the Kegel-Wielandt-Theorem.

The following result (already conjectured in [28]) has been verified in 1993 by E. Fisman [14]:

> (*)*Let the finite group $G = G_1G_2\cdots G_r$ be a product of the pairwise permutable subgroups $G_1, \cdots, G_r$. If G_iG_j is soluble for all $1 \le i, j \le r$, then G is soluble.*

In [8] is shown:

> *If $G = G_1G_2$ with two soluble and mutually sf-permutable subgroups, then G is soluble.*

Therefore, using (*), the following generalization of the Kegel-Wielandt-Theorem is true:

> 6′) *Let $G = G_1G_2\cdots G_r$ be the product of the soluble and pairwise mutually sf-permutable subgroups. Then G is soluble.*

Still better would be if it were possible to verify:

> **Conjecture** 6″) *If the G_i are soluble and G_i is permutable with the Carter subgroups of G_j for all $1 \leq i, j \leq r$, then G is soluble.*

(Also here, like in every potential solubility criterion for the product of pairwise permutable subgroups, the restriction to two factors is sufficient).

If H and K are two totally s-permutable subgroups of a group and if $x \in H$ and $y \in K$, then $<x,y>=<x><y>=<y><x>$ is a supersoluble subgroup by Huppert's theorem. This leads to the following definition which tries to establish certain connections between two subgroups:

> *Let $\mathcal{L}$ be a nonempty class of groups. Two subgroups H and K of a group are $\mathcal{L}$-connected, if for every $x \in H$ and every $y \in K$, the subgroup $<x, y>$ is an $\mathcal{L}$-group.*

So total s-permutable subgroups are $\mathcal{SS}$-connected for the class $\mathcal{L} = \mathcal{SS}$ of supersoluble groups. $\mathcal{A}$-connection of two subgroups for the class $\mathcal{L} = \mathcal{A}$ of abelian groups means exactly their elementwise permutability. Whereas the total s-permutability of the factors in result 4) can certainly not be weakened to "the G_i are pairwise $\mathcal{SS}$-connected" (see [27]), in [8] and [10] is proved, however:

> *Let the finite group $G = G_1G_2 \cdots G_r$ be the product of the pairwise permutable subgroups $G_1, \cdots, G_r$. Suppose, the saturated formation $\mathcal{F}$ contains the class $\mathcal{N}$ of all nilpotent groups. If the G_i are $\mathcal{N}$-connected $\mathcal{F}$-groups, then also $G \in \mathcal{F}$.*

In recent years, interest in problems on factorizable groups has increased notably due to the appearance of new powerful methods, developed while solving the classification problem for finite simple groups. However, the knowledge of the simple groups does not automatically give an answer to many important problems, including the question of groups with factorization.

Consider the class $\mathcal{S}$ of all soluble groups. In view of the fact that the finite simple groups are 2-generated, in [10] is conjectured:

> **Conjecture**) *Let $G = G_1G_2...G_r$ be a product of the pairwise permutable subgroups $G_1, \cdots G_r$. If the G_i are $\mathcal{S}$-connected soluble subgroups, then G is soluble.*

A particular case of this conjecture is solved in [10]:

> *Let $\mathcal{T}$ denote the class of groups having Sylow tower for the prime numbers arranged in decreasing order. If the G_i are pairwise $\mathcal{T}$-connected supersoluble subgroups, then G is a $\mathcal{T}$-group. In particular, G is soluble.*

Meanwhile (in 1998), using classification theorems of simple groups, in [13] is to be published the general proof of the above conjecture.

Using similar ideas, in [11] is proved:

Let p be an odd prime number, $G = HK$ a group, such that H is 2-nilpotent and K is a Sylow-p-subgroup of G. Then G is soluble.

References

[1] Asaad, M. and Shaalan, A. *On the supersolvability of finite groups.* Archiv der Mathematik, **53** (1989), 318 -326.

[2] Baer, R. *Classes of finite groups and their properties.* Illinois Math. J. **1** (1957), 115 - 187.

[3] Ballester-Bolinches, A. and Perez-Ramos. M.D. *A question of R. Maier concerning formations.* Journal of Algebra **182** (1996), 738 -747.

[4] Ballester-Bolinches, A., Pedraza-Aguilera, M. C. and Perez-Ramos. M.D. *Finite groups which are products of pairwise totally permutable subgroups.* To appear in the Proc. Edinburgh Math. Soc.

[5] Burnside, W. *On groups of order $p^\alpha q^\beta$.* Proc. London Math. Society (2), **2** (1904), 432 - 437.

[6] Carocca, A. *Grupos p-solúveis de p-comprimento menor ou igual a um.* Dissertação de Mestrado. Univ. de Brasília, (1990).

[7] Carocca, A. *p-supersolvability of factorized finite groups.* Hokkaido Math. J. **21** (1992), 395 - 403.

[8] Carocca, A. *Produtos de subgrupos mutuamente f-permutáveis em grupos finitos.* Tese de Doutorado. Univ. de Brasília, (1992).

[9] Carocca, A. *A note on the product of $\mathcal{F}$-subgroups in a finite group.* Proc. Edinburgh Math. Soc. **39** (1996), 37 - 42.

[10] Carocca, A. *Completeness properties of certain formations.* Mat. Contemporânea **6** (1994), 1-6.

[11] Carocca, A. and Matos, H. *Some solvability criteria for finite groups.* Hokkaido Math. J. **26** (1997), 157 - 161.

[12] Carocca, A. *On factorized finite groups in which certain subgroups of the factors permute.* To appear in Archiv der Math.

[13] Carocca, A. *Solvability of factorized finite groups.* Submitted.

[14] Fisman, E. *Products of solvable subgroups of a finite group.* Archiv der Mathematik **61** (1993), 201 - 205.

[15] Hall, P. *A note on soluble groups.* J. London Math. Soc. **3** (1928), 98 - 105.

[16] Hall, P. *On the Sylow systems of a soluble group.* Proc. London Math. Soc. (2) **43** (1937), 316 - 323.

[17] Hall, P. *A characteristic property of soluble groups.* J. London Math. Soc. **12** (1937), 198 - 200.

[18] Hall, P. and Higman, G. *On the p-length of p-soluble groups and reduction theorems for Burnside's problem.* Proc. London Math. Soc. **6** (1956), 1 - 40.

[19] Huppert, B. *Über das Produkt von paarweise vertauschbaren zyklischen Gruppen.* Math. Zeitschrift **58** (1953), 243 - 264.

[20] Itô, N. *Über das Produkt von zwei abelschen Gruppen.* Math. Zeitschrift **62** (1955), 400 - 401.

[21] Kegel, O. H. *Produkte nilpotenter Gruppen.* Archiv der Mathematik **12** (1961), 90 - 93.

[22] Maier, R. *Endliche metanilpotente Gruppen.* Archiv der Mathematik **23** (1972), 139 - 144.

[23] Maier, R. *Faktorisierte p-auflösbare Gruppen.* Archiv der Mathematik **27** (1976), 576 - 583.

[24] Maier, R. *Über die 2-Nilpotenz faktorisierbarer endlicher Gruppen.* Archiv der Mathematik **27** (1976), 480 - 483.

[25] Maier, R. *Um problema da teoria dos subgrupos subnormais.* Bol. Soc. Brasil. Mat. **8** (1977), 127 - 130.

[26] MAIER, R. *Subnormalidade em produtos.* Trabalho de Matemática 167, Universidade de Brasília (1980).

[27] Maier, R. *A completeness property of certain formations.* Bull. London Math. Soc. **24** (1992), 540 - 544.

[28] Wielandt, H. *Über das Produkt von paarweise vertauschbaren nilpotenten Gruppen* . Math. Zeitschrift **55** (1951), 1 - 7.

[29] Wielandt, H. *Über Produkte von nilpotenten Gruppen.* Illinois J. Math. **2** (1958), 611 - 618.

[30] Wielandt, H. *Vertauschbarkeit von Untergruppen und Subnormalität.* Math. Zeitschrift **133** (1973), 275 - 276.

[31] Wielandt, H. *Subnormalität in faktorisierten endlichen Gruppen.* J. Algebra **69** (1982), 305 - 311.

SINGLY GENERATED RADICALS ASSOCIATED WITH VARIETIES OF GROUPS

CARLES CASACUBERTA*, JOSÉ L. RODRÍGUEZ*[1] and DIRK SCEVENELS†

*Departament de Matemàtiques, Universitat Autònoma de Barcelona, E–08193 Bellaterra, Spain

†Centre de Recerca Matemàtica, Apartat 50, E–08193 Bellaterra, Spain

Abstract

To every variety of groups $\mathcal{W}$ one can associate an idempotent radical $\mathcal{P}_\mathcal{W}$ by iterating the verbal subgroup. The basic example is the perfect radical, which is the intersection of the transfinite derived series. We prove that each such radical $\mathcal{P}_\mathcal{W}$ is generated by a single locally free group F, in the sense that, for every group G, the subgroup $\mathcal{P}_\mathcal{W}G$ is generated by images of homomorphisms $F \to G$.

Our motivation comes from algebraic topology. In fact we show that every variety of groups $\mathcal{W}$ determines a localization functor in the homotopy category, which kills the radical $\mathcal{P}_\mathcal{W}$ of the fundamental group while preserving homology with certain coefficients.

0 Introduction

Radicals have been broadly studied in abelian categories; see e.g. [15, VI.1]. However, they have received less attention in the category of groups, where the fundamentals of radical theory were first investigated by Kurosh; see [13, I.3]. We say that a functor R is a radical if RG is a normal subgroup of G and $R(G/RG) = 1$ for all groups G.

Radicals are useful in localization theory. Given a family Φ of group homomorphisms $\varphi_\alpha: A_\alpha \to B_\alpha$, a group K is called Φ-local if the induced map of sets $\mathrm{Hom}(\varphi_\alpha, K): \mathrm{Hom}(B_\alpha, K) \to \mathrm{Hom}(A_\alpha, K)$ is bijective for all α. Under mild assumptions (e.g., if Φ is a set, or also if each φ_α is surjective), every group G admits a Φ-localization $G \to LG$, which is initial among homomorphisms from G into Φ-local groups. The main examples are localizations at sets of primes and homological localizations [2].

As we explain in Section 2, there is a bijective correspondence between radicals and surjective localizations in the category of groups, and there is also a bijective correspondence between idempotent radicals (that is, such that $RRG = RG$ for all G) and localizations with respect to homomorphisms whose target is the trivial group (such localizations will be called reductions). In particular, the projection onto an arbitrary variety of groups determines a radical R, and to this radical we can associate by standard methods an idempotent radical and hence a reduction. The basic example is $RG = [G, G]$, the commutator subgroup, whose associated idempotent radical $R^\infty G$ is the perfect radical.

[1]The two first-named authors were supported by DGICYT grant PB94–0725

Our motivation comes from [1], where a universal locally free group $\mathcal{F}$ was constructed with the property that localization with respect to the homomorphism $\mathcal{F} \to 1$ has the effect of dividing out the perfect radical. In Theorem 3.3 below we generalize the construction of Example 5.3 in [1] so that it applies to any other idempotent radical associated with a variety.

Furthermore, every radical R associated with a variety gives rise to a localization functor in the homotopy category of CW-complexes. When applied to a space X, this functor kills the subgroup $R\pi_1(X)$ of the fundamental group and preserves homology with certain coefficients. Quillen's plus-construction [12] is the special case corresponding to the perfect radical.

1 Radicals in group theory

We shall work in the category $\mathcal{G}$ of groups. A *radical* R is a subfunctor of the identity (i.e., a functor assigning to each group G a subgroup RG in such a way that every homomorphism $G \to K$ induces $RG \to RK$ by restriction), with the property that RG is normal in G and $R(G/RG) = 1$ for all groups G. A radical R is said to be *idempotent* if $RRG = RG$ for all groups G. These notions are standard in abelian categories (see e.g. [15, Ch. VI]), although the terminology varies slightly depending on the authors; cf. [7], [8], [9], [13].

Example 1.1 The best-known example of a (nonidempotent) radical is the commutator subgroup $RG = [G, G]$. Two idempotent examples are the perfect radical (i.e., the largest perfect subgroup, where a group G is called *perfect* if $[G, G] = G$), and the torsion radical (i.e., the smallest normal subgroup $\tau(G)$ such that $G/\tau(G)$ is torsion-free). In fact, for each set of primes J there is a J-torsion radical, and there is also a largest J-perfect subgroup, where a group G is called J-perfect if the first mod p homology group $H_1(G; \mathbb{Z}/p)$ is zero for $p \in J$.

Given any radical R, the class of groups G such that $RG = G$ is closed under quotients and free products, and the class of groups G such that $RG = 1$ is closed under subgroups and cartesian products. The proof is the same as the one given in [15, VI.1.2] for abelian categories. It follows that, if two groups G and K satisfy $RG = G$ and $RK = 1$, then $\mathrm{Hom}(G, K)$ is trivial.

Proposition 1.2 *Let R be any radical in the category of groups. Then every group G contains a unique subgroup $R^\infty G$ maximal with the property that $RR^\infty G = R^\infty G$. Moreover, $R^\infty G$ is normal in G and it is contained in RG.*

Proof Let $R^\infty G$ be the product of all subgroups H of G such that $RH = H$. Then $R^\infty G$ is a quotient of the free product of all such subgroups H, and this family of subgroups is closed under conjugation. Hence, $RR^\infty G = R^\infty G$ and $R^\infty G$ is normal. Finally, since R is a subfunctor of the identity, we have $RR^\infty G \subseteq RG$, that is, $R^\infty G \subseteq RG$. □

It follows, similarly as in [15, VI.1.6], that R^∞ is also a radical and it is in fact the largest idempotent radical which is a subfunctor of R. There is another standard way of constructing R^∞ from R by transfinite induction. Namely, if α is a successor ordinal, define $R^\alpha = RR^{\alpha-1}$, with $R^0 = R$, and if α is a limit ordinal, then let R^α be the intersection of R^β for all $\beta < \alpha$. Finally, take $R^\infty G$ to be $R^\alpha G$ for the smallest α such that $R^{\alpha+1}G = R^\alpha G$.

Proposition 1.3 *Every family of groups $\mathcal{C}$ determines a radical R, by defining RG to be the intersection of the kernels of all epimorphisms $f: G \to C$ where C is in $\mathcal{C}$.*

We omit the details since this is a classical construction. Note that every radical R arises this way, by taking $\mathcal{C}$ to be the class of groups G such that $RG = 1$.

There is another important source of radicals. Namely, the same argument as in [9, 2.2] proves the following.

Proposition 1.4 *Let $\mathcal{G}$ be the category of groups and $\mathcal{D}$ any category. Suppose given an adjoint pair of functors $L: \mathcal{G} \to \mathcal{D}$ and $E: \mathcal{D} \to \mathcal{G}$, where L is the left adjoint. Then we obtain a radical by defining RG to be the kernel of the unit map $G \to ELG$, for each group G.*

2 Localizations

A full subcategory $\mathcal{D}$ of the category of groups is called *reflective* if the inclusion $E: \mathcal{D} \to \mathcal{G}$ has a left adjoint L. In this case, for every group G we have a group homomorphism $G \to LG$ which is initial among homomorphisms from G into groups in $\mathcal{D}$. Then the groups in $\mathcal{D}$ are called *L-local* and the homomorphisms $G \to K$ inducing isomorphisms $LG \cong LK$ are called *L-equivalences.* The functor L will be called a *localization* or a *reflection* onto the subcategory $\mathcal{D}$. It will be called an *epireflection* if, for all groups G, the localization homomorphism $G \to LG$ is surjective.

If L is any reflection in the category of groups, and RG denotes the kernel of the localization homomorphism $G \to LG$, then it follows from Proposition 1.4 that R is a radical. In fact, there is a bijective correspondence between radicals and epireflections in the category of groups.

Given any family of group homomorphisms $\varphi_\alpha: A_\alpha \to B_\alpha$, we can consider the subcategory $\mathcal{D}$ of all groups D which are *orthogonal* to φ_α for all α, that is, such that the induced map

$$\mathrm{Hom}(\varphi_\alpha, D): \mathrm{Hom}(B_\alpha, D) \to \mathrm{Hom}(A_\alpha, D)$$

is bijective for all α. If the family $\{\varphi_\alpha\}$ is a set, then we can form the free product φ of all homomorphisms in the family, and it follows by standard methods that the orthogonal subcategory $\mathcal{D}$ is reflective. For further details and applications to homotopy theory, see [1], [4], [5], [14]. If φ is a surjective homomorphism, then the φ-localization functor L_φ is an epireflection. In the special case when φ is of the form $A \to 1$ for some group A, the orthogonal groups are called *A-reduced.* Thus, a group G is A-reduced if and only if $\mathrm{Hom}(A, G)$ is trivial. In this case, the φ-localization of a group G will be called *A-reduction* and denoted by $G//A$. It is the largest quotient of G admitting no nontrivial homomorphisms from A.

If the given class of homomorphisms $\{\varphi_\alpha\}$ is proper (i.e., not a set), then we cannot infer in general that the orthogonal subcategory $\mathcal{D}$ is reflective. However, there is a special situation where it can be proved that the subcategory orthogonal to a (possibly proper) class of homomorphisms is reflective.

Proposition 2.1 *The subcategory $\mathcal{D}$ orthogonal to any class of group epimorphisms is reflective.*

Proof Let TG be the intersection of all kernels of epimorphisms from G onto groups in $\mathcal{D}$. By Proposition 1.3, T is a radical, and hence $G \to G/TG$ is a reflection. Thus, it suffices to check that a group G belongs to $\mathcal{D}$ if and only if $TG = 1$. If G is in $\mathcal{D}$, then the identity homomorphism $G \to G$ is an epimorphism onto a group in $\mathcal{D}$ and hence TG is indeed trivial. To prove the converse, suppose that $TG = 1$ and let $\gamma: G \to \widehat{G}$ be the inverse limit of all epimorphisms from G onto groups in $\mathcal{D}$. Our assumption ensures that γ is injective. Since $\mathcal{D}$ is closed under subgroups and inverse limits, it follows that G is in $\mathcal{D}$. □

In the special case where the targets of all homomorphisms φ_α are trivial, the associated localization will be called a *reduction*. This terminology is consistent with our previous use of the same word. The same arguments as in [14, Theorem 2.7] lead to the following characterization.

Theorem 2.2 *Let L be a localization in the category of groups and let $\mathcal{D}$ be the class of L-local groups. Then L is an epireflection if and only if $\mathcal{D}$ is closed under subgroups, and L is a reduction if and only if $\mathcal{D}$ is closed under subgroups and formation of extensions.*

Finally, we prove that the bijection between epireflections and radicals makes reductions correspond with idempotent radicals.

Theorem 2.3 *There is a bijective correspondence between idempotent radicals and reduction functors in the category of groups.*

Proof We show that, given an idempotent radical R, the functor $LG = G/RG$ is a reduction. Consider the family of all homomorphisms $A \to 1$ where $LA = 1$, and let L' denote the corresponding reduction. From the fact that all such homomorphisms are L-equivalences it follows that there is a natural transformation $L' \to L$. Conversely, let G be L'-local. Then we have $LRG = RG/RRG = 1$ and this implies that the homomorphism $RG \to 1$ is an L'-equivalence. Therefore $\mathrm{Hom}(RG, G)$ is trivial, from which it follows that $RG = 1$ and G is L-local. Thus we have proved that $L' = L$, as desired.

The kernel R of any reduction with respect to a class $\varphi_\alpha: A_\alpha \to 1$ can be described as a possibly transfinite direct limit, as in [4, Theorem 3.2], where the first step is the subgroup generated by the images of all homomorphisms from A_α for all α. From this description one sees that R is in fact an idempotent radical. □

3 Varieties of groups

Let $\mathcal{W}$ be the variety of groups defined by a set of words W. That is, W is a set of elements of the free group F_∞ on a countably infinite set of generators $\{x_1, x_2, x_3, \ldots\}$, and $\mathcal{W}$ is the family of groups G with the property that every homomorphism $f: F_\infty \to G$ satisfies $f(w) = 1$ for all $w \in W$; see [11]. For an arbitrary group G, the verbal subgroup WG of G is the subgroup generated by all the images of words in W under homomorphisms $F_\infty \to G$.

Proposition 3.1 *A group G is in the variety $\mathcal{W}$ if and only if G is orthogonal to the natural homomorphism $\varphi: F_\infty \to F_\infty/WF_\infty$.*

Proof A group G is orthogonal to φ if and only if every homomorphism $f: F_\infty \to G$ satisfies $f(WF_\infty) = 1$. But this condition is equivalent to $WG = 1$, and this means that G is in the variety $\mathcal{W}$. □

Let φ be as in Proposition 3.1. Then the φ-localization functor L_φ is the projection onto the variety $\mathcal{W}$, sending each group G onto G/WG. Thus, the verbal subgroup is a radical, which we denote by the same letter W. It is not idempotent in general; indeed, WWG need not be equal to WG.

The groups G such that $WG = G$ will be called *$\mathcal{W}$-perfect*. That is, G is $\mathcal{W}$-perfect if and only if $L_\varphi G = 1$. This notion specializes to ordinary perfect groups when WG is the commutator subgroup of G.

It follows from Proposition 1.2 that every group G has a largest $\mathcal{W}$-perfect subgroup. We call it the $\mathcal{W}$-perfect radical of G, and denote it by $\mathcal{P}_\mathcal{W}G$. This radical $\mathcal{P}_\mathcal{W}$ is idempotent. Thus, we can consider the reduction functor assigning to each group G the quotient $G/\mathcal{P}_\mathcal{W}G$; cf. Theorem 2.3. We say that a group F *generates* the radical $\mathcal{P}_\mathcal{W}$ if $G//F = G/\mathcal{P}_\mathcal{W}G$ for all groups G. Our main result in this section (Theorem 3.3) states that each radical $\mathcal{P}_\mathcal{W}$ is generated by some locally free group.

In the case where W consists of the word x^m alone, where m is a nonnegative integer, the $\mathcal{W}$-perfect radical of a group G is the largest subgroup H such that $H = H^m$, where H^m denotes the subgroup generated by all m-powers of elements of H. We call this radical the *Burnside radical* of exponent m. The Burnside radical of exponent 0 is the trivial subgroup and the Burnside radical of exponent 1 is the whole group. Note that the Burnside radical of exponent m coincides with the radical generated by $\mathbb{Z}[1/m]$ on commutative groups, but not on other groups in general.

Recall from [11, 12.12] that every word w is equivalent to a power word x^m together with a commutator word c, in the sense that w is a law in a group G if and only if the words x^m and c are both laws in G. (A commutator word is any element of $[F_\infty, F_\infty]$.) Thus, given a variety $\mathcal{W}$, we can assume without loss of generality that $\mathcal{W}$ is defined by a set of words of the form

$$W = \{x^m, c_1, c_2, c_3, \ldots\},$$

where each c_i is a commutator word and m is a nonnegative integer, called the exponent of the variety.

If a variety $\mathcal{W}$ is defined by commutator words only, then the $\mathcal{W}$-perfect radical is contained in the ordinary perfect radical. Indeed, the inclusion $WG \subseteq [G,G]$ yields an epimorphism $G/WG \to G/[G,G]$ and hence all $\mathcal{W}$-perfect groups are perfect. However, the inclusion of the $\mathcal{W}$-perfect radical into the perfect radical can be proper, as the following example shows.

Example 3.2 Let $\mathcal{W}$ be the variety defined by the word $c = [x,y]^m$, where m is any integer greater than 2. Then there exist perfect groups which are not $\mathcal{W}$-perfect; it suffices to pick any perfect group G such that $G \neq G^m$.

We next prove that, for every variety $\mathcal{W}$, the reduction $G \to G/\mathcal{P}_{\mathcal{W}}G$ coincides with F-reduction (that is, localization with respect to $F \to 1$), for some locally free group F. To this aim, we shall generalize the construction described in Example 5.3 of [1].

Theorem 3.3 *Let $\mathcal{W}$ be any variety of groups. Then there exists a locally free, $\mathcal{W}$-perfect group F such that, for all groups G, the radical $\mathcal{P}_{\mathcal{W}}G$ is generated by images of homomorphisms $F \to G$.*

Proof Let $W = \{w_1, w_2, \ldots\}$ be a set of words defining the variety $\mathcal{W}$. In order to simplify the notation, we will assume, as we may (by reordering the words in W and inserting the trivial word as many times as needed), that w_j is a word on a subset of the generators $x_1, \ldots, x_j$ of the free group F_∞. Thus we write $w_j = w_j(x_1, \ldots, x_j)$.

We shall construct a countable, locally free group $F_{\mathbf{n}}$ for each sequence $\mathbf{n} = (n_1, n_2, n_3, \ldots)$ of positive integers, and define F to be the free product of the groups $F_{\mathbf{n}}$ for all increasing sequences $\mathbf{n}$. The group $F_{\mathbf{n}}$ is defined as the colimit of a directed system $(F_{\mathbf{n},r}, \varphi_r)$ of free groups and homomorphisms. For $r = 0$, the group $F_{\mathbf{n},0}$ is infinite cyclic with a generator x_0. For $r \geq 1$, the group $F_{\mathbf{n},r}$ is the free group on the symbols

$$x_r(\delta_1, \ldots, \delta_r; \varepsilon_1, \ldots, \varepsilon_r; i_1, \ldots, i_r),$$

where $1 \leq \varepsilon_k \leq \delta_k \leq n_k$ and $1 \leq i_k \leq n_k$, for $k = 1, \ldots, r$. The homomorphism $\varphi_r \colon F_{\mathbf{n},r} \to F_{\mathbf{n},r+1}$ is determined by letting

$$\varphi_r \left(x_r(\delta_1, \ldots, \delta_r; \varepsilon_1, \ldots, \varepsilon_r; i_1, \ldots, i_r) \right)$$

be the product

$$\prod_{i_{r+1}=1}^{n_{r+1}} \bar{w}_1(i_{r+1})\, \bar{w}_2(i_{r+1}) \cdots \bar{w}_{n_{r+1}}(i_{r+1})$$

in which $\bar{w}_j(i_{r+1})$ denotes the value of the word w_j on the symbols

$$x_{r+1}(\delta_1, \ldots, \delta_r, j; \varepsilon_1, \ldots, \varepsilon_r, \varepsilon_{r+1}; i_1, \ldots, i_r, i_{r+1}),$$

where ε_{r+1} runs from 1 to j.

The image of φ_r is contained in the verbal subgroup of $F_{\mathbf{n},r+1}$. Hence, $F_{\mathbf{n}} = WF_{\mathbf{n}}$ for each sequence $\mathbf{n}$, so that $F = WF$ as well; i.e., F is $\mathcal{W}$-perfect. Since every epimorphic image of a $\mathcal{W}$-perfect group is $\mathcal{W}$-perfect, it follows that the image of every homomorphism $F \to G$ is contained in the $\mathcal{W}$-perfect radical of G. Thus the argument will be complete if we show that, for every element $x \in \mathcal{P}_{\mathcal{W}}G$, there is an increasing sequence $\mathbf{n}$ and a homomorphism $F_{\mathbf{n}} \to G$ whose image contains the element x. To see this, pick the minimum n_1 such that x can be written as a product of n_1 values (or less) of some of the words $w_1, \ldots, w_{n_1}$ (possibly repeated and in any order), and choose one such decomposition of x to continue the process. Consider all the elements of G which appear in the chosen expression of x as arguments in the words w_j; pick the minimum n_2 which is greater than or equal to the lengths of the expressions of these elements as products of values of words in W, and greater than or equal to the subindices of the words involved. Replace n_2 with $1 + n_1$ if necessary, in order that the sequence $\mathbf{n}$ be increasing. By continuing this way, one obtains a sequence $\mathbf{n} = (n_1, n_2, \ldots)$ of positive integers and a homomorphism $\psi: F_{\mathbf{n}} \to G$ sending x_0 to x. (In order to illustrate how ψ is defined, suppose e.g. that $x = w_3(a,b,c)\,w_2(d,e)$, so that $n_1 = 3$. Then ψ sends $x_1(3;1;1) \mapsto a$, $x_1(3;2;1) \mapsto b$, $x_1(3;3;1) \mapsto c$, $x_1(2;1;2) \mapsto d$, $x_1(2;2;2) \mapsto e$, and it sends all the other generators $x_1(\delta;\varepsilon;i)$ of $F_{\mathbf{n},1}$ to 1. Then one proceeds similarly by choosing decompositions of a, b, c, d, e as products of values of words in W, and so on.) □

4 Applications to homotopy theory

For a map $f: A \to B$ between CW-complexes, a space X is called *f-local* if the induced map of function spaces

$$\mathrm{map}(f,X): \mathrm{map}(B,X) \to \mathrm{map}(A,X)$$

is a weak homotopy equivalence. Each map f determines a localization functor L_f in the homotopy category of CW-complexes; see [6]. Thus, for every CW-complex X there is a map $X \to L_f X$ which is homotopy initial among maps from X into f-local spaces.

Let $\mathcal{W}$ be any variety of groups. Let F be the locally free $\mathcal{W}$-perfect group constructed in the proof of Theorem 3.3 (or any other locally free group generating the same radical). Since F is a direct limit of free groups, its classifying space $K(F,1)$ is a homotopy colimit of wedges of circles and hence it is two-dimensional. It then follows from [5, Theorem 2.1] that localization with respect to $f: K(F,1) \to *$ is *π_1-compatible*; that is,

$$\pi_1(L_f X) \cong L_\varphi \pi_1(X) \qquad \text{for all spaces } X,$$

where φ denotes the homomorphism $F \to 1$ induced by f on the fundamental group; thus, the localization functor L_φ has the efect of dividing out the $\mathcal{W}$-perfect radical. Therefore, the functor L_f assigns to each space X a space for which the $\mathcal{W}$-perfect radical of the fundamental group is trivial. The following theorem ensures that such localizations are not trivial themselves, since they preserve homology with certain coefficients. The steering example is Quillen's plus-construction; cf. [1].

Recall once more that, by [11, 12.12], every variety $\mathcal{W}$ can be defined by a power word together with commutator words.

Theorem 4.1 *Let $\mathcal{W}$ be any variety of groups of exponent $m \geq 0$. Let F be any locally free group generating the $\mathcal{W}$-perfect radical. Consider the map $f\colon K(F,1) \to *$. Then, for each space X, the natural map $X \to L_f X$ kills the $\mathcal{W}$-perfect radical from the fundamental group of X, and it induces an isomorphism in homology with coefficients in $\mathbb{Z}/m$.*

Proof By assumption, we have $F = WF$; that is, every element of F can be written as a product of commutators and power words of the form a^m with $a \in F$. If $m = 0$, then F is perfect and, since it is is locally free, it is in fact acyclic. Therefore, the map $f\colon K(F,1) \to *$ is an integral homology equivalence and hence all maps $X \to Y$ inducing a homotopy equivalence $L_f X \simeq L_f Y$ are integral homology equivalences. In particular, the natural map $X \to L_f X$ is an integral homology equivalence for all spaces X. If $m \geq 2$, then the abelianization of F is a group A such that $A = mA$; that is, A is p-divisible for all primes p dividing m. Hence, $H_1(F;\mathbb{Z}/m) = 0$ and, since F is locally free, F is mod m acyclic. In other words, the map f is a mod m homology equivalence. It then follows, as above, that the natural map $X \to L_f X$ is a mod m homology equivalence for all X. □

Plus-constructions for homology with mod m coefficients have long been known; see [3, VII.6] or [10]. These occur in our framework, up to homotopy, by choosing the variety defined by the words x^m and $[x,y]$; what they kill is the J-perfect radical of the fundamental group, where J is the set of prime divisors of m. The word x^m alone yields a localization which kills the Burnside radical of exponent m from the fundamental group, while preserving homology with mod m coefficients. This localization does not alter, for example, spaces whose fundamental group is a finite perfect group of exponent m.

References

[1] A. J. Berrick and C. Casacuberta, A universal space for plus-constructions, *Topology*, to appear.

[2] A. K. Bousfield, Homological localization towers for groups and π-modules, Mem. Amer. Math. Soc. 10, no. 186, Providence, 1977.

[3] A. K. Bousfield and D. M. Kan, Homotopy Limits, Completions and Localizations, Lecture Notes in Math. 304, Springer-Verlag, Berlin Heidelberg New York, 1972.

[4] C. Casacuberta, Anderson localization from a modern point of view, in: The Čech Centennial; A Conference on Homotopy Theory, Contemp. Math. 181, Amer. Math. Soc., Providence, 1995, 35–44.

[5] C. Casacuberta and J. L. Rodríguez, On towers approximating homological localizations, *J. London Math. Soc.* 56 (1997), 645–656.

[6] E. Dror Farjoun, Cellular Spaces, Null Spaces and Homotopy Localization, Lecture Notes in Math. 1622, Springer-Verlag, Berlin Heidelberg New York, 1996.

[7] M. Dugas, T. H. Fay, and S. Shelah, Singly cogenerated annihilator classes, *J. Algebra* 109 (1987), 127–137.

[8] T. H. Fay, E. P. Oxford, and G. L. Walls, Singly generated socles and radicals, in: Abelian Group Theory, Lecture Notes in Math. 1006, Springer-Verlag, Berlin Heidelberg New York, 1983, 671–684.

[9] P. Jara, A. Verschoren, and C. Vidal, Localization and Sheaves: a Relative Point of View, Pitman Research Notes in Math. Series 339, Longman, Essex, 1995.

[10] W. Meier and R. Strebel, Homotopy groups of acyclic spaces, Quart. J. Math. Oxford Ser. (2) 32 (1981), 81–95.
[11] H. Neumann, Varieties of Groups, Ergeb. Math. Grenzgeb. 37, Springer-Verlag, Berlin Heidelberg New York, 1967.
[12] D. Quillen: Cohomology of groups, Actes Congrès Int. Math. Nice, vol. 2, 1970, 47–51.
[13] D. J. S. Robinson, Finiteness Conditions and Generalized Soluble Groups Part 1, Ergeb. Math. Grenzgeb. 62, Springer-Verlag, Berlin Heidelberg New York, 1972.
[14] J. L. Rodríguez and D. Scevenels, Universal epimorphic equivalences for group localizations, preprint, 1998.
[15] B. Stenström, Rings of Quotients, Grundlehren Math. Wiss. 217, Springer-Verlag, Berlin Heidelberg New York, 1975.

THE WORD PROBLEM IN GROUPS OF COHOMOLOGICAL DIMENSION 2

DONALD J. COLLINS* and CHARLES F. MILLER III†

*School of Mathematical Sciences, Queen Mary and Westfold College, University of London, London E1 4NS, England
†Department of Mathematics, University of Melbourne, Parkville 3052, Australia

Abstract

We show that the finitely presented groups with unsolvable word problem given by the Boone-Britton construction have cohomological dimension 2. More precisely we show these groups can be obtained from a free group by successively forming HNN-extensions where the associated subgroups are finitely generated free groups. Also the presentations obtained for these groups are aspherical. Using this we show there is no algorithm to determine whether a presentation is aspherical. There is no algorithm to determine whether a finite 2-complex is aspherical.

1 Introduction

Fundamental algorithms in combinatorial group theory (the original due to Nielsen [6]) enable one to decide membership in a finitely generated subgroup of a free group. It follows that the free product of two free groups with finitely generated amalgamation has a solvable word problem. Similarly, an HNN-extension of a free group with finitely generated associated subgroups has a solvable word problem. (However, such groups can have unsolvable conjugacy problem and the problem of deciding whether an arbitrary pair of them are isomorphic is unsolvable - see [7].) More generally, the fundamental group of a finite graph of groups whose edge and vertex groups are all finitely generated and free is finitely presented and has a solvable word problem.

Using the Mayer-Vietoris sequence for the (co)homology of a graph of groups [3], it is easy to check that such groups have cohomological dimension ≤ 2. In fact, if the vertex groups of a graph of groups have cohomological dimension ≤ 2 and the edge groups are all free, then its fundamental group will have cohomological dimension ≤ 2.

In particular, groups obtained from free groups by repeatedly forming HNN-extensions and amalgamated free products where the associated or amalgamated subgroups are always free will have cohomological dimension ≤ 2.

The purpose of this article is to establish the following:

Theorem 1.1 *There exists a finitely presented group $\mathcal{B}$ of cohomological dimension 2 having unsolvable word problem. Indeed, $\mathcal{B}$ can be obtained from a free group by applying three successive HNN-extensions where the associated subgroups are finitely generated free groups.*

The group in question is one constructed by Boone as later modified by Boone, Britton and the authors. The construction is described in Rotman's textbook [9]. The verification that the group has the required properties is somewhat technical and requires detailed understanding of the proof in [9]. In fact we discovered this result some years ago and have even mentioned it in print (see for example [8] p.29). As some interesting applications have recently been found [1], it seems timely to publish a proof.

We will write down a presentation $\mathcal{P}_{\mathcal{B}}$ of $\mathcal{B}$ which exhibits its structure as a successive HNN-extension with free associated subgroups. Each of the relators of $\mathcal{P}_{\mathcal{B}}$ is non-empty, reduced and not conjugate to another relator or its inverse; that is, $\mathcal{P}_{\mathcal{B}}$ is *concise* in the sense of [2]. No relator of $\mathcal{P}_{\mathcal{B}}$ is a proper power. In [2] such presentations are shown to be aspherical in a number of senses. In particular, a presentation is *aspherical* means the standard 2-complex associated with it is topologically aspherical. By applying [2] we conclude the following:

Corollary 1.2 *The finite presentation $\mathcal{P}_{\mathcal{B}}$ for the group $\mathcal{B}$ having unsolvable word problem is aspherical, combinatorially aspherical, diagramatically aspherical and Cohen-Lyndon aspherical.*

Applying one of the constructions used in proving a theorem of Adian and Rabin showing one cannot recognise the trivial group (as for instance in [8] Theorem 3.3 and Lemma 3.6), we obtain a collection of presentations Π_w indexed by words w in the generators of $\mathcal{B}$ such that:

1. each Π_w is concise, has more relators than generators, and no relator is a proper power;
2. if $w \neq_{\mathcal{B}} 1$, then the group presented by Π_w is infinite and the presentation Π_w is aspherical, combinatorially aspherical, diagramatically aspherical and Cohen-Lyndon aspherical;
3. if $w =_{\mathcal{B}} 1$, then the group presented by Π_w is the trivial group and Π_w is not aspherical in any of these senses.

Most of this follows easily from [2] and the construction [8] which applies ordinary free products with free groups and amalgamated free products with free amalgamated subgroups in the case $w \neq_{\mathcal{B}} 1$. In case $w =_{\mathcal{B}} 1$, the standard finite 2-dimensional CW-complex K_w associated with Π_w is simply connected, has a single 0-cell, and has more 2-cells than 1-cells and so is not aspherical. Hence if $w =_{\mathcal{B}} 1$, then Π_w is not aspherical in any of the above senses. Since $\mathcal{B}$ has unsolvable word problem, we conclude the following:

Corollary 1.3 *There is no algorithm to determine of an arbitrary finite presentation Π of a group whether or not Π is any of aspherical, combinatorially aspherical, diagramatically aspherical or Cohen-Lyndon aspherical.*

Corollary 1.4 *There is no algorithm to determine of an arbitrary finite 2-dimensional CW-complex K whether or not K is aspherical.*

The principal difficulty in the proof of the main theorem is showing that a certain subgroup is free. Our proof of this relies on an argument due to Post [4] which involves the deterministic nature of a Turing machine T. Computations in a Turing machine can be viewed as a directed graph with instantaneous descriptions as vertices and operations in the machine as edges. Since T is deterministic, the component of the halting state of T is actually a tree. It is this fact which underlies Post's argument.

2 The structure of the group $\mathcal{B} = \mathcal{B}(T)$

The construction of Boone's group begins with a Turing machine T having an unsolvable halting problem. A construction of Markov and Post is then applied to obtain a finitely presented semigroup $\Gamma(T)$ of the form

$$\Gamma(T) = \langle q, q_0, \ldots, q_N, s_0, \ldots, s_M \mid F_i q_{i_1} G_i = H_i q_{i_2} K_i, i \in I \rangle$$

where the F_i, G_i, H_i, K_i are positive s-words and $q_{i_j} \in \{q, q_0, \ldots, q_N\}$. By an *s-word* we mean a word on the symbols $s_0, \ldots, s_M$ and their inverses. A word is *positive* if it contains no s_i^{-1} symbols. Of course the above presentation for $\Gamma(T)$ is a semigroup presentation and the symbols s_i^{-1} are not present in $\Gamma(T)$. We will give more details concerning $\Gamma(T)$ later.

We use $X \equiv Y$ to mean the words X and Y are identical (letter by letter). If $X \equiv s_{b_1}^{e_1} \cdots s_{b_m}^{e_m}$ is an s-word, we define $X^\# \equiv s_{b_1}^{-e_1} \cdots s_{b_m}^{-e_m}$. Note that $X^\#$ is not the same as X^{-1}. Also, if X and Y are s-words, then $(X^\#)^\# \equiv X$ and $(XY)^\# = X^\# Y^\#$.

The group $\mathcal{B} = \mathcal{B}(T)$ constructed (essentially) by Boone is then presented as follows:

generators: $q, q_0, \ldots, q_N, s_0, \ldots, s_M, r_i, i \in I, x, t, k$;
relations: for all $i \in I$ and all $b = 0, \ldots, M$,

$$\left.\begin{array}{l} \left.\begin{array}{l} \left.\begin{array}{l} x s_b = s_b x^2 \end{array}\right] \Delta_1 \\ r_i s_b = s_b x r_i x \\ r_i^{-1} F_i^\# q_{i_1} G_i r_i = H_i^\# q_{i_2} K_i \end{array}\right] \Delta_2 \\ t r_i = r_i t \\ t x = x t \end{array}\right] \Delta_3$$

$$k r_i = r_i k$$
$$k x = x k$$
$$k(q^{-1} t q) = (q^{-1} t q) k$$

The subsets $\Delta_1 \subset \Delta_2 \subset \Delta_3$ of the relations each define a presentation of a group $\mathcal{B}_i$ generated by the symbols appearing in the Δ_i. Also let $\mathcal{B}_0$ denote the infinite cyclic group generated by x, and let Q denote the free group with basis $\{q, q_0, \ldots, q_N\}$. The following is Lemma 12.11 in [9]:

Lemma 2.1 *In the chain*

$$\mathcal{B}_0 \le \mathcal{B}_1 \le \mathcal{B}_1 * Q \le \mathcal{B}_2 \le \mathcal{B}_3 \le \mathcal{B}$$

each group is an HNN-extension of its predecessor; moreover, the free product $\mathcal{B}_1 * Q$ is an HNN-extension of $\mathcal{B}_0$.

We want to look at $\mathcal{B}_3$ somewhat differently. Let $\mathcal{A}$ be the group with presentation

$$\mathcal{A} = \langle x, s_0, \ldots, s_M, q, q_0, \ldots, q_N, t \mid t^{-1}xt = x, s_b^{-1}xs_b = x^2, b = 0, \ldots, M \rangle.$$

Then $\mathcal{A}$ is an HNN-extension of $\mathcal{B}_0 = \langle x \mid \rangle$ with all the listed generators other than x as stable letters. In particular, the associated subgroups are either cyclic or trivial and hence are finitely generated free groups. We now have the following easy fact.

Lemma 2.2 *In the chain*

$$\mathcal{B}_0 \leq \mathcal{A} \leq \mathcal{B}_3$$

each group is an HNN-extension of its predecessor; moreover, the associated subgroups are finitely generated free groups.

Proof Let $\mathcal{F}$ denote the free group on the stable letters of $\mathcal{A}$ and $\phi : \mathcal{A} \to \mathcal{F}$ the retraction sending stable letters to themselves and x to 1. One of the associated subgroups for r_i is generated by the $M + 3$ elements $\{F_i^{\#} q_{i_1} G_i, t, s_0 x, \ldots, s_M x\}$. The image of this subgroup under ϕ is easily seen to be the (free) subgroup of $\mathcal{F}$ generated by $\{q_{i_1}, t, s_0, \ldots, s_M\}$ which has rank $M + 3$. Hence the associated subgroup is free on the given generators. Similar considerations show the other associated subgroup for r_i is free. This completes the proof □

Lemma 2.3 *The elements* $\{x, r_i, i \in I\}$ *freely generate a free subgroup of* $\mathcal{B}_3$.

Proof Consider $\mathcal{B}_3$ as an HNN-extension of $\mathcal{A}$ and adopt the notation in the previous proof. If w is a (non-empty) freely reduced word in x and the r_i and if $w = 1$ in $\mathcal{B}_3$ then Britton's Lemma implies w contains a subword of the form $r_i^{-e} x^n r_i^e$ and x^n belongs to one of the associated subgroups of r_i depending on the sign of e. But $x \in \ker \phi$ while $\ker \phi$ intersects the associated subgroups in the identity. Thus $n = 0$ and w is not freely reduced which is a contradiction. □

What is needed to finish the proof of the main theorem is the following improvement of the previous lemma.

Lemma 2.4 (Main Lemma) *The elements* $\{q^{-1}tq, x, r_i, i \in I\}$ *freely generate a free subgroup of* $\mathcal{B}_3$. *Hence in the chain*

$$\mathcal{B}_0 \leq \mathcal{A} \leq \mathcal{B}_3 \leq \mathcal{B}$$

each group is an HNN-extension of its predecessor; moreover, the associated subgroups are finitely generated free groups.

The main theorem is an immediate consequence of this result. We are going to prove this "Main Lemma" by contradiction. We begin with the following.

Lemma 2.5 *Assume there is some non-empty freely reduced word* W *in the generators* $\{q^{-1}tq, x, r_i, i \in I\}$ *such that* $W = 1$ *in* $\mathcal{B}_3$. *Then there are non-empty freely reduced words* L_1 *and* L_2 *on* $\{x, r_i, i \in I\}$ *such that* $L_1 q L_2 = q$ *in* $\mathcal{B}_3$.

Proof By Lemma 2.3, W must contain $q^{-1}tq$ or $q^{-1}t^{-1}q$ and hence must involve t. Now $\mathcal{B}_3$ is the HNN-extension of $\mathcal{B}_2$ by the stable letter t having associated subgroup freely generated by $\{x, r_i, i \in I\}$ and t commutes with these elements. So by Britton's Lemma, W must contain a subword $t^{-e}qL_2q^{-1}t^e$ such that $qL_2q^{-1} = L_1^{-1}$ in $\mathcal{B}_2$, where L_1 and L_2 are words in the associated subgroup. Hence $L_1qL_2 = q$ in $\mathcal{B}_2$ as desired. □

3 Reduction to proofs in $\Gamma(T)$

A word Σ in $\mathcal{B}_3$ is *special* if $\Sigma \equiv X^{\#}q_jY$ where X and Y are positive s-words and $q_j \in \{q, q_0, \ldots, q_N\}$. If $\Sigma \equiv X^{\#}q_jY$ is special, then we define $\Sigma^* \equiv XqY$, which is a word in the semigroup $\Gamma(T)$.

By a *proof* in $\Gamma(T)$ is meant a sequence of words

$$w_1 \to w_2 \to \ldots \to w_n$$

on the generators where $w_j \to w_{j+1}$ means that w_j yields w_{j+1} by a single application of a relation of $\Gamma(T)$. We say this is a *forward* application if the left-hand side of the relation is replaced by the right-hand side and a *backward* application if the right-hand side is replaced by the left-hand side. A proof is *reversal-free* if it does not contain an application $w_{j-1} \to w_j$ of a relation immediately followed by an application $w_j \to w_{j+1}$ of the same relation in the opposite direction. (Note that when w_j contains only a single letter from $\{q, q_0, \ldots, q_N\}$ such a reversal would mean $w_{j-1} \equiv w_{j+1}$.)

The inductive proof of Lemma 12.15 in [9] actually shows the following more detailed result:

Lemma 3.1 *Suppose that Σ_1 and Σ_2 are special words and that L_1 and L_2 are freely reduced words on $\{x, r_i, i \in I\}$. If the equation*

$$L_1\Sigma_1L_2 = \Sigma_2$$

holds in $\mathcal{B}_2$, then for some choice of $e_i = \pm 1$ and integers m_i, n_i,

$$L_1 \equiv x^{m_p}r_{i_p}^{-e_p}\cdots x^{m_1}r_{i_1}^{-e_1}x^{m_0},$$

$$L_2 \equiv x^{n_0}r_{i_1}^{e_1}x^{n_1}\cdots r_{i_p}^{e_p}x^{n_p},$$

and the sequence

$$(i_1, e_1), \ldots, (i_p, e_p)$$

determines a reversal-free proof that $\Sigma_1^* = \Sigma_2^*$ in $\Gamma(T)$. Under this correspondence (i_j, e_j) corresponds to a forward application of the j-th relation if $e_j = +1$ and a backward application of the j-th relation if $e_j = -1$.

Proof The only point not covered by the argument given in [9] is the fact that the resulting proof in $\Gamma(T)$ is reversal-free. We need to show that if $i_j = i_{j+1}$, then $e_j = e_{j+1}$. To obtain a contradiction we may without loss of generality assume that

$p = 2$ and $i = i_1 = i_2$ and $e_1 = -e_2$. We consider the case $e_1 = 1$ (the case $e_1 = -1$ is similar). The left hand side of the equation $L_1\Sigma_1 L_2 = \Sigma_2$ is transformed to the right by two successive pinches,

$$\begin{aligned} L_1\Sigma_1 L_2 &\equiv x^{m_2} r_i x^{m_1} r_i^{-1} x^{m_0} X_1^{\#} q_j Y_1 x^{n_0} r_i x^{n_1} r_i^{-1} x^{n_2} \\ &\equiv x^{m_2} r_i x^{m_1} r_i^{-1} x^{m_0} u(s_b) F_i^{\#} q_{i_1} G_i v(s_b) x^{n_0} r_i x^{n_1} r_i^{-1} x^{n_2} \\ &= x^{m_2} r_i x^{m_1} r_i^{-1} u(s_b x) F_i^{\#} q_{i_1} G_i v(s_b x) r_i x^{n_1} r_i^{-1} x^{n_2} \\ &= x^{m_2} r_i x^{m_1} u(s_b x^{-1}) H_i^{\#} q_{i_2} K_i v(s_b x^{-1}) x^{n_1} r_i^{-1} x^{n_2} \\ &= x^{m_2} r_i u(s_b x^{-1}) H_i^{\#} q_{i_2} K_i v(s_b x^{-1}) r_i^{-1} x^{n_2} \\ &= x^{m_2} u(s_b x) F_i^{\#} q_{i_1} G_i v(s_b x) x^{n_2} \\ &= X_1 q_j Y_1 \equiv \Sigma_2 \end{aligned}$$

first by r_1 then by r_1^{-1}. Here $X_1^{\#} \equiv u(s_b) F_i^{\#}$ and $Y_1 \equiv G_i v(s_b)$, where $u(s_b)$ and $v(s_b)$ are s-words. The word $u(s_b x)$ is the word obtained from u by replacing each s_b by $s_b x$, and similarly for v.

Also all equations in this calculation except for the pinches are either identities or are consequences of the relations in $\mathcal{B}_1 * Q$. In particular, in $\mathcal{B}_1 * Q$

$$x^{m_1} u(s_b x^{-1}) H_i^{\#} q_{i_2} K_i v(s_b x^{-1}) x^{n_1} = u(s_b x^{-1}) H_i^{\#} q_{i_2} K_i v(s_b x^{-1})$$

and so $x^{m_1} u(s_b x^{-1}) H_i^{\#} = u(s_b x^{-1}) H_i^{\#}$ and $K_i v(s_b x^{-1}) x^{n_1} = K_i v(s_b x^{-1})$ in $\mathcal{B}_1$. Thus $x^{m_1} = x^{n_1} = 1$, and so $m_1 = n_1 = 0$ and L_1 and L_2 are not freely reduced. This is a contradiction and the proof of the lemma is complete. □

Combining this lemma with Lemma 2.5 we obtain the following result which reduces the Main Lemma to a fact about proofs in $\Gamma(T)$:

Corollary 3.2 *Assume there is some non-empty freely reduced word W in the generators $\{q^{-1}tq, x, r_i, i \in I\}$ such that $W = 1$ in $\mathcal{B}_3$. Then there is a non-trivial, reversal-free proof in $\Gamma(T)$*

$$q \to w_2 \to \ldots w_{n-1} \to q.$$

In the next section we will show that, because T is deterministic, there is no such non-trivial, reversal-free proof.

4 Reversals and the determinism of T

We need to look at the construction due to Post [4] of $\Gamma(T)$ from the Turing machine T in some detail. Suppose the Turing machine T has alphabet $s_0, s_1, \ldots, s_M$ and internal states $q_0, q_1, \ldots, q_N$ with q_1 as the start state and q_0 as the unique halting state.

Let $\gamma(T)$ be the semigroup with presentation

$$\gamma(T) = \langle h, s_0, s_1, \ldots, s_M, q, q_0, q_1, \ldots, q_N \mid R(T) \rangle$$

where the relations $R(T)$ are

$$q_i s_j = q_l s_k \text{ if } q_i s_j s_k q_l \in T.$$

and for all $b = 0, 1, \ldots, M$:

$$\begin{array}{rcll} q_i s_j s_b & = & s_j q_l s_b & \text{if } q_i s_j R q_l \in T, \\ q_i s_j h & = & s_j q_l s_0 h & \text{if } q_i s_j R q_l \in T, \\ s_b q_i s_j & = & q_l s_b s_j & \text{if } q_i s_j L q_l \in T, \\ h q_i s_j & = & h q_l s_0 s_j & \text{if } q_i s_j L q_l \in T, \end{array}$$

$$\begin{array}{rcl} q_0 s_b & = & q_0 \\ s_b q_0 h & = & q_0 h \\ h q_0 h & = & q \end{array}$$

We recall that intuitively the symbol h marks the ends of the tape the Turing machine T is reading.

The semigroup $\Gamma(T)$ used in the previous sections is obtained from $\gamma(T)$ by regarding the h as the last s-letter and reindexing these s-letters so that $h = s_M$. (So $\Gamma(T)$ is just $\gamma(T)$ in a slightly revised notation.)

A word of $\gamma(T)$is *h-special* if it has the form huq_jvh where u and v are positive s-words and $q_j \in \{q, q_0, q_1, \ldots, q_N\}$. Since only the last relation $hq_0h = q$ creates or destroys h, we observe the following:

Lemma 4.1 *If w_1 and w_2 are words of $\gamma(T)$ and neither if them is the word q, and if $w_1 \to w_2$ is an application of a relation, then w_1 is h-special if and only if w_2 is h-special.*

Lemma 4.2 *Let $w \equiv huq_jvh$ be an h-special word of $\gamma(T)$. Then at most one of the relations of $\gamma(T)$ has a forward application to w.*

Proof Clearly we may assume $q_j \not\equiv q$. If $q_j \not\equiv q_0$ the conclusion is immediate from the deterministic nature of the Turing machine T. So suppose $q_j \equiv q_0$. If u and v are empty then there is a forward application of $hq_0h = q$, but clearly this is the only possibility. If u is non-empty and v is empty, then just one of the relations $s_b q_0 h = q_0 h$ has a forward application. Finally if v is non-empty, just one of the relations $q_0 s_b = q_0$ has a forward application. This completes the proof. □

Lemma 4.3 *Let $w \equiv huq_jvh$ be an h-special word of $\gamma(T)$. If*

$$w \equiv w_1 \to w_2 \to \cdots \to w_{n-1} \to q$$

is a proof in $\gamma(T)$, then either this proof contains a reversal or all the applications of relations are forward.

Proof Clearly the application $w_{n-1} \to q$ is forward and $w_{n-1} \equiv hqh$. Assume that not all applications of relations in the given proof are forward. Suppose that $w_{j-1} \to w_j$ is the last backward application of a relation in the proof. Then $w_j \to w_{j+1}$ and all subsequent applications are forward. But $w_{j-1} \to w_j$ backward implies that $w_j \to w_{j-1}$ is a forward application of the same relation.

Now w_{n-1} is h-special and none of $w_j, \ldots, w_{n-1}$ could be q since only a backward application applies to q. So each of $w_j, \ldots, w_{n-1}$ must be h-special. Hence at most one relation has a forward application to w_j by the previous result. Thus $w_{j+1} \equiv w_{j-1}$ and hence $w_{j-1} \to w_j \to w_{j+1}$ is a reversal. This proves the lemma. □

Corollary 4.4 *Every non-trivial proof in* $\Gamma(T)$ *of the form*

$$q \to w_2 \to \cdots w_{n-1} \to q$$

contains a reversal.

Proof Clearly $w_2 \equiv hq_0h$. So $w_2 \to \cdots w_{n-1} \to q$ either contains a reversal (as claimed), or consists of all forward applications. Assume all the applications $w_2 \to \cdots w_{n-1} \to q$ are forward. But the only forward application of a relation to $w_2 \equiv hq_0h$ is $hq_0h \to q$ so the given proof begins with a reversal. This proves the corollary. □

Corollaries 3.2 and 4.4 together immediately imply the Main Lemma 2.4, and so the proof of our theorem is complete.

References

[1] G. Baumslag, M. R. Bridson, C. F. Miller III and H. Short, *Fibre products, non-positive curvature and decision problems*, in preparation.

[2] I. M. Chiswell, D. J. Collins and J. Huebschmann, *Aspherical group presentations*, Math. Zeit. **178** (1981), 1-36.

[3] K. S. Brown, "Cohomology of groups", Graduate Texts in Mathematics 87, Springer-Verlag, Berlin-Heidelberg-New York (1982).

[4] E. L. Post, *Recursive unsolvability of a problem of Thue*, J. Symbolic Logic, **12** (1947), 1-11.

[5] R. C. Lyndon and P. E. Schupp, "Combinatorial group theory", Springer-Verlag, Berlin-Heidelberg-New York (1977).

[6] W. Magnus, A. Karrass, and D. Solitar, "Combinatorial group theory", Interscience-Wiley, New York (1966), also revised edition Dover (1976).

[7] C. F. Miller III, " On group-theoretic decision problems and their classification", Annals of Mathematics Studies, No. 68, Princeton University Press (1971).

[8] C. F. Miller III, *Decision problems for groups: survey and reflections* - in Algorithms and Classification in Combinatorial Group Theory (eds. G Baumslag and C. F. Miller III), MSRI Publications No. 23, Springer-Verlag (1992), 1-59.

[9] J. J. Rotman, "An introduction to the theory of groups", Fourth Edition, Graduate Texts in Mathematics 148, Springer-Verlag, Berlin-Heidelberg-New York (1995).

POLYCYCLIC-BY-FINITE GROUPS: FROM AFFINE TO POLYNOMIAL STRUCTURES

KAREL DEKIMPE[1]

Katholieke Universiteit Leuven, Campus Kortrijk, B-8500 Kortrijk, Belgium

1 Introduction

This paper sketches the mathematics which arose from a question posed by John Milnor in 1977, concerning the existence of certain affinely flat manifolds. Although the original question is quite geometrical, several results have been obtained by translating Milnor's problem into the language of specific (discrete and Lie) group and Lie algebra representations. These representations are called affine structures. The examples, constructed in the beginning of the 90's, show that not all reasonable groups admit an affine structure, and gave a new boost to this research topic. Besides the investigation (without many results up till now) of the exact nature of the groups that (do not) allow an affine structure, we began to study what might replace the missing affine structures. Very recently we have shown that there is a notion of a polynomial structure, which can be seen as an alternative to the inadequate notion of an affine structure, and which exists on any polycyclic-by-finite group.

In this paper, we first give a historical survey of the most important results on affine structures and later we explain the ideas behind the origin and the existence of polynomial structures. The paper contains no new results, but the old results are put together in a nicer way and presented perhaps more clearly, without to many technical details. We end the paper with a small list of open problems.

2 Affine structures: geometrically and algebraically

The affine structures on a polycyclic-by-finite group Γ originate from differential geometry. Without going too much into detail, I believe that it is worthwhile to sketch this geometrical background ([27], [31]).

An n-dimensional manifold M is said to be equipped with an affinely flat structure, if there is given a system of charts $f_i : U_i \to \mathbb{R}^n$ (i belonging to a family I) covering M, such that the transition functions are given by affine maps of $\mathbb{R}^n$. This means that for all $i, j \in I$, with $U_i \cap U_j \neq \emptyset$, there is a unique affine map, consisting of a linear part $A \in \mathrm{Gl}(n, \mathbb{R})$ and a translational part $a \in \mathbb{R}^n$, such that the change of local coordinates is given by

$$f_j \circ f_i^{-1} : f_i(U_i \cap U_j) \to f_j(U_i \cap U_j) : x \mapsto Ax + a.$$

(Remark that $x \in \mathbb{R}^n$ stands for a column vector $x = (x_1, x_2, \ldots, x_n)^T$.)

[1]Postdoctoral Fellow of the Fund for Scientific Research - Flanders (F.W.O.)

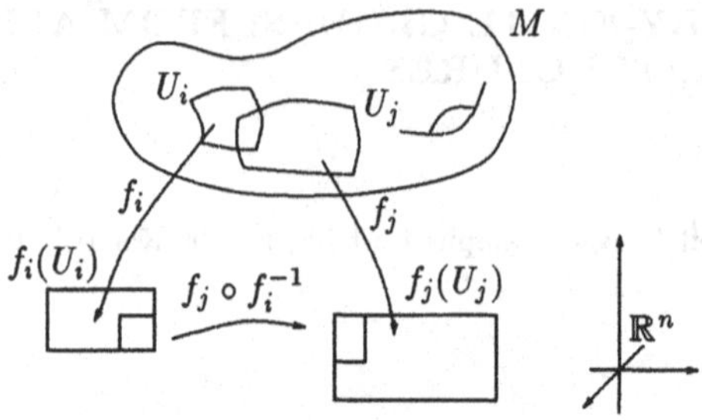

One says that the manifold M is an affinely flat manifold.

As the change of coordinates is given by means of affine maps, the manifold M inherits the usual affine connection of $\mathbb{R}^n$. This implies that a geodesic in M is a curve $\gamma :]a, b[\subseteq \mathbb{R} \to M : t \mapsto \gamma(t)$, such that in local coordinates, γ is just a straight line traversed at constant speed. So, a curve γ is a geodesic, if and only if for any index $i \in I$ there exist real numbers $a_1, a_2, \ldots, a_n \in \mathbb{R}$, which are not all zero, and real numbers $b_1, b_2, \ldots, b_n$, such that

$$f_i(\gamma(t)) = (a_1 t + b_1, a_2 t + b_2, \ldots, a_n t + b_n)^T.$$

The affinely flat manifold M is said to be complete if and only if every geodesic γ can be extended to a geodesic which is defined on the whole time axis $\mathbb{R}$.

The easiest way to construct a complete affinely flat manifold is achieved by taking the quotient manifold $\Gamma\backslash\mathbb{R}^n$, where Γ is a group acting freely and properly discontinuously and via affine motions on $\mathbb{R}^n$. We remark here that although there are many (non-equivalent) definitions of the notion of "properly discontinuous action", most of these notions do coincide if the space where the group acts on is locally compact, like $\mathbb{R}^n$. The definition we propose is that an action on a space X is properly discontinuous if for any compact set $K \subseteq X$, the set $\{\gamma \in \Gamma \mid {}^{\gamma}K \cap K \neq \emptyset\}$ is finite.

In 1955, Auslander and Markus showed that the construction above yields all the complete affinely flat manifolds.

Theorem 2.1 ([3], see also [31, page 45]) *The connected, complete, affinely flat manifolds are exactly the quotients $\Gamma\backslash\mathbb{R}^n$, where Γ acts properly discontinuously, freely and via affine motions on $\mathbb{R}^n$.*

One of the easiest and most important examples is obtained by considering the action of the free abelian group of rank k, $\mathbb{Z}^k$, on a space $\mathbb{R}^n$ ($n \geq k$), where the action is given by translation on the first k components, i.e.

$$\forall z = (z_1, \ldots, z_k)^T \in \mathbb{Z}^k, \ \forall r = (r_1, \ldots, r_n)^T \in \mathbb{R}^n :$$
$${}^{z}r = (r_1 + z_1, \ldots, r_k + z_k, r_{k+1}, \ldots, r_n)^T.$$

The resulting manifolds are cylinders and in case $k = n$, we obtain an affinely flat torus.

In 1977, John Milnor ([27]) proved that every torsion-free polycyclic-by-finite group Γ can be realised as the fundamental group of a connected, complete, affinely flat manifold. In other words, Milnor showed that any torsion-free polycyclic-by-finite group Γ admits a properly discontinuous and affine action on some space $\mathbb{R}^n$.

On the other hand, it was already known that any such a group Γ could be realised as the fundamental group of a connected, *compact* manifold (whose universal covering space is some space $\mathbb{R}^k$). Therefore, Milnor combined these two offerings into the following question, which became widely known as Milnor's conjecture:

> Let Γ be any torsion-free polycyclic-by-finite group. Is it true that Γ can be realized as the fundamental group of a connected, compact, affinely flat manifold?

In other words, Milnor questioned the existence of a properly discontinuous action of Γ on some space $\mathbb{R}^n$, which has both the properties of being affine and having a compact quotient. It is this kind of actions which are called affine structures. Let us denote the group of invertible affine transformations of $\mathbb{R}^n$ by $\mathrm{Aff}(\mathbb{R}^n)$.

Definition 2.2 Let Γ be a polycyclic-by-finite group. An affine structure on Γ is given by a morphism $\rho : \Gamma \to \mathrm{Aff}(\mathbb{R}^n)$, which lets Γ act properly discontinuously and with compact quotient on $\mathbb{R}^n$.

Using this notion, we can reformulate Milnor's question as follows:

> Is it true that any torsion-free polycyclic-by-finite group Γ admits an affine structure?

We note here that a properly discontinuous action of a polycyclic-by-finite group on some space $\mathbb{R}^n$ has a compact quotient if and only if $n = h(\Gamma)$, where $h(\Gamma)$ denotes the Hirsch length of the group Γ.

Before we start talking about what is known about this question, let us first of all explain why we are so interested in the class of polycyclic-by-finite groups.

In 1964, Auslander published a paper ([1]) in which he proved that the fundamental group of any connected, complete, affinely flat manifold is polycyclic-by-finite. Unfortunately, his proof contains a gap. Moreover, Margulis ([25], [26]) constructed examples of connected, complete affinely flat manifolds having a free (non-abelian) fundamental group. This shows that even the statement of Auslander's theorem was incorrect. However, the examples provided by Margulis are non-compact and the statement of Auslander remains still open in the compact case. Therefore, one often speaks of Auslander's conjecture:

> **Auslander's conjecture:** The fundamental group of a connected, compact, complete affinely flat manifold is a polycyclic-by-finite group.

At the time of writing the (affirmative) answer to this conjecture is only known in very small dimensions. Nevertheless, most people believe that the conjecture holds and therefore one should consider the class of polycyclic-by-finite groups in the study of the compact, affinely flat manifolds.

3 Answers to Milnor's question

As a first approach to Milnor's question, many people restricted themselves to the class of finitely generated nilpotent(-by-finite) groups. The class of finitely generated torsion-free nilpotent groups is particularly interesting since in this case there are many equivalent formulations of Milnor's question.

In order to explain this, we fix a finitely generated torsion-free nilpotent group N of Hirsch length n. It is well known that there exists a unique connected and simply connected nilpotent Lie group G (of dimension n) containing N as a lattice. (By a lattice we mean a discrete and cocompact subgroup). Associated to this Lie group G, there is a Lie algebra $\mathfrak{g}$ and the exponential map $\exp : \mathfrak{g} \to G$ is a global diffeomorphism in this case.

Assume that the nilpotent Lie group G admits a simply transitive and affine action on some space $\mathbb{R}^n$. It then follows that the restriction of this action to the lattice N is an affine structure on N. Conversely, Fried, Goldman and Hirsch ([18, Theorem 7.1]), showed that any affine structure on N is obtained as the restriction of a simply transitive affine action of the Lie group G. Therefore, the study of affine structures on a torsion-free nilpotent group N can be translated to the study of the simply transitive affine actions of its Mal'cev completion.

Throughout this paper, we will, as usual, regard the group of affine motions $\mathrm{Aff}(\mathbb{R}^n)$ as a subgroup of $\mathrm{Gl}(\mathbb{R}^{n+1})$, by mapping the transformation with linear part $A \in \mathrm{Gl}(\mathbb{R}^n)$ and translational part $a \in \mathbb{R}^n$ to the matrix $\begin{pmatrix} A & a \\ 0 & 1 \end{pmatrix}$.

A Lie group homomorphism $\tilde{\rho} : G \to \mathrm{Aff}(\mathbb{R}^n)$ can be described by specifying both the linear part, say $\tilde{\rho}_l$, and the translational part, say $\tilde{\rho}_t$ of the map. It is obvious that such a morphism

$$\tilde{\rho} : G \to \mathrm{Aff}(\mathbb{R}^n) : g \mapsto \begin{pmatrix} \tilde{\rho}_l(g) & \tilde{\rho}_t(g) \\ 0 & 1 \end{pmatrix}$$

represents a simply transitive affine action of G on $\mathbb{R}^n$ if and only if $\tilde{\rho}_t : G \to \mathbb{R}^n$ is a bijective map. In [29], [18] it is shown that for a simply transitive action $\tilde{\rho}$ of the nilpotent Lie group G, the image $\tilde{\rho}(G)$ (or equivalently $\tilde{\rho}_l(G)$) entirely consists of unipotent matrices.

Remark 3.1 If $\tilde{\rho} : G \to \mathrm{Aff}(\mathbb{R}^n)$ is a (Lie) group homomorphism, then $\tilde{\rho}_l : G \to \mathrm{Gl}(\mathbb{R}^n)$ is also a (Lie) group homomorphism. Via $\tilde{\rho}_l$, the vector space $\mathbb{R}^n$ becomes a G-module. In contrast to $\tilde{\rho}_l$, the map $\tilde{\rho}_t : G \to \mathbb{R}^n$ is not a homomorphism. However, one can easily check that it is a 1-cocycle with respect to the G-module structure of $\mathbb{R}^n$ which is given by $\tilde{\rho}_l$ (i.e. $\tilde{\rho}_t(g_1g_2) = {}^{g_1}\tilde{\rho}_t(g_2) + \tilde{\rho}_t(g_1) = \tilde{\rho}_l(g_1)\tilde{\rho}_t(g_2) + \tilde{\rho}_t(g_1)$).

It is now very obvious where to search for the concept on the Lie algebra level, which should play the same role as the simply transitive actions. Indeed, the affine Lie algebra (the Lie algebra of the affine group) is given as

$$\mathfrak{aff}(\mathbb{R}^n) = \left\{ \begin{pmatrix} A & a \\ 0 & 0 \end{pmatrix} \parallel A \in \mathfrak{gl}(\mathbb{R}^n),\ a \in \mathbb{R}^n \right\}.$$

As a simply transitive action of a Lie group corresponds to a morphism with a bijective translational part, it is natural to consider the Lie algebra morphisms

$$\varphi : \mathfrak{g} \to \mathfrak{aff}(\mathbb{R}^n) : X \mapsto \begin{pmatrix} \varphi_l(X) & \varphi_t(X) \\ 0 & 0 \end{pmatrix}$$

where $\varphi_t : \mathfrak{g} \to \mathbb{R}^n$ is a bijective map. Such a Lie algebra morphism for which the translational part is a bijective map, will be called an affine structure of the nilpotent Lie algebra $\mathfrak{g}$. However, in contrast to the group case where $\tilde{\rho}(G)$ consists of unipotent matrices, the bijectivity of the map φ_t does not imply a nice structure on the matrices of $\varphi(\mathfrak{g})$. Therefore, one has to introduce the stronger notion of a complete affine structure of a Lie algebra, which is an affine structure satisfying the fact that $\varphi(\mathfrak{g})$ consists on nilpotent matrices.

Remark 3.2 If $\varphi : \mathfrak{g} \to \mathfrak{aff}(\mathbb{R}^n)$ is a Lie algebra morphism, then $\varphi_l : \mathfrak{g} \to \mathfrak{gl}(\mathbb{R}^n)$ is also a Lie algebra morphism. Via φ_l, the vector space $\mathbb{R}^n$ becomes a $\mathfrak{g}$-module. In contrast to φ_l, the map $\varphi_t : \mathfrak{g} \to \mathbb{R}^n$ is not a morphism. It is however a 1-cocycle (derivation) with respect to the $\mathfrak{g}$-module structure of $\mathbb{R}^n$ which is given by φ_l (i.e. $\varphi_t([X,Y]) = \varphi_l(X)\varphi_t(Y) - \varphi_l(Y)\varphi_t(X)$).

The notion of a complete affine structure of a Lie algebra and that of a simply transitive affine action of the Lie group are not only analogous because they are defined in an analogous way, but they are really equivalent. If $\rho : \mathfrak{g} \to \mathfrak{aff}(\mathbb{R}^n)$ is a complete affine structure of the Lie algebra $\mathfrak{g}$, we can construct the commutative diagram

$$\begin{array}{ccc} G & \xrightarrow{\tilde{\rho}} & \mathrm{Aff}(\mathbb{R}^n) \\ \uparrow \exp & & \uparrow \exp \\ \mathfrak{g} & \xrightarrow[\varphi]{} & \mathfrak{aff}(\mathbb{R}^n) \end{array}$$

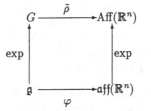

to obtain a Lie group morphism $\tilde{\rho} : G \to \mathrm{Aff}(\mathbb{R}^n)$. It turns out that the action of G on $\mathbb{R}^n$ is simply transitive. Conversely, given a simply transitive affine action of the nilpotent Lie group G on $\mathbb{R}^n$, by means of a morphism $\tilde{\rho} : G \to \mathrm{Aff}(\mathbb{R}^n)$, one can construct the commutative diagram (log exists on unipotent matrices)

$$\begin{array}{ccc} G & \xrightarrow{\tilde{\rho}} & \mathrm{Aff}(\mathbb{R}^n) \\ \downarrow \log & & \downarrow \log \\ \mathfrak{g} & \xrightarrow[\varphi = d\tilde{\rho}]{} & \mathfrak{aff}(\mathbb{R}^n) \end{array}$$

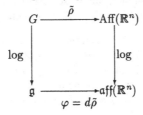

leading to a Lie algebra morphism $\varphi : \mathfrak{g} \to \mathfrak{aff}(\mathbb{R}^n)$. Again, the obtained φ is a complete affine structure of $\mathfrak{g}$. For more details on this procedure, we refer the reader to [18] and the references therein. As a conclusion, we formulate the following theorem:

Theorem 3.3 *Let N be a finitely generated torsion-free nilpotent group with Mal'cev completion G. Denote the Lie algebra of G by $\mathfrak{g}$. Then, there are one-to-one correspondences between the following three sets of objects*

1. *The affine structures on the group N.*
2. *The simply transitive affine actions of G.*
3. *The complete affine structures of $\mathfrak{g}$.*

It is now time to give the historical evolution of the answers to Milnor's question.

The first result concerning affine structures on nilpotent groups seems to go back to Elie Cartan, who proved that any finitely generated torsion-free 2-step nilpotent group admits an affine structure. Although, this result is quoted by many authors, nobody seems to be able to give an exact reference to this result and nor do I.

In 1974, i.e. 3 years before Milnor's question was formulated, Scheuneman ([28]) constructed a complete affine structure on any 3-step nilpotent Lie algebra. As a consequence, one can conclude that any finitely generated torsion-free 3-step nilpotent group has an affine structure. It then lasted until

1983, when K.B. Lee ([21]) could improve Scheuneman's result a little bit by proving the existence of an affine structure on any finitely generated virtually 3-step nilpotent group. K.B. Lee presents a proof of this fact by using some cohomology techniques (similar to the ones we mention in Section 5) on the (discrete) group level. Alternative proofs of the fact that the virtually 3-step nilpotent groups admit an affine structure are given in [8] (group level) and [16] (mainly on the Lie algebra level).

From then onwards, besides the announcement of several faulty results each claiming a general "yes" as answer to Milnor's question, no real progress was made.

In 1992, Benoist ([4]) announced a counter-example to Milnor's question. More precisely, Benoist described an 11-dimensional 10-step nilpotent Lie algebra admitting no complete affine structure (even worse, admitting no 12-dimensional faithful nilpotent matrix representation).

In 1995, the details of Benoist's counter-example were published ([5]) and in the mean time this counter-example was checked by an alternative method and generalized to a family of counter-examples (all of them still 10-step nilpotent and 11-dimensional) by Burde and Grunewald ([7]). It follows that the general answer to Milnor's question is *no; there are even finitely generated, torsion-free nilpotent groups not admitting any affine structure.*

In 1996, Burde could reduce both the dimension (to 10) as the nilpotency class (to 9) of the counter-example to Milnor's question by one. We note here that the techniques used thus far to obtain counter-examples cannot any longer be used to produce counter-examples of lower nilpotency class.

In some sense, the work of Benoist, Burde and Grunewald settles Milnor's problem completely. On the other hand, the problem is now open more than ever in the following two aspects.

1. Since it is not true that any finitely generated torsion-free nilpotent group (polycyclic-by-finite group) admits an affine structure, one should now ask: which groups do admit an affine structure. Even in the nilpotent case, there is still a big gap between the best known positive results (up till class 3) and the best known negative results (class 9). It is for instance not known whether any finitely generated, torsion-free 4-step nilpotent group admits an affine structure. Here, I should mention the papers [10] (on the Lie algebra level) and [16] (both on the Lie algebra as the group level), where it is shown that a rather broad class of (virtually) 4-step nilpotent groups admits an affine structure.

2. The other question one can ask, and this is the one we will focus on in the rest of this paper, is

What kind of structure is there, when there is a lack of affine structures?

It was already known for a long time that any torsion-free polycyclic-by-finite group acts properly discontinuously and via smooth maps on some space $\mathbb{R}^n$, with compact quotient. Analogously to the affine case, one can refer to such an action as a smooth structure. So the question one should ask is whether there is something in between affine structures and smooth structures. The beginning of the answer to this question is given in the next section.

4 The case of virtually nilpotent groups

In this section we will describe a rather easy construction of properly discontinuous and cocompact actions of virtually nilpotent groups on some space $\mathbb{R}^n$. It will turn out that this action has some very nice properties.

Let us fix a finitely generated virtually nilpotent group E, which need not be torsion-free. It follows that there is a short exact sequence of groups

$$1 \longrightarrow N \longrightarrow E \xrightarrow{p} F \longrightarrow 1, \tag{1}$$

where N is a torsion-free nilpotent group and F is a finite group. Choose a normalised section $s : F \to E$. The section s induces a map (not a morphism) $\psi : F \to \mathrm{Aut}(N)$ as follows:

$$\psi : F \to \mathrm{Aut}(N) : x \mapsto \varphi(x), \text{ with } \forall n \in N : \ \psi(x)(n) = s(x)ns(x)^{-1}.$$

The section s also induces a second map $c : F \times F \to N$ defined by

$$\forall x, y \in F : \ s(x)s(y) = c(x,y)s(xy).$$

The map c (or the pair (c,ψ)) is often called a non–abelian 2-cocycle. The maps c and ψ determine completely the group E. We refer the reader to [6, Capter IV] for more details on this subject.

Now, let G be the Mal'cev completion of N. A well known result, due to Mal'cev [24], states that any automorphism of N can be extended in a unique way to a continuous automorphism of G. In other words there is a canonical embedding $i : \mathrm{Aut}(N) \to \mathrm{Aut}(G)$. Using this embedding, we can define a new non–abelian 2–cocycle $(\tilde{c}, \tilde{\psi})$:

$$\tilde{\psi} = i \circ \psi \text{ and } \tilde{c} : F \times F \to G : (x,y) \mapsto c(x,y) \in N \hookrightarrow G.$$

determining an extension $\tilde{E}$ of G by F. It follows that there is a commutative diagram, with exact rows

$$\begin{array}{ccccccccc}
1 & \longrightarrow & N & \longrightarrow & E & \longrightarrow & F & \longrightarrow & 1\\
 & & \downarrow & & \downarrow & & \downarrow{=} & & \\
1 & \longrightarrow & G & \longrightarrow & \tilde{E} & \longrightarrow & F & \longrightarrow & 1
\end{array}$$

The reason for considering the group $\tilde{E}$ in stead of E will follow from the following theorem:

Theorem 4.1 ([9, page 33], [22]) *Let G be a connected and simply connected nilpotent Lie group. Then, any extension of G by a finite group F splits.*

Note that this theorem is a straightforward generalization of the fact that any extension of the vector group $\mathbb{R}^n$ by a finite group F splits.

The above theorem shows that the extension $\tilde{E}$ is equivalent to a semi-direct product $G \rtimes_\varphi F$, where $\varphi : F \to \mathrm{Aut}(G)$ denotes the action of F on G. Finally we note that there is a morphism σ with finite kernel

$$\sigma = 1_G \times \varphi : G \rtimes_\varphi F \to G \rtimes \mathrm{Aut}(G) : (g,x) \mapsto (g, \varphi(x)).$$

The groups and morphisms we constructed so far, can be summarized in the following commutative diagram with exact rows.

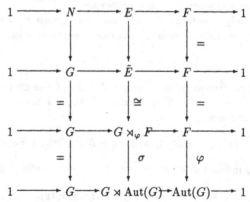

$$\begin{array}{ccccccccc}
1 & \longrightarrow & N & \longrightarrow & E & \longrightarrow & F & \longrightarrow & 1\\
 & & \downarrow & & \downarrow & & \downarrow{=} & & \\
1 & \longrightarrow & G & \longrightarrow & \tilde{E} & \longrightarrow & F & \longrightarrow & 1\\
 & & {=}\downarrow & & \downarrow{\cong} & & \downarrow{=} & & \\
1 & \longrightarrow & G & \longrightarrow & G \rtimes_\varphi F & \longrightarrow & F & \longrightarrow & 1\\
 & & {=}\downarrow & & \downarrow{\sigma} & & \downarrow{\varphi} & & \\
1 & \longrightarrow & G & \longrightarrow & G \rtimes \mathrm{Aut}(G) & \longrightarrow & \mathrm{Aut}(G) & \longrightarrow & 1
\end{array}$$

As a conclusion, we find that there is a morphism $\tau : E \to G \rtimes \mathrm{Aut}(G)$ with finite kernel, which is the identity on N. So, via this morphism τ, N is embedded as a discrete cocompact subgroup of G.

It is our aim to construct a properly discontinuous action of E on $\mathbb{R}^n$ with compact quotient. There is however a very nice model action. Indeed, the group $G \rtimes \mathrm{Aut}(G)$ acts on G in the following natural way

$$\forall (g,\alpha) \in G \rtimes \mathrm{Aut}(G),\ \forall h \in G : \ {}^{(g,\alpha)}h = g\alpha(h). \tag{2}$$

It is well known that the group G is diffeomorphic to a space $\mathbb{R}^n$, where n denotes the dimension of G, or equivalently the Hirsch length of N. After having fixed such a diffeomorphism $\nu : G \to \mathbb{R}^n$, i.e. ν is a global chart on G, we can say that the group $G \rtimes \mathrm{Aut}(G)$ acts on the space $\mathbb{R}^n$. This action, which is nothing else then the coordinate expression for the natural action (2), is explicitly given by

$$\forall (g,\alpha) \in G \rtimes \mathrm{Aut}(G),\ \forall x \in \mathbb{R}^n : \ {}^{(g,\alpha)}x = \nu(g\alpha(\nu^{-1}(x))).$$

It is obvious that a subgroup of $G \rtimes \mathrm{Aut}(G)$ will act properly discontinuously on $\mathbb{R}^n$ if and only if it acts properly discontinuously on G. Looking at $\tau(E)$, we see that this group is a finite extension of the lattice N of G. As any lattice has the property of acting properly discontinuously on G and with compact quotient via the natural action (2), it follows immediately that the action of $\tau(E)$ on G and so on $\mathbb{R}^n$ is also properly discontinuous, having a compact quotient space. We can now define an action of the original group E on $\mathbb{R}^n$, by taking

$$\forall e \in E,\ \forall x \in \mathbb{R}^n : \ {}^{e}x = {}^{\tau(e)}\,x.$$

As τ has a finite kernel, this action of E on $\mathbb{R}^n$ will be properly discontinuous too. In this way, we constructed a smooth structure on every virtually nilpotent group E.

In the above construction, we did not exploit the fact that there are several diffeomorphisms $\nu : G \to \mathbb{R}^n$. It seems natural to think that the nicer the choice of this ν is, the nicer the action of E on $\mathbb{R}^n$ will be. There are two commonly used global charts on a connected, simply connected nilpotent Lie group, called the Mal'cev coordinates of the first and the second kind respectively ([24]). In this paper we will only use the coordinate systems of the first kind. Let G be a n-dimensional connected and simply connected nilpotent Lie group with Lie algebra $\mathfrak{g}$. After having fixed a basis $(A_1, A_2, \ldots, A_n)$ of $\mathfrak{g}$, any element $X \in \mathfrak{g}$ can be uniquely decomposed as a sum $X = x_1A_1 + x_2A_2 + \cdots + x_nA_n$. So, there is an identification

$$co : \mathfrak{g} \to \mathbb{R}^n : X = x_1A_1 + x_2A_2 + \cdots + x_nA_n \mapsto co(X) = (x_1, x_2, \ldots, x_n)^T.$$

We already mentioned that in case of simply connected nilpotent Lie groups, the exponential map is an analytical bijection (with inverse log). Therefore, we can define the following global chart on G:

$$\nu : G \to \mathbb{R}^n : g \mapsto co(\log(g)).$$

It is this map ν which is called a Mal'cev coordinate system of the first kind on G. In terms of these coordinates of the first kind, the action of $G \rtimes \mathrm{Aut}(G)$ on $\mathbb{R}^n$, now becomes very nice.

Theorem 4.2 *Let G be a connected and simply connected c-step nilpotent Lie group, equipped with a Mal'cev coordinate system $\nu : G \to \mathbb{R}^n$ of the first kind. Let $G \rtimes \mathrm{Aut}(G)$ act on $\mathbb{R}^n$, via the coordinate expression for the natural action of $G \rtimes \mathrm{Aut}(G)$ on G. Then, for any $(g,\alpha) \in G \rtimes \mathrm{Aut}(G)$, there exists a polynomial map $p : \mathbb{R}^n \to \mathbb{R}^n$, depending on g and α such that*

$$\forall (x_1, x_2, \ldots, x_n)^T \in \mathbb{R}^n : {}^{(g,\alpha)}(x_1, x_2, \ldots, x_n)^T = p(x_1, x_2, \ldots, x_n).$$

Moreover, the polynomial map p is of total degree $\leq \mathrm{Max}\{c-1, 1\}$.

Proof (see also [8], [9], [15]) We will prove this theorem in two steps. As $(g,\alpha) = (g,1)(1,\alpha)$, we will first investigate the action of $(1,\alpha)$ on $\mathbb{R}^n$ and afterwards the action of $(g,1)$. By definition, we have that

$${}^{(1,\alpha)}(x_1, x_2, \ldots, x_n)^T = \nu(\alpha(\nu^{-1}(x_1, x_2, \ldots, x_n)^T)).$$

Recall that the differential of α, denoted by $d\alpha$, is the unique Lie algebra automorphism of $\mathfrak{g}$ satisfying $\alpha(\exp(X)) = \exp(d\alpha(X))$ for all $X \in \mathfrak{g}$. It follows that there is a commutative diagram

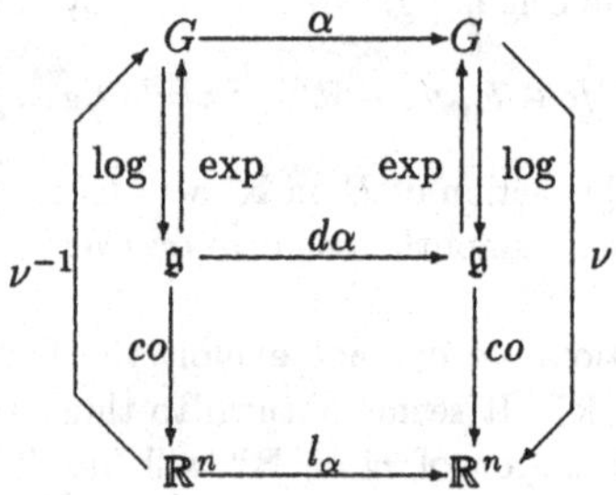

where l_α is the linear map representing $d\alpha$ with respect to the fixed basis $(A_1, A_2, \ldots, A_n)$. From the diagram one now immediately deduces that the action of $(1,\alpha)$ is given by the linear map l_α, i.e.

$${}^{(1,\alpha)}(x_1, x_2, \ldots, x_n)^T = l_\alpha(x_1, x_2, \ldots, x_n).$$

For the second step, we consider the action of $(g,1)$ on $\mathbb{R}^n$, which is given by

$$\begin{aligned} {}^{(g,1)}(x_1, x_2, \ldots, x_n)^T &= \nu(g\nu^{-1}(x_1, x_2, \ldots, x_n)^T) \\ &= co(\log(g\exp(x_1A_1 + x_2A_2 + \cdots + x_nA_n))). \end{aligned}$$

The Campbell-Baker-Hausdorff formula now states that

$$\log(g\exp(x_1A_1 + x_2A_2 + \cdots + x_nA_n)) =$$

$$\log(g) + x_1A_1 + \cdots + x_nA_n + \sum_{j=2}^{c} q_j(\log(g), x_1A_1 + \cdots + x_nA_n)$$

where $q_j(\log(g), x_1A_1 + \cdots + x_nA_n)$ is a rational linear combination of j-fold Lie-brackets in $\log(g)$ and $x_1A_1 + \cdots + x_nA_n$. By the bilinearity of the Lie bracket, one sees that $co(q_j(\log(g), x_1A_1 + \cdots + x_nA_n))$ can be expressed by a polynomial of degree $\leq j-1$ in the variables $x_1, x_2, \ldots, x_n$. Using this, it is now obvious that the action of $(g, 1)$ on $\mathbb{R}^n$ is expressed by means of a polynomial map $p_g : \mathbb{R}^n \to \mathbb{R}^n$ which is of total degree $\leq \mathrm{Max}(1, c-1)$ in the variables $x_1, x_2, \ldots, x_n$.

The proof now finishes by realizing that the map p in the statement is the composition $p_g \circ l_\alpha$. □

Corollary 4.3 *Let E be a finitely generated group containing a nilpotent subgroup N of nilpotency class $\leq c$. Then there exists a properly discontinuous action of E on some space $\mathbb{R}^n$, with compact quotient, such that the action of any element $e \in E$ is expressed by means of a polynomial map $p : \mathbb{R}^n \to \mathbb{R}^n$, which is of total degree* $\leq \mathrm{Max}\{c-1, 1\}$.

As a comment to the above corollary, we wish to make the following two remarks.

1. If E is a finitely generated virtually 2-step nilpotent group, then the above construction yields an affine structure on E.

2. For more general groups, the above construction shows what can lie between affine and smooth structures. Indeed, the actions we constructed for general virtually nilpotent groups are expressed by means of polynomial maps. Therefore, we will call such actions polynomial structures. We give the precise definition in a few moments.

In the sequel, we will denote by $\mathrm{P}(\mathbb{R}^n)$ the group of polynomial diffeomorphisms of $\mathbb{R}^n$. So, the group $\mathrm{P}(\mathbb{R}^n)$ consists of all bijective maps $\mu : \mathbb{R}^n \to \mathbb{R}^n$, which are expressed by means of polynomials and such that the inverse is also expressed by polynomials. The multiplication in $\mathrm{P}(\mathbb{R}^n)$ is of course the composition of maps. The reader who is not familiar with the group $\mathrm{P}(\mathbb{R}^n)$ should reflect some moments about this group. We remark here for instance that $\mathrm{P}(\mathbb{R}) = \mathrm{Aff}(\mathbb{R}^1)$, while $\mathrm{P}(\mathbb{R}^n)$, for $n \geq 2$, is no longer a finite dimensional Lie group. A prototype example for an element μ of $\mathrm{P}(\mathbb{R}^2)$ is given as follows

$$\mu : \mathbb{R}^2 \to \mathbb{R}^2 : \begin{pmatrix} x \\ y \end{pmatrix} \mapsto \begin{pmatrix} x + p(y) \\ y \end{pmatrix}$$

where $p : \mathbb{R} \to \mathbb{R}$, can be any polynomial map.

Definition 4.4 Let Γ be a group. A **polynomial structure** on Γ is a group morphism $\rho : \Gamma \to \mathrm{P}(\mathbb{R}^n)$ for some n, such that via ρ, the group Γ acts properly discontinuous and with compact quotient on $\mathbb{R}^n$. If moreover, s is an integer such that $\forall \gamma \in \Gamma$, $\rho(\gamma)$ is expressed by polynomials of degree $\leq s$, we say that ρ is a polynomial structure of degree $\leq s$.

As an affine structure is in fact a polynomial structure of degree 1 and a smooth structure can be seen as being described by "polynomials of infinite degree" (think of a Taylor series expansion), it is natural to consider, from that point of view, polynomial structures as really being situated between the affine and the smooth actions.

We can summarize this section as follows:

> Any finitely generated, virtually c-step nilpotent group admits a polynomial structure of degree $\leq \mathrm{Max}(c-1,1)$.

5 Polynomial structures on polycyclic-by-finite groups

In the previous section we saw that any finitely generated virtually nilpotent group admits a polynomial structure of finite degree. We will now explain how it was possible to extend this result to any polycyclic-by-finite group.

The idea behind the construction of a polynomial structure on a general polycyclic-by-finite group, uses the fact that all polycyclic-by-finite groups can be build up in a very nice iterative way. It is well known (see e.g. [30]) that any infinite polycyclic-by-finite group has a non-trivial free abelian normal subgroup. In other words, for any polycyclic-by-finite group Γ there exists a finite ascending series Γ_* of normal subgroups Γ_i (say $0 \leq i \leq c+1$) of Γ:

$$\Gamma_* : \; \Gamma_0 = 1 \subseteq \Gamma_1 \subseteq \Gamma_2 \subseteq \cdots \subseteq \Gamma_{c-1} \subseteq \Gamma_c \subseteq \Gamma_{c+1} = \Gamma \tag{3}$$

with

$$\Gamma_i/\Gamma_{i-1} \cong \mathbb{Z}^{k_i} \text{ for } 1 \leq i \leq c \text{ and some } k_i \in \mathbb{N}_0 \text{ and } \Gamma/\Gamma_c \text{ is finite.}$$

Such a sequence Γ_* will be called a torsion-free filtration (of length c) of Γ.

It is our intention to build up a polynomial structure for Γ by starting with the smallest quotient Γ/Γ_c of Γ_* and then gradually use the polynomial structure on the group Γ/Γ_{i+1} to obtain a polynomial structure on the bigger group Γ/Γ_i.

To explain our construction, we introduce the vector space $\mathrm{P}(\mathbb{R}^K,\mathbb{R}^k)$, where K and k are positive integers, of polynomial mappings of $\mathbb{R}^K$ to $\mathbb{R}^k$. So any element of $\mathrm{P}(\mathbb{R}^K,\mathbb{R}^k)$ is given by k polynomials in K indeterminates. Note that though $\mathrm{P}(\mathbb{R}^K)$ is a subset of $\mathrm{P}(\mathbb{R}^K,\mathbb{R}^K)$, it is not a subgroup, since the group operation in $\mathrm{P}(\mathbb{R}^K,\mathbb{R}^K)$ is simply the addition, and not composition, of the polynomials considered. On the other hand, we note here that $\mathrm{P}(\mathbb{R}^K,\mathbb{R}^k)$ contains the group $\mathbb{R}^k$ as the subgroup of constant mappings. The nice thing about $\mathrm{P}(\mathbb{R}^K,\mathbb{R}^k)$ is that it can be made into a $(\mathrm{Gl}(\mathbb{R}^k)\times \mathrm{P}(\mathbb{R}^K))$-module by defining

$$\forall p \in \mathrm{P}(\mathbb{R}^K,\mathbb{R}^k),\ \forall g \in \mathrm{Gl}(\mathbb{R}^k),\ \forall h \in \mathrm{P}(\mathbb{R}^K):\ {}^{(g,h)}p = g \circ p \circ h^{-1}.$$

Moreover, the resulting semi-direct product group $\mathrm{P}(\mathbb{R}^K,\mathbb{R}^k) \rtimes (\mathrm{Gl}(\mathbb{R}^k)\times \mathrm{P}(\mathbb{R}^K))$ can be seen as a subgroup of $\mathrm{P}(\mathbb{R}^{k+K})$ as follows:
$\forall p \in \mathrm{P}(\mathbb{R}^K,\mathbb{R}^k)$, $\forall g \in \mathrm{Gl}(\mathbb{R}^k)$, $\forall h \in \mathrm{P}(\mathbb{R}^K)$: $\forall x \in \mathbb{R}^k$, $\forall y \in \mathbb{R}^K$:

$$(p,g,h)\begin{pmatrix} x \\ y \end{pmatrix} = \begin{pmatrix} g(x)+p(h(y)) \\ h(y) \end{pmatrix}$$

We are now ready to define the central notion of our construction, namely that of a canonical type polynomial structure. This structure can be seen as the immediate generalization of the canonical type affine structures used already in [21], [20], [23], [11] and [10]. The definition of such a canonical type polynomial structure can be formulated in many equivalent ways. The definition we present here is the one which is closest to the iterative idea of the whole construction.

Definition 5.1 A canonical type polynomial structure of a polycyclic-by-finite group with respect to a given torsion-free filtration is completely characterized (by induction on the length of the torsion-free filtration) by the following two properties:

1. The trivial representation $\rho : F \to \mathrm{P}(\mathbb{R}^0) = 1$ of a finite group F, is a canonical type polynomial structure with respect to the torsion-free filtration $F_* : F_0 = 1 \subseteq F_1 = F$. (Note that in this case $c = 0$).

2. A group morphism $\rho : \Gamma \to \mathrm{P}(\mathbb{R}^K)$ of a polycylic-by-finite group Γ is called of canonical type with respect to the torsion-free filtration Γ_* as in (3) if it fits in a commutative diagram

$$\begin{array}{ccccccc}
1 \longrightarrow \mathbb{Z}^{k_1} \cong \Gamma_1 & \longrightarrow & \Gamma & \longrightarrow & \Gamma/\Gamma_1 & \longrightarrow & 1 \\
\downarrow i = \rho_{|\Gamma_1} & & \downarrow \rho & & \downarrow \psi \times \bar{\rho} & & \\
1 \longrightarrow \mathrm{P}(\mathbb{R}^{K_2}, \mathbb{R}^{k_1}) & \longrightarrow & \mathrm{P}(\mathbb{R}^{K_2}, \mathbb{R}^{k_1}) \rtimes (\mathrm{Gl}(\mathbb{R}^{k_1}) \times \mathrm{P}(\mathbb{R}^{K_2})) & \longrightarrow & \mathrm{Gl}(\mathbb{R}^{k_1}) \times \mathrm{P}(\mathbb{R}^{K_2}) & \longrightarrow & 1
\end{array}$$

where

(a) $K_2 = K - k_1$.

(b) $\bar{\rho} : \bar{\Gamma} = \Gamma/\Gamma_1 \to \mathrm{P}(\mathbb{R}^{K_2})$ is a canonical type polynomial structure with respect to the torsion-free filtration

$$\bar{\Gamma}_* = \Gamma_*/\Gamma_1 : \bar{\Gamma}_0 = 1 \subseteq \bar{\Gamma}_1 = \Gamma_2/\Gamma_1 \subseteq \cdots \subseteq \bar{\Gamma}_{c-1} = \Gamma_c/\Gamma_1 \subseteq \bar{\Gamma}_c = \bar{\Gamma} = \Gamma/\Gamma_1.$$

(c) $i = \rho_{|\Gamma_1} : \Gamma_1 = \mathbb{Z}^{k_1} \to \mathrm{P}(\mathbb{R}^{K_2}, \mathbb{R}^{k_1})$ maps Γ_1 into the group of constant mappings $\mathbb{R}^{k_1}$, in such a way that $i(\Gamma_1)$ is a discrete and cocompact subgroup of $\mathbb{R}^{k_1}$. In other words, $i(\Gamma_1)$ spans $\mathbb{R}^{k_1}$ as a vector space.

(d) $\psi : \Gamma/\Gamma_1 \to \mathrm{Gl}(\mathbb{R}^{k_1})$ is a group morphism.

It is important to note that a canonical type polynomial structure is indeed a polynomial structure, i.e. determines a properly discontinuous action with compact quotient.

From the definition itself, it follows that a canonical type polynomial structure, if it exists, is built up in an iterative way. Let us fix a polycyclic-by-finite group Γ and a torsion-free filtration (3) Γ_* of Γ. Moreover, we also assume that there is a polynomial structure $\bar{\rho} : \bar{\Gamma} \mapsto \mathrm{P}(\mathbb{R}^{K_2})$ which is of canonical type with respect to the torsion-free filtration $\bar{\Gamma}$ as in the definition above. The question we have to ask ourselves now is whether the polynomial structure $\bar{\rho}$ can be lifted to the whole group Γ.

This problem can be solved in a way which is very analogous to the way which we constructed a polynomial structure on every virtually nilpotent group. Indeed, in the case of a virtually nilpotent group we worked our way up, starting from the extension $1 \to N \to E \to F \to 1$ towards the extension $1 \to G \to G \rtimes \mathrm{Aut}(G) \to \mathrm{Aut}(G) \to 1$. Here, we start from the short exact sequence $1 \to \mathbb{Z}^{k_1} = \Gamma_1 \to \Gamma \to \bar{\Gamma} \to 1$ and we wish to arrive at the exact sequence $1 \to \mathrm{P}(\mathbb{R}^{K_2}, \mathbb{R}^{k_1}) \to \mathrm{P}(\mathbb{R}^{K_2}, \mathbb{R}^{k_1}) \rtimes (\mathrm{Gl}(\mathbb{R}^{k_1}) \times \mathrm{P}(\mathbb{R}^{K_2})) \to \mathrm{Gl}(\mathbb{R}^{k_1}) \times \mathrm{P}(\mathbb{R}^{K_2}) \to 1$. As a first step, we choose the embedding $i : \mathbb{Z}^{k_1} \to \mathrm{P}(\mathbb{R}^{K_2}, \mathbb{R}^{k_1})$. One can show that the exact choice of i is not really important (this follows from the fact that any two embeddings of $\mathbb{Z}^{k_1}$ in $\mathbb{R}^{k_1}$ as discrete subgroups, are conjugated to each other by means of an affine map) and therefore, we can think of i as being the map sending each element $z \in \mathbb{Z}^{k_1}$ to the constant polynomial in $\mathrm{P}(\mathbb{R}^{K_2}, \mathbb{R}^{k_1})$ mapping everything onto z. Recall that an extension $1 \to \mathbb{Z}^{k_1} \to \Gamma \to \bar{\Gamma} \to 1$ induces a morphism $\psi : \bar{\Gamma} \to \mathrm{Aut}(\mathbb{Z}^{k_1})$, so that $\mathbb{Z}^{k_1}$ is a $\bar{\Gamma}$-module. The extension itself is then given by an element of $H^2(\bar{\Gamma}, \mathbb{Z}^{k_1})$. As $\mathrm{Aut}(\mathbb{Z}^{k_1}) \subseteq \mathrm{Gl}(\mathbb{R}^{k_1})$, we can also regard ψ as being a morphism from $\bar{\Gamma}$ to $\mathrm{Gl}(\mathbb{R}^{k_1})$. In this way, $\mathrm{P}(\mathbb{R}^{K_2}, \mathbb{R}^{k_1})$ can be seen as a $\bar{\Gamma}$-module via the map $\psi \times \bar{\rho} : \bar{\Gamma} \to \mathrm{Gl}(\mathbb{R}^{k_1}) \times \mathrm{P}(\mathbb{R}^{K_2})$. It is important to realize that $\mathbb{Z}^{k_1}$, to be precise $i(\mathbb{Z}^{k_1})$, becomes a submodule of $\mathrm{P}(\mathbb{R}^{K_2}, \mathbb{R}^{k_1})$. The above observations, allow us to conclude that there is a commutative diagram

$$\begin{array}{ccccccc}
1 \longrightarrow & \mathbb{Z}^{k_1} & \longrightarrow & \Gamma & \longrightarrow & \bar{\Gamma} & \longrightarrow 1 \\
& \downarrow i & & \downarrow \tilde{i} & & \downarrow = & \\
1 \longrightarrow & \mathrm{P}(\mathbb{R}^{K_2}, \mathbb{R}^{k_1}) & \longrightarrow & \tilde{\Gamma} & \longrightarrow & \bar{\Gamma} & \longrightarrow 1
\end{array}$$

Here, $\tilde{\Gamma}$ corresponds to the element in $H^2(\bar{\Gamma}, \mathrm{P}(\mathbb{R}^{K_2}, \mathbb{R}^{k_1}))$ which is the image under the induced morphism $i_* : H^2(\bar{\Gamma}, \mathbb{Z}^{k_1}) \to H^2(\bar{\Gamma}, \mathrm{P}(\mathbb{R}^{K_2}, \mathbb{R}^{k_1}))$ of the cohomology class, determining Γ.

The main point in the construction we did for nilpotent groups, was the fact that the group $\tilde{E}$ we obtained, could be written as a semi-direct product. Assume that in the new situation also, the extension $1 \to \mathrm{P}(\mathbb{R}^{K_2}, \mathbb{R}^{k_1}) \to \tilde{\Gamma} \to \bar{\Gamma} \to 1$ splits. Then, we can complete the above commutative diagram to obtain

$$\begin{array}{ccccccc}
1 \longrightarrow & \mathbb{Z}^{k_1} & \longrightarrow & \Gamma & \longrightarrow & \bar{\Gamma} & \longrightarrow 1 \\
& \downarrow i & & \downarrow \tilde{i} & & \downarrow = & \\
1 \longrightarrow & \mathrm{P}(\mathbb{R}^{K_2}, \mathbb{R}^{k_1}) & \longrightarrow & \tilde{\Gamma} & \longrightarrow & \bar{\Gamma} & \longrightarrow 1 \\
& \downarrow = & & \theta \downarrow \cong & & \downarrow = & \\
1 \longrightarrow & \mathrm{P}(\mathbb{R}^{K_2}, \mathbb{R}^{k_1}) & \longrightarrow & \mathrm{P}(\mathbb{R}^{K_2}, \mathbb{R}^{k_1}) \rtimes \bar{\Gamma} & \longrightarrow & \bar{\Gamma} & \longrightarrow 1 \\
& \downarrow = & & \downarrow \sigma & & \downarrow \psi \times \bar{\rho} & \\
1 \longrightarrow & \mathrm{P}(\mathbb{R}^{K_2}, \mathbb{R}^{k_1}) & \longrightarrow & \mathrm{P}(\mathbb{R}^{K_2}, \mathbb{R}^{k_1}) \rtimes (\mathrm{Gl}(\mathbb{R}^{k_1}) \times \mathrm{P}(\mathbb{R}^{K_2})) & \longrightarrow & \mathrm{Gl}(\mathbb{R}^{k_1}) \times \mathrm{P}(\mathbb{R}^{K_2}) & \longrightarrow 1
\end{array}$$

In the diagram above, $\sigma : \mathrm{P}(\mathbb{R}^{K_2}, \mathbb{R}^{k_1}) \rtimes \bar{\Gamma} \to \mathrm{P}(\mathbb{R}^{K_2}, \mathbb{R}^{k_1}) \rtimes (\mathrm{Gl}(\mathbb{R}^{k_1}) \times \mathrm{P}(\mathbb{R}^{K_2})$ is the morphism which maps $(p, \bar{\gamma})$ to $(p, \psi(\bar{\gamma}), \bar{\rho}(\bar{\gamma}))$, for all $p \in \mathrm{P}(\mathbb{R}^{K_2}, \mathbb{R}^{k_1})$ and $\bar{\gamma} \in \bar{\Gamma}$.

It is now obvious that, in case the construction can be carried out (so in case the second extension splits), the map $\rho = \sigma \circ \theta \circ \bar{i}$ is a canonical type polynomial structure of Γ with respect to the torsion-free filtration Γ_*. If we could show that $H^2(\bar{\Gamma}, \mathrm{P}(\mathbb{R}^{K_2}, \mathbb{R}^{k_1})) = 0$, then any extension of $\mathrm{P}(\mathbb{R}^{K_2}, \mathbb{R}^{k_1})$ by $\bar{\Gamma}$ splits, so certainly also the one we are considering. It follows that the existence problem is solved, once you know the vanishing of a second cohomology group. In [13] (see also [14]), we proved even a more general statement, namely

Theorem 5.2 ([13]) *Let $\bar{\Gamma}$ be a polycyclic-by-finite group equipped with a canonical type polynomial structure $\bar{\rho} : \bar{\Gamma} \to \mathrm{P}(\mathbb{R}^n)$ and let $\psi : \bar{\Gamma} \to \mathrm{Aut}(\mathbb{Z}^k)$ be any morphism, then*

$$\forall i > 0 : \quad H^i(\bar{\Gamma}, \mathrm{P}(\mathbb{R}^n, \mathbb{R}^k)) = 0.$$

(The $\bar{\Gamma}$-module structure of $\mathrm{P}(\mathbb{R}^n, \mathbb{R}^k)$ is given by $\psi \times \bar{\rho}$.)

As an immediate consequence of the above theorem, we can deduce that any polycyclic-by-finite group admits a canonical type polynomial structure. But there is even more information available. Indeed, the vanishing of $H^1(\bar{\Gamma}, \mathrm{P}(\mathbb{R}^{K_2}, \mathbb{R}^{k_1}))$ implies that the splitting of the second last layer in the diagram above is unique up to conjugation with an element of $\mathrm{P}(\mathbb{R}^{K_2}, \mathbb{R}^{k_1})$. This, combined with the fact that the choice of $i : \Gamma_1 = \mathbb{Z}^{k_1} \to \mathbb{R}^{k_1} \hookrightarrow \mathrm{P}(\mathbb{R}^{K_2}, \mathbb{R}^{k_1})$ is determined up to a nice conjugation leads to the following theorem:

Theorem 5.3 ([13]) *Let Γ be any polycyclic-by-finite group equipped with a torsion-free filtration Γ_*. Then, Γ admits a canonical type polynomial structure with respect to Γ_*. Moreover, any two canonical type polynomial structures of Γ, with respect to the same torsion-free filtration Γ_*, are conjugated to each other by means of a polynomial diffeomorphism.*

6 Properties and open problems of polynomial structures

There are two properties of canonical type polynomial structures which deserve our attention:

Theorem 6.1 *Let $\rho : \Gamma \to \mathrm{P}(\mathbb{R}^K)$ be a canonical type polynomial structure of an infinite polycyclic-by-finite group Γ (with respect to some torsion-free filtration), then*

1. *there is a non-trivial free abelian subgroup of Γ (i.e. Γ_1) acting as pure translations on $\mathbb{R}^K$;*
2. *there exists a positive integer s, such that for all $\gamma \in \Gamma$, the degree of the polynomial diffeomorphism $\rho(\gamma)$ is $\leq s$.*

The first property is interesting in the light of a (false) conjecture of Auslander concerning the existence of translations in affine structures on nilpotent groups (see [2], [29] and [17]).

The second property is even more interesting, because it states that we really obtained the best possible alternative to Milnor's question. Indeed, we did not only obtain a properly discontinuous action expressed by means of polynomial functions but we also know that there is an upper-bound on the degree of the polynomial maps involved. We should however warn the reader that the existence of this upper-bound follows immediately from the fact that there is a canonical type polynomial structure, but there is no information at all concerning the exact value of such an upper bound. Indeed, even for the three-dimensional Heisenberg group, one can construct a canonical type polynomial structure of arbitrary (but bounded) degree.

On the other hand, using a different approach to the existence problem of polynomial structures on polycyclic-by-finite groups we could prove the following theorem in [12].

Theorem 6.2 *Let Γ be any polycyclic-by-finite group of Hirsch length h, then Γ has a (characteristic) subgroup of finite index admitting a polynomial structure of degree* $\leq \mathrm{Max}\{h-2,1\}$.

The idea behind the proof of this theorem is rather simple. It is known that any polycyclic-by-finite group Γ admits a (characteristic) subgroup of finite index Γ' which is splittable ([30], [19]). Being splittable implies that Γ' embeds into a semi-direct product group $N \rtimes T$, where N is a finitely generated torsion-free nilpotent subgroup of the same Hirsch length as Γ' and T is a free abelian group acting faithfully on N. It follows that there is a sequence of injective morphisms

$$\Gamma' \to N \rtimes T \to N \rtimes \mathrm{Aut}(N) \to G \rtimes \mathrm{Aut}(G) \qquad (4)$$

where G is the Mal'cev completion of N. Note that the dimension of G coincides with the Hirsch length of Γ' and so equals h. The nilpotency class of G can then at most be $h-1$. Recall that we showed in a previous section that $G \rtimes \mathrm{Aut}(G)$ acts on $\mathbb{R}^h$ via polynomial diffeomorphisms of degree $\leq \mathrm{Max}\{h-2,1\}$. A careful analysis of the morphisms (4) above, shows that the restriction of the action of Γ' is properly discontinuous and has compact quotient as desired.

The above observations make it natural to introduce the following concept.

Definition 6.3 Let Γ be a polycyclic-by-finite group of Hirsch length $h(\Gamma)$. The **affine defect of** Γ, denoted by $d(\Gamma)$ is defined as

$$\mathrm{Min}\left\{ s \in \mathbb{N} \,\|\, \begin{array}{l} \Gamma \text{ acts properly discontinuous and via polynomial} \\ \text{diffeomorphisms of degree} \leq s+1 \text{ on } \mathbb{R}^{h(\Gamma)} \end{array} \right\}$$

Note that the affine defect of a group is 0 if and only if that group admits an affine structure. In [15], we showed that the affine defect of the known counter-examples to Milnor's question is exactly 1, the smallest possible value. These (and other) observations lead to the first problem.

Problem 1 Is it true that any finitely generated nilpotent group of Hirsch length n and nilpotency class $n-1$ has an affine defect which is ≤ 1?

Hinted by Theorem 6.2 one can also ask:

Problem 2 Let Γ be a any polycyclic-by-finite group of Hirsch length $h \geq 2$. Is $d(\Gamma) \leq h - 2$? More general, we can ask if there exists a linear (or polynomial) function $\mu : \mathbf{N} \to \mathbf{N}$, such that for any Γ of Hirsch length h, we have that $d(\Gamma) \leq \mu(\Gamma)$?

We also recall here the three open problems listed in [13].

Problem 3 Are any two polynomial structures on a given polycyclic-by-finite group conjugated to each other by means of a polynomial diffeomorphism?

Problem 4 Is it true that any polynomial structure on a polycyclic-by-finite group is of bounded degree?

Problem 5 Let E be any group admitting a polynomial structure. Is E polycyclic-by-finite?

To finish, I want to add two more problems, which have been occupying my mind for a while.

Problem 6 Let M and N be finitely generated nilpotent groups, where M is a central extension of N. Is it true that $d(M) \leq d(N) + 1$?

Problem 7 Let Γ be a polycyclic-by-finite group containing Γ' as a subgroup of finite index. Is $d(\Gamma) = d(\Gamma')$?

References

[1] Auslander, L. *The structure of complete locally affine manifolds.* Topology, 1964, 3 Suppl. 1., 131–139.

[2] Auslander, L. *Simply Transitive Groups of Affine Motions.* Amer. J. Math., 1977, 99 (4), 809–826.

[3] Auslander, L. and Markus, L. *Holonomy of Flat Affinely Connected Manifolds.* Ann. of Math., 1955, 62 (1), 139–151.

[4] Benoist, Y. *Une nilvariété non affine.* C. R. Acad. Sci. Paris Sér. I Math., 1992, 315, 983–986.

[5] Benoist, Y. *Une nilvariété non affine.* J. Differential Geom., 1995, 41, 21–52.

[6] Brown, K. S. *Cohomology of groups*, volume 87 of *Grad. Texts in Math.* Springer–Verlag New York Inc., 1982.

[7] Burde, D. and Grunewald, F. *Modules for certain Lie algebras of maximal class.* J. Pure Appl. Algebra, 1995, 99, 239–254.

[8] Dekimpe, K. *The construction of affine structures on virtually nilpotent groups.* Manuscripta Math., 1995, 87, 71 – 88.

[9] Dekimpe, K. *Almost-Bieberbach Groups: Affine and Polynomial Structures*, volume 1639 of *Lect. Notes in Math.* Springer–Verlag, 1996.

[10] Dekimpe, K. and Hartl, M. *Affine Structures on 4-step Nilpotent Lie Algebras.* 1994. To appear in J. Pure Appl. Algebra.

[11] Dekimpe, K. and Igodt, P. *Computational aspects of affine representations for torsion free nilpotent groups via the Seifert construction.* J. Pure Appl. Algebra, 1993, 84, 165–190.

[12] Dekimpe, K. and Igodt, P. *Polynomial structures on polycyclic groups.* 1995. Preprint, to appear in Trans. Amer. Math. Soc.

[13] Dekimpe, K. and Igodt, P. *Polycyclic-by-finite groups admit a bounded-degree polynomial structure.* 1996. Prepint, to appear in Inventiones Math.

[14] Dekimpe, K. and Igodt, P. *Chaque groupe polycyclique-par-fini admet une structure polynomial de degré borné.* C. R. Acad. Sci. Paris Sér. I Math., 1997, 324, 303–306.

[15] Dekimpe, K., Igodt, P. and Lee, K. B. *Polynomial structures for nilpotent groups.* Trans. Amer. Math. Soc., 1996, 348, 77–97.

[16] Dekimpe, K. and Malfait, W. *Affine structures on a class of virtually nilpotent groups.* Topology Appl., 1996, 73 (2), 97–119.

[17] Fried, D. *Distality, completeness and affine structures.* J. Differential Geom., 1986, 24, 265–273.

[18] Fried, D., Goldman, W. and Hirsch, M. *Affine manifolds with nilpotent holonomy.* Comment. Math. Helv., 1981, 56, 487–523.

[19] Grunewald, F. J., Pickel, P. F., and Segal, D. *Polycyclic groups with isomorphic finite quotients.* Ann. of Math., 1980, 111, 155–195.

[20] Igodt, P. and Lee, K. B. *On the uniqueness of certain affinely flat infra-nilmanifolds.* Math. Zeitschrift, 1990, 204, 605–613.

[21] Lee, K. B. *Aspherical manifolds with virtually 3–step nilpotent fundamental group.* Amer. J. Math., 1983, 105, 1435–1453.

[22] Lee, K. B. and Raymond, F. *Rigidity of almost crystallographic groups.* Contemporary Math. A. M. S., 1985, 44, 73–78.

[23] Lee, K. B. and Raymond, F. *Examples of solvmanifolds without certain affine structure.* Illinois J. Math., 1990, 37, 69–77.

[24] Mal'cev, A. I. *On a class of homogeneous spaces.* Transl. Amer. Math. Soc.., 1951, 39, 1–33.

[25] Margulis, G. A. *Free properly discontinuous groups of affine transformations.* Dokl. Akad. Nauk SSSR, 1983, 272, 937–940.

[26] Margulis, G. A. *Complete affine locally flat manifolds with a free fundamental group.* J. Soviet Math., 1987, 134, 129–134.

[27] Milnor, J. *On fundamental groups of complete affinely flat manifolds.* Adv. Math., 1977, 25, 178–187.

[28] Scheuneman, J. *Affine structures on three-step nilpotent Lie algebras.* Proc. Amer. Math. Soc., 1974, 46 (3), 451–454.

[29] Scheuneman, J. *Translations in certain groups of affine motions.* Proc. Amer. Math. Soc., 1975, 47 (1), 223–228.

[30] Segal, D. *Polycyclic Groups.* Cambridge University Press, 1983.

[31] Wolf, J. A. *Spaces of constant curvature.* Publish or Perish Inc., Berkeley, 1977.

ON GROUPS WITH RANK RESTRICTIONS ON SUBGROUPS

MARTYN R. DIXON*, MARTIN J. EVANS* and HOWARD SMITH†

*Department of Mathematics, University of Alabama, Tuscaloosa, AL 35487-0350, U.S.A.
†Department of Mathematics, Bucknell University, Lewisburg, PA 17837, U.S.A.

1 Introduction

In this article we give a brief survey of recent work of the authors and others. We recall that if r is a fixed positive integer then a group G has *(Prüfer) rank* r if every finitely generated subgroup of G can be generated by r elements and r is the least such integer. Throughout this paper we shall say that a group G has rank r if, in the above sense, it has rank at most r. We shall be primarily concerned with those groups which have the property that certain of their subgroups have finite rank, although in the last section we discuss what might be thought of as the dual problem.

The types of question we have in mind are these: suppose that $\mathcal{X}$ is some class of groups and let G be a group in which every proper subgroup belongs to $\mathcal{X}$. What can be said about G? Is G necessarily in $\mathcal{X}$? Can we classify those groups G that are not in $\mathcal{X}$? What if we relax the condition on G and only suppose that certain distinguished subgroups of G belong to $\mathcal{X}$?

Such questions are, of course, quite old. One example, and perhaps the most famous, is Schmidt's problem: if G is a group and all proper subgroups of G are finite then what can be said about G? For many years the only infinite groups G that were known to have this property were the Prüfer p-groups. However, Ol'šanskii and Rips later constructed 2-generator infinite simple groups with all proper subgroups cyclic of prime order (the so-called Tarski monsters). In fact these examples of Ol'šanskii and Rips are often obstructions to obtaining answers to the questions we wish to consider. For this reason we make the following standard definition. A group G is said to be *locally graded* if every non-trivial finitely generated subgroup of G has a non-trivial finite image. Natural examples of locally graded groups are the free groups, locally finite groups, residually finite groups, and locally soluble groups. Thus the class of locally graded groups is quite extensive whilst it does not contain the pathological examples of Ol'šanskii and Rips.

For the sake of completeness we note that it is a fairly easy matter to prove the following well-known result which classifies the locally graded groups with all proper subgroups finite.

Lemma 1 *Let G be a locally graded group and suppose that all proper subgroups of G are finite. Then either G is finite or $G \cong C_{p^\infty}$ for some prime p.*

The only thing that need be said here is that we can quickly reduce this to the locally finite case and then appeal to a theorem of Hall and Kulatilaka [10] which says that an infinite locally finite group has an infinite abelian subgroup.

We shall be particularly concerned in Sections 2 and 3 with groups in which certain distinguished subgroups are of finite rank. For instance, in Section 2 we discuss groups in which all locally soluble subgroups have finite rank. Our results here are concerned with locally (soluble-by-finite) groups. These groups form an important subclass of the class of locally graded groups. In Section 3 we discuss the reverse situation, so to speak. We shall be interested in, among other things, groups with all proper insoluble subgroups of finite rank. The class of groups we consider has a somewhat technical definition which is given later. The result (Theorem 9) we obtain gives a complete classification of such groups. We also use a remarkable theorem of Ol'šanskii to construct an interesting example.

Recently a new line of inquiry, which has perhaps been motivated by the work on groups with subgroups satisfying certain properties, has been opened (see [1], [14] and [23]). In Section 4 we consider the class of those groups which are residually of rank r, for some fixed integer r, and give a synthesis of results appearing in [8] and [9]. The results we obtain are perhaps not quite as decisive as those in earlier sections. One difficulty is again the possibility that Tarski monsters can appear, but this time in the factor groups. For example a non-abelian free group is residually (finite-p of rank 9), but has Tarski groups as images.

Our notation, where not explained, is that in standard use.

2 Groups with locally soluble subgroups of finite rank

The theory of groups all of whose abelian subgroups satisfy some finiteness condition on their ranks is well documented. In particular, in [25], Šunkov showed that a locally finite group all of whose abelian subgroups have finite rank itself has finite rank and in [16] Merzljakov showed that a locally soluble group in which all abelian subgroups have bounded rank also has finite rank. We refer the reader to the excellent survey paper [22] where further results of this nature are obtained.

In this section we discuss some of our recent results for groups satisfying similar rank conditions. Our notation for closure operations is that used in [20]. We denote by Λ the set of closure operations $\{\mathbf{L}, \mathbf{R}, \acute{\mathbf{P}}, \grave{\mathbf{P}}\}$; thus a class $\mathcal{C}$ of groups is Λ-closed if it is closed under the formation of local systems, subcartesian products and both ascending and descending series. In [4], Černikov introduced the class $\mathcal{X}$ obtained by taking the Λ-closure of the class of periodic locally graded groups. It is straightforward to show that every $\mathcal{X}$-group is locally graded and the authors know of no example of a locally graded group which is not in the class $\mathcal{X}$. We call a group G *almost locally soluble* if it has a locally soluble subgroup of finite index.

The main result of [4] is the following.

Theorem 1 *Let G be an $\mathcal{X}$-group of finite rank. Then G is almost locally soluble.*

By Šunkov's theorem, mentioned above, a locally finite group all of whose abelian subgroups have finite rank is almost locally soluble. Theorem 1 therefore generalizes the theorem of Šunkov and also the now well-known result of Lubotzky and Mann [14] that a residually finite group of finite rank is almost locally soluble.

In [5] we gave a partial generalization of the theorem of Černikov which runs as follows.

Theorem 2 *Let G be a locally (soluble-by-finite) group and suppose that every locally soluble subgroup of G has finite rank. Then G has finite rank and is almost locally soluble.*

Here is a sketch proof of this theorem. Note first that we only need show that G has finite rank, since Černikov's theorem will do the rest. The property of "having finite rank" is a property of countably recognizable character. Thus to show that our group is of finite rank we may assume that G is countable. So we let G be a counterexample to our theorem and then write G as the ascending union $\cup_{i\geq 1}G_i$ of finitely generated soluble-by-finite subgroups. If R_i is the soluble radical of G_i then $F_i = G_i/R_i$ is a finite semisimple group and the hypotheses on G show that each R_i is a soluble minimax group. Moreover the group $R = \langle R_i \mid i \geq 1\rangle$ is easily seen to be locally soluble and hence has finite rank r, say. Since G is a counterexample, the orders of the F_i have to be unbounded.

Let us consider this sequence $\{F_i\}_{i\geq 1}$ more closely. We denote the rank of the socle of F_i by n_i. Clearly, F_i is a section of F_{i+1} for each i and Propositions 2.1 and 2.4 of [5] show that there is no bound on the ranks of the socles of the F_i. The classification of finite simple groups is involved here.

Next we recall that the periodic subgroups of $GL(r,\mathbb{Q})$ are all finite of order bounded by some number $k = f(r)$ (see 9.33 of [27]). Also, it is well-known that in a locally soluble group R of rank r some term $R^{(d)}$ of the derived series is a periodic hypercentral group, where $d = d(r)$ (see, for example, Lemma 10.39 of [20]). By discarding certain of the G_i, and reindexing if necessary, we may assume that $n_i > kd$ and that $n_{i+1} > n_i$, for all i. It is then possible to show using Proposition 3.3 of [5] that each G_i contains a normal locally finite subgroup A_i of rank at least $n_i - kd$. The product A of the groups A_i is a locally finite group of infinite rank and Šunkov's theorem then gives a contradiction, which proves the result.

Using this result we can extend the theorem of Šunkov mentioned earlier to the class of locally (soluble-by-finite) groups. Before giving this extension (Theorem 3 below) let us consider the possibility of proving results of the form 'If all abelian subgroups of G have finite rank, then G has finite rank', when G is assumed to lie in some natural class of groups. We recall that a group G is called *radical* if it is the terminus of its upper Hirsch-Plotkin series. Baer and Heineken showed in [2] that a radical group with all abelian subgroups of finite rank itself has finite rank. We remark that in obtaining such theorems it is often necessary to restrict attention to groups which are the terminus of some type of ascending series with factors which are, at worst, locally nilpotent. One reason for this is that Merzljakov [16] has given an example of a torsionfree locally soluble group which has infinite rank, but whose abelian subgroups all have finite rank. Although this example is not difficult, the details are a little bit technical so we shall only give a brief description. Let A be a free abelian group of finite rank s which is contained as a subgroup of finite index in a group H, say $|H/A| = m$. Let k be a natural number. Then A has a subgroup B

such that A/B has order k and for some n there is a faithful representation of A/B as a subgroup of $GL(n, \mathbb{Z})$. Hence there is a map from A to $GL(n, \mathbb{Z})$ with kernel B. There is an induced representation of H in $GL(mn, \mathbb{Z})$. If the corresponding module for H is denoted by C then C is free abelian of rank mn and since B acts trivially on C the group BC is a free abelian normal subgroup of $C \rtimes H$ of rank mns. The trick now is to iterate this construction infinitely often, which can be done, and in this way Merzljakov's group can be constructed. Of course to verify that it has the correct properties takes some work, and one has to be a little bit more careful with the choice of B than has been indicated here.

The following result from [5] shows where the problem lies: the groups of Merzljakov have abelian subgroups of finite but unbounded ranks.

Theorem 3 *Let G be a locally (soluble-by-finite) group with all abelian torsion subgroups of finite rank and all torsionfree abelian subgroups of bounded rank. Then G has finite rank and is almost locally soluble.*

Thus obstructions to proving rank theorems for locally soluble groups in general lie with the torsionfree subgroups. It would be interesting to have further examples of locally soluble groups of infinite rank, all of whose abelian subgroups have finite rank.

Next we recall that a group is said to have *finite abelian subgroup rank* if the rank of every abelian p-subgroup is finite for all primes p and if the torsionfree-rank of every abelian subgroup is finite. The class of groups with finite abelian subgroup rank is not closed under taking homomorphic images so that it is sometimes more useful to consider groups with *finite abelian section rank*; such groups have been discussed in [22]. Of course if G is a group with all abelian subgroups of finite rank then G has finite abelian subgroup rank. The main result of [2] gives the structure of radical groups with finite abelian subgroup rank. The structure of locally finite groups with finite abelian subgroup rank is also known. If G is locally finite and has finite abelian subgroup rank then G has min-p for all primes p and a theorem of Belyaev [3] then shows that G is almost locally soluble.

We wish to prove a complement to Theorem 3 for locally (soluble-by-finite) groups of finite abelian section rank. Certainly we can no longer expect to deduce that our groups have finite rank. We say that a group has *bounded torsionfree abelian subgroup rank* if there is actually a bound on the ranks of the torsionfree abelian subgroups. If N is a soluble group with finite abelian subgroup rank and bounded torsionfree rank r_0, Theorem 10(iii) of [22] shows that there exists an integer d depending on r_0 only such that $N^{(d)}$ is periodic. Theorem 10(iv) of [22] then shows that there is an integer $h(r_0)$ such that each torsionfree abelian factor of N has rank at most $h(r_0)$. These two observations imply the following modification of Proposition 3.3 of [5]:

Proposition 1 *Let M be a finitely generated soluble-by-finite minimax group such that the soluble radical N of M has torsionfree rank r_0. Suppose that $N^{(d)}$ is periodic and let $k = k(r_0)$ denote the order of a largest finite subgroup of $GL(h(r_0), \mathbb{Q})$. Suppose that M/N has socle of rank n where $n > kd$. Then*

(i) *$M/N^{(d)}$ has a normal Černikov subgroup with an image which is a direct product of simple groups and of rank at least $n - kd$.*

(ii) *M contains a normal locally finite subgroup of rank at least $n - kd$.*

This enables us to prove the following theorem.

Theorem 4 *Let G be a locally (soluble-by-finite) group with finite abelian subgroup rank. Suppose that G has bounded torsionfree abelian subgroup rank. Then G is almost locally soluble.*

Proof By Lemma 3.5 of [5] we may assume that G is countable. Let $G = \cup_{i\geq 1} G_i$ where G_i is a finitely generated soluble-by-finite group and $G_i \leq G_{i+1}$ for all $i \geq 1$. Let R_i be the soluble radical of G_i for each i. Using 1.K.2 of [12] we may assume that the orders of the subgroups $F_i = G_i/R_i$ are unbounded. Naturally F_i embeds in F_{i+1} and $\{F_i\}_{i\geq 1}$ is a sequence of finite semisimple groups. Propositions 2.1 and 2.4 of [5] then show that the ranks of the socles of the groups F_i are unbounded.

Now R_i is a finitely generated soluble group with finite abelian subgroup rank. Hence R_i has finite abelian section rank by Theorem 7 of [22]. A theorem of Robinson [21] shows that R_i is minimax and hence has torsionfree rank at most r_0, where r_0 is the torsionfree rank of G.

If the socle of G_i/R_i has rank n_i where $n_i > kd$ and $n_{i+1} > n_i$ for all i then Proposition 1 shows that G_i contains a normal locally finite subgroup A_i with an image which is a direct product of finite simple groups and of rank at least $n_i - kd$. Let $A = \langle A_1, A_2, \ldots \rangle$. Then A is a locally finite group with finite abelian subgroup rank and Belyaev's theorem shows that A is almost locally soluble. If B is a normal locally soluble subgroup of A such that $|A : B| = m < \infty$ then every finite section of A has a soluble subgroup of index at most m. This is impossible in a group A_i for which $n_i - kd > m$, a contradiction which proves the result. □

If G is a free group then all abelian, and indeed all locally soluble, subgroups of G have rank 1. Thus a locally graded group with all locally soluble subgroups of finite rank need not have finite rank. We pose one question to end this section, among the many that could be asked.

Question 1 If G is a periodic locally graded group and all locally soluble subgroups of G have finite rank then does G have finite rank?

3 Groups with all insoluble subgroups of finite rank

In view of Theorem 2, where we obtained a result when we have a rank condition on the locally soluble subgroups, a natural question arises. What can we say if we assume that our group has a rank condition on its non-locally soluble subgroups? The theorems mentioned in Section 2 show that the locally soluble and, indeed, the abelian subgroups of a group G exert a strong influence on the structure of the group, especially when such subgroups exhibit some finiteness condition on their ranks. We might therefore expect that the non-abelian subgroups also exert such an influence and in this section we show that this is indeed the case, at least in the class $\mathcal{X}$ defined in Section 2.

If $\mathcal{Z}$ is a class of groups then we let $\mathcal{Z}^*$ denote the class of groups in which every proper subgroup is either of finite rank or in $\mathcal{Z}$. In [6] and [7] we discuss the cases $\mathcal{Z} = \mathcal{N}_c$ and $\mathcal{Z} = \mathcal{S}_d$ respectively, where $\mathcal{N}_c$ is the class of nilpotent groups of class at most c and $\mathcal{S}_d$ is the class of groups which are soluble of derived length at most d.

Suppose G is in the class $\mathcal{N}_1^*$, so all proper subgroups of G are abelian or of finite rank. Clearly groups which themselves are abelian or of finite rank fall into this class and we might reasonably hope that these are the only groups in $\mathcal{N}_1^*$. Note that the Tarski monsters do not cause any problems here since they are of finite rank. However, using the following remarkable result of Ol'šanskii [18] one can show that we are being overly optimistic.

Theorem 5 *Let $\{G_i\}_{i\in I}$ be a countable set of non-trivial finite or countably infinite groups G_i without involutions. Suppose that $|I| \geq 2$ and that n is a sufficiently large odd number. Suppose further that $G_i \cap G_j = 1$ for $i \neq j$. Then there is a countable simple group G which contains a copy of G_i for each $i \in I$ with the following properties:*

(i) *If $x, y \in G$ with $x \in G_i \setminus \{1\}, y \notin G_i$ for some $i \in I$ then G is generated by x and y.*

(ii) *Every proper subgroup of G is either a cyclic group of order dividing n or is contained in a subgroup conjugate to some G_i.*

We content ourselves with just one application of this theorem.

Example 1 There is a 2-generator countably infinite simple group G with no proper non-abelian subgroups, which has abelian subgroups of infinite rank. In particular G has infinite rank.

Proof Let G_1 and G_2 be free abelian groups of infinite rank and apply the Ol'šanskii construction to the groups G_1 and G_2. The group G obtained has no proper non-abelian subgroups and hence is in the class $\mathcal{N}_1^*$. However the group clearly has infinite rank and is simple. □

In [6] we proved the first result along the lines mentioned at the start of this section.

Theorem 6 *Let c be a positive integer and suppose that $G \in \mathcal{X} \cap \mathcal{N}_c^*$. Then either $G \in \mathcal{N}_c$ or G has finite rank. In either case G is almost locally soluble.*

This result follows from the following theorem in which $\mathcal{N}$ denotes the class of all nilpotent groups.

Theorem 7 *Let $G \in \mathcal{X} \cap (\mathbf{L}\mathcal{N})^*$. Then either $G \in \mathbf{L}\mathcal{N}$ or G has finite rank. In either case G is almost locally soluble.*

If F is a finitely generated subgroup of G in Theorem 7 then, being a locally graded group, F has a proper normal subgroup N of finite index such that F/N is a finite non-trivial group. If N is nilpotent then F is nilpotent-by-finite. If N has finite rank then F is a finitely generated $\mathcal{X}$-group of finite rank so is soluble-by-finite by Černikov's theorem. It follows that G is locally (soluble-by-finite). The proof of Theorem 7 requires a discussion of the almost locally soluble and locally finite cases and then an argument similar to that used in Theorem 4 is used to mesh these two cases together.

Of course there is no corresponding result if we replace $\mathcal{N}_c$ by $\mathcal{N}$ here since the Heineken-Mohamed groups [11] are locally nilpotent, non-nilpotent and of infinite rank, but have all proper subgroups nilpotent. However the existence of such groups is essentially the only reason that we cannot replace $\mathcal{N}_c$ by $\mathcal{N}$ in Theorem 7. For we show in [7] that if $G \in \mathcal{X} \cap \mathcal{N}^*$ then either G has finite rank or every proper subgroup of G is nilpotent. The class of groups with all proper subgroups nilpotent has been extensively studied in [17] and [24], which are useful references. Such groups are either locally nilpotent or finitely generated. However Theorem 7 shows that a finitely generated group with every proper subgroup nilpotent which is not in the class $\mathbf{L}\mathcal{N}$ must be of finite rank, so the groups arising here are locally nilpotent if they are not of finite rank. By contrast, Theorem 7 of [7] shows that hypercentral groups in the class $\mathcal{N}^*$ are nilpotent or of finite rank, as we would hope. The proof of this result again depends on structural results concerning groups with all proper subgroups nilpotent.

At this point it is natural to ask if one can obtain similar results for groups in the class $\mathcal{S}_d^*$. One potential obstacle to such a classification is the existence of locally finite simple groups of infinite rank in which every proper subgroup is finite or metabelian. For example, the group $PSL(2, \mathbb{F})$ has this property if $\mathbb{F}$ is a locally finite field with no proper infinite subfields (see [19]). In order to settle this question we will require some information concerning soluble groups with all proper subgroups soluble of derived length d. We remark in passing that the proof of Theorem 7 requires a well-known result of Schmidt which states that a finite group with all proper subgroups nilpotent is soluble. For groups with all proper subgroups soluble of bounded derived length we have the following theorem the proof of which is precisely the one given by Zaicev in [29].

Theorem 8 *Let G be a finitely generated soluble group, and N a normal subgroup of G such that G/N is infinite. Suppose that every proper subgroup H of G that contains N and has finite index in G is of derived length at most d. Then G has derived length at most d.*

Using this result we can show, for example, that if $\mathcal{Y}$ is the Λ-closure of the class of locally soluble groups and if $G \in \mathcal{Y} \cap \mathcal{S}_d^*$ for some d then either G is soluble of derived length at most d or G is of finite rank and locally soluble. We can also obtain the more general:

Theorem 9 *Let $G \in \mathcal{X} \cap (\mathbf{L}\mathcal{S})^*$. Then either*

(i) *$G \in \mathbf{L}\mathcal{S}$ or*

(ii) *G has finite rank or*

(iii) *G is isomorphic to one of* $SL(2,\mathbb{F}), PSL(2,\mathbb{F})$ *or* $Sz(\mathbb{F})$ *for some infinite locally finite field* $\mathbb{F}$ *in which every proper subfield is finite.*

Here $Sz(\mathbb{F})$ denotes the Suzuki group over the field $\mathbb{F}$. We note that each of the groups in (iii) actually is in the class under consideration.

Finally in this section we mention the work [13]. There the authors consider the class of groups G with the property that every finitely generated subgroup of G which is not nilpotent of class at most c is at most r-generated, for fixed c and r. They prove that a locally (soluble-by-finite) group with this property is either nilpotent of class c or of finite rank. There is no bound in terms of r for the rank here. No such result holds in the class of groups in which every finitely generated subgroup is either r-generator or soluble of derived length at most d. In [7] we give an example of a 3-generator soluble group of derived length 3 which has infinite rank and in which every proper subgroup is either 3-generator or of derived length at most 2.

4 Groups which are residually of finite rank

In the previous sections we discussed groups in which the *subgroups* satisfy some kind of condition on their ranks. Now we consider the dual situation of groups having factor groups with finite rank.

The theorem of Lubotzky and Mann [14] mentioned earlier stating that a residually finite group of finite rank is almost locally soluble is virtually the dual of Šunkov's theorem [25] that a locally finite group of finite rank is almost locally soluble. As we have seen, Theorem 1 simultaneously generalizes both these results. The work of Lubotzky and Mann uses the theory of pro-p groups, which Černikov's argument avoids. In [15] Mann and Segal go further and show that a finitely generated residually finite group in which *all* finite quotients have rank at most r is almost soluble minimax and hence is of finite rank. This theorem is strictly a result about finitely generated groups since there exist residually finite groups containing a non-abelian free subgroup in which every finite quotient has rank at most 3. However, we shall give an interesting variant of this result for groups that are not finitely generated as Theorem 14 below. In [23] another result of a similar nature is obtained which we now state. Here, and throughout this section, if $\mathcal{P}$ is a class (or property) of groups then we denote the class of residually $\mathcal{P}$-groups by res $\mathcal{P}$.

Theorem 10 *Let G be a finitely generated group that is res(finite soluble of rank r). Then G has a normal nilpotent subgroup Q such that G/Q is a subdirect product of finitely many linear groups over fields. If, moreover, every finite quotient of G is soluble then G is almost nilpotent-by-abelian.*

This result can be used to obtain the theorem of Mann and Segal mentioned above. We have obtained a partial generalization of Theorem 10 in [8] which runs as follows.

Theorem 11 *Let G be a group that is residually (of rank r and locally soluble), for some integer r. Then there are subgroups M, N of G with $M \triangleleft N \triangleleft G$ such that*

(i) *M is hyperabelian and locally nilpotent,*

(ii) *N/M is residually (linear of r-bounded degree) and*

(iii) *G/N is soluble of r-bounded derived length.*

If a group G is locally soluble and residually of rank r, then the well-known results on the structure of soluble linear groups and Theorem 11 together imply that G is locally nilpotent-by-(soluble of r-bounded derived length). Furthermore, Amberg and Sysak [1] have shown that a locally soluble group which is residually of rank r is hyperabelian. With these facts at hand, it is easy to prove the following nice generalization of the result that, for groups of finite rank, the properties radical, locally soluble and hyperabelian coincide.

Theorem 12 *Let G be a group that is residually of rank r. The following are equivalent.*

(i) *G is locally soluble.*

(ii) *G is hyperabelian.*

(iii) *G is radical.*

The Amberg-Sysak result mentioned above suggests several questions and in [8] we give a number of examples which indicate the limitations of a general theory of groups which are res(rank r). For example a locally finite group which is res(rank r) need not be hyperfinite. Certain wreath products show this. A somewhat more restrictive definition often enables us to say more. If $\mathcal{P}$ is a property (or class) of groups we say that a group is res*($\mathcal{P}$) if it has a countable descending series of normal subgroups N_i intersecting in the identity such that each G/N_i is a $\mathcal{P}$-group. It is easily seen that a locally finite res*(rank r)-group has finite rank and in [8] we show that a locally (soluble-by-finite) group which is res*(rank r) is almost locally soluble. This result may be viewed as dual to the last assertion of Theorem 2. Note, however, that a locally (soluble-by-finite) group with res*(rank r) need not have finite rank as is shown by the group of p-adic integers. Indeed, not even finitely generated soluble groups with res*(rank r) need be of finite rank since we showed in [8] that $\mathbb{Z} \wr \mathbb{Z}$ is res*(finite and of rank 2).

In his important paper [26], J. Tits proved that a finitely generated linear group either is almost soluble or contains a non-abelian free subgroup. In [9], by using results of Wilson [28], we obtained a Tits alternative for groups that are residually of rank r.

Theorem 13 *Let G be a res (locally (soluble-by-finite) of rank r) group. Then either G is locally (soluble-by-finite) or G contains a non-abelian free subgroup.*

We can obtain a stronger conclusion if we assume that G is res*(locally (soluble-by-finite) of rank r). Namely, either G is almost locally soluble or G contains a non-abelian free subgroup (see Corollary A of [9]). We are now in a position to give the variant of the Mann-Segal theorem we mentioned at the beginning of this section.

Theorem 14 *Let G be a residually finite group and suppose that every finite quotient of G has rank r. Then either G is almost locally soluble or G contains a non-abelian free subgroup.*

Proof We suppose throughout that G contains no non-abelian free subgroups. Let H be a countable subgroup of G and let $h_1, h_2, \ldots$ be a list of the non-trivial elements of H. For each $i \geq 1$ there exists $N_i \triangleleft G$ such that G/N_i is finite and $h_i \notin N_i$. For each $j \geq 1$, let $M_j = \cap_{i=1}^{j} N_j$ and note that G/M_j is finite of rank r and $h_1, h_2, \ldots, h_j \notin M_j$. It follows that $H/(H \cap M_j)$ is finite of rank r for each $j \geq 1$. Since $\cap_{j=1}^{\infty}(H \cap M_j) = 1$, we deduce that H is res*(finite of rank r) and it follows from the remark preceding this theorem that H is almost locally soluble. Thus every countable subgroup of G is almost locally soluble and so by Lemma 3.5 of [5] G itself is almost locally soluble. The proof is complete. □

Finally we mention a Tits alternative for finitely generated groups. A proof can be found in [9].

Theorem 15 *Let G be a finitely generated res (soluble of rank r) group. Then either G is nilpotent-by-abelian-by-finite or G contains a non-abelian free subgroup.*

References

[1] B. Amberg and Y. P. Sysak, Locally soluble products of two minimax subgroups, in *Groups-Korea '94* (edited by A. C. Kim and D. Johnson, de Gruyter, Berlin 1995), 9–14.

[2] R. Baer and H. Heineken, Radical groups with finite abelian subgroup rank, *Illinois J. Math.* **16**(1972), 533–580.

[3] V. V. Belyaev, Locally finite groups with Černikov Sylow p-subgroups, *Algebra i Logika* **20**(1981), 605–619; English transl. in *Algebra and Logic* **20**(1981), 393–402.

[4] N. S. Černikov, A theorem on groups of finite special rank, *Ukrain. Mat. Zh.* **42**(1990), 962–970; English transl. in *Ukrainian Math. J.* **42**(1990), 855–861.

[5] M. R. Dixon, M. J. Evans and H. Smith, Locally (soluble–by–finite) groups of finite rank, *J. Algebra* **182**(1996),756–769.

[6] M. R. Dixon, M. J. Evans and H. Smith, Locally (soluble-by-finite) groups with all proper non-nilpotent subgroups of finite rank, *J. Pure Appl. Algebra*, to appear.

[7] M. R. Dixon, M. J. Evans and H. Smith, Locally (soluble-by-finite) groups with all proper insoluble subgroups of finite rank, *Arch. Math.* **68**(1997), 100–109.

[8] M. R. Dixon, M. J. Evans and H. Smith, On groups that are residually of finite rank, *Israel J. Math.*, to appear.

[9] M. R. Dixon, M. J. Evans and H. Smith, A Tits alternative for groups that are residually of bounded rank, *Israel J. Math.*, to appear.

[10] P. Hall and C. R. Kulatilaka, A property of locally finite groups, *J. London Math. Soc.* **39**(1964), 235–239.

[11] H. Heineken and I. J. Mohamed, A group with trivial centre satisfying the normalizer condition, *J. Algebra* **10**(1968), 368–376.

[12] O. H. Kegel and B. A. F. Wehrfritz, *Locally Finite Groups* (North-Holland Publishing Company, Amsterdam 1973).

[13] P. Longobardi, M. Maj and H. Smith, A finiteness condition on non-nilpotent subgroups, *Comm. Algebra* **24**(1996), 3567–3588.

[14] A. Lubotzky and A. Mann, Residually finite groups of finite rank, *Math. Proc. Cambridge Philos. Soc.* **106**(1989), 385–388.

[15] A. Mann and D. Segal, Uniform finiteness conditions in residually finite groups, *Proc. London Math. Soc. (3)* **61**(1990), 529–545.

[16] Yu. I. Merzljakov, Locally soluble groups of finite rank, *Algebra i Logika* **3**(1964), 5–16; erratum *ibid* **8**(1969), 686–690.

[17] M. F. Newman and J. Wiegold, Groups with many nilpotent subgroups, *Arch. Math.* **15**(1964), 241–250.

[18] A. Yu. Ol'šanskii, *Geometry of Defining Relations in Groups* (Kluwer Academic Publishers, Dordrecht, Boston, London; Mathematics and its Applications, vol 70(1989)).

[19] J. Otal and J. M. Peña, Infinite locally finite groups of type $PSL(2, K)$ or $Sz(K)$ are not minimal under certain conditions, *Publ. Math.* **32**(1988), 43–47.

[20] D. J. S. Robinson, *Finiteness Conditions and Generalized Soluble Groups vols. 1 and 2* (Springer-Verlag, Berlin, Heidelberg, New York; Ergebnisse der Mathematik und ihrer Grenzgebiete. Band 62 and 63 1972).

[21] D. J. S. Robinson, On the cohomology of soluble groups of finite rank, *J. Pure Appl. Algebra* **6**(1975), 155–164.

[22] D. J. S. Robinson, A new treatment of soluble groups with finiteness conditions on their abelian subgroups, *Bull. London Math. Soc.* **8**(1976), 113–129.

[23] D. Segal, A footnote on residually finite groups, *Israel J. Math.* **94**(1996), 1–5.

[24] H. Smith, Groups with few non-nilpotent subgroups, *Glasgow J. Math.* to appear.

[25] V. P. Šunkov, On locally finite groups of finite rank, *Algebra i Logika* **10**(1971), 199–225; English transl. in *Algebra and Logic* **10**(1971), 127–142.

[26] J. Tits, Free subgroups of linear groups, *J. Algebra* **20**(1972), 250–270.

[27] B. A. F. Wehrfritz, *Infinite Linear Groups* (Springer-Verlag, New York, Heidelberg, Berlin; Ergebnisse der Mathematik und ihrer Grenzgebiete Band 76 1973).

[28] J. S. Wilson, Two generator conditions for residually finite groups, *Bull. London Math. Soc.* **23**(1991), 239–248.

[29] D. I. Zaicev, Stably solvable groups, *Izv. Akad. Nauk SSSR Ser. Mat.* **33**(1969), 765–780; English transl. in *Math. USSR-Izv.* **3**(1969), 723–736.

ON DISTANCES OF MULTIPLICATION TABLES OF GROUPS

ALEŠ DRÁPAL[1]

Department of Algebra, Charles University, Sokolovská 83, 18600 Prague 8, Czech Republic

Abstract

If $G(*)$ and $G(\circ)$ are two different groups on the same set, then their distance is defined as the number of pairs $(a, b) \in G \times G$ with $a * b \neq a \circ b$. Questions about minimal possible distances lead to different problems as isomorphic groups are or are not admitted. The paper presents many facts known about these problems and establishes the minimal distance of an elementary-abelian 2-group from non-isomorphic groups of the same order.

Introduction

The purpose of this paper is to report on facts known about (Hamming) distances of finite groups and to prove some new results in this area.

If $G(\circ)$ and $G(*)$ are two groups on G, then their distance $dist(G(\circ), G(*))$ is defined to be the number of pairs $(a, b) \in G \times G$ with $a \circ b \neq a * b$. For a fixed group $G(\circ)$ define $\delta(G(\circ))$ to be the minimum of $dist(G(\circ), G(*))$, where $G(*)$ runs through all groups on G that are different from $G(\circ)$. Similarly, if $G(*)$ runs through all groups on G that are not isomorphic to $G(\circ)$, then the minimum of $dist(G(\circ), G(*))$ will be denoted by $\nu(G(\circ))$.

The problem of determining $\delta(G(\circ))$ and $\nu(G(\circ))$ goes back to a book of Lászlo Fuchs [7]. If $|G| \geq 51$ holds, then $\delta(G(\circ))$ is known—see Section 1.

To determine $\nu(G(\circ))$ is far more difficult, and up to now the exact value of $\nu(G(\circ))$ was known only for a very few groups of low order. In Section 2 we shall show that $\nu(E_{2^k})$ equals 2^{2k-2} for all elementary Abelian groups E_{2^k} with $k \geq 2$.

The problem of distances of multiplication tables of groups is close to the problem of distances of Latin squares that satisfy the quadrangle criterion. By a result of J. Denes [1] (which has been rendered more precise by S. Frische in [6], see also [2] and [3, p. 315]), the minimal distance of any two different such Latin squares of order $n \geq 2$ is $2n$, with the exception of $n \in \{4, 6\}$. Latin squares that satisfy the quadrangle criterion arise from multiplication tables of groups by permutation of rows and columns. The distance $2n$ can thus be achieved in several ways, for example by exchanging two rows.

Some of the results in Sections 1 and 3 that concern groups of low order were obtained with the aid of computer programs written by Petr Vojtěchovský.

[1]Partially supported by the Grant Agency of Czech Republic, grant number 201/96/0312

1 Isomorphic groups

Theorem 1.1 *Let $G(\circ)$ be a finite group of order $n \geq 51$. If n is odd, then $\delta(G(\circ))$ is $6n-18$. If $G(\circ)$ is a generalized dihedral group of order $n = 4k+2$, then $\delta(G(\circ))$ is $6n-20$, and $\delta(G(\circ)) = 6n-24$ holds in the remaining cases.*

The above theorem was proved in [5]. If $n \geq 51$ and $G(*)$ satisfies $\delta(G(\circ)) = dist(G(\circ), G(*))$, then there always exists an isomorphism $f : G(\circ) \simeq G(*)$ such that f is a transposition of two elements of G.

It seems that there are many groups of order less than 51, for which the above theorem continues to holds. To find them all, one is inclined to use computer. However, the naive algorithm consumes too much time when $n \geq 14$, and so some additional theory is necessary to determine $\delta(G)$ for all groups $G = G(\cdot)$ of order $n \leq 50$.

If $n \geq 5$, then the formula of 1.1 gives an upper bound for $\delta(G)$. We shall now list all orders $n \geq 5$, for which a smaller distance is known to the present author (the distance appears in parentheses and always applies to the cyclic group of the respective order): 6 (8), 7 (18), 8 (16), 9 (18), 10 (24), 12 (32), 14 (48), 15 (50), 16 (64), 18 (72), 21 (98). With exception of the cases $n = 7$ and $n = 9$ one always can obtain the distance in the list by the method of Proposition 1.2 (see below).

If H and K are groups and $\lambda : H \to K$ is a mapping, let m_λ denote the number of pairs $(h_1, h_2) \in H \times H$ with $\lambda(h_1h_2) \neq \lambda(h_1)\lambda(h_2)$, and put $m_{H,K} = \min\{m_\lambda;\ \lambda : H \to K$ is not a homomorphism$\}$.

If $\phi : G(*) \simeq G(\circ)$ is an isomorphism, then $dist(G(\circ), G(*))$ is given by number of pairs $(g_1, g_2) \in G \times G$ with $\phi(g_1 \circ g_2) \neq \phi(g_1) \circ \phi(g_2)$, which means that $dist(G(\circ), G(*))$ equals m_ϕ.

Let $G = G(\cdot)$ be a group, and put $m_G = \min\{m_\phi;\ \phi : G \to G$ is a permutation, but not a homomorphism$\}$. One immediately observes $\delta(G) \leq m_G$, and, in fact, it is natural to expect $\delta(G) = m_G$ in most cases. However, there are exceptions to this rule, and they include the groups of order 4, the quaternion group Q_8 and the elementary group E_8.

Proposition 1.2 *Let G be a finite group, $K \leq G$, $H \leq Z(G)$, $\alpha : G \to K$ an epimorphism and $\lambda : K \to H$ a mapping. Assume $H \leq Ker(\alpha)$ and define $\phi : G \to G$ so that $\phi(g) = g \cdot \lambda\alpha(g)$ for all $g \in G$. Then ϕ is a permutation and we have*

$$m_\phi = |G : K|^2 \cdot m_\lambda.$$

Proof Assume $\phi(g) = \phi(g')$. Then $g' = gh$ for some $h \in H$, and we have $\phi(g') = g \cdot h \cdot \lambda(\alpha(h)\alpha(g)) = hg\lambda(\alpha(g)) = h\phi(g)$. This gives $h = 1$, g' equals g and ϕ is a permutation.

Furthermore, for all $g, g' \in G$ the equality $\phi(g) \cdot \phi(g') = \phi(g \cdot g')$ holds if and only if $\lambda(\alpha(g)) \cdot \lambda(\alpha(g')) = \lambda(\alpha(g \cdot g')) = \lambda(\alpha(g) \cdot \alpha(g'))$ is true, and the rest is clear. □

Corollary 1.3 *Let G be a finite Abelian group with $K \times H \leq G$. Then the inequality $|G : K|^2 \cdot m_{K,H} \geq m_G$ holds.*

Proof Denote by π the projection $K \times H \to K$, and by γ an endomorphism of G with $\gamma(H) \leq H$ and $\gamma(G) = K \times H$. Put $\alpha = \pi\gamma$. The conditions $\alpha(G) = K$ and $H \leq Ker(\alpha)$ are then clearly satisfied. □

In general, computation of $m_{K,H}$ is, of course, no easier than computation of m_G. However, sometimes the subgroups H and K are so small that we are able to obtain $m_{K,H}$ by a complete enumeration of all mappings $K \to H$.

2 Distances and involutions

First recall a well-known and easy fact (see, e.g. [2]).

Proposition 2.1 *Let G be a finite group with non-empty subsets A and B. If $|A| = |B| = |AB|$ holds, then there exists a (unique) subgroup H of G such that A is a left coset of H and B is a right coset of H.*

For the rest of this section G denotes a finite set of $n \geq 4$ elements, n is even, and $\circ$ and $*$ denote two different group operations on G. For $A \subseteq G$ and $B \subseteq G$ let $d(A, B)$ denote the number of pairs $(a, b) \in A \times B$ with $a \circ b \neq a * b$. Put $d = dist(G(\circ), G(*))$ and observe that d equals $d(G, G)$.

If H is a subgroup of $G(\circ)$, then its set of left and right cosets will be denoted by $L^\circ(H)$ and $R^\circ(H)$, respectively. Similarly, $L^*(K)$ and $R^*(K)$ refer to cosets of $K \leq G(*)$.

Now fix a two-element subgroup H of $G(\circ)$.

Lemma 2.2 *If $4d < n^2$, then there exist a two-element subgroup K of $G(*)$, a coset $A \in L^\circ(H) \cap L^*(K)$ and a coset $B \in R^\circ(H) \cap R^*(K)$ such that $a \circ b = a * b$ holds for all $(a, b) \in A \times B$.*

Proof By 2.1, we just need to find $A \in L^\circ(H)$ and $B \in R^\circ(H)$ with $d(A, B) = 0$. There are $n^2/4$ pairs $(A, B) \in L^\circ(H) \times R^\circ(H)$. If $d(A, B) \geq 1$ holds for all of them, then we obtain $d \geq n^2/4$. □

Lemma 2.3 *Let K be a two-element subgroup of $G(*)$ and let B be contained in $R^\circ(H) \cap R^*(K)$. If A is in $L^\circ(H) \setminus L^*(K)$, then $d(A, B) \geq 2$.*

Proof We have $B = \{b, h \circ b\} = \{b, k * b\}$, with $h \in H$ and $k \in K$, and $A = \{a, a \circ h\} = \{a, a * x\}$, $x \in G$. If $d(A, B) \leq 1$, then the equalities $a * b = a \circ b$, $(a \circ h) \circ b = (a \circ h) * b$ and $a \circ (h \circ b) = a * (h \circ b)$ can be assumed to be true. Then $a * x * b = (a \circ h) * b = a \circ h \circ b = a * (h \circ b) = a * k * b$ implies $k = x$, and A is in $L^*(K)$ as well. □

Proposition 2.4 *If $4d < n^2$ is true, then for every two-element subgroup H of $G(\circ)$ there exists exactly one two-element subgroup K of $G(*)$ that satisfies*

$$4\,|L^\circ(H) \cap L^*(K)| > |G|.$$

Proof We wish to find a subgroup K of $G(*)$ such that more than half of the left cosets of H are also left cosets of K. If K_1 and K_2 are two such subgroups, then they have a common coset, and hence are equal.

Suppose now that no such group K exists. To get a contradiction with the assumed inequality $4d < n^2$, it suffices to show $2d(G,B) \geq n$ for all $B \in R^\circ(H)$. Choose an arbitrary $B \in R^\circ(H)$. Suppose first that B is not in $R^*(T)$ for all $T \leq G(*)$. By 2.1, $d(A,B) \geq 1$ holds for all $A \in L^\circ(H)$, and $d(G,B) \geq |L^\circ(H)| = n/2$ follows immediately.

Now consider the case $B \in R^*(T)$, where T is a subgroup of $G(*)$, and denote $|L^\circ(H) \cap L^*(T)|$ by t. By 2.3, $d(G,B) \geq 2(n/2 - t) = n - 2t$. We have assumed $4t \leq n$, and $d(G,B) \geq n/2$ is therefore true also in this case. □

Remark 2.5 To prove 2.6 and 2.7 below, we do not need more than Proposition 2.4. However, this proposition can be strengthened in the following way:

(i) the subgroup K also satisfies $4\,|R^\circ(H) \cap R^*(K)| > |G|$;

(ii) if H is normal in $G(\circ)$, then K is normal in $G(*)$ as well; and

(iii) in such a case the number of cosets contained in $(G(\circ)/H) \cap (G(*)/K)$ exceeds $(|G|/4)(1 + 3^{-\frac{1}{2}}) \approx 0.39 \cdot |G|$.

Theorem 2.6 *If $G(*)$ and $G(\circ)$ are two groups on a finite set G, and if the inequality $4\,dist(G(\circ), G(*)) < |G|^2$ holds, then $G(*)$ and $G(\circ)$ contain the same number of involutions.*

Proof By 2.5, each involution in $G(\circ)$ determines a unique involution in $G(*)$. As Proposition 2.5 operates also from $G(*)$ to $G(\circ)$, we see that we really have a bijection between the respective sets of involutions. □

Corollary 2.7 *If $k \geq 2$, then $\nu(E_{2^k}) = 2^{2k-2}$ holds.*

Proof By 2.6, $\nu(E_{2^k}) \geq 2^{2k-2}$ takes place. It is well known that by changing the multiplication table of Z_4 at $\{1,3\} \times \{1,3\}$ one can obtain a multiplication table of an elementary abelian group. Denote this group by V; we have $dist(V, Z_4) = 4$. The density of differences in multiplication tables does not change when both groups are multiplied by a common factor, and hence $dist(V \times E_{2^k}, Z_4 \times E_{2^k}) = 2^{2k+2}$ holds for all $k \geq 0$. □

3 Conjectures and problems

Conjecture 3.1 Let G be a finite group of order n and p the least prime dividing n. Then $\nu(G) \geq n^2(p-1)/2p$.

If n is divisible by p^2, then it is easy [5] to construct groups $G(\circ)$ and $G(*)$ of order n with $dist(G(\circ), G(*)) = n^2(p-1)/2p$.

A weaker form of 3.1 just states $\nu(G) \geq n^2/4$. The bound $\nu(G) > n^2/9$ proved in [5] seems to be the only general lower bound of $\nu(G)$ that is currently known.

It seems to be an open question if there exist two groups G_1 and G_2 of the same order and with $\nu(G_1) \neq \nu(G_2)$.

In a joint paper [4], Donovan, Oates-Williams and Praeger investigated the distances of $G(*)$ and $G(\circ)$ under the assumption that they have a common subgroup. They showed, among other things, that for every $k \geq 4$ and every pair of groups $(A, B) \in \{(D_{2^k}, SD_{2^k}), (D_{2^k}, Q_{2^k}), (Q_{2^k}, SD_{2^k})\}$ one can define $G(\circ)$ and $G(*)$ in such a way that $G(\circ) \simeq A$, $G(*) \simeq B$, $dist(G(\circ), G(*)) = 2^{2k-2}$, $G(\circ)$ and $G(*)$ have a common two-element normal subgroup H, and the cosets of H in $G(\circ)$ coincide with cosets of H in $G(*)$.

We shall now state a conjecture which is far weaker than 3.1, but which might be attacked by methods generalizing the approach used in Section 2.

Conjecture 3.2 If $G(*)$ and $G(\circ)$ are groups on a finite set G of n elements, and $dist(G(*), G(\circ)) < n^2/4$ holds, then these groups have the same number of 2-Sylow subgroups and their 2-Sylow subgroups are isomorphic.

Working along the lines of the proof of 2.4, the present author was able to show that if $dist(G(*), G(\circ)) < n^2/4$ holds, then there exists a bijection between subgroups of order 4, and this bijection sends a subgroup of $G(*)$ to an isomorphic subgroup of $G(\circ)$, and preserves normality. However, the proof is far more technical than that of 2.4 and a further generalization need not be easy.

Note that 3.2 implies 3.1 for 2-groups. If 3.2 holds, then one can ask which pairs (A, B) of groups of order 2^k can have their multiplication tables rearranged in such a way that they differ only at 2^{2k-2} entries. If $k = 3$, then this is possible exactly for the pairs $(Z_8, Z_4 \times Z_2)$, $(Z_4 \times Z_2, E_8)$, $(Z_4 \times Z_2, D_8)$, $(Z_4 \times Z_2, Q_8)$, (E_8, D_8), (D_8, Q_8). One can regard the list of preceding pairs as a relation on isomorphism types of groups of order 8. It is obvious that the equivalence closure of this relation consists of only one class. Is this true for all $k \geq 2$?

Added in proof: Conjecture 3.2 has recently been proved by the author. The proof (which is quite long) constructs the necessary isomorphism explicitly. Subsquares of 2-groups are compared in the proof in a similar way as in Section 2. However, if the order of a 2-group is greater than 4, then there need not exist any coinciding Latin subsquares, and hence one has to relate subgroups of $G(\circ)$ and $G(*)$ by measuring the partial overlaps of the respective subsquares.

References

[1] J. Dénes: *On a problem of L. Fuchs*, Acta Sci. Math. (Szeged), 23(1962), 237–241.

[2] J. Dénes and A. D. Keedwell: Latin Squares and their Applications, Akadémiai Kiadó, Budapest, 1974.

[3] J. Dénes and A. D. Keedwell: Latin Squares: New Developments in the theory and Applications, North Holland, Amsterdam, 1991.

[4] D. Donovan, S. Oates-Williams and Ch. E. Praeger: *On the distance between distinct group Latin squares*, J. Comb. Designs, 5(1997), 235–248.

[5] A. Drápal: *How far apart can the group multiplication table be?*, Europ. J. Combinatorics, 13(1992), 335–343.

[6] S. Frische, Lateinische Quadrate, Diploma Thesis, Vienna, 1988.

[7] L. Fuchs: Abelian Groups, Akadémiai Kiadó, Budapest, 1958.

THE DADE CONJECTURE FOR THE MCLAUGHLIN GROUP

GUDULA ENTZ and HERBERT PAHLINGS

Lehrstuhl D für Mathematatik, RWTH Aachen, Templergraben 64, 52072 Aachen, Germany

Abstract

The Dade Conjectures, which relate the numbers of irreducible ordinary or projective characters with given defects and given inertia groups of certain local subgroups of a finite group are verified for the sporadic simple McLaughlin Group McL.

AMS (MOS) subject classification Numbers: 20C15 (20C20 20C25).
Keywords: Dade Conjectures, characters, blocks, defects.

1 Introduction

In the Representation Theory of Finite Groups there is a large number of open problems and long-standing conjectures. Many of these originate from a famous lecture [4] of R. Brauer, in which he described the subject by listing its most interesting and natural open problems and which stimulated an enormous amount of further research. About 10 years later an observation of J. McKay [17] on character degrees of finite groups led to a series of conjectures, notably the Alperin-McKay-Conjecture, and the Alperin Weight Conjecture [1], which has been reformulated in several ways, see for example [15]. The latest in this series seem to be the conjectures of Dade given in [6], [7], and [8]. Any of these conjectures implies the Alperin Weight Conjecture. The strongest form of Dade's conjectures, the "Inductive Conjecture", which implies all others, has the big advantage, that it can be shown to hold for all finite groups provided that it could be verified for all finite simple groups. It has already been verified for a number of simple groups, including all the Mathieu groups, the first three Janko groups, the Held Group, the Suzuki groups $Sz(q)$ and the linear groups $PSL_2(q)$, also quite generally for blocks with cyclic defect groups, see [2], [3], [8], [9], [14]. For the McLaughlin sporadic simple group the Inductive Conjecture is equivalent to the Projective Dade Conjecture and the Invariant Conjecture, the latter being a refinement of the Ordinary Conjecture. These conjectures are verified using the computer algebra system GAP (see [21]). For the Ordinary Conjecture and the Projective Conjecture this has been done by a somewhat different method by the first author [10].

2 The conjectures

To describe the conjectures we fix a finite group G, a prime p and a p-block B of G. A p-chain C of G of length $|C| = n$ is any strictly increasing chain

$$C : P_0 < P_1 < ... < P_n$$

of p-subgroups of G. The group G (and also any extension group A, with $G \triangleleft A$) operates on such chains by conjugation and the stabilizer $N_G(C)$ of C is just the intersection of the normalizers of the P_i in G. The chain C is called "elementary" if all its members P_i are elementary abelian, C is called "normal" if all P_i are normal subgroups in P_n and C is called a "radical p-chain" if for all i

$$P_i = O_p(N_G(C_i)),$$

where as usual $O_p(G)$ denotes the largest normal p-subgroup of G and C_i is the subchain

$$C_i : P_0 < P_1 < \ldots < P_i$$

The set of all elementary or normal p-chains C or p-chains C consisting of radical p-subgroups (i.e. satisfying $P_i = O_p(N_G(P_i))$) for $i > 0$ of G starting with $P_0 = \{1\}$ will be denoted by $\mathcal{E}(G)$ or $\mathcal{N}(G)$ or $\mathcal{U}(G)$, respectively, the set of all radical p-chains C of G beginning with $P_0 = O_p(G)$ by $\mathcal{R}(G)$. We choose sets of representatives of the G-orbits on these sets of chains and denote these by $\mathcal{E}(G)/G$, $\mathcal{N}(G)/G$, $\mathcal{R}(G)/G$ or $\mathcal{U}(G)/G$, respectively.

It follows from a result of Knörr and Robinson ([15], Lemma 3.2), that for any chain C in $\mathcal{E}(G)$ or $\mathcal{N}(G)$ or $\mathcal{R}(G)$ or $\mathcal{U}(G)$ and any p-block b of $N_G(C)$ the "induced" block b^G is defined. The (p-) defect of an irreducible character χ of a finite group G is $d(\chi) = \nu_p(|G|) - \nu_p(\chi(1))$, where for an integer n as usual $\nu_p(n) = m$ if p^m divides n but p^{m+1} doesn't. The defect $d(B)$ of a block B of G is the maximum of the defects of its characters. For any chain C in $\mathcal{E}(G)$ or $\mathcal{N}(G)$ or $\mathcal{R}(G)$ or $\mathcal{U}(G)$ as above we define $k(C, B, d)$ to be the number of irreducible characters χ of $N_G(C)$ with $d(\chi) = d$ belonging to (p-) blocks b of $N_G(C)$ with $b^G = B$. Then the "Ordinary Dade Conjecture" can be stated like this:

The ordinary Dade conjecture If $O_p(G) = \{1\}$ and $d(B) > 0$ then

$$\sum_{C \in \mathcal{R}(G)/G} (-1)^{|C|}\, k(C, B, d) = 0$$

for any d.

Although all the conjectures of Dade's are stated in terms of radical p-subgroups, it follows from a result of Knörr and Robinson [15], that one can replace $\mathcal{R}(G)$ in all of these by $\mathcal{E}(G)$ or $\mathcal{N}(G)$ or $\mathcal{U}(G)$. In fact it appears that for large groups $\mathcal{E}(G)/G$ is easier to compute than $\mathcal{R}(G)/G$ or $\mathcal{U}(G)/G$ although there are cases, in which $\mathcal{R}(G)/G$ or $\mathcal{U}(G)/G$ is significantly smaller than $\mathcal{E}(G)/G$.

For the Invariant Conjecture one has to embed G as a normal subgroup of an extension group A (for a simple group G it suffices to consider $A = Aut(G)$). This group also operates on the chains $\mathcal{E}(G)$ or $\mathcal{N}(G)$ or $\mathcal{R}(G)$ or $\mathcal{U}(G)$ and one has to compute their normalizers in A. For each such chain C the normalizer $N_G(C)$ is a normal subgroup in $N_A(C)$, the factor group being a subgroup of A/G (for a simple group this is $Out(G)$). Hence $N_A(C)$ operates on $Irr(N_G(C))$ and each irreducible character χ of $N_G(C)$ has an inertia group (i.e. stabilizer) in $N_A(C)$ and hence an inertia factor $N_{A/G}(\chi)$ in A/G. We define for each subgroup H of A/G and each chain C as above $k(C,B,d,H)$ to be the number of irreducible characters χ of $N_G(C)$ with $d(\chi) = d$ and $N_{A/G}(\chi) = H$ which belong to (p-) blocks b of $N_G(C)$ with $b^G = B$. One then has

The invariant Dade conjecture If $O_p(G) = \{1\}$ and $d(B) > 0$ then

$$\sum_{C \in \mathcal{R}(G)/G} (-1)^{|C|}\, k(C,B,d,H) = 0$$

for any d and any subgroup H of of A/G.

For the next conjecture, we also have to consider projective characters. For this we consider coverings of G i.e. exact sequences

$$1 \to Z \to G^* \to G \to 1$$

with Z a cyclic subgroup of the Schur multiplier of G embedded into the center $Z(G^*)$ and the commutator subgroup $(G^*)'$ of the finite group G^*. Furthermore for a faithful linear character ζ of Z and any subgroup U of G^* containing Z we let $\mathcal{B}l(U|\zeta)$ denote the set of all (p-) blocks of U which contain irreducible characters χ lying above ζ, which means that $\chi|Z = \chi(1)\cdot\zeta$. For any $B \in \mathcal{B}l(G^*|\zeta)$ and any integer d let $k(U,B,d,\zeta)$ denote the number of those irreducible characters χ lying above ζ with $d(\chi) = d$ which belong to blocks b of $\mathcal{B}l(U|\zeta)$ such that b^{G^*} is defined and equal to B and $k(C,B,d,\zeta) = k(N_{G^*}(C),B,d,\zeta)$. With this notation one has

The projective Dade conjecture: If $O_p(G) = \{1\}$ and $B \in \mathcal{B}l(G^*|\zeta)$ with defect $> \nu_p(|Z|)$ one has for any integer d

$$\sum_{C \in \mathcal{R}(G)/G} (-1)^{|C|}\, k(C,B,d,\zeta) = 0.$$

3 The McLaughlin group

Since the Schur multiplier of $G = McL$ has order 3 and the outer automorphism group has order 2 with the outer automorphism inverting the central elements of the non-trivial Schur-cover of McL the Inductive Conjecture of Dade's is equivalent to the Invariant Conjecture and the Projective Conjecture for G (see [8]).

We now turn to the verification of these conjectures for G. The McLaughlin group G has a faithful permutation representation of degree 275, so it is quite feasible to compute within the group using for example the GAP-system (see [21]). For instance given the generating permutations (for example from [22]) it is possible to compute the conjugacy classes of elements of McL and thus a set of representatives for the conjugacy classes of subgroups of order p. The relevant primes p here are $p = 2$ and $p = 3$ and $p = 5$, the Sylow subgroups for the other primes being cyclic. A set of representatives of the G-conjugacy classes of elementary abelian p-subgroups is then recursively constructed as follows: if the representatives of such

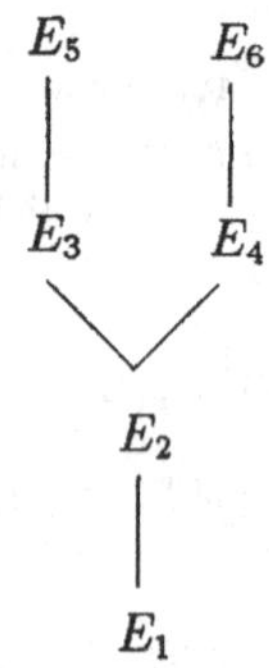

Figure 1

subgroups of order p^i are already known, one computes for each of these representatives P_i its centralizer $C_G(P_i)$, finds the conjugacy classes of subgroups of order p in $C_G(P_i)$ as above, and discards all those which are contained in P_i. The resulting subgroups of order p^{i+1} have to be tested for conjugacy within G, retaining just a set of representatives for these classes.

In fact the authors have written programs in GAP which compute the poset of conjugacy classes of elementary abelian p-subgroups of a permutation group and also all elementary and radical p-chains as well as chains of radical p-subgroups up to conjugacy together with their normalizers. These programs have been tested to work for groups of degree up a few thousands and of order up to 10^{12}. The programs computing radical p-subgroups and radical p-chains first compute the lattice of subgroups of a Sylow-p-subgroup of G using the "Lattice"-program in GAP, so there is some restriction on the size of the Sylow-p-subgroups, which should probably be not larger than, say 2^{10} or 3^7 for reasonable timings.

On the other hand, the table of marks of McL has been computed by G. Pfeiffer and is available in GAP, see [19], [16]. It is a trivial matter to determine from this table the poset of conjugacy classes of elementary p-subgroups for each prime p. Here one uses the usual partial order on the set of conjugacy classes $[U]$ of subgroups U of G, defining $[U] \leq [V]$ if and only if V contains a subgroup which is conjugate to U in G. The table of marks of McL in GAP contains also information about the conjugacy class of the normalizers for each subgroup of McL and also for each conjugacy class of subgroups of McL a system of generators for a representative. So also the poset of conjugacy classes of radical p-subgroups can be determined with a minimum of computations. We now record our results for the relevant primes 2, 3 and 5.

3.1 Case $p = 2$

Here it is much easier to work with elementary subgroups. We may choose representatives $E_1, ..., E_6$ of the conjugacy classes of non-trivial elementary 2-subgroups of G with inclusions as shown in Fifure 1. This can also be deduced (without further computation) from [11], Lemma 5.1 and Lemma 5.3.

This picture describes also the poset of conjugacy classes $[E_i]$ of non-trivial

Order	Name of group	Normalizer in McL	Order of normalizer
2	E_1	$2.A_8$	$40320 = 2^7.3^2.5.7$
4	E_2	$2^{2+4}.3^2.2$	$1152 = 2^7.3^2$
8	E_3	$2^{3+1}.L_2(7)_a$	$2688 = 2^7.3.7$
8	E_4	$2^{3+1}.L_2(7)_b$	$2688 = 2^7.3.7$
16	E_5	$(2^4 : A_7)_a$	$40320 = 2^7.3^2.5.7$
16	E_6	$(2^4 : A_7)_b$	$40320 = 2^7.3^2.5.7$

Table 1

elementary 2-subgroups of G – see Table 1; in particular E_5 (resp. E_6) contains no conjugate of E_4 (resp. E_3).

In Table 1 we use the notation of the ATLAS [5] and suffixes a and b to distinguish between groups which are conjugate in $\mathrm{Aut}(G)$ but not in G. The normalizers of the groups E_1, E_5 and E_6 are, in fact maximal subgroups of G and can also be found in the ATLAS. Looking at the orders of the normalizers, one immediately observes, that $N_G(E_5)$ (resp. $N_G(E_6)$) operates transitively on the 15 subgroups (hyperplanes) of E_5 (resp. E_6) of order 8 and hence by Brauer's Permutation Lemma (see [13], p. 536) also on the 15 subgroups (points) of order 2. Likewise $N_G(E_3)$ (resp. $N_G(E_4)$) operates transitively on the 7 subgroups (hyperplanes) of E_3 (resp. E_4) and hence also on the 7 subgroups (points) of order 2. Hence $N_G(E_5)$ (resp. $N_G(E_6)$) operates transitively on the 35 subgroups (lines) of E_5 (resp. E_6) of order 4. Finally $N_G(E_2)$ operates transitively on the 3 subgroups of E_2 of order 2, since it has trivial center.

The upshot of these observations is, that every normalizer $N_G(E_i)$ operates flag-transitively on E_i ($1 \leq i \leq 6$), so that we can conclude that every elementary chain of G is conjugate in G to one all of whose members belong to the set $\{\{1\}, E_1, E_2, E_3, E_4, E_5, E_6\}$. Hence we have a complete overview of $\mathcal{E}(G)/G$ and also we can also easily give the normalizers of the chains. Since the latter are what we need, we order the chains accordingly. Here we abbreviate a chain $\{1\} < E_1 < E_2 < E_6$ for example by $C_{1,2,6}$ and the trivial chain of length zero by C_0. But in fact, we don't have to compute all the normalizers of all the chains: since $N_G(E_3) \leq N_G(E_5)$ and $N_G(E_4) \leq N_G(E_6)$ (see also [11], Lemma 5.3) and also $N_{Aut(G)}(E_3) \leq N_{Aut(G)}(E_5)$ and $N_{Aut(G)}(E_4) \leq N_{Aut(G)}(E_6)$ it is clear that the normalizers of any chain ending in E_3 (resp. E_4) are the same as the normalizers of the corresponding chain with E_5 (resp. E_6) appended, and these pairs give rise to terms which cancel in the alternating sums occurring in Dade's conjectures. This means that we can omit from our consideration all chains containing E_3 or E_4. So there are only very few chains left to consider; they are shown in Table 2.

Observe that the last two normalizers are perfect groups and different from the groups $2^{3+1}.L_2(7)_{a/b}$ appearing in the first table.

It is an easy matter to compute the character tables of the N_i and of the groups $N_{Aut(McL)}(C)$ using GAP [21], in which the Dixon-Schneider algorithm (see [20], [12]) is implemented, and also the fusions of their conjugacy classes in G. The character tables of G, $Aut(G)$, N_1 and also $N_{Aut(McL)}(N_1)$ can be found in the ATLAS [5]. We print the tables of N_2, N_5, $N_{1,2}$, and $N_{1,5}$ in the appendix using

Chains C	$(-1)^{\|C\|}$	$N_{McL}(C)$	$N_{Aut(McL)}(C)$
C_1	$-$	$N_1 = 2.A_8$	$2.S_8$
C_2	$-$	$N_2 = 2^{2+4} : (3 \times 3) : 2$	$2^{2+4}.(S_3 \times S_3)$
$C_{2,5}, C_{2,6}$	$++$	N_2	N_2
C_5	$-$	$N_5 = (2^4 : A_7)_a$	N_5
C_6	$-$	$N_6 = (2^4 : A_7)_b$	N_6
$C_{1,2}$	$+$	$N_{1,2} = 2^{2+4}.S_3$	$2^{2+4}.(S_3 \times 2)$
$C_{1,2,5}, C_{1,2,6}$	$--$	$N_{1,2}$	$N_{1,2}$
$C_{1,5}$	$+$	$N_{1,5} = 2^{1+3}.L_2(7)_a$	$N_{1,5}$
$C_{1,6}$	$+$	$N_{1,6} = 2^{1+3}.L_2(7)_b$	$N_{1,6}$

Table 2

the same format as in [5]. The lower parts of these tables, which give the projective characters, are not relevant at the moment. Observe that $Out(G)$ has order 2 and the irreducible characters of G or N_i with inertia-factor $Out(G)$ are marked by a ":" at the end of the rows in the tables of the ATLAS [5] and also in the appendix whereas pairs of conjugate characters with trivial inertia factor $\{1\}$ are grouped together with a "|". Of course all irreducible characters of N_5, N_6, $N_{1,5}$, $N_{1,6}$ have trivial inertia factors and have no mark. Although the chains C_2 and $C_{2,5}, C_{2,6}$ have the same normalizer in G their normalizers in $Aut(G)$ are different and likewise for $C_{1,2}$ and $C_{1,2,5}, C_{1,2,6}$. This makes a difference in the alternating sums in the Invariant Conjecture, but no difference in the Ordinary and Projective Conjecture.

McL has just the principal 2-block B_0 except for blocks of defect 0 and one block B_1 of defect 1 (consisting of χ_{10} and χ_{11} in the notation of the ATLAS [5] with inertia factors $Out(G)$), which we wouldn't really need to consider in view of the fact that the conjectures are settled for blocks with cyclic defect groups as mentioned in the introduction. All the N_i have just one 2-block except for N_1, which has also a block b of defect 1 with $b^G = B_1$.

The results are summarized in Table 3, where for greater clarity zeros have been replaced by dots.

Here the chains which give no non-zero contribution to $k(C, B_0, d, Out(G))$ have been omitted. In both tables all columns add up to 0, which shows that the Invariant Conjecture and hence also the Ordinary Conjecture holds for the principal block. For B_1 one just has to observe that

$$k(C_0, B_1, 1, Out(G)) = k(C_1, B_1, 1, Out(G)) = 2.$$

As one can see the Invariant Conjecture holds true for McL and $p = 2$.

Having verified the Ordinary Conjecture, we have to consider for the Projective Conjecture proper projective characters, i.e. characters of $3.G$ lying above one of the non-trivial linear characters ζ of $Z(3.G)$. Thus we need also the character tables of the preimages N_i^* of the N_i in $3.McL$. Observe that we don't have a nice permutation representation for $3.McL$, so that calculating elementary 2-subgroups and their normalizers in $3.McL$ in order to apply again the Dixon Schneider algorithm would be cumbersome. But in fact it is possible to compute these character

	$d \le 2$	$d = 3$	$d = 4$	$d = 5$	$d = 6$	$d = 7$
$k(C_0, B_0, d, \{1\})$	.	.	.	.	4	4
$-k(C_1, B_0, d, \{1\})$	.	.	−6	.	.	−4
$-k(C_2, B_0, d, \{1\})$	.	.	.	.	−4	−4
$k(C_{2,5}, B_0, d, \{1\})$	.	.	1	2	6	8
$k(C_{2,6}, B_0, d, \{1\})$	.	.	1	2	6	8
$-k(C_5, B_0, d, \{1\})$	.	.	−1	.	−6	−8
$-k(C_6, B_0, d, \{1\})$	.	.	−1	.	−6	−8
$k(C_{1,2}, B_0, d, \{1\})$	.	.	2	.	.	4
$-k(C_{1,2,5}, B_0, d, \{1\})$	.	.	−4	−2	−2	−8
$-k(C_{1,2,6}, B_0, d, \{1\})$	.	.	−4	−2	−2	−8
$k(C_{1,5}, B_0, d, \{1\})$	.	.	6	.	2	8
$k(C_{1,6}, B_0, d, \{1\})$	.	.	6	.	2	8

	$d \le 2$	$d = 3$	$d = 4$	$d = 5$	$d = 6$	$d = 7$
$k(C_0, B_0, d, Out(G))$	.	1	1	2	2	4
$-k(C_1, B_0, d, Out(G))$	.	−1	−2	−2	−2	−4
$-k(C_2, B_0, d, Out(G))$	.	.	−1	−2	−2	−4
$k(C_{1,2}, B_0, d, Out(G))$	.	.	2	2	2	4

Table 3

	$d \le 2$	$d = 3$	$d = 4$	$d = 5$	$d = 6$	$d = 7$
$k(C_0, B_0(\zeta), d, \zeta)$	.	1	5	2	2	8
$-k(C_1, B_0(\zeta), d, \zeta)$	.	−1	−8	−2	−2	−8
$k(C_2, B_0(\zeta), d, \zeta)$	.	.	1	2	2	8
$-k(C_5, B_0(\zeta), d, \zeta)$	.	.	−3	.	−2	−8
$-k(C_6, B_0(\zeta), d, \zeta)$	.	.	−3	.	−2	−8
$-k(C_{1,2}, B_0(\zeta), d, \zeta)$	.	.	−4	−2	−2	−8
$k(C_{1,5}, B_0(\zeta), d, \zeta)$	.	.	6	.	2	8
$k(C_{1,6}, B_0(\zeta), d, \zeta)$	.	.	6	.	2	8

Table 4

tables together with their fusions into $3.McL$ just using the known character table of $3.McL$ and those of N_i using standard techniques, described in [18] and [16]. This has been done by T. Breuer (Aachen) and the character tables are now available in the GAP system. The projective characters of N_1 are in the ATLAS, those of N_2, N_5 are printed in ATLAS-format in the appendix. The preimages of $N_{1,2}$ and $N_{1,5}$ are direct products with the center of $3.McL$.

Let $B_0(\zeta)$ and $B_1(\zeta)$ denote the two blocks of $3.G$ lying above ζ with defect 7 and 1, respectively. We get the information in Table 4.

Here we have omitted the chains $C_{2,5}, C_{2,6}$ which have the same normalizer in G as C_2, and $C_{1,2,5}, C_{1,2,6}$ with the same normalizer as $C_{1,2}$ and have adjusted the signs for C_2 and $C_{1,2}$ instead.

For $B_1(\zeta)$ one sees that $k(C_0, B_1(\zeta), 1, \zeta) = k(C_1, B_1(\zeta), d, \zeta) = 2$ and the other normalizers give no non-zero contribution. Thus the Projective Conjecture holds

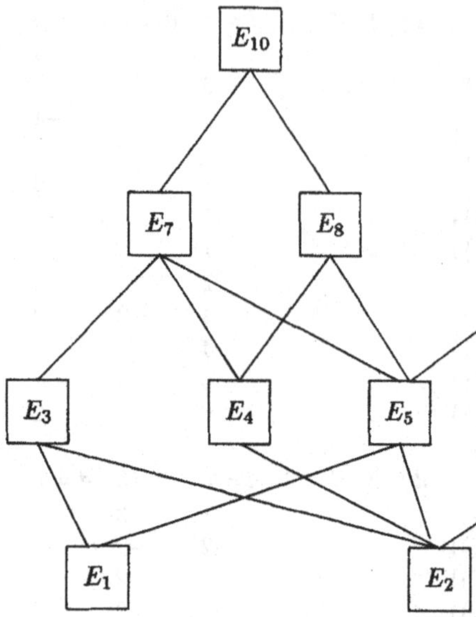

Figure 2

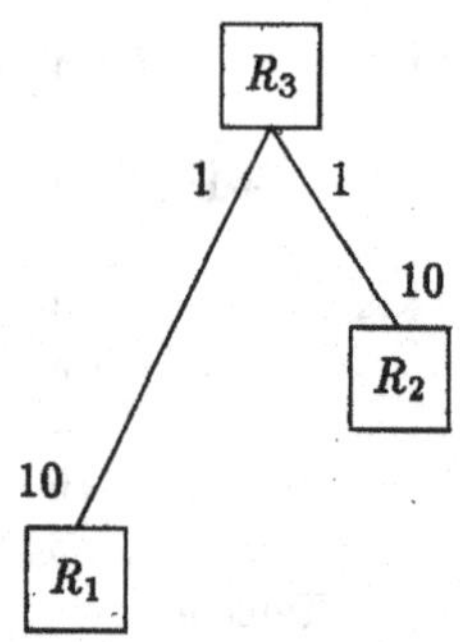

Figure 3

for McL and $p = 2$.

3.2 Case $p = 3$

For the prime $p = 3$ the situation is quite different. There are 10 conjugacy classes of non-trivial elementary abelian 3-subgroups of $G = McL$. The poset of these is shown in Figure 2, where the boxes stand for conjugacy classes.

In this case one could not derive immediately $\mathcal{E}(G)/G$ from this poset, in fact there are not 8 conjugacy classes of elementary chains of length 4 but 10 . On the other hand there are only 3 non-trivial radical 3-subgroups of G up to conjugacy and the poset of their conjugacy is shown in Figure 3.

Here R_1, R_2 and a Sylow-3-subgroup R_3 are radical 3-subgroups; see Table 5.

Order	Name	$N_{McL}(R_i)$	$\|N_G(R_i)\|$	$N_{Aut(McL)}(R_i)$
3^4	R_1	$M_1 = 3^4 : M_{10}$	58320	$3^4 : (M_{10} \times 2)$
3^5	R_2	$M_2 = 3_+^{1+4} : 2S_5$	58320	$3_+^{1+4} : 4S_5$
3^6	R_3	$M_3 = 3^4.3^2.Q_8$	5832	$3^4.3^2.Q_8.2$

Table 5

	$d \leq 2$	$d=3$	$d=4$	$d=5$	$d=6$
$k(C_0, B_0, d, \{1\})$	.	2	.	2	4
$-k(C_1, B_0, d, \{1\})$	.	.	.	-2	-4
$-k(C_2, B_0, d, \{1\})$	.	-2	-2	-2	-4
$k(C_{1,3}, B_0, d, \{1\})$	.	.	2	2	4

	$d \leq 2$	$d=3$	$d=4$	$d=5$	$d=6$
$k(C_0, B_0, d, Out(G))$	.	1	3	1	8
$-k(C_1, B_0, d, Out(G))$	.	.	-3	-1	-8
$-k(C_2, B_0, d, Out(G))$	.	-1	-4	-1	-8
$k(C_{1,3}, B_0, d, Out(G))$	.	.	4	1	8

Table 6

One sees from the table of marks that R_3 contains exactly one subgroup conjugate to R_1 and one conjugate to R_2 whereas R_1 and R_2 are contained in 10 conjugates of R_3 each as indicated in the above figure. From this it follows that $\mathcal{U}(G)/G$ contains just two chains of length 2 namely $C_{1,3} = (\{1\} \leq R_1 \leq R_3)$ and $C_{2,3} = (\{1\} \leq R_2 \leq R_3)$, with normalizer M_3 so that M_3 is the normalizer of two chains of length 2 and one of length 1 resulting in a positive contribution in the alternating sums to be considered . The character tables of the groups M_i are easily computed using GAP as in the case $p = 2$ and are listed in the appendix in ATLAS-format (together with the projective charcters which are needed below). From these one finds the results shown in Table 6.

This verifies the Invariant Conjecture (and hence also the Ordinary Conjecture) for the prime $p = 3$.

$3.McL$ has, apart form the principal 3-block B_0^*, only blocks of defect 1 (which are irrelevant for the Projective Conjecture). It turns out that the preimages M_i^* of the M_i are all non-split extensions of $Z(3.McL)$ and their character tables and fusions into $3.McL$ can be deduced from the character tables of the M_i. All these extensions have just one 3-block. The results are in the appendix. From this we get the information in Table 7.

Hence the Dade Conjectures are verified for McL and $p = 3$.

3.3 Case $p = 5$

Here it is again easier to work with radical p-subgroups. In fact there is only one non-trivial radical p-subgroup R of $G = McL$ up to conjugacy, as can be seen immediately form the table of marks. This is (of course) a Sylow-5-subgroup with

	$d \leq 3$	$d=4$	$d=5$	$d=6$	$d=7$
$k(C_0, B_0^*, d, \zeta)$	.	6	12	.	.
$-k(C_1, B_0^*, d, \zeta)$	.	-6	-15	.	.
$-k(C_2, B_0^*, d, \zeta)$	.	.	-12	.	.
$k(C_{1,3}, B_0^*, d, \zeta)$	.	.	15	.	.

Table 7

	$d=1$	$d=2$	$d=3$
$k(C_0, B_0, d, \{1\})$	.	4	4
$-k(C, B_0, d, \{1\})$	.	-4	-4

	$d=1$	$d=2$	$d=3$
$k(C_0, B_0, d, Out(G))$	.	2	9
$-k(C, B_0, d, Out(G))$	.	-2	-9

	$d=1$	$d=2$	$d=3$
$k(C_0, B_0(\zeta), d, \zeta)$	.	6	13
$-k(C, B_0(\zeta), d, \zeta)$	.	-6	-13

Table 8

Order	Name of group	Normalizer in McL	Normalizer in $Aut(McL)$
5^3	R	$5_+^{1+2}:3:8$	$5_+^{1+2}:3:8.2$

as can be read off from the ATLAS [5].

Except for blocks of defect zero G has only the principal 5-block B_0 and $3.McL$ has only one block $B_0(\zeta)$ lying over ζ which is not of defect zero. We have, with $C = (\{1\} \leq R)$, the information shown in Table 8.

This shows that the Dade Conjectures holds for McL and $p=5$.

Acknowledgement Thanks are due to Prof. E. C. Dade for pointing out an error in a previous version of this paper. Prof. Dade also informed us that the conjectures had been verified independently by John Murray. We wish also to thank Thomas Breuer for calculating several of the relevant character tables and also for his help in bringing these tables into ATLAS- format.

4 Appendix

$N_2 \cong 2^{2+4}{:}(3 \times 3){:}2 \le McL$

																			McL.2:								
McL:	1A	2A	2A	2A	4A	3B	3B	3B	3B	6B	6B	6B	2A	4A	4A	4A	8A		2B	2B	4B	6C	6C	8B	8C	8C	12C
		3																									
	1152	84	96	96	32	36	36	36	9	12	12	12	16	16	8	8	8		48	24	48	6	3	16	16	8	6
p power		A	A	A	A	A	A	A	A	AA	BC	CB	A	A	B	C	A		A	A	A	AE	CF	A	A	A	AE
p' part		A	A	A	A	A	A	A	A	AA	BC	CB	A	A	A	A	A		A	A	A	AE	CF	A	A	A	AE
ind	1A	2A	2B	2C	4A	3A	3B	3C	3D	6A	6B	6C	2D	4B	4C	4D	8A	fu	2E	2F	4E	6D	6E	8B	8C	8D	12A
+	1	1	1	1	1	1	1	1	1	1	1	1	1	1	1	1	1	:	1	1	1	1	1	1	1	1	1
+	1	1	1	1	1	1	1	1	1	1	1	1	-1	-1	-1	-1	-1	:	-1	1	-1	-1	1	1	1	-1	-1
+	2	2	2	2	2	-1	-1	-1	2	-1	-1	-1	0	0	0	0	0	:	2	0	2	-1	0	0	0	2	-1
+	2	2	2	2	2	2	-1	-1	-1	2	-1	-1	0	0	0	0	0	:	0	2	0	0	-1	2	2	0	0
+	2	2	2	2	2	-1	2	-1	-1	-1	2	-1	0	0	0	0	0	|	0	0	0	0	0	0	0	0	0
+	2	2	2	2	2	-1	-1	2	-1	-1	-1	2	0	0	0	0	0	|									
+	3	3	-1	3	-1	0	0	3	0	0	0	-1	-1	-1	1	-1	1	|	0	0	0	0	0	0	0	0	0
+	3	3	3	-1	-1	0	3	0	0	0	-1	0	-1	-1	-1	1	1	|									
+	3	3	-1	3	-1	0	0	3	0	0	0	-1	1	1	-1	1	-1	|	0	0	0	0	0	0	0	0	0
+	3	3	3	-1	-1	0	3	0	0	0	-1	0	1	1	1	-1	-1	|									
+	6	6	-2	6	-2	0	0	-3	0	0	0	1	0	0	0	0	0	|	0	0	0	0	0	0	0	0	0
+	6	6	6	-2	-2	0	-3	0	0	0	1	0	0	0	0	0	0	|									
+	9	9	-3	-3	1	0	0	0	0	0	0	0	-1	-1	1	1	-1	:	3	-3	3	0	0	1	1	-1	0
+	9	9	-3	-3	1	0	0	0	0	0	0	0	1	1	-1	-1	1	:	3	3	3	0	0	-1	-1	-1	0
+	12	-4	0	0	0	3	0	0	0	-1	0	0	2	-2	0	0	0	:	2	0	-2	-1	0	2	-2	0	1
+	12	-4	0	0	0	3	0	0	0	-1	0	0	-2	2	0	0	0	:	2	0	-2	-1	0	-2	2	0	1
+	24	-8	0	0	0	-3	0	0	0	1	0	0	0	0	0	0	0	:	4	0	-4	1	0	0	0	0	-1
ind	1	2	2	2	4	3	3	3	3	6	6	6	2	4	4	4	8	fus	2	2	4	6	6	8	8	8	12
	3	6	6	6	12					6	6	6	6	12	12	12	24										
	3	6	6	6	12					6	6	6	6	12	12	12	24										
o2	3	3	3	3	3	0	0	0	0	0	0	0	1	1	1	1	1	*	+								
o2	3	3	3	3	3	0	0	0	0	0	0	0	-1	-1	-1	-1	-1	*	+								
o2	3	3	-1	3	-1	0	0	0	0	0	0	2	-1	-1	1	-1	1	|	+								
o2	3	3	3	-1	-1	0	0	0	0	0	2	0	-1	-1	-1	1	1	*									
o2	3	3	-1	3	-1	0	0	0	0	0	0	2	1	1	-1	1	-1	|	+								
o2	3	3	3	-1	-1	0	0	0	0	0	2	0	1	1	1	-1	-1	*									
o2	6	6	-2	6	-2	0	0	0	0	0	0	-2	0	0	0	0	0	|	+								
o2	6	6	6	-2	-2	0	0	0	0	0	-2	0	0	0	0	0	0	*									
o2	9	9	-3	-3	1	0	0	0	0	0	0	0	1	1	-1	-1	1	*	+								
o2	9	9	-3	-3	1	0	0	0	0	0	0	0	-1	-1	1	1	-1	*	+								
o2	12	-4	0	0	0	0	0	0	0	2	0	0	-2	2	0	0	0	*	+								
o2	12	-4	0	0	0	0	0	0	0	2	0	0	2	-2	0	0	0	*	+								
o2	24	-8	0	0	0	0	0	0	0	-2	0	0	0	0	0	0	0	*	+								

$N_2 \cong 2^{2+4}{:}(3 \times 3){:}2$

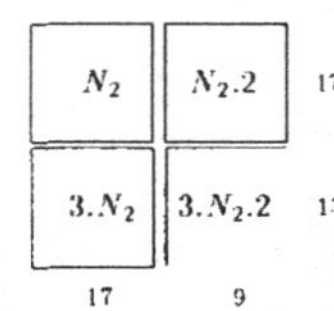

$N_5, N_6 \cong 2^4{:}A_7$

McL:	1A	2A	2A	4A	3B	3B	6B	4A	8A	5B	6B	7A	14A	7B	14B
;	@	@	@	@	@	@	@	@	@	@	@	@	@	@	@
		2													
	40320	688	96	32	36	36	12	8	8	5	12	14	14	14	14
p power		A	A	A	A	A	BA	B	A	A	AB	A	AA	A	BA
p' part		A	A	A	A	A	BA	A	A	A	AB	A	AA	A	BA
ind	1A	2A	2B	4A	3A	3B	6A	4B	8A	5A	6B	7A	14A	7B	14B
+	1	1	1	1	1	1	1	1	1	1	1	1	1	1	1
+	6	6	2	2	3	0	0	0	0	1	-1	-1	-1	-1	-1
o	10	10	-2	-2	1	1	1	0	0	0	1	b7	b7	**	**
o	10	10	-2	-2	1	1	1	0	0	0	1	**	**	b7	b7
+	14	14	2	2	2	-1	-1	0	0	-1	2	0	0	0	0
+	14	14	2	2	-1	2	2	0	0	-1	-1	0	0	0	0
+	15	15	-1	-1	3	0	0	-1	-1	0	-1	1	1	1	1
+	21	21	1	1	-3	0	0	-1	-1	1	1	0	0	0	0
+	35	35	-1	-1	-1	-1	-1	1	1	0	-1	0	0	0	0
+	15	-1	3	-1	0	3	-1	1	-1	0	0	1	-1	1	-1
o	45	-3	-3	1	0	0	0	1	-1	0	0	b7	-b7	**	**
o	45	-3	-3	1	0	0	0	1	-1	0	0	**	**	b7	-b7
+	90	-6	6	-2	0	0	0	0	0	0	0	-1	1	-1	1
+	105	-7	-3	1	0	3	-1	-1	1	0	0	0	0	0	0
+	120	-8	0	0	0	-3	1	0	0	0	0	1	-1	1	-1
ind	1	2	2	4	3	3	6	4	8	5	6	7	14	7	14
	3	6	6	12			6	12	24	15	6	21	42	21	42
	3	6	6	12			6	12	24	15	6	21	42	21	42
o2	6	6	2	2	0	0	0	0	0	1	2	-1	-1	-1	-1
o2	15	15	-1	-1	0	0	0	-1	-1	0	2	1	1	1	1
o2	15	15	3	3	0	0	0	1	1	0	0	1	1	1	1
o2	21	21	1	1	0	0	0	-1	-1	1	-2	0	0	0	0
o2	21	21	-3	-3	0	0	0	1	1	1	0	0	0	0	0
o2	24	24	0	0	0	0	0	0	0	-1	0	b7	b7	**	**
o2	24	24	0	0	0	0	0	0	0	-1	0	**	**	b7	b7
o2	15	-1	3	-1	0	0	2	1	-1	0	0	1	-1	1	-1
o2	45	-3	-3	1	0	0	0	1	-1	0	0	b7	-b7	**	**
o2	45	-3	-3	1	0	0	0	1	-1	0	0	**	**	b7	-b7
o2	90	-6	6	-2	0	0	0	0	0	0	0	-1	1	-1	1
o2	105	-7	-3	1	0	0	2	-1	1	0	0	0	0	0	0
o2	120	-8	0	0	0	0	-2	0	0	0	0	1	-1	1	-1

$3^4{:}M_{10}$

McL:	1A	3A	3B	2A	6B	6A	3B	3B	3B	9A	9B	4A	5B	4A	12A	12A	8A	8A
;	@	@	@	@	@	@	@	@	@	@	@	@	@	@	@	@	@	@
		2																
	58320	916	972	144	36	36	81	81	81	27	27	8	5	12	12	12	8	8
p power		A	A	A	BA	AA	A	A	A	A	A	A	A	A	BB	BB	A	A
p' part		A	A	A	BA	AA	A	A	A	A	A	A	A	A	AB	AB	A	A
ind	1A	3A	3B	2A	6A	6B	3C	3D	3E	9A	B**	4A	5A	4B	12A	B*	8A	B**
+	1	1	1	1	1	1	1	1	1	1	1	1	1	1	1	1	1	1
+	1	1	1	1	1	1	1	1	1	1	1	1	1	-1	-1	-1	-1	-1
+	9	9	9	1	1	1	0	0	0	0	0	1	-1	1	1	1	-1	-1
+	9	9	9	1	1	1	0	0	0	0	0	1	-1	-1	-1	-1	1	1
+	10	10	10	2	2	2	1	1	1	1	1	-2	0	0	0	0	0	0
o	10	10	10	-2	-2	-2	1	1	1	1	1	0	0	0	0	0	i2	-i2
o	10	10	10	-2	-2	-2	1	1	1	1	1	0	0	0	0	0	-i2	i2
+	16	16	16	0	0	0	-2	-2	-2	-2	-2	0	1	0	0	0	0	0
+	20	-7	2	4	-2	1	2	2	2	-1	-1	0	0	2	-1	-1	0	0
+	20	-7	2	4	-2	1	2	2	2	-1	-1	0	0	-2	1	1	0	0
-	20	-7	2	-4	2	-1	2	2	2	-1	-1	0	0	0	r3	-r3	0	0
-	20	-7	2	-4	2	-1	2	2	2	-1	-1	0	0	0	-r3	r3	0	0
+	60	6	-3	4	1	-2	6	-3	-3	0	0	0	0	0	0	0	0	0
+	60	6	-3	4	1	-2	-3	6	-3	0	0	0	0	0	0	0	0	0
+	60	6	-3	4	1	-2	-3	-3	6	0	0	0	0	0	0	0	0	0
o	80	-28	8	0	0	0	-1	-1	-1	**	-b27	0	0	0	0	0	0	0
o	80	-28	8	0	0	0	-1	-1	-1	-b27	**	0	0	0	0	0	0	0
+	180	18	-9	-4	-1	2	0	0	0	0	0	0	0	0	0	0	0	0
ind	1	3	3	2	6	6	3	3	3	9	9	4	5	4	12	12	8	8
	3	3		6	6	6						12	15	12	12	12	24	24
	3	3		6	6	6						12	15	12	12	12	24	24
o2	36	9	0	4	-2	1	0	0	0	0	0	0	1	2	-1	-1	0	0
o2	36	9	0	4	-2	1	0	0	0	0	0	0	1	-2	1	1	0	0
o2	36	9	0	-4	2	-1	0	0	0	0	0	0	1	0	r3	-r3	0	0
o2	36	9	0	-4	2	-1	0	0	0	0	0	0	1	0	-r3	r3	0	0
o2	45	-9	0	-3	0	3	0	0	0	0	0	1	0	1	1	1	-1	-1
o2	45	-9	0	-3	0	3	0	0	0	0	0	1	0	-1	-1	-1	1	1
o2	45	-9	0	5	2	-1	0	0	0	0	0	1	0	1	1	1	1	1
o2	45	-9	0	5	2	-1	0	0	0	0	0	1	0	-1	-1	-1	-1	-1
o2	90	-18	0	2	2	2	0	0	0	0	0	-2	0	0	0	0	0	0
o2	90	-18	0	-2	-2	-2	0	0	0	0	0	0	0	0	0	0	-i2	i2
o2	90	-18	0	-2	-2	-2	0	0	0	0	0	0	0	0	0	0	i2	-i2
o2	144	36	0	0	0	0	0	0	0	0	0	0	-1	0	0	0	0	0

$N_{1,2} \cong 2^{2+4}.S_3 \leq McL$

																		McL.2:									
McL:	1A	2A	2A	4A	2A	2A	4A	3B	6B	6B	6B	2A	4A	4A	4A	8A		2B	2B	4B	6C	8B	8C	8B	8C	8C	12C
		3	1																								
	384	84	92	32	32	32	16	12	12	12	12	16	16	8	8	8		48	8	48	6	16	16	8	8	8	6
p power		A	A	A	A	A	B	A	AA	AB	AB	A	A	D	C	A		A	A	A	AF	A	A	B	B	A	AF
p' part		A	A	A	A	A	A	A	AA	AB	AB	A	A	A	A	A		A	A	A	AF	A	A	A	A	A	AF
ind	1A	2A	2B	4A	2C	2D	4B	3A	6A	6B	C**	2E	4C	4D	4E	8A	fus	2F	2G	4F	6D	8B	8C	8D	8E	8F	12A
+	1	1	1	1	1	1	1	1	1	1	1	1	1	1	1	1	:	1	1	1	1	1	1	1	1	1	1
+	1	1	1	1	1	1	1	1	1	1	1	-1	-1	-1	-1	-1	:	1	-1	1	1	-1	-1	-1	-1	1	1
+	2	2	2	2	2	2	2	-1	-1	-1	-1	0	0	0	0	0	:	2	0	2	-1	0	0	0	0	2	-1
+	3	3	3	3	-1	-1	-1	0	0	0	0	-1	-1	1	1	-1	:	3	-1	3	0	-1	-1	1	1	-1	0
+	3	3	3	3	-1	-1	-1	0	0	0	0	1	1	-1	-1	1	:	3	1	3	0	1	1	-1	-1	-1	0
+	3	3	3	-1	-1	3	-1	0	0	0	0	-1	-1	-1	1	1	|	0	0	0	0	0	0	0	0	0	0
+	3	3	3	-1	3	-1	-1	0	0	0	0	-1	-1	1	-1	1	|										
+	3	3	3	-1	-1	3	-1	0	0	0	0	1	1	1	-1	-1	|	0	0	0	0	0	0	0	0	0	0
+	3	3	3	-1	3	-1	-1	0	0	0	0	1	1	-1	1	-1	|										
+	6	6	6	-2	-2	-2	2	0	0	0	0	0	0	0	0	0	:	0	2	0	0	-2	-2	0	0	0	0
+	4	4	-4	0	0	0	0	1	1	-1	-1	-2	2	0	0	0	:	2	0	-2	-1	-2	2	0	0	0	1
+	4	4	-4	0	0	0	0	1	1	-1	-1	2	-2	0	0	0	:	2	0	-2	-1	2	-2	0	0	0	1
+	8	8	-8	0	0	0	0	-1	-1	1	1	0	0	0	0	0	:	4	0	-4	1	0	0	0	0	0	-1
+	8	-8	0	0	0	0	0	2	-2	0	0	0	0	0	0	0	:	0	0	0	0	0	0	2	-2	0	0
o	8	-8	0	0	0	0	0	-1	1	i3	-i3	0	0	0	0	0	|	0	0	0	0	0	0	0	0	0	0
o	8	-8	0	0	0	0	0	-1	1	-i3	i3	0	0	0	0	0	|										

$N_{1,5}, N_{1,6} \cong 2^{1+3}.L_2(7) \leq McL$

McL:	1A	2A	2A	2A	4A	4A	3B	6B	6B	6B	4A	8A	7A	7B	14A	14B
		2														
	2688	688	192	32	32	16	12	12	12	12	8	8	14	14	14	14
p power		A	A	A	A	B	A	AA	AB	AB	C	A	A	A	AA	BA
p' part		A	A	A	A	A	A	AA	AB	AB	A	A	A	A	AA	BA
ind	1A	2A	2B	2C	4A	4B	3A	6A	6B	C**	4C	8A	7A	B*	14A	B*
+	1	1	1	1	1	1	1	1	1	1	1	1	1	1	1	1
o	3	3	3	-1	-1	-1	0	0	0	0	1	1	b7	**	b7	**
o	3	3	3	-1	-1	-1	0	0	0	0	1	1	**	b7	**	b7
+	6	6	6	2	2	2	0	0	0	0	0	0	-1	-1	-1	-1
+	7	7	7	-1	-1	-1	1	1	1	1	-1	-1	0	0	0	0
+	7	7	-1	-1	3	-1	1	1	-1	-1	-1	1	0	0	0	0
+	7	7	-1	3	-1	-1	1	1	-1	-1	1	-1	0	0	0	0
+	8	8	8	0	0	0	-1	-1	-1	-1	0	0	1	1	1	1
+	14	14	-2	2	2	-2	-1	-1	1	1	0	0	0	0	0	0
+	21	21	-3	-3	1	1	0	0	0	0	1	-1	0	0	0	0
+	21	21	-3	1	-3	1	0	0	0	0	-1	1	0	0	0	0
+	8	-8	0	0	0	0	2	-2	0	0	0	0	1	1	-1	-1
o	8	-8	0	0	0	0	-1	1	i3	-i3	0	0	1	1	-1	-1
o	8	-8	0	0	0	0	-1	1	-i3	i3	0	0	1	1	-1	-1
o	24	-24	0	0	0	0	0	0	0	0	0	0	**	b7	**	-b7
o	24	-24	0	0	0	0	0	0	0	0	0	0	b7	**	-b7	**

$3^4.3^2.Q_8$

McL:	1A	3A	3B	3A	3B	3B	3B	3B	9B	9A	6A	2A	6A	6B	4A	4A	12A	12A	12A	12A	4A
;	@	@	@	@	@	@	@	@	@	@	@	@	@	@	@	@	@	@	@	@	@
		2																			
	5832	916	243	324	162	81	81	81	27	27	36	72	36	18	4	12	12	12	12	12	12
p power		A	A	A	A	A	A	A	A	A	AA	A	CA	DA	A	A	BB	BB	AC	AC	A
p' part		A	A	A	A	A	A	A	A	A	AA	A	CA	DA	A	A	CB	CB	AC	AC	A
ind	1A	3A	3B	3C	3D	3E	3F	3G	9A	B**	6A	2A	6B	6C	4A	4B	12A	B*	12C	D*	4C
+	1	1	1	1	1	1	1	1	1	1	1	1	1	1	1	1	1	1	1	1	1
+	1	1	1	1	1	1	1	1	1	1	1	1	1	1	1	-1	-1	-1	-1	-1	-1
+	1	1	1	1	1	1	1	1	1	1	1	1	1	1	-1	1	1	1	-1	-1	-1
+	1	1	1	1	1	1	1	1	1	1	1	1	1	1	-1	-1	-1	-1	1	1	1
-	2	2	2	2	2	2	2	2	2	2	-2	-2	-2	-2	0	0	0	0	0	0	0
+	8	8	8	8	8	-1	-1	-1	-1	-1	0	0	0	0	0	0	0	0	0	0	0
+	2	2	2	-1	-1	2	2	2	-1	-1	2	2	-1	-1	0	2	-1	-1	0	0	0
+	2	2	2	-1	-1	2	2	2	-1	-1	2	2	-1	-1	0	-2	1	1	0	0	0
-	2	2	2	-1	-1	2	2	2	-1	-1	-2	-2	1	1	0	0	-r3	r3	0	0	0
-	2	2	2	-1	-1	2	2	2	-1	-1	-2	-2	1	1	0	0	r3	-r3	0	0	0
o	8	8	8	-4	-4	-1	-1	-1	-b27	**	0	0	0	0	0	0	0	0	0	0	0
o	8	8	8	-4	-4	-1	-1	-1	**	-b27	0	0	0	0	0	0	0	0	0	0	0
+	24	24	-3	0	0	6	-3	-3	0	0	0	0	0	0	0	0	0	0	0	0	0
+	24	24	-3	0	0	-3	6	-3	0	0	0	0	0	0	0	0	0	0	0	0	0
+	24	24	-3	0	0	-3	-3	6	0	0	0	0	0	0	0	0	0	0	0	0	0
+	18	-9	0	-6	3	0	0	0	0	0	-1	2	2	-1	0	0	0	0	-1	-1	2
+	18	-9	0	-6	3	0	0	0	0	0	-1	2	2	-1	0	0	0	0	1	1	-2
-	18	-9	0	-6	3	0	0	0	0	0	1	-2	-2	1	0	0	0	0	r3	-r3	0
-	18	-9	0	-6	3	0	0	0	0	0	1	-2	-2	1	0	0	0	0	-r3	r3	0
+	36	-18	0	6	-3	0	0	0	0	0	-2	4	-2	1	0	0	0	0	0	0	0
-	36	-18	0	6	-3	0	0	0	0	0	2	-4	2	-1	0	0	0	0	0	0	0
ind	1	3	3	3	3	3	3	3	9	9	6	2	6	6	4	4	12	12	12	12	4
	3	3		3	3						6	6	6	6	12	12	12	12	12	12	12
	3	3		3	3						6	6	6	6	12	12	12	12	12	12	12
o2	9	9	0	-3	-3	0	0	0	0	0	1	1	1	1	1	1	1	1	1	1	1
o2	9	9	0	-3	-3	0	0	0	0	0	1	1	1	1	1	-1	-1	-1	-1	-1	-1
o2	9	9	0	-3	-3	0	0	0	0	0	1	1	1	1	-1	1	1	1	-1	-1	-1
o2	9	9	0	-3	-3	0	0	0	0	0	1	1	1	1	-1	-1	-1	-1	1	1	1
o2	18	-9	0	6	-3	0	0	0	0	0	-1	2	2	-1	0	0	0	0	1	1	-2
o2	18	-9	0	6	-3	0	0	0	0	0	-1	2	2	-1	0	0	0	0	-1	-1	2
o2	18	-9	0	6	-3	0	0	0	0	0	1	-2	-2	1	0	0	0	0	-r3	r3	0
o2	18	-9	0	6	-3	0	0	0	0	0	1	-2	-2	1	0	0	0	0	r3	-r3	0
o2	18	18	0	-6	-6	0	0	0	0	0	-2	-2	-2	-2	0	0	0	0	0	0	0
o2	18	18	0	3	3	0	0	0	0	0	-2	-2	1	1	0	0	-r3	r3	0	0	0
o2	18	18	0	3	3	0	0	0	0	0	-2	-2	1	1	0	0	r3	-r3	0	0	0
o2	18	18	0	3	3	0	0	0	0	0	2	2	-1	-1	0	2	-1	-1	0	0	0
o2	18	18	0	3	3	0	0	0	0	0	2	2	-1	-1	0	-2	1	1	0	0	0
o2	36	-18	0	-6	3	0	0	0	0	0	-2	4	-2	1	0	0	0	0	0	0	0
o2	36	-18	0	-6	3	0	0	0	0	0	2	-4	2	-1	0	0	0	0	0	0	0

$3_+^{1+4}{:}2S_5$

McL:	1A	3A	3B	2A	6A	4A	12A	3A	3B	9B	9A	6B	6A	5A	15A	15B	10A	30A	30B	4A	8A	8A	12A	12A
;	@	@	@	@	@	@	@	@	@	@	@	@	@	@	@	@	@	@	@	@	@	@	@	@
		29	2	7	3			3	1															
	58320	160	43	20	60	24	12	24	62	27	27	18	36	30	30	30	30	30	30	12	8	8	12	12
p power		A	A	A	AA	A	AA	A	A	A	A	DA	CA	A	AA	AA	AA	AAA	BAA	A	A	A	CB	CB
p' part		A	A	A	AA	A	AA	A	A	A	A	DA	CA	A	AA	AA	AA	AAA	BAA	A	A	A	CB	CB
ind	1A	3A	3B	2A	6A	4A	12A	3C	3D	9A	B**	6B	6C	5A	15A	B**	10A	30A	B**	4B	8A	B*	12B	C*
+	1	1	1	1	1	1	1	1	1	1	1	1	1	1	1	1	1	1	1	1	1	1	1	1
+	1	1	1	1	1	1	1	1	1	1	1	1	1	1	1	1	1	1	1	-1	-1	-1	-1	-1
+	6	6	6	6	6	-2	-2	0	0	0	0	0	0	1	1	1	1	1	1	0	0	0	0	0
+	4	4	4	4	4	0	0	1	1	1	1	1	1	-1	-1	-1	-1	-1	-1	2	0	0	-1	-1
+	4	4	4	4	4	0	0	1	1	1	1	1	1	-1	-1	-1	-1	-1	-1	-2	0	0	1	1
+	5	5	5	5	5	1	1	-1	-1	-1	-1	-1	-1	0	0	0	0	0	0	1	-1	-1	1	1
+	5	5	5	5	5	1	1	-1	-1	-1	-1	-1	-1	0	0	0	0	0	0	-1	1	1	-1	-1
-	4	4	4	-4	-4	0	0	-2	-2	-2	-2	2	2	-1	-1	-1	1	1	1	0	0	0	0	0
-	4	4	4	-4	-4	0	0	1	1	1	1	-1	-1	-1	-1	-1	1	1	1	0	0	0	r3	-r3
-	4	4	4	-4	-4	0	0	1	1	1	1	-1	-1	-1	-1	-1	1	1	1	0	0	0	-r3	r3
-	6	6	6	-6	-6	0	0	0	0	0	0	0	0	1	1	1	-1	-1	-1	0	r2	-r2	0	0
-	6	6	6	-6	-6	0	0	0	0	0	0	0	0	1	1	1	-1	-1	-1	0	-r2	r2	0	0
+	80	80	-1	0	0	0	0	8	8	-1	-1	0	0	0	0	0	0	0	0	0	0	0	0	0
o	80	80	-1	0	0	0	0	-4	-4	-b27	**	0	0	0	0	0	0	0	0	0	0	0	0	0
o	80	80	-1	0	0	0	0	-4	-4	**	-b27	0	0	0	0	0	0	0	0	0	0	0	0	0
+	18	-9	0	2	-1	2	-1	-6	3	0	0	-1	2	-2	1	1	2	-1	-1	0	0	0	0	0
o	36	-18	0	-4	2	0	0	6	-3	0	0	-1	2	1	b15	**	1	b15	**	0	0	0	0	0
o	36	-18	0	-4	2	0	0	6	-3	0	0	-1	2	1	**	b15	1	**	b15	0	0	0	0	0
o	54	-27	0	6	-3	-2	1	0	0	0	0	0	0	-1	-b15	**	1	b15	**	0	0	0	0	0
o	54	-27	0	6	-3	-2	1	0	0	0	0	0	0	-1	**	-b15	1	**	b15	0	0	0	0	0
+	72	-36	0	8	-4	0	0	-6	3	0	0	-1	2	2	-1	-1	-2	1	1	0	0	0	0	0
-	72	-36	0	-8	4	0	0	-6	3	0	0	1	-2	2	-1	-1	2	-1	-1	0	0	0	0	0
+	90	-45	0	10	-5	2	-1	6	-3	0	0	1	-2	0	0	0	0	0	0	0	0	0	0	0
-	108	-54	0	-12	6	0	0	0	0	0	0	0	0	-2	1	1	-2	1	1	0	0	0	0	0
ind	1	3	3	2	6	4	12	3	3	9	9	6	6	5	15	15	10	30	30	4	8	8	12	12
	3	3		6	6	12	12	3	3			6	6	15	15	15	30	30	30	12	24	24	12	12
	3	3		6	6	12	12	3	3			6	6	15	15	15	30	30	30	12	24	24	12	12
o2	9	9	0	1	1	1	1	-3	-3	0	0	1	1	-1	-1	-1	1	1	1	1	-1	-1	1	1
o2	9	9	0	1	1	1	1	-3	-3	0	0	1	1	-1	-1	-1	1	1	1	-1	1	1	-1	-1
o2	18	-9	0	2	-1	2	-1	6	-3	0	0	-1	2	-2	1	1	2	-1	-1	0	0	0	0	0
o2	36	36	0	4	4	0	0	-3	-3	0	0	1	1	1	1	1	-1	-1	-1	-2	0	0	1	1
o2	36	36	0	4	4	0	0	-3	-3	0	0	1	1	1	1	1	-1	-1	-1	2	0	0	-1	-1
o2	36	36	0	-4	-4	0	0	-3	-3	0	0	-1	-1	1	1	1	1	1	1	0	0	0	r3	-r3
o2	36	36	0	-4	-4	0	0	-3	-3	0	0	-1	-1	1	1	1	1	1	1	0	0	0	-r3	r3
o2	36	36	0	-4	-4	0	0	6	6	0	0	2	2	1	1	1	1	1	1	0	0	0	0	0
o2	36	-18	0	-4	2	0	0	-6	3	0	0	-1	2	1	b15	**	1	b15	**	0	0	0	0	0
o2	36	-18	0	-4	2	0	0	-6	3	0	0	-1	2	1	**	b15	1	**	b15	0	0	0	0	0
o2	45	45	0	5	5	1	1	3	3	0	0	-1	-1	0	0	0	0	0	0	-1	-1	-1	-1	-1
o2	45	45	0	5	5	1	1	3	3	0	0	-1	-1	0	0	0	0	0	0	1	1	1	1	1
o2	54	-27	0	6	-3	-2	1	0	0	0	0	0	0	-1	-b15	**	1	b15	**	0	0	0	0	0
o2	54	-27	0	6	-3	-2	1	0	0	0	0	0	0	-1	**	-b15	1	**	b15	0	0	0	0	0
o2	54	54	0	6	6	-2	-2	0	0	0	0	0	0	-1	-1	-1	1	1	1	0	0	0	0	0
o2	54	54	0	-6	-6	0	0	0	0	0	0	0	0	-1	-1	-1	-1	-1	-1	0	r2	-r2	0	0
o2	54	54	0	-6	-6	0	0	0	0	0	0	0	0	-1	-1	-1	-1	-1	-1	0	-r2	r2	0	0
o2	72	-36	0	8	-4	0	0	6	-3	0	0	-1	2	2	-1	-1	-2	1	1	0	0	0	0	0
o2	72	-36	0	-8	4	0	0	6	-3	0	0	1	-2	2	-1	-1	2	-1	-1	0	0	0	0	0
o2	90	-45	0	10	-5	2	-1	-6	3	0	0	1	-2	0	0	0	0	0	0	0	0	0	0	0
o2	108	-54	0	-12	6	0	0	0	0	0	0	0	0	-2	1	1	-2	1	1	0	0	0	0	0

References

[1] J. L. Alperin, *Weights for finite groups*, in "The Arcata Conference on Representations of Finite Groups", Proc. Sympos. Pure Math., Vol. 47, Amer. Math. Soc., Providence, RI, pp. 369–379, 1987.

[2] J. An, *The Alperin and Dade conjectures for the simple Held Group*, J. Algebra **189** (1997), 34–57.

[3] J. An, M. Conder, *The Alperin and Dade conjectures for the simple Mathieu groups*, Comm. Algebra, **23** (1995), 2797–2823.

[4] R. Brauer, *Representations of finite groups*, Lect. on Modern Math., Vol. **1** , 133–175, Wiley, New York 1963.

[5] J. Conway, R. Curtis, S. Norton, R. Parker, R. Wilson, *An Atlas of Finite Groups*. Oxford: Clarendon Press, 1987.

[6] E. C. Dade, *Counting characters in blocks I*, Invent. Math. **109** (1992), 187–272.

[7] E. C. Dade, *Counting characters in blocks II*, J. Reine Angew. Math. **448** (1994), 97–190.

[8] E. C. Dade, *Counting characters in blocks 2.9*, in "Representation Theory of Finite Groups" ed. R. Solomon, pp. 45–59, 1997.

[9] E. C. Dade, *Counting characters in blocks with cyclic defect group I*, J. Algebra **186** (1996), 934–969.

[10] G. Entz, *Dades Vermutung für gewöhnliche und projektive Charaktere*, Diplomarbeit Aachen, 1997, unpublished.

[11] L. Finkelstein, *The maximal subgroups of Conway's Group C_3 and McLaughlin's Group*, J. Algebra **25** (1973), 58–89.

[12] A. Hulpke, *Zur Berechnung von Charaktertafeln*, Diplomarbeit Aachen 1993, unpublished.

[13] B. Huppert, *Endliche Gruppen I*, Berlin, Heidelberg, New York 1967.

[14] S. Kotlica, *Verification of Dade's conjecture for Janko group J_3*, J. Algebra **187** (1997), 579–619.

[15] R. Knörr, G. Robinson, *Some remarks on a conjecture of Alperin*, J. London Math. Soc. (2) **39** (1989), 48–60.

[16] K. Lux, H. Pahlings, *Computational aspects of representation theory of finite groups* in Representation Theory of Finite Groups and Finite-Dimensional Algebras, ed. G. O. Michler, C. M. Ringel, pp. 37–64, 1991.

[17] J. McKay, *Irreducible representations of odd degree*, J. Algebra **20** (1972), 416–418.

[18] J. Neubüser, H. Pahlings, W. Plesken, *CAS; design and use of a system for the handling of characters of finite groups*, in: *Computational Group Theory* (ed. M. D. Atkinson) pp. 195–247, London: Academic Press 1984.

[19] G. Pfeiffer, *The subgroups of M_{24} or how to compute a table of marks*, Experimental Math. **6** (1997), 247–270.

[20] G.J. Schneider, *Dixon's character table algorithm revisited*, J. Symbolic Comput. **9** (1990), 601–606.

[21] M. Schönert et al., *GAP - Groups, Algorithms, and Programming, Lehrstuhl D für Mathematik*, RWTH Aachen, fifth edition, 1995.

[22] R.A. Wilson, *An Atlas of group representations*, http://www.mat.bham.ac.uk./R.A.Wilson/mtx-atlas, 1996.

AUTOMORPHISM GROUPS OF CERTAIN NON-QUASIPRIMITIVE ALMOST SIMPLE GRAPHS

XIN GUI FANG*, GEORGE HAVAS*[1] and JIE WANG†

*Centre for Discrete Mathematics and Computing, Department of Computer Science and Electrical Engineering, The University of Queensland, Queensland 4072, Australia
† Department of Mathematics, Peking University, Beijing, 100871, P. R. China

1 Introduction

A permutation group on a set Ω is said to be *quasiprimitive* if each of its nontrivial normal subgroups is transitive on Ω. By the structure theorem in [18, Section 2] quasiprimitive permutation groups may be divided into 8 disjoint types (AS, TW, HA, HS, HC, SD, CD, PA) in a way which is helpful for applications to graph theory (see the description in [3]). The main type considered in this paper is AS: a quasiprimitive group G is said to be of type AS if $S \leq G \leq \mathrm{Aut}(S)$, for some nonabelian simple group S.

Let Γ be a finite simple undirected graph with vertex set $V\Gamma$ and edge set $E\Gamma$ and let G be a group of automorphisms of Γ. The graph Γ is said to be *G-quasiprimitive* if and only if G is quasiprimitive on $V\Gamma$. In particular, an AutΓ-quasiprimitive graph is said to be *quasiprimitive.* The graph Γ is said to be *G-locally-primitive* if G is transitive on $V\Gamma$ and the stabilizer G_α is primitive on $\Gamma(\alpha)$, where $\Gamma(\alpha)$ denotes the set of all vertices adjacent to α. The graph Γ is said to be *$(G,2)$-arc transitive* if G is transitive on the 2-arcs of Γ.

Praeger in [19] raises the questions: (1) for a G-quasiprimitive graph Γ, under what conditions can we be certain that AutΓ is quasiprimitive on $V\Gamma$? (2) If AutΓ and G are quasiprimitive with the same quasiprimitive type, is it possible that $soc(G) \neq soc(\mathrm{Aut}\Gamma)$, and if so what are the possibilities for these socles?

We look first at question (1). It is shown in [7] that, for a connected G-quasiprimitive graph Γ with G of type AS, if Γ is G-locally-primitive then either Γ is quasiprimitive or AutΓ has a restricted structure. Baddeley in [2, Section 6] constructs the first example of connected non-quasiprimitive graph Γ which is $(G,2)$-arc transitive, for some quasiprimitive group G of type TW, and he comments there that such graphs seem difficult to construct. An infinite family of $(\mathrm{L}_2(q),2)$-arc transitive graphs with valency 4 is constructed in [12], and Li [14] proves that the full automorphism groups are isomorphic to $Z_2 \times \mathrm{L}_2(q)$, which gives the first infinite family of non-quasiprimitive graphs Γ such that $\mathrm{L}_2(q)$ acts quasiprimitively on $V\Gamma$. Recently an infinite family of $\mathrm{U}_3(q)$-quasiprimitive transitive graphs of valency 9 has been constructed by the authors of this paper in [8]; their full automorphism groups are isomorphic to $Z_3.G$, a semidirect product, where $Z_3 = \mathrm{C}_{\mathrm{Aut}\Gamma}(\mathrm{U}_3(q))$ and $\mathrm{U}_3(q) \leq G \leq \mathrm{Aut}(\mathrm{U}_3(q))$. From this we obtain a new infinite family of non-quasiprimitive graphs Γ admitting a quasiprimitive group of type AS acting transitively on the vertices and the 2-arcs of Γ. Observing all non-quasiprimitive

[1]Partially supported by the Australian Research Council.

examples with G quasiprimitive given above, we find that $\mathrm{Aut}\Gamma = C.G$ with C some small nontrivial cyclic group, which motivates us to suggest a set of conditions under which we may guarantee that either Γ is non-quasiprimitive or $\mathrm{Aut}\Gamma$ has a restricted structure. In this paper we concentrate on G-quasiprimitive graphs for G of type AS. The set of conditions we employ is the following:

Condition $\mathcal{P}$ *Let Γ be a connected, S-locally-primitive graph with S a nonabelian simple group. Set $G = \mathrm{Aut}(S) \cap \mathrm{Aut}\Gamma$ and $C = \mathrm{C}_{\mathrm{Aut}\Gamma}(S)$. Γ and G are said to satisfy Condition $\mathcal{P}$ if $C \neq 1$ and $\mathrm{Aut}\Gamma$ has no quasiprimitive subgroup of type AS containing $C.G$ as its maximal subgroup.*

Remarks on Condition $\mathcal{P}$ (a) Using [7, Theorem 1.2] we shall prove that, for a connected S-locally-primitive graph Γ with S a nonabelian simple group with Γ and G satisfying Condition $\mathcal{P}$, then any quasiprimitive overgroup Y of G in $\mathrm{Aut}\Gamma$ must be of type AS (see Theorem 1.1). From this we conclude that Condition $\mathcal{P}$ does guarantee that either Γ is non-quasiprimitive or $\mathrm{Aut}\Gamma$ has a non-quasiprimitive subgroup Y such that $C.G$ is maximal in Y.

(b) For an arbitrary group C, it is in general quite difficult to verify that Γ and G satisfy Condition $\mathcal{P}$. However, if C is a nilpotent group and if p is a prime divisor of $|C|$, then $C.G$ is a maximal p-local subgroup of Y whenever there exists such a quasiprimitive group Y of type AS in $\mathrm{Aut}\Gamma$. Note that all maximal p-local subgroups of almost simple groups have been determined (see, for example, [1, 5, 13, 16]). So we can find out which Γ and G satisfy Condition $\mathcal{P}$ using a case-by-case check.

The main results of this paper are the following.

Theorem 1.1 *Suppose that Γ and G satisfy Condition $\mathcal{P}$. Then either Γ is a non-quasiprimitive graph; or $\mathrm{Aut}\Gamma$ is a quasiprimitive group of type AS and $\mathrm{Aut}\Gamma$ contains a non-quasiprimitive subgroup Y such that $C.G$ is maximal in Y. Further, for any intransitive minimal normal subgroup N of Y, then: N centralizes S; or $N \cap C = 1$ and $N = Z_p^n$ for some prime p; and $S = S(q)$ is a simple group of Lie type over a field of order $q = p^e$, and further, S and n/e are as in Table 1. In particular, if S lies in the first column of Table 1 or $S = E_7(q)$, then N is the unique intransitive minimal normal subgroup of $\mathrm{Aut}\Gamma$ not centralized by S.*

For Γ and G given as in Theorem 1.1, let Γ^* denote the quotient graph modulo the C-orbits on $V\Gamma$ (obtained by taking C-orbits as vertices and joining two C-orbits by an edge if there is at least one edge in Γ joining a point in the first C-orbit to a point in the second one). By [17], Γ is a cover of Γ^*. As an application of Theorem 1.1 we have the following.

Theorem 1.2 *Let Γ and G satisfy Condition $\mathcal{P}$. Suppose that C is a cyclic group with prime order and that $\mathrm{Aut}\Gamma$ has no a subgroup $N{:}S$ given as in Table 1. If $\mathrm{Aut}\Gamma^*$ is quasiprimitive of type AS with socle S, then Γ is non-quasiprimitive and $\mathrm{Aut}\Gamma = C.G$.*

S	values for n/e	S	values for n/e
$B_l(q), l = 3$ or 4	$[8,9]$ or $[8,16]$	$E_7(q)$	$[56,63]$
$C_l(q), l = 3$ or 4	$[8,9]$ or $[9,16]$	$A_l^{\pm}(q),\ l > 1$	$[l+1, l(l+1)/2]$
$D_l^{\pm}(q), l = 4$ or 5	$[8,12]$ or $[16,20]$	$B_l(q),\ l > 4$	$[2l+1, l^2]$
${}^3D_4(q),\ q$ odd	12	$C_l(q),\ l \neq 1,3,4$	$[2l, l^2]$
$E_6^{\pm}(q)$	$[27,36]$	$D_l^{\pm}(q),\ l \neq 1,2,4,5$	$[2l, l(l-1)]$

Table 1
where $[i,j]$ stands $i \leq n/e \leq j$, for distinct integers i and j with $i < j$.

Remark on Theorem 1.2 For each graph given as in [8, 14], we know $C = Z_3$ or Z_2 and that AutΓ has no subgroup $N.S$ given as in Table 1. It is also easy to prove that $S \leq \text{Aut}\Gamma^* \leq \text{Aut}(S)$ (up to isomorphism). By Theorem 1.2, to show Aut$\Gamma = C.G$, it remains only to verify that Γ and G satisfy Condition $\mathcal{P}$, which can be completed by using the method mentioned in Remark (b) on Condition $\mathcal{P}$. Indeed, this is not difficult to check for the graphs given in [8, 14].

Theorems 1.1 and 1.2 are proved in the next section. In the final section we consider Question (2) raised at the beginning of this paper and give some examples of G-quasiprimitive graphs such that AutΓ and G have same quasiprimitive type but $soc(\text{Aut}\Gamma) \neq soc(G)$.

2 Proof of Theorems 1.1 and 1.2

The first lemma is used in the proof of Theorem 1.1. Its proof follows immediately from [17] and the connectivity of Γ.

Lemma 2.1 *Let Γ be a connected, nonbipartite, Y-locally-primitive graph. Then each intransitive normal subgroup of Y is semiregular.*

Suppose now that G is an almost simple group with socle S and let G and Γ be given as in Theorem 1.1. Write $X :=$ AutΓ and $C := \mathrm{C}_X(S)$. Since S is locally primitive on $V\Gamma$, S is transitive on the arcs of Γ, and so S is not regular on $V\Gamma$. It follows that C is not transitive on $V\Gamma$ (see, for example, [21] or [9, Proposition 2.4]).

Proof [of Theorem 1.1] First we show that, for each overgroup K of S in X, if K is quasiprimitive then K is of type AS. Suppose to the contrary that K is not of type AS. By [7, Theorem 1.2], either $\Gamma = K_8$ with $S = \text{PSL}(2,7)$ and $K = \text{AGL}(3,2)$; or K is of type PA with $soc(K) = S_1 \times S_2 \cong S \times S$ and S is a diagonal subgroup of $soc(K)$. In the former case, by [6], $S_\alpha = Z_7{:}Z_3$ and S_α is a maximal subgroup, and hence S acts primitively on $V\Gamma$. It follows from [21] that $C = 1$, which contradicts Condition $\mathcal{P}$. In the latter case, since $S < soc(K)$, Γ is also $soc(K)$-locally-primitive. Since $S_i \lhd soc(K)$ and $S_i \cong S$, for $i = 1,2$, the both S_1 and S_2 are transitive by Lemma 2.1. It follows that S_1 and S_2 are regular, which is impossible since $|S_1| = |S| > |V\Gamma|$. So K is of type AS. In particular, take $K = X$, then X is of type AS.

Suppose now that X is quasiprimitive. Since X is of type AS, by Condition $\mathcal{P}$, $G.C$ is not maximal in X, and hence there is a proper subgroup Y of X containing $G.C$ such that $G.C$ is maximal in Y. If Y is quasiprimitive then Y is of type AS, which contradicts Condition $\mathcal{P}$. Thus Y is not quasiprimitive.

Let N be an intransitive minimal normal subgroup of Y with $N \not\leq C$. Since Γ is also a Y-locally-primitive graph, by Lemma 2.1, N is semiregular on $V\Gamma$. If $N < C.G$ then CN/C is isomorphic to a normal subgroup of G, which implies that either $N \leq C$ or S isomorphic to a subgroup of CN/C. The former case contradicts $N \not\leq C$. While in the latter case we conclude that $|N|$ is divisible by $|S|$, which contradicts the fact that N is semiregular. Thus $N \not\leq C.G$ and so $Y = \langle N, C.G\rangle$ (since $C.G$ is maximal in Y). Then arguing as in the proof of [7, Theorem 1.1] we conclude that N is an elementary abelian p-group, for some prime p. If $N \cap C \neq 1$, then $N \cap C$ is normalized by N and $C.G$, respectively. Thus $N \cap C$ is a nontrivial normal subgroup of Y. Since $N \cap C \leq C$ and N is a minimal normal subgroup of Y, $N \cap C = N$ and hence $N \leq C$, which is not the case. So $N \cap C = 1$. It follows that S acts faithfully by conjugation on N. Thus S has a faithful projective p-modular representation of degree n. A similar argument as in the proof of [10, Theorem 1.1] shows that S and n/e are given as in Table 1. For example, we look at the case where $S = A_l(q)$ with $q = p_1^e$ for some prime p_1. If $p_1 \neq p$ then $n/e \geq (q-1)/(2, q-1)$ by [13, Table 5.3.A], and hence

$$|N| \geq q^{(q-1)/(2,q-1)} \geq |A_l(q)| > |V\Gamma|,$$

which is impossible. So $p_1 = p$. Now let $R_p(S)$ denote the minimal dimension of a faithful, irreducible, projective KS-module, where K is an algebraically closed field of characteristic p. By [13, Table 5.4.C], $R_p(A_l(q)) = l+1$, which implies that $n/e \geq l+1$. On the other hand, since a Sylow p-subgroup of $A_l(q)$ has order $q^{l(l+1)/2}$ and since $|S|$ is divisible by $|N|$, $n/e \leq l(l+1)/2$. So n/e lies in $[l+1, l(l+1)/2]$. A similar argument deals with other nonabelian simple groups S.

Finally, for S in the first column of Table 1 or $E_7(q)$, if there is another intransitive minimal normal subgroup K not centralized by S, then a similar argument as above implies that $K \cong Z_p^m$ with m/e in the second column of Table 1 or $[56, 63]$. Set $W = NK$. Now W is a normal p subgroup of X with $|W| = p^{n+m}$. Since p^{n+m} is greater than the p-part of $|S|$, W must be transitive on $V\Gamma$ by Lemma 2.1. Recall that S is one of the following groups: $B_l(q)$ or $C_l(q)$ with $l \in \{3,4\}$, $D_l^{\pm}(q)$ with $l = 4$ or 5, ${}^3D_4(q)$ with q odd or $E_7(q)$. From [11, Corollary 2] it follows that $S = \mathrm{PSp}(4,3)$ and $|V\Gamma| = 27$, and so $X \leq \mathrm{AGL}(3,3)$. However it is trivial to see that $\mathrm{PSp}(4,3)$ is not isomorphic to a subgroup of $\mathrm{AGL}(3,3)$, which is a contradiction. So N is the unique intransitive minimal normal subgroup of X not centralized by S. □

Proof [of Theorem 1.2] We claim first that AutΓ is not quasiprimitive. If this is not the case, by Theorem 1.1, AutΓ contains a non-quasiprimitive subgroup Y such that $C.G$ is maximal in Y. Let N be an intransitive minimal normal subgroup Y.

If $N \not\leq C$, by Theorem 1.1, we conclude that $N{:}S$ would be in Table 1, which is not the case. So $N \leq C$ and hence $N = C = Z_{p_1}$. Since $G \cong (C.G)/C < Y/C$ and Y/C is isomorphic to a subgroup of $\mathrm{Aut}\Gamma^*$, $S \leq Y/C \leq \mathrm{Aut}(S)$ (up to isomorphism). It follows that $Y \cong C.G_1$, for some group G_1 with $S \leq G_1 \leq \mathrm{Aut}(S)$. On the other hand, from the definition of G we know that G is the maximal group among the groups H with the properties that $S \leq H \leq \mathrm{Aut}(S)$ and $C.H \leq \mathrm{Aut}\Gamma$, which implies that $G_1 \cong G$. Thus $Y = C.G$, which is a contradiction. So $\mathrm{Aut}\Gamma$ is non-quasiprimitive.

Let N be an intransitive minimal normal subgroup of $\mathrm{Aut}\Gamma$. If $N \not\leq C$ then S acts nontrivially by conjugation on N. Now N is semiregular on $V\Gamma$ by Lemma 2.1. Then arguing as in the proof of Theorem 1.1, we conclude that $N = Z_p^n$, for some prime p and integer $n > 1$, and $N{:}S$ lies in Table 1, which is a contradiction. So $N \leq C$, and hence $N = C$. Now $C \lhd \mathrm{Aut}\Gamma$ and $\mathrm{Aut}\Gamma/C$ is a subgroup of $\mathrm{Aut}\Gamma^*$. Thus $\mathrm{Aut}\Gamma \cong C.G_1^*$, for some subgroup G_1^* of $\mathrm{Aut}\Gamma^*$ with $S \leq G_1^* \leq \mathrm{Aut}(S)$ (up to isomorphism). From the definition of G it follows that $G_1^* \cong G$ and hence $\mathrm{Aut}\Gamma = C.G$. □

3 Some examples of quasiprimitive graphs

In this section we always assume that G is an almost simple group and Γ is a connected G-arc transitive graph with valency d_Γ. By [20] there exists a 2-element $g \in G$ with the properties: $g \notin \mathrm{N}_G(H)$, $g^2 \in H$ and $\langle H, g\rangle = G$, such that $\Gamma \cong \Gamma^* := \Gamma(G, H, HgH)$ with $d_\Gamma = |H : H \cap H^g|$, where Γ^* is defined by

$$V\Gamma^* = \{Hx \mid x \in G\}, \quad E\Gamma^* = \{\{Hx, Hy\} \mid x, y \in G, xy^{-1} \in HgH\}. \qquad (1)$$

All connected regular G-arc transitive graphs considered in this section will be defined in terms of a subgroup H and a 2-element g as in (1).

Now we give some examples of G-quasiprimitive graphs. The first example gives all G-quasiprimitive graphs with a prime power number of vertices, for G a non-abelian simple group. The proof of Example 3.1 follows immediately from [11, Corollary 2] and [6].

Example 3.1 Let Γ be a finite connected graph and suppose that G is a transitive subgroup of $\mathrm{Aut}\Gamma$ with G a nonabelian simple group. If $|V\Gamma|$ is a prime power, then either Γ is a complete graph or $G = \mathrm{PSU}(4,2)$ and $|V\Gamma| = 27$. Further, G and $soc(\mathrm{Aut}\Gamma)$ are given in Table 2.

The graphs on the first five lines of Table 2 are all complete, but the graph for $\mathrm{PSU}(4,2)$ is not. The graphs on lines two to five provide one infinite family and three particular graphs with the property that $soc(G) \neq soc(\mathrm{Aut}\Gamma)$.

Example 3.2 Let Γ be a connected G-arc transitive graph of valency d_Γ, where $(G, |V\Gamma|)$ is one of $(\mathrm{P\Gamma L}(2,8), 36)$, $(A_9, 120)$ and $(M_{11}, 55)$. Then $\mathrm{Aut}\Gamma$ is an almost simple group. Moreover, the pair $(G, soc(\mathrm{Aut}G)$ is $(\mathrm{PSL}_2(8), A_9)$, $(A_9, \Omega_8^+(2))$ or (M_{11}, A_{11}), respectively.

G	$\lvert V\Gamma\rvert$	$soc(\mathrm{Aut}\Gamma)$	comments on G_α
A_{p^a}	p^a	A_{p^a}	$G_\alpha \cong A_{p^a-1}$
$\mathrm{PSL}(n,q)$	$\frac{q^n-1}{q-1} = p^a$	A_{p^a}	the stabilizer of a line or hyperplane
$\mathrm{PSL}(2,11)$	11	A_{11}	$G_\alpha \cong A_5$
M_{23}	23	A_{23}	$G_\alpha \cong M_{22}$
M_{11}	11	A_{11}	$G_\alpha \cong M_{10}$
$\mathrm{PSU}(4,2)$	27	$\mathrm{PSU}(4,2)$	$G_\alpha \cong 2^4{:}A_5$

Table 2

G	d_Γ	$\lvert V\Gamma\rvert$	G_α	$soc(G)$	$\mathrm{Aut}\Gamma$
$\mathrm{P\Gamma L}(2,8)$	14 or 21	36	7:6	$\mathrm{PSL}(2,8)$	S_9
A_9	56 or 63	120	$\mathrm{PSL}(2,8).2$	A_9	$\Omega_8^+(2)$
M_{11}	18 or 36	55	$3^2{:}Q_8.2$	M_{11}	S_{11}

Table 3

Proof For $\alpha \in V\Gamma$ write $H = G_\alpha$. For $(G, |V\Gamma|)$ given as above, by [6] we have the structure of H as given in Table 3 (see Column 4). Note that H is a maximal subgroup of G, for G and G_α given as in Table 3. So $\langle H, g\rangle = G$, for any 2-element $g \in G\backslash H$, and hence the graph $\Gamma = \Gamma(G, H, HgH)$ is a connected G-arc transitive graph with valency $d_\Gamma = |H : H \cap H^g|$. Look first at $G = \mathrm{P\Gamma L}(2,8)$ and $H = R{:}T$ with $R \cong Z_7$ and $T \cong Z_6$. Computation shows that T is self-normalized in G. Thus $|H \cap H^g|$ is either 2 or 3, and so $d_\Gamma = 21$ or 14. Moreover, since a Sylow 2-subgroup of G is elementary abelian all suitable 2-elements g have order 2. If $d_\Gamma = 21$ then $H \cap H^g = 2$. Using the MAGMA program (see Figure 1) we obtain both all connected $(\mathrm{P\Gamma L}(2,8))$-arc transitive graphs with valency 21 and their full automorphism groups. Using a similar method we obtain Table 3, and the theorem follows from Columns 4 and 5 of Table 3. □

Remark on Example 3.2 Recall the definition of 2-closure of a finite permutation group: if G is a finite permutation group on a set Ω the 2-closure $G^{(2)}$ of G is the largest subgroup of $\mathrm{Sym}(\Omega)$ containing G which has the same orbits as G in the induced action on $\Omega \times \Omega$. For each group G in Example 3.2, Table 3 shows that $Aut\Gamma \cong G^{(2)}$, and all pairs of $(G, \mathrm{Aut}\Gamma)$ in the example occur in [15, Table 1].

Acknowledgement We are grateful to John Cannon for assistance in setting up the MAGMA programs for calculating general arc-transitive graphs.

```
H := PGammaL(2,8);
A :=SylowSubgroup(H, 7);
M :=Normalizer(H, A);
for x in M do
  if Order(x) eq 3 then
     for y in M do
        if Order(y) eq 2 then
        M6:=sub<H | x,y>;
          if Order(M6) eq 6 then
          B:=M6; break;
          end if;
        end if;
     end for;
   end if;
end for;

B3:=SylowSubgroup(B,3);
B2:=SylowSubgroup(B,2);
A21:=sub<H |A, B3>;
N2:=Normalizer(H,B2);

phi, G := CosetAction(H, M);
T := Stabilizer(G, 1);
A21 := phi(A21);
B2 := phi(B2);
NB2 := Normalizer(G, B2);

Grphs := [];
for g in NB2 do
   if Order(g) eq 2 and sub< G | T, g > eq G then
      print "Success with", g;
      found := true;
      Nbs := { 1^(g*x) : x in A21 };
      Gr := Graph< Support(G) | <1, Nbs>^G >;
      print Gr;
      print (AutomorphismGroup(Gr));
      print Order(AutomorphismGroup(Gr));

      Append(~Grphs, Gr);
   end if;
end for;
```

Figure 1: MAGMA program for $\mathrm{P\Gamma L}(2,8)$-arc transitive graphs of valency 21

References

[1] M. Aschbacher, On the maximal subgroups of the finite subgroups of the finite classical groups, *Invent. Math.* **76** (1984), 469–514.

[2] R. Baddeley, Two-arc transitive graphs and twisted wreath products, *J. Algebraic Combin.* **2** (1993), 215–237.

[3] R. Baddeley and C.E. Praeger, On primitive overgroups of quasiprimitive permutation groups, Technical Report No. 1996/4, University of Leicester, 1996.

[4] W. Bosma, J. Cannon and C. Playoust, The Magma algebra system I: The user language, *J. Symbolic Comput.* **24** (1997), 235–265.

[5] A.M. Cohen, M.W. Liebeck, J. Saxl and G.M. Seitz, The local maximal subgroups of exceptional groups of Lie type, finite and algebraic, *Proc. London Math. Soc.* (3) **64** (1992), 21–48.

[6] J.H. Conway, R.T. Curtis, S.P. Norton, R.A. Parker and R.A. Wilson, *Atlas of Finite Groups*, Clarendon Press, Oxford, 1985.

[7] X.G. Fang, G. Havas and C.E. Praeger, On the automorphism group of quasiprimitive graphs admitting arc-transitive actions of almost simple groups, *Technical Report* **405**, The University of Queensland, 1997.

[8] X.G. Fang, G. Havas and J. Wang, A Family of Non-Quasiprimitive Graphs Admitting a Quasiprimitive 2-Arc Transitive Group Action, *Technical Report* **406**, The University of Queensland, 1997.

[9] X.G. Fang and C.E. Praeger, Finite two-arc transitive graphs admitting a Suzuki simple group, *Comm. Algebra* (to appear).

[10] X.G. Fang and C.E. Praeger, On graphs admitting arc-transitive actions of almost simple groups, *J. Algebra* (to appear).

[11] R.M. Guralnick, Subgroups of prime power index in a simple group, *J. Algebra* **81** (1983), 304–311.

[12] A. Hassani, L. Nochefranca and C.E. Praeger, Two-arc transitive graphs admitting a two-dimensional projective linear group, *Preprint*, 1995.

[13] P. Kleidman and M. Liebeck, *The subgroup structure of the finite classical groups*, Cambridge University Press, Cambridge, 1990.

[14] C.H. Li, A family of quasiprimitive 2-arc transitive graphs which have non-quasiprimitive full automorphism groups, *European J. Combin.* (to appear).

[15] M.W. Liebeck, C.E. Praeger and J. Saxl, On the 2-closures of finite permutation groups, *J. London Math. Soc.* (2) **37** (1988), 241–252.

[16] M.W. Liebeck, J. Saxl, G.M. Seitz, Subgroups of maximal rank in finite exceptional groups of Lie type, *Proc. London Math. Soc.* (3) **65** (1992), 297–325.

[17] C.E. Praeger, Imprimitive symmetric graphs, *Ars Combin.*, **19A** (1985), 149–163.

[18] C.E. Praeger, An O'Nan-Scott Theorem for finite quasiprimitive permutation groups and an application to 2-arc transitive graphs, *J. London Math. Soc.* (2) **47** (1993), 227–239.

[19] C.E. Praeger, Finite quasiprimitive graphs, *Surveys in combinatorics*, 65–85, London Math. Soc. Lecture Note Ser., **241**, Cambridge Univ. Press, Cambridge, 1997.

[20] G.O. Sabidussi, Vertex-transitive graphs, *Monatsh. Math.* **68** (1964), 426–438.

[21] H. Wielandt, *Finite permutation groups*, Academic Press, New York, 1964.

SUBGROUPS OF THE UPPER-TRIANGULAR MATRIX GROUP WITH MAXIMAL DERIVED LENGTH AND A MINIMAL NUMBER OF GENERATORS

S.P. GLASBY

Department of Mathematics and Computing Science, The University of the South Pacific, PO Box 1168, Suva, Fiji

1 Introduction

Let $\mathbb{F}$ be a field and let $U_n(\mathbb{F})$ (or U_n) denote the group of $n \times n$ upper-triangular matrices over $\mathbb{F}$ with 1's on the main diagonal and 0's below. If $2^{d-1} < n \leq 2^d$, then U_n has derived length d and has a subgroup generated by $n-1$ elements which also has derived length d (see [3]). We show in Theorem 2.2 that U_n has a 3-generated subgroup with derived length d. In Theorem 3.4 we show that U_n has a 2-generated subgroup of derived length d if and only if $\frac{21}{32}2^d < n \leq 2^d$. It follows that the proportion, $\pi(N)$, of $n \leq N$ such that U_n has a 2-generated subgroup of maximal derived length satisfies $\frac{11}{21} < \pi(N) \leq 1$, $\liminf \pi(N) = \frac{11}{21}$ and $\limsup \pi(N) = \frac{11}{16}$. Theorems 2.2 and 3.4 are constructive in the sense that the generating matrices are explicitly given by recursive formulas.

We shall now introduce some notation and state some well-known properties of U_n (see [3]). The kth term of the lower central series for U_n, denoted $\gamma_k(U_n)$, comprises the matrices $(a_{i,j}) \in U_n$ with $a_{i,j} = 0$ if $0 < j - i < k$. Furthermore, the kth term in the derived series for U_n is $U_n^{(k)} = \gamma_{2^k}(U_n)$.

In the sequel we shall assume that $d = \lfloor \log_2(n-1) \rfloor + 1$ and consider subgroups G of U_n where $G^{(d-1)}$ is not trivial. Let $1 \leq i < j \leq n$ and let $X_{i,j} \in U_n$ be the matrix obtained by adding row j of the identity matrix, I, to row i (so its (i,j)th entry is 1). Then

$$[X_{i,j}, X_{k,\ell}] = X_{i,j}^{-1} X_{k,\ell}^{-1} X_{i,j} X_{k,\ell}$$

equals I if $j < k$, and equals $X_{i,\ell}$ if $j = k$. In order to show that U_n has derived length d for all n satisfying $2^{d-1} < n \leq 2^d$, it suffices to show that $\langle X_{1,2}, X_{2,3}, \ldots, X_{n-1,n} \rangle$ has derived length d when $n = 2^{d-1}+1$. The latter can be proved using induction on d based on the following reasoning

$$\begin{aligned} X_{1,9} &= [X_{1,5}, X_{5,9}] \\ &= [\,[X_{1,3}, X_{3,5}], [X_{5,7}, X_{7,9}]\,] \\ &= [\,[\,[X_{1,2}, X_{2,3}], [X_{3,4}, X_{4,5}]\,], [\,[X_{5,6}, X_{6,7}], [X_{7,8}, X_{8,9}]\,]\,]. \end{aligned}$$

At the heart of this proof is a binary tree with d layers and $2^d - 1$ vertices. The vertices at layer k are the elements $X_{1+(i-1)2^{k-1}, 1+i2^{k-1}}$ of $U_n^{(k-1)}$. If $j = 1 + (i-1)2^{k-1}$, then the vertices $X_{j,j+2^{k-1}}$ and $X_{j+2^{k-1}, j+2^k}$ of layer k are joined to $X_{j,j+2^k}$ on the next layer.

2 3-generated subgroups

The idea behind the proof of Theorem 2.2 is to "re-cycle" vertices of the above binary tree. For example, the four matrices $X_{1,2}, X_{2,3}, X_{3,4}, X_{4,5}$ are not needed to show that $X_{1,5} \in U_5^{(2)}$: three matrices suffice as

$$[[X_{1,2}, X_{2,3}X_{3,4}], [X_{2,3}X_{3,4}, X_{4,5}]] = [X_{1,3}X_{1,4}, X_{3,5}] = X_{1,5}.$$

The graph at the heart of the proof of Theorem 2.2 has fewer vertices than the complete bipartite binary tree with $2^d - 1$ vertices. It has d layers with 3 vertices per layer, where the vertices of layer k correspond to elements of $G^{(k-1)}$. Let A, B, C be the matrices corresponding to the vertices of layer k. Then the commutators $[B,C]$, $[C,A]$, $[A,B]$ correspond to the vertices of layer $k+1$. Thus the edges between layers k and $k+1$ form a bipartite graph K, and the full graph is obtained by joining $d-1$ copies of K end-to-end. Our objective is to inductively construct three layer 1 matrices, so that at least one of the layer d matrices is non-trivial.

Let F be the free group $\langle x_1, x_2, x_3 \mid \ \rangle$ of rank 3. The following lemma was much harder to conceive than to prove.

Lemma 2.1 *Let d be a positive integer, and let $n = 2^{d-1} + 1$. Then there exist matrices $A_n, B_n, C_n \in U_n$ and a word $w_n(x_1, x_2, x_3) \in F^{(d-1)}$ such that*

$$w_n(A_n, B_n, C_n) = X_{1,n}, \ w_n(B_n, C_n, A_n) = I, \text{ and } w_n(C_n, A_n, B_n) = I. \quad (1)$$

Proof The proof uses induction on d. When $d = 1$, take $w_2(x_1, x_2, x_3) = x_1$ and $A_2 = X_{1,2}$, $B_2 = C_2 = I$. (More generally, if $r^3 + s^3 + t^3 - 3rst \neq 0$ in $\mathbb{F}$ where $r, s, t \in \mathbb{Z}$, then we may take $w_2 = x_1^r x_2^s x_3^t$ and find $A_2, B_2, C_2 \in U_2$ such that (1) holds.) Suppose that $A_n, B_n, C_n \in U_n$ and $w_n \in F^{(d-1)}$ satisfy (1). We shall construct appropriate $A_{2n-1}, B_{2n-1}, C_{2n-1}$ and w_{2n-1}. Now $n = 2^{d-1} + 1$ and $2n - 1 = 2^d + 1$. There is a surjective homomorphism

$$\pi \ U_{2n-1} \to U_n \times U_n \quad \text{given by} \quad \pi(A) = (\lambda(A), \rho(A)),$$

where $\lambda(A)$ is the upper-left $n \times n$ submatrix of A, and $\rho(A)$ is the lower-right $n \times n$ submatrix of A. Note that $\lambda(A)$ and $\rho(A)$ overlap at the (n,n)th entry of A, which is a 1.

Choose $A_{2n-1}, B_{2n-1}, C_{2n-1} \in U_{2n-1}$ such that

$$\pi(A_{2n-1}) = (A_n, B_n), \qquad \pi(B_{2n-1}) = (B_n, C_n), \qquad \pi(C_{2n-1}) = (C_n, A_n).$$

Clearly A_{2n-1}, B_{2n-1} and C_{2n-1} are not uniquely defined. (A different choice may be obtained by multiplying by an element of $\ker(\pi) \cong \mathbb{F}^{(n-1)^2}$.) Define w_{2n-1} by $w_{2n-1}(x_1, x_2, x_3) = [\, w_n(x_1, x_2, x_3), w_n(x_3, x_1, x_2)\,]$. Clearly, $w_{2n-1} \in F^{(d)}$.

Consider $w_{2n-1}(A_{2n-1}, B_{2n-1}, C_{2n-1})$. Now

$$\begin{aligned} \pi(\, w_n(A_{2n-1}, B_{2n-1}, C_{2n-1})\,) &= w_n(\pi(A_{2n-1}), \pi(B_{2n-1}), \pi(C_{2n-1})) \\ &= (\, w_n(A_n, B_n, C_n), w_n(B_n, C_n, A_n)\,) \\ &= (X_{1,n}, I). \end{aligned}$$

Similarly,

$$\begin{aligned}\pi(w_n(B_{2n-1}, C_{2n-1}, A_{2n-1})) &= (I, I) \quad \text{and} \\ \pi(w_n(C_{2n-1}, A_{2n-1}, B_{2n-1})) &= (I, X_{1,n}).\end{aligned}$$

Now $\pi(X_{1,n}) = (X_{1,n}, I)$ and $\pi(X_{n,2n-1}) = (I, X_{1,n})$. (Here we can tell from the context whether $X_{1,n}$ lies in U_n or U_{2n-1}.) Therefore

$$\begin{aligned}w_n(A_{2n-1}, B_{2n-1}, C_{2n-1}) &= X_{1,n}Z_1, \\ w_n(B_{2n-1}, C_{2n-1}, A_{2n-1}) &= Z_2, \quad \text{and} \\ w_n(C_{2n-1}, A_{2n-1}, B_{2n-1}) &= X_{n,2n-1}Z_3\end{aligned}$$

where $Z_1, Z_2, Z_3 \in \ker(\pi)$. Since $\ker(\pi)$ is Abelian, and is centralized by both $X_{1,n}$ and $X_{n,2n-1}$, it follows that

$$\begin{aligned}w_{2n-1}(A_{2n-1}, B_{2n-1}, C_{2n-1}) &= [X_{1,n}Z_1, X_{n,2n-1}Z_3] \\ &= [X_{1,n}, X_{n,2n-1}] = X_{1,2n-1}, \\ w_{2n-1}(B_{2n-1}, C_{2n-1}, A_{2n-1}) &= [Z_2, X_{1,n}Z_1] = I, \\ w_{2n-1}(C_{2n-1}, A_{2n-1}, B_{2n-1}) &= [X_{n,2n-1}Z_3, Z_2] = I.\end{aligned}$$

This completes the induction and the proof. □

Recall the observation that $U_n(\mathbb{F})$ has derived length $\lfloor \log_2(n-1) \rfloor + 1$, and the subgroup $\langle X_{1,2}, X_{2,3}, \ldots, X_{n-1,n} \rangle$ has $n-1$ generators and the same derived length.

Theorem 2.2 *The group $U_n(\mathbb{F})$ of $n \times n$ upper-triangular matrices over a field $\mathbb{F}$ with all eigenvalues 1, has a 3-generated subgroup whose derived length is $d = \lfloor \log_2(n-1) \rfloor + 1$. Furthermore, if $n \le \frac{5}{8}2^d$ then $U_n(\mathbb{F})$ has no 2-generated subgroup of derived length d.*

Proof Let $d = \lfloor \log_2(n-1) \rfloor + 1$ and let $m = 2^{d-1} + 1$. Then both U_m and U_n have derived length d. By Lemma 2.1, U_m has a 3-generated subgroup with derived length d, and hence so does U_n, as U_m is isomorphic to a subgroup of U_n.

If $d < 3$, then there are no integers in the range $\frac{1}{2}2^d < n \le \frac{5}{8}2^d$. Suppose $d \ge 3$ and $G = \langle A, B \rangle$ is a 2-generated subgroup of U_n where $\frac{1}{2}2^d < n \le \frac{5}{8}2^d$. Then $\gamma_2(G)/\gamma_3(G) = \langle [A, B]\gamma_3(G) \rangle$ is cyclic, and therefore

$$G^{(2)} = [\gamma_2(G), \gamma_2(G)] = [\gamma_2(G), \gamma_3(G)] \subseteq \gamma_5(G).$$

A simple induction shows $G^{(d-1)} \subseteq \gamma_{5 \cdot 2^{d-3}}(G)$ for $d \ge 3$, and hence

$$G^{(d-1)} \subseteq \gamma_{5 \cdot 2^{d-3}}(G) \subseteq \gamma_n(G) \subseteq \gamma_n(U_n) = \{I\}.$$

Therefore G has derived length less than d. □

In the above proof, there were choices for A_2, B_2, C_2 and for the subsequent generators A_n, B_n, C_n where $n = 2^{d-1}+1$. However, once A_2, B_2 and C_2 were specified, the $(i,i+1)$ entries of A_n, B_n, C_n $(d>1)$ were determined, but the (i,j) entries with $j-i>1$ could be arbitrary. It should not surprise the reader that different choices for A_2, B_2 and C_2 can yield different subgroups $\langle A_n, B_n, C_n\rangle$.

We shall give an example of a 2-generated group $G = \langle A, B\rangle$ of U_n that shows that both the derived length and the order can depend on $\mathbb{F}$. Let G be the subgroup $\langle A, B\rangle$ of U_6 where $A = X_{1,2}X_{5,6}$ and $B = X_{2,3}X_{3,4}^{-1}X_{4,5}$, and suppose that $\mathrm{char}(\mathbb{F}) = p$ is prime. It follows from $[[[B,A],B],[B,A]] = X_{1,6}^{2}$ and $[[[B,A],A],[B,A]] = I$ that G is metabelian if $p=2$, and has derived length 3 if $p>2$. Furthermore, $|G| = p^7$ if $p = 2,3$ and $|G| = p^6$ if $p>3$. In the latter case G has maximal class (see [1, p.61]).

3 2-generated subgroups

Suppose that $\frac{5}{8}2^d < n \le 2^d$. It is natural to ask whether $U_n(\mathbb{F})$ has a 2-generated subgroup of derived length d. If $U_m(\mathbb{F})$ has a 2-generated subgroup of derived length d, then so too does $U_n(\mathbb{F})$ all n satisfying $m \le n \le 2^d$. In this section we show that the smallest value of m for which U_m has a 2-generated subgroup of derived length d is $m = \lfloor \frac{21}{32}2^d \rfloor + 1$. This is clearly the case if $0 \le d < 3$. Henceforth assume that $d \ge 3$.

Let $F = \langle a, b \mid \ \rangle$ denote a free group of rank 2. Then $\gamma_r(F)/\gamma_{r+1}(F)$ is an Abelian group, for each positive integer r, which is freely generated by the basic commutators of weight k (see [2]). Thus a typical element of $\gamma_2(F)/\gamma_4(F)$ has the form $[b,a]^i[b,a,a]^j[b,a,b]^k\gamma_4(F)$, where $[b,a,a]$ and $[b,a,b]$ denote left-normed commutators, i.e. $[[b,a],a]$ and $[[b,a],b]$ respectively. We shall need three lemmas in the sequel. Lemmas 3.1 and 3.2 are standard so we omit their proofs.

Lemma 3.1 *Let $x, x' \in \gamma_r(F)$ and $y, y' \in \gamma_s(F)$ where $x \equiv x' \mod \gamma_{r+1}(F)$ and $y \equiv y' \mod \gamma_{s+1}(F)$. Then $[x,y] \equiv [x',y'] \mod \gamma_{r+s+1}(F)$.*

Applying Lemma 3.1 to $[[b,a]^i[b,a,a]^j[b,a,b]^k, [b,a]^\ell]$ shows that

$$[[b,a,a],[b,a]]\gamma_6(F) \quad \text{and} \quad [[b,a,b],[b,a]]\gamma_6(F)$$

generate $F^{(2)}\gamma_6(F)/\gamma_6(F)$.

Lemma 3.2 *Let $T_{r,n}(\tau_1,\ldots,\tau_{n-r})$ denote a coset of $\gamma_{r+1}(U_n)$ comprising matrices $(t_{i,j})$ satisfying $t_{i,j} = 0$ if $1 \le j-i < r$, $t_{i,j} = \tau_i$ if $j-i = r$, and $t_{i,j}$ arbitrary if $j-i > r$. Then $[T_{r,n}(\alpha_1,\ldots,\alpha_{n-r}), T_{s,n}(\beta_1,\ldots,\beta_{n-s})]$ is contained in $T_{r+s,n}(\alpha_1\beta_{1+r} - \alpha_{1+s}\beta_1, \ldots, \alpha_{n-r-s}\beta_{n-s} - \alpha_{n-r}\beta_{n-r-s})$.*

How might we go about finding matrices $A, B \in U_n$ such that $\langle A, B\rangle$ has derived length d? Motivated by the previous section we suspect that the $(i,i+1)$ entries of A and B are important. Let $A \in T_{1,n}(\alpha_1,\ldots,\alpha_{n-1})$ and $B \in T_{1,n}(\beta_1,\ldots,\beta_{n-1})$

where the α_i and the β_j are regarded as variables. An evaluation homomorphism from the polynomial ring

$$P = \mathbb{Z}[\alpha_1, \ldots, \alpha_{n-1}, \beta_1, \ldots, \beta_{n-1}]$$

to $\mathbb{F}$ gives rise to a group homomorphism $\phi : U_n(P) \to U_n(\mathbb{F})$. We shall find a word $c_{n-1}(a,b) \in \gamma_{n-1}(F) \cap F^{(d-1)}$ and values for the α_i and β_j in $\mathbb{F}$ such that $c_{n-1}(\phi(A), \phi(B)) = X_{1,n}$ or $X_{1,n}^{-1}$.

The first case not excluded by Theorem 2.2, or already excluded, is $n = 6$. Let $c_5(a,b) = [\,[b,a,a],[b,a]\,]$. By repeated application of Lemma 3.2 the $(1,6)$ entry of $c_5(A,B)$ is

$$\begin{aligned}
[c_5(A,B)]_{1,6} &= [\,[B,A,A],[B,A]\,]_{1,6} \\
&= [B,A,A]_{1,4}[B,A]_{4,6} - [B,A]_{1,3}[B,A,A]_{3,6} \\
&= -\alpha_1\alpha_2\beta_3\alpha_4\beta_5 + \alpha_1\alpha_2\beta_3\beta_4\alpha_5 + 3\alpha_1\beta_2\alpha_3\alpha_4\beta_5 - 4\alpha_1\beta_2\alpha_3\beta_4\alpha_5 \\
&\quad +\alpha_1\beta_2\beta_3\alpha_4\alpha_5 - 2\beta_1\alpha_2\alpha_3\alpha_4\beta_5 + 3\beta_1\alpha_2\alpha_3\beta_4\alpha_5 - \beta_1\alpha_2\beta_3\alpha_4\alpha_5
\end{aligned}$$

We make some remarks about this polynomial. First each monomial summand has five variables. The variables have distinct subscripts and contain three α's and two β's. The polynomial has integer coefficients and $[B,A,A]$ contributes two α_i and one β_j to the first three variables, or to the last three variables of each monomial summand. Similarly, $[B,A]$ contributes an α_i and a β_j to the first two variables, or to the last two variables of each monomial summand. Thus, even without computing $[c_5(A,B)]_{1,6}$, we know that $\alpha_1\alpha_2\alpha_3\beta_4\beta_5$ is not a summand. Setting $\alpha_1 = \alpha_2 = \beta_3 = \alpha_4 = \beta_5 = 1$ and $\beta_1 = \beta_2 = \alpha_3 = \beta_4 = \alpha_5 = 0$ shows that $[c_5(\phi(A), \phi(B))]_{1,6} = -1$ and hence $c_5(\phi(A), \phi(B)) = X_{1,6}^{-1}$. This proves that $\langle \phi(A), \phi(B) \rangle$ is a 2-generated subgroup of $U_6(\mathbb{F})$ of derived length 3 for all fields $\mathbb{F}$.

Many of the above remarks generalize *mutatis mutandis* to other words in $\gamma_{n-1}(F) \cap F^{(d-1)}$. We shall use the following lemma repeatedly.

Lemma 3.3 (Multiplication Lemma) *With the above notation, suppose that $w \in \gamma_r(F)$, $w' \in \gamma_s(F)$, and $[w(A,B)]_{1,1+r}$ and $[w'(A,B)]_{1,1+s}$ have monomial summands m and m' respectively. If $r \geq s$, and no monomial summand of $[w'(A,B)]_{1,1+s}$ divides m, then $m\psi_r(m')$ is a monomial summand of $[\,[w(A,B), w'(A,B)]\,]_{1,1+r+s}$ where $\psi_r(m')$ is the polynomial obtained from m' by adding r to each subscript.*

Proof By Lemma 3.2, $[\,[w(A,B), w'(A,B)]\,]_{1,1+r+s}$ equals

$$[w(A,B)]_{1,1+r}[w'(A,B)]_{1+r,1+r+s} - [w'(A,B)]_{1,1+s}[w(A,B)]_{1+s,1+s+r}$$

and $m\psi_r(m')$ divides the first term. However, since no monomial summand of $[w'(A,B)]_{1,1+s}$ divides m, it follows that $m\psi_r(m')$ is a monomial summand of $[\,[w(A,B), w'(A,B)]\,]_{1,1+r+s}$ as desired. □

By Theorem 2.2, the next case of interest is when $n = 11$. Mimicking the $n = 6$ case, we seek a word $c_{10}(a,b) \in \gamma_{10}(F) \cap F^{(3)}$ such that the polynomial

$[c_{10}(A,B)]_{1,11}$ has a monomial summand with coefficient ± 1. We then assign the value of 1 to the variables in this summand, and zero to the variables not in the summand. Since $F^{(2)}\gamma_6(F)/\gamma_6(F)$ has two generators, it follows from Lemma 3.1 that $F^{(3)}\gamma_{11}(F)/\gamma_{11}(F) = \langle c_{10}(a,b)\gamma_{11}(F)\rangle$ is cyclic where

$$c_{10}(a,b) = [[[b,a,b],[b,a]], c_5(a,b)] = [[[b,a,b],[b,a]],[[b,a,a],[b,a]]].$$

We abbreviate the phrase "m is a monomial summand of p" by "$m \in p$". Now

$$\begin{aligned} m_5 &= \beta_1\beta_2\alpha_3\alpha_4\beta_5 \in [[[B,A,B],[B,A]]]_{1,6} \quad \text{and} \\ m_5' &= \alpha_1\alpha_2\beta_3\beta_4\alpha_5 \in [c_5(A,B)]_{1,6}. \end{aligned}$$

Hence by Lemma 3.3

$$m_{10} = m_5\psi_5(m_5') = \beta_1\beta_2\alpha_3\alpha_4\beta_5\alpha_6\alpha_7\beta_8\beta_9\alpha_{10} \in [c_{10}(A,B)]_{1,11}.$$

Setting $\beta_1 = \beta_2 = \alpha_3 = \cdots = \alpha_{10} = 1$ and $\alpha_1 = \alpha_2 = \beta_3 = \cdots = \beta_{10} = 0$ shows that U_{11} has a 2-generated subgroup of derived length 4.

Theorem 3.4 *Let $d = \lfloor \log_2(n-1)\rfloor + 1$. Then U_n has a 2-generated subgroup of derived length d if and only if $\frac{21}{32}2^d < n \le 2^d$.*

Proof Suppose that U_n has a 2-generated subgroup G of derived length d. It follows from Theorem 2.2 that $\frac{5}{8}2^d < n \le 2^d$. However, if $0 \le d < 5$ then $\lfloor \frac{5}{8}2^d \rfloor = \lfloor \frac{21}{32}2^d \rfloor$. Hence $\frac{21}{32}2^d < n \le 2^d$ for $d < 5$. Suppose now that $d \ge 5$. We showed in the preamble to this theorem that $F^{(3)}\gamma_{11}(F)/\gamma_{11}(F)$ is cyclic. Hence by Lemma 3.1, $F^{(4)} \subseteq \gamma_{21}(F)$. For $d \ge 5$, a simple induction shows that $F^{(d-1)} \subseteq \gamma_{21\cdot 2^{d-5}}(F)$. Since $G^{(d-1)} \subseteq \gamma_{21\cdot 2^{d-5}}(G)$ and $\gamma_n(G) = \{I\}$ it follows that $21\cdot 2^{d-5} < n \le 2^d$ as desired.

Conversely, suppose $\frac{21}{32}2^d < n \le 2^d$. If $d = 0,1,2,3,4$, then the values of $n = \lfloor \frac{21}{32}2^d \rfloor + 1$ are $1,2,3,6,11$ respectively. In each of these cases we have shown that U_n has a 2-generated subgroup of derived length d. Suppose henceforth that $d \ge 5$. We shall give a recursive procedure for constructing a 2-generated subgroup of U_n. It suffices to do this for $n = 21\cdot 2^{d-5}+1$.

We use induction on d. The initial case when $d = 5$ and $n = 22$ requires the most lengthy calculations. Note that the hypothesis in Lemma 3.3 that no monomial summand of $[w'(A,B)]_{1,1+s}$ divides m is easily verified in the case when the first s variables of m have a different number of α's than one (and hence every) summand of $[w'(A,B)]_{1,1+s}$. A lengthy argument which repeatedly uses this observation and the Multiplication Lemma shows that

$$\begin{aligned} m_{21} &= -\alpha_1\alpha_2\alpha_3\beta_4\beta_5\alpha_6\psi_6(m_5)\psi_{11}(m_{10}) \in c_{21}(a,b) \\ m_{21}' &= \alpha_1\alpha_2\beta_3\beta_4\beta_5\alpha_6\psi_6(m_5)\psi_{11}(m_{10}) \in c_{21}'(a,b) \\ m_{21}'' &= -\beta_1\beta_2\beta_3\alpha_4\alpha_5\beta_6\psi_6(m_5)\psi_{11}(m_{10}) \in c_{21}''(a,b) \end{aligned}$$

where

$$\begin{aligned} c_{21}(a,b) &= [[[[b,a,a,a],[b,a]], c_5(a,b)], c_{10}(a,b)] \\ c_{21}'(a,b) &= [[[[b,a,a,b],[b,a]], c_5(a,b)], c_{10}(a,b)] \\ c_{21}''(a,b) &= [[[[b,a,b,b],[b,a]], c_5(a,b)], c_{10}(a,b)]. \end{aligned}$$

This proves the result for $d = 5$ because the polynomial $[c_{21}(A, B)]_{1,22}$ has a monomial summand with coefficient ± 1. The number of α's in m_{21}, m''_{21}, m'_{21} is congruent to 0, 1, 2 modulo 3 respectively, and so by the Multiplication Lemma

$$\begin{aligned} m'_{21}\psi_{21}(m''_{21}) \in d_{21}(a,b) &= [c'_{21}(a,b), c''_{21}(a,b)] \\ m''_{21}\psi_{21}(m_{21}) \in d'_{21}(a,b) &= [c''_{21}(a,b), c_{21}(a,b)] \\ m_{21}\psi_{21}(m'_{21}) \in d''_{21}(a,b) &= [c_{21}(a,b), c'_{21}(a,b)]. \end{aligned}$$

The argument may be applied repeatedly as the number of α's occurring in $m'_{21}\psi_{21}(m''_{21})$, $m''_{21}\psi_{21}(m_{21})$, $m_{21}\psi_{21}(m'_{21})$ is congruent to 0, 1, 2 modulo 3 respectively. This completes the inductive proof. □

Acknowledgment. I would like to thank Peter Pleasants and the referee for their helpful comments. This research was partially supported by URC grant 6294-1341-70766-15.

References

[1] N. Blackburn, *On a special class of p-groups*, Acta Math. **100** (1958), 45–92.
[2] M. Hall, *The theory of groups*, Macmillan, 1964.
[3] B. Huppert, *Endliche Gruppen I*, Springer-Verlag, 1967.

ON p-PRONORMAL SUBGROUPS OF FINITE p-SOLUBLE GROUPS

M. GÓMEZ-FERNÁNDEZ

Departamento de Matemática e Informática, Universidad Pública de Navarra, Campus de Arrosadía. 31006 Pamplona, Spain

Throughout this note we will denote by p a fixed prime number. All groups considered will be finite.

In the theory of groups it is well known that the formula "subnormal + pronormal = normal". In this note, we define an embedding property of subgroups such that the previous formula with p-subnormal instead subnormal is also true. We call this property, which is stronger than pronormality, p-pronormality.

We give tests for p-pronormality that will be used in inductive proofs. By means of these we can show that the introduced concept is essentially new, in the sense that p-pronormality is not a particular case of the already known property of $\mathfrak{F}$-pronormality for any saturated formation $\mathfrak{F}$.

Recall that if G is a group, P a Sylow subgroup of G and $H \leq G$ it is said that P reduces into H if $P \cap H$ is a Sylow subgroup of H.

Definition 1 Let G be a group and H a subgroup of G. Then H is said to be *p-pronormal* in G if each Sylow p-subgroup P of G reduces into an unique conjugate subgroup of H in G; i.e. if $P \cap H \in \mathrm{Syl}_p(H)$ and $P \cap H^g \in \mathrm{Syl}_p(H^g)$, then $g \in N_G(H)$.

Notice that the Sylow p-subgroups of a normal subgroup of a group G are p-pronormal subgroups in G. Moreover it follows from an argument by Mann that every p-pronormal subgroup is also pronormal (see [2, Ch. I, proof of Theorem 6.6]).

We collect some basic results from p-pronormality.

Proposition 2 *Let G be a group and H a p-pronormal subgroup of G.*

i) *If K is a subgroup of G such that $H \leq K$, then H is p-pronormal in K.*

ii) *If N is a normal subgroup of G, then HN/N is p-pronormal in G/N.*

iii) *If N is a normal subgroup of H and H/N is p-pronormal in G/N, then H is p-pronormal in G.*

iv) *If N is a normal subgroup of G, then HN is p-pronormal in G.*

v) *If H is p-subnormal in G, then H is a normal subgroup of G.*

vi) *$N_G(H)$ is both p-pronormal and selfnormalizing in G.*

Proof These are straightforward consequences of the definition. □

An important property of p-pronormal subgroups is the following:

Proposition 3 *Let H be a p-pronormal subgroup of a group G. If $P \in \mathrm{Syl}_p(G)$ and P reduces into H, then $N_G(P) \leq N_G(H)$. In particular, if H is a subgroup such that $|G : H|$ is a p'-number, H is p-pronormal in G if and only if there exists $P \in \mathrm{Syl}_p(G)$ verifying $N_G(P) \leq N_G(H)$.*

Proof Suppose that there exists $g \in N_G(P) \backslash N_G(H)$. Hence P reduces into two different conjugates of H, a contradiction. Thus $N_G(P) \leq N_G(H)$.

Now we assume that $N_G(P) \leq N_G(H)$ for some $P \in \mathrm{Syl}_p(G)$. We suppose also that Q, $Q^g \in \mathrm{Syl}_p(G)$ reduce into H. Since $|G : H|$ is a p'-number, thus $\mathrm{Syl}_p(H) \subseteq \mathrm{Syl}_p(G)$.In this case there exists $h \in H$ such that $Q^g = Q^h$.

On the other hand Q, $P \in \mathrm{Syl}_p(N_G(H))$, then $Q = P^x$ with $x \in N_G(H)$. It follows that $P^{xg} = Q^g = Q^h = P^{xh}$ and $xhg^{-1}x^{-1} \in N_G(P) \leq N_G(H)$. Hence $g \in N_G(H)$, and so H is p-pronormal in G, as required. □

It is easy to see, from the previous proposition, that if H is a p-pronormal subgroup of a group G then $O_p(G) \leq N_G(H)$.

The pronormality criterion, due to Gaschütz (cf. [2, Ch. I, Prop. 6.4]), suggests the next result.

Proposition 4 *Let H be a subgroup of a group G, and let N be a normal subgroup of G such that $O_p(N) = O^{p'}(N)$. Then the following statements are equivalent:*

i) *H is a p-pronormal subgroup of G,*

ii) *H is p-pronormal in $N_G(HN)$ and HN is p-pronormal in G.*

Proof From Proposition 2 (i) and (iv), it is clear that statement (i) implies (ii).

Therefore suppose that statement (ii) holds. Let $g \in G$, and let P a Sylow p-subgroup of G such that P and P^g reduce into H. Because N is a p-closed normal of G, $O_p(N)$ is the unique Sylow p-subgroup of N and $O_p(N)$ is normal in G. Whence, it is easy to see that P and P^g reduces into HN. Since HN is p-pronormal in G, we have $g \in N_G(HN)$.

Using that there exists $Q \in \mathrm{Syl}_p(N_G(HN))$ such that $P \cap H = Q \cap H$, we have $P^g \cap HN = (Q \cap HN)^g$. It follows that $Q^g \cap H = P^g \cap H$ and clearly Q, Q^g reduce into H. Now, because we suppose that (ii) holds, we conclude that $g \in N_{N_G(HN)}(H) \leq N_G(H)$. Consequently H is p-pronormal in G. □

Notice that, in particular, Proposition 4 holds for subgroups that are either p-groups or p'-groups. Hence, working with p-soluble groups, we can use this result in inductive processes.

In our next results, we study sufficient conditions for p-pronormality of subgroups in finite groups. The first one is suggested by Wood (cf. [6]). The second one will be crucial to characterize the p-pronormal saturated formations.

Proposition 5 *Let N be a normal subgroup of a group G, and H a subgroup of N. Then H is p-pronormal in G if and only if the following two conditions are satisfied:*

i) *H is p-pronormal in N;*

ii) $G = N_G(H)N$.

Proof If H is p-pronormal in G, by Proposition 2 (i), statement (i) holds. Furthermore, it is easy to prove that H is a pronormal subgroup of G. Then $G = N_G(H)N$ (see [2, I, 6.3]).

Conversely, suppose that the statements (i) and (ii) hold. Assume that there exist P, $P^g \in \mathrm{Syl}_p(G)$ such that both reduce into H. By assumption we have $g = nx$ with $x \in N$ and $n \in N_G(H)$. Then

$$P^g \cap H = P^x n \cap H = (P^x \cap H)^n,$$

so that $P^x \cap H$ is a Sylow p-subgroup of H. Therefore $P \cap N$ and $P^x \cap N$ are Sylow p-subgroups of N reducing into H. It follows that $x \in N_G(H)$, so that also g belongs to $N_G(H)$, and H is p-pronormal in G. □

Theorem 6 *Let G be a p-soluble group and H a subgroup of G. The subgroup H is p-pronormal in G if and only if the following two conditions are satisfied:*

i) *H is p-pronormal in every proper subgroup K of G such that $H \leq K$;*

ii) *there exists $P \in \mathrm{Syl}_p(G)$ such that $N_G(P) \leq N_G(H)$.*

Proof If H is p-pronormal in G, by Propositions 2 (i) and 3, the claim is true.

To prove that H is a p-pronormal subgroup, we assume that there exists a counterexample G of minimal order: there exists a proper subgroup H of G such that H is not p-pronormal in G despite H verifies conditions (i) and (ii).

Firstly, notice that by (ii) it is clear that P reduces into $N_G(H)$. If P^g reduces into $N_G(H)$ too, then $P^g \in \mathrm{Syl}_p(N_G(H))$. Whence there exists $x \in N_G(H)$ such that $P^g = P^x$. It implies that $gx^{-1} \in N_G(P) \leq N_G(H)$ and then $g \in N_G(H)$.

Consider any minimal normal subgroup N of G. Since the claim is true for the subgroup HN/N in the group G/N, we deduce that HN is a p-pronormal subgroup of G. If $N_G(HN) < G$, we apply Proposition 4 to conclude that H is p-pronormal in G, a contradiction. Consequently $G = N_G(HN)$ for all minimal normal subgroup N of G.

Assume that there exists a minimal normal subgroup N of G such that $G = HN$. If N is a p'-subgroup, by Proposition 3 we have that H is p-pronormal in G. If N is a p-subgroup, then H is maximal in G. Moreover $H = N_G(H)$ and $N \leq P \leq N_G(P) \leq N_G(H) = H$. Hence $G = H$. In both cases we obtain a contradiction; therefore HN is a proper normal subgroup of G.

Now, we take a minimal normal p-subgroup N of G. Then H is p-subnormal in HN and subnormal in G. By (i) we have H is normal in any proper subgroup of G in which is contained. In others words, $N_G(H)$ is the unique maximal subgroup of G that contains H. Suppose that HN is contained in a maximal normal T of G. Hence $H \leq T$ and so $T \leq N_G(H)$. Since $|G : N_G(H)|$ is a p'-number, thus G/T is a chief p'-factor of G. It implies that T and $N_G(H)$ contains all the Sylow p-subgroups of G. Using first remark we deduce that $N_G(H) = G$, contradiction. Then $O_p(G) = 1$.

We take a minimal normal p'-subgroup N of G and denote $M = HN$. Since M is normal in G we can build $L = M\langle g\rangle$ with $g \in G \setminus N_G(H)$ such that P^g reduces into H.

Suppose that $L < G$. By assumption H is p-pronormal in L. Moreover $P \cap H \in \mathrm{Syl}_p(M)$ and then $P \cap H = P \cap M$. Analogously $P^g \cap H \in \mathrm{Syl}_p(H) \subseteq \mathrm{Syl}_p(M)$; thus $P^g \cap H = P^g \cap M = (P \cap M)^g = (P \cap H)^g$. On the other hand there exists $Q \in \mathrm{Syl}_p(L)$ such that $P \cap H = Q \cap H$. It implies that $Q \cap H \in \mathrm{Syl}_p(M)$, and so $Q \cap H = Q \cap M$. Then $P^g \cap H = (P \cap H)^g = (Q \cap H)^g = (Q \cap M)^g = Q^g \cap M$. It follows that $Q^g \cap M \in \mathrm{Syl}_p(H)$. Since $Q^g \cap M \leq Q^g \cap H$ thus $Q^g \cap M = Q^g \cap H$. Because H is p-pronormal in L and Q, Q^g both reduce into H, we have that $g \in N_L(H) \leq N_G(H)$; a contradiction. Then $L = G$.

In this case G/M is a cyclic group; and evidently PM/M is normal in G/M. Then $G/M = N_{G/M}(PM/M) = N_G(P)M/M$. Moreover

$$G = N_G(P)M = N_G(P)HN = N_G(H)N.$$

By Proposition 5, we conclude that H is p-pronormal in G. This is the final contradiction. The minimal counterexample does not exist and the claim is true. □

Now, we turn our attention to the study of the class of soluble groups in which their projectors are p-pronormal subgroups.

Definition 7 Let $\mathfrak{V}$ be a universe and $\mathfrak{X}$ a Schunk class. We say that $\mathfrak{X}$ is a *p-pronormal Schunck class*, if for any $G \in \mathfrak{V}$, $\mathrm{Proy}_{\mathfrak{X}}(G)$ is a set of p-pronormal subgroups of G. Analogously, we define a *p-pronormal saturated formation* $\mathfrak{F}$.

In the sequel every class of groups considered is a class of soluble groups.

Proposition 8 *Let G be a non simple primitive group with abelian socle. If M is a core free maximal subgroup of G, then M is p-pronormal in G if and only if there exists $P \in \mathrm{Syl}_p(G)$ such that $N_G(P) \leq M$.*

Proof Suppose that M is p-pronormal in G. Notice that if $|G : M|$ is a power of p, then M is p-subnormal in G. By Proposition 2 (v) it follows that M is normal in G. Thus $M = 1$ and G is simple group, contradiction. Then p does not divide to $|G : M|$. Moreover, from Proposition 3, there exists $P \in \mathrm{Syl}_p(G)$ such that $N_G(P) \leq N_G(M) = M$.

Conversely, since $|G : M|$ is a p'-number, we can apply the Proposition 3 and we deduce that the claim is true. □

Recall that if $\mathfrak{H}$ is a Schunck class, define the avoidance class of $\mathfrak{H}$

$$a(\mathfrak{H}) = (G \in \mathfrak{V} : H \cap \mathrm{Soc}(G) = 1 \text{ for all } H \in \mathrm{Proy}_{\mathfrak{H}}(G)).$$

Lemma 9 *Let $\mathfrak{H}$ be a p-pronormal Schunck class. If $G \in b(\mathfrak{H})$, then either G is a simple group or $G \in a(\mathfrak{S}_p\mathfrak{S}_{p'})$.*

Proof Consider $G \in b(\mathfrak{H})$ such that G is not a simple group. Then G is a primitive group that can be factorized $G = MN$, with $N = \mathrm{Soc}(G)$ and $M \in \mathfrak{H}$ a maximal subgroup of G.

Furthermore $M \in \mathrm{Proy}_{\mathfrak{H}}(G)$, and thus M is p-pronormal in G. By Theorem 6 there exists $P \in \mathrm{Syl}_p(G)$ such that $H = N_G(P) \leq M$. Then we have $H \cap N = 1$. Since $H \in \mathrm{Proy}_{\mathfrak{S}_p\mathfrak{S}_{p'}}(G)$ we conclude that $G \in a(\mathfrak{S}_p\mathfrak{S}_{p'})$. □

Proposition 10 *Let $\mathfrak{H}$ be a p-pronormal Schunck class, then*

$$\mathfrak{H} = \mathfrak{Q}^{\pi} \cap \mathfrak{H}_0,$$

where $\mathfrak{H}_0$ denotes a Schunk class such that $\mathfrak{S}_p\mathfrak{S}_{p'} << \mathfrak{H}_0$ and $\mathfrak{Q}^{\pi}$ denotes the class of π-perfect groups, with $\pi = \mathbb{P}\setminus$ Char($\mathfrak{H}$).

Proof We use the following notation:

1. $b_0(\mathfrak{H}) = (G \in b(\mathfrak{H}) : G$ is not simple$)$, $h(b_0(\mathfrak{H})) = \mathfrak{H}_0$;
2. $b_1(\mathfrak{H}) = (G \in b(\mathfrak{H}) : G$ is simple$)$, $h(b_1(\mathfrak{H})) = \mathfrak{H}_1$.

Notice that $b(\mathfrak{H})$ is a Schunck Q-boundary and $b(\mathfrak{H}) = b(\mathfrak{H}_0) \cup b(\mathfrak{H}_1)$ with $b(\mathfrak{H}_0) \cap b(\mathfrak{H}_1) = \emptyset$. It follows that $b(\mathfrak{H}_0)$ and $b(\mathfrak{H}_1)$ are Schunck Q-boundaries and $\mathfrak{H}_0 \cap \mathfrak{H}_1 = \mathfrak{H}$.

From Lemma 9, we have $b(\mathfrak{H}_0) \subseteq a(\mathfrak{S}_p\mathfrak{S}_{p'})$. We apply [2, Lemma 1.5 , VI] and we deduce that $\mathfrak{S}_p\mathfrak{S}_{p'} << \mathfrak{H}_0$. On the other hand, we consider $\pi = \{p \in \mathbb{P} : C_p \in b(\mathfrak{H}_1)\}$. Since $b(\mathfrak{H}_1)$ consists of abelian simple groups, from [2, Th. 4.2, Ch. III], we conclude that $\mathfrak{H}_1$ is a normal Schunck class.

Now using the characterization of Blessenohl and Gaschütz (cf. [2, Ch. III, Th. 4.4]), we deduce $\mathfrak{H}_1 = \mathfrak{Q}^{\pi}$ as required. □

Theorem 11 *Let $\mathfrak{F}$ be a saturated formation. Then $\mathfrak{F}$ is a p-pronormal saturated formation if and only one of following statements holds:*

i) $\mathfrak{F} = (1)$,

ii) $\mathfrak{F} = \mathfrak{S}_p$, *or*

iii) $\mathfrak{S}_p\mathfrak{S}_{p'} << \mathfrak{F}$.

Proof Assume that $\mathfrak{F}$ is a p-pronormal saturated formation. We denote by $\pi = \mathrm{Char}(\mathfrak{F})$ and we consider three possible cases.

(1) Suppose $p \notin \pi$. Suppose that there exists also a prime $q \in \pi$. Let C be the cyclic group of order q. Then we can build the semidirect product $G = [V]C$, with V an irreducible and faithful C-module over GF(p). Since $p \notin \pi$, thus $G \notin \mathfrak{F}$. By [2, Ch. III, Prop. 3.23(a)] we have C is a $\mathfrak{F}$-projector of G. Then, by hypothesis, C is p-pronormal in G. Moreover C is a normal subgroup of G because C is a p'-group. Since $(|C|, |V|) = 1$, thus $C \leq C_G(V) = V$, contradiction. Consequently $\pi = \emptyset$ and then $\mathfrak{F} = (1)$.

(2) If $\pi = \{p\}$ it is clear that $\mathfrak{F} = \mathfrak{S}_p$.

(3) We claim that if $|\pi| \geq 2$, then $\pi = \mathbb{P}$. Consider $\{p, q\} \subseteq \pi$ with $p \neq q$, and suppose that there exists $r \in \mathbb{P}$ such that $r \notin \pi$. Denote by H the cyclic

group of order q. Then for all $p, r \neq q$ there exist irreducible and faithful H-modules V_p, V_r, over $\mathrm{GF}(p)$ and $\mathrm{GF}(r)$ respectively. Whence we construct the group $G = [V_p \times V_r]H$. Since $r \notin \pi$ thus $G \notin \mathfrak{F}$.

We take the primitive group $T = [V_p]H$. Suppose that $T \notin \mathfrak{F}$, since $H \in \mathfrak{F}$ we can argue as in (1) to deduce that H is normal in G. This contradicts the fact that $\mathrm{core}_T(H) = 1$. Then $T \in \mathfrak{F}$. Notice that $G = V_r T$ thus arguing as in (1) again, we obtain that T is p-pronormal in G.

Since $\mathrm{Syl}_p(G) = \{V_p\}$, from Proposition 3, we can deduce that $G = N_G(V_p) \leq N_G(T)$. Thus T is normal in G and $T \cap V_r = 1$. Moreover $T \leq C_G(V_r) = V_p \times V_r$, contradiction. Consequently $\pi = \mathbb{P}$.

Therefore by Lemma 9 and Proposition 10 $\mathfrak{F} = \mathfrak{Q}^\pi \cap \mathfrak{F}_0$. In this case $\pi = \emptyset$ then $\mathfrak{Q}^\pi = \mathfrak{S}$. Consequently $\mathfrak{S}_p\mathfrak{S}_{p'} << \mathfrak{F}$.

Conversely it is clear that (1) and $\mathfrak{S}_p$ are p-pronormal saturated formations. In order to prove that if $\mathfrak{F}$ verifies (iii), then $\mathfrak{F}$ is a p-pronormal saturated formation, we consider $G \in \mathfrak{F}$ of minimal order such that for certain $H \in \mathrm{Proy}_{\mathfrak{F}}(G)$, and H is not p-pronormal in G.

If K is a proper subgroup of G such that $H \leq K$, then $H \in \mathrm{Proy}_{\mathfrak{F}}(K)$ and H is p-pronormal in K, by minimality of G.

On the other hand since $\mathfrak{S}_p\mathfrak{S}_{p'} << \mathfrak{F}$, thus $N_G(P) \leq H \leq N_G(H)$. By Theorem 6, we have that H is p-pronormal in G, a contradiction. Then the minimal counterexample does not exit and the claim is true. □

Our next objective is to prove that it does not exist any saturated formation $\mathfrak{F}$ in which the concepts of $\mathfrak{F}$-pronormality and p-pronormality are equivalent. We take as definition of $\mathfrak{F}$-pronormality the characterizations of this property that appear in [4] and [5].

Proposition 12 (i) *Let G be a group, then every p-pronormal subgroup of G is $\mathfrak{S}_p\mathfrak{S}_{p'}$-pronormal in G.*

(ii) *There exist examples of groups in which some $\mathfrak{S}_p\mathfrak{S}_{p'}$-pronormal subgroups are not p-pronormal subgroups.*

Proof To prove (i), we assume that there exists a counterexample G of minimal order. Then we take H a p-pronormal subgroup of G such that H is not $\mathfrak{S}_p\mathfrak{S}_{p'}$-pronormal in G.

Suppose that N is a minimal normal subgroup of G such that $N \leq H$. From Proposition 2 (ii) H/N is p-pronormal in G/N. It follows, by minimality of G, that H/N is $\mathfrak{S}_p\mathfrak{S}_{p'}$-pronormal in G/N. Since H is $\mathfrak{S}_p\mathfrak{S}_{p'}$-pronormal in $N_G(H)$, we apply [5, 1.22] and we deduce that H is $\mathfrak{S}_p\mathfrak{S}_{p'}$-pronormal in G, contradiction. Then $\mathrm{Core}_G(H) = 1$.

Moreover, for every minimal normal subgroup N of G, H is a proper subgroup of HN. Notice that, from Proposition 2 (ii) and minimality of G, we obtain that HN/N is $\mathfrak{S}_p\mathfrak{S}_{p'}$-pronormal in G/N.

If $N_G(HN)$ is a proper subgroup of G, using the minimality of G again, H is $\mathfrak{S}_p\mathfrak{S}_{p'}$-pronormal in $N_G(HN)$. By [5, 1.22], we have H is p-pronormal in G, a contradiction. Then $G = N_G(HN) = N_G(H)N$; i.e. HN is a normal subgroup of G.

Assume that HN is a proper subgroup of G, then H is $\mathfrak{S}_p\mathfrak{S}_{p'}$-pro–normal in HN. From [5, 1.23] we deduce that H is $\mathfrak{S}_p\mathfrak{S}_{p'}$-pronormal in G, contradiction. Therefore $G = HN$. It follows that H is a core free maximal subgroup of G, and so G is a primitive group. Moreover, from Proposition 3, we obtain that $O_p(G) = 1$. It implies that N is a p'-group.

If H is a $\mathfrak{S}_p\mathfrak{S}_{p'}$-normal maximal subgroup of G, then $G/\operatorname{core}_G(H) = H \in \mathfrak{S}_p\mathfrak{S}_{p'}$. Furthermore $G \in \mathfrak{S}_{p'}$ and so H is p-subnormal and p-pronormal in G. It follows that H is a normal subgroup of G. Consequently $H = 1$, and so H is $\mathfrak{S}_p\mathfrak{S}_{p'}$-pronormal in G, contradiction. Therefore H is a $\mathfrak{S}_p\mathfrak{S}_{p'}$-critical maximal subgroup of G. It implies that $H = N_G(H)$ contains a $\mathfrak{S}_p\mathfrak{S}_{p'}$-normalizer of G. Now, we apply [5, 1.31] and we deduce that H is $\mathfrak{S}_p\mathfrak{S}_{p'}$-pronormal in G; a contradiction. Then does not exist a counter–example to the assertion (i).

(ii) Consider a prime $q \neq p$ and let C the cyclic group of order p. We take the group $H = [V]C$, with V an irreducible and faithful C-module over GF(q). Notice that there exists an irreducible and faithful H-module W over GF(p). We construct the semidirect product $G = [W]H$.

On the other hand $C \in \mathfrak{S}_p\mathfrak{S}_{p'}$ and C is a $\mathfrak{S}_p\mathfrak{S}_{p'}$-critical maximal subgroup of H. Moreover, H is a $\mathfrak{S}_p\mathfrak{S}_{p'}$-critical maximal of G. Then C is a $\mathfrak{S}_p\mathfrak{S}_{p'}$-normalizer of G. Since H is selfnormalizing in G, we deduce from [5, 1.31] that H is $\mathfrak{S}_p\mathfrak{S}_{p'}$-pronormal in G. But $O_p(G) = W$ is not contained in H. Then, using Proposition 4, we obtain that H is not p-pronormal in G. □

Theorem 13 *It does not exist any saturated formation $\mathfrak{F}$ satisfying the property:*
(*) *for each group G, if $H \leq G$, then H is a $\mathfrak{F}$-pronormal subgroup of G if and only if H is a p-pronormal subgroup of G.*

Proof Assume that $\mathfrak{F}$ is a saturated formation that satisfies property (*).

Step 1. $\mathfrak{N} \subseteq \mathfrak{F}$, in other words Char($\mathfrak{F}$) is the set of all prime numbers. Suppose that there exists a prime q such that $q \notin$ Char($\mathfrak{F}$). In this case $\mathfrak{F} \subseteq \mathfrak{S}_{q'}$ and then, each group X verifies that $O^{q'}(X) \leq X^{\mathfrak{F}}$. For every prime r such that $q \neq r \neq p$ we construct the group $G = [V]C$, with C the cyclic group of order r and V an irreducible and faithful C-módule over $GF(q)$.

Notice that for every $g \in G$, $\langle C, C^g\rangle$ either is C, if $g \in N_G(C) = C$, or G if $g \notin C$. Moreover $C^g = C^x$ for certain $x \in V = O^{q'}(\langle C, C^g\rangle) \leq \langle C, C^g\rangle^{\mathfrak{F}}$.

By [5, 1.18] we deduce that C is $\mathfrak{F}$-pronormal in G. Thus C is p-pronormal in G because $\mathfrak{F}$ satisfies property (*). But C is p'-group and then C would be normal in G, contradiction. It follows that for every prime q the claim is true.

Step 2. $\mathfrak{S}_p\mathfrak{S}_{p'} \subseteq \mathfrak{F}$. Consider a group $G \in \mathfrak{S}_p\mathfrak{S}_{p'} \setminus \mathfrak{F}$ of minimal order. Then $G \in b(\mathfrak{F})$ and G is a primitive group factorized by $G = MN$, with M $\mathfrak{F}$-projector of G and N the unique minimal normal of G. Notice that M contains an $\mathfrak{F}$-normalizer of G. Furthermore M is an $\mathfrak{F}$-normalizer of G. Since $\mathfrak{N} \subseteq \mathfrak{F}$, from step 1, applying [5, 1.31] we obtain that M is $\mathfrak{F}$-pronormal in G, and so M is p-pronormal in G.

If N is a p-group, then M is p-subnormal in G. If N is a p'-group, it implies that $G \in \mathfrak{S}_{p'}$ and thus M is p-subnormal in G too. In both cases, from Proposition 2 (iv), M is a normal subgroup of G. Consequently $M = 1$ and G is a cyclic group. It follows that $G \in \mathfrak{N} \subseteq \mathfrak{F}$, contradiction. Then $\mathfrak{S}_p\mathfrak{S}_{p'} \subseteq \mathfrak{F}$.

Step 3. $\mathfrak{F} \subseteq \mathfrak{S}_p\mathfrak{S}_{p'}$. Consider a group $G \in \mathfrak{F} \setminus \mathfrak{S}_p\mathfrak{S}_{p'}$ of minimal order. Then $G \in b(\mathfrak{S}_p\mathfrak{S}_{p'})$; i.e. G is a primitive group factorized by $G = MN$, with M a $\mathfrak{S}_p\mathfrak{S}_{p'}$-projector of G and $N = O_{p'}(G)$ unique minimal normal of G.

If $P \in \mathrm{Syl}_p(G)$ such that $P \leq M$, then $M = N_G(P)$. We apply Proposition 8 and we deduce that M is p-pronormal en G. Since $\mathfrak{F}$ satisfies (*), the subgroup M is $\mathfrak{F}$-pronormal in G. It follows, from [5, 1.31], that M contains a $\mathfrak{F}$-normalizer of G. Now since $G \in \mathfrak{F}$ thus $G = M$, contradiction. Then the claim is true.

Step 4. Conclusion. Suppose that there exists a saturated formation $\mathfrak{F}$ satisfying the property (*). By steps 2 and 3, $\mathfrak{F} = \mathfrak{S}_p\mathfrak{S}_{p'}$. We apply Proposition 12 to deduce that the formation $\mathfrak{S}_p\mathfrak{S}_{p'}$ not satisfy property (*). Then the formation $\mathfrak{F}$ satisfying the Property (*) does not exist. □

Acknowledgements

The author would to like to thank Professor L. M. Ezquerro for his very helpful comments and suggestions while this article was being written.

References

[1] S. V. Aleshin. "Finite automata and the Burnside problem for periodic groups", *Mat. Zametki* **11**(1972), 319-328.

[2] D. Blessenohl and W. Gaschütz, "Über normale Schunck-und Fitting–klassen", *Math. Z.* **118**(1970), 1-8.

[3] K. Doerk and T. O. Hawkes, *Finite Soluble Groups*, De Gruyter Expositions in Mathematics 4, Berlin 1992.

[4] B. Huppert, *Endliche Gruppen I*, Springer-Verlag, 1967.

[5] N. Müller, *$\mathfrak{F}$-pronormale Untergruppen endlichen auflösbarer Gruppen*, Diplomarbeit Johannes Gutenberg-Universität, Mainz , 1991.

[6] M. C. Pedraza, *Subgrupos $\mathfrak{F}$-pronormales e inmeros normalmente de grupos finitos resolubles*, Memoria de Licenciatura. Dto de Algebra; U. De Valencia, (1993) (Spanish).

[7] G. J. Wood, "On pronormal subgroups of finite soluble groups", *Arch. Math. (Basel)* **25** (1974), 578-588.

ON THE SYSTEM OF DEFINING RELATIONS AND THE SCHUR MULTIPLIER OF PERIODIC GROUPS GENERATED BY FINITE AUTOMATA

R. I. GRIGORCHUK[1]

Steklov Mathematical Institute, Gubkina Street 8, Moscow 117966, Russia

Abstract

Proofs are given that the 3-generator 2-group G of intermediate growth constructed in [Gr 80] is not finitely presented. A method is suggested for constructing an independent system of relations for this group and for other periodic groups generated by finite automata, for example the Gupta and Sidki groups [Gu1]. It is proved that $H_2(G, \mathbb{Z}) \cong \mathbb{Z}_2^\infty$ and that the system of defining relations for G constructed by Lysionok in [Ly] is independent. Thus we answer a question of G. Baumslag in [Ba] for this particular group.

1 Introduction

Infinite finitely generated periodic groups (also called Burnside groups) play a special role in group theory. The question of their existence constituted the essence of the so called general Burnside problem which was resolved by Golod [Go] using the Golod-Shafarevich construction [Sa].

Subsequently papers containing examples of Burnside groups appeared, written by P. S. Novikov and Adyan [No], Aleshin [Al], Sushansky [Su], Grigorchuk [Gr 80], Gupta and Sidki [Gu1] and other authors.

The groups from the papers [Al, Go, Gr 80, Gu1, Su] are residually finite but they do not have finite exponent. In contrast, the free periodic Burnside groups $B(m,n) = \langle a_1, \ldots, a_m \mid x^n = 1 \rangle$, $m \geq 2$, $n = 2k+1 \geq 665$ considered in [No] have finite exponent but are not residually finite. The fact that these two properties are not compatible follows from a theorem of Zelmanov [Ze 90, Ze 91].

An arbitrary countably generated residually finite p-group (where p is a prime) can be embedded into the group of automorphisms of a homogeneous rooted tree T_n, $n = p^k$ with branching number n (where k is an arbitrary natural number).

Any such automorphism is determined by a labelling of the vertices by elements of the symmetric group S_n of degree n (the elements of S_n determine the action of the automorphism on the edges incident to the given vertex). Moreover, in the embedding theorem mentioned above, one can limit oneself to a subgroup Aut T_n consisting of automorphisms which act at each vertex via a power of a fixed cyclic permutation ε of the set of n edges.

Among the groups with elements of the type described, we highlight the class GGS of periodic groups generated by finite automata and considered in [Al, Su,

[1]The author acknowledges the support of the Russian Foundation for Fundamental Research, grant 96-01-00974 and INTAS, grant 94-3420.

Gr 80, Gu1] and other publications (the notation for the class is taken from [Ba]).

Without giving a precise definition here of this class we note that the groups in the class GGS are generated by automorphisms $a, b, \ldots, d$ of the tree T_n, $n = p^k$, where p is a prime and a acts as a cyclic permutation ε of the edges incident to the root vertex, and the automorphisms $b, \ldots, d$ are characterized by the fact that the non-trivial part of their action is in a neighbourhood of the right-most side infinite path connecting the root vertex to infinity.

Examples of such groups include both of the groups constructed in [Gr 80]. The first group (which we will denote by G and which will be the main object of study in this paper) is generated by four automorphisms a, b, c, d of the tree T_2 defined by their labellings in Diagram 1 (1 designates the identity permutation; the part of the tree T_2 not given in the diagram is labelled with '1's).

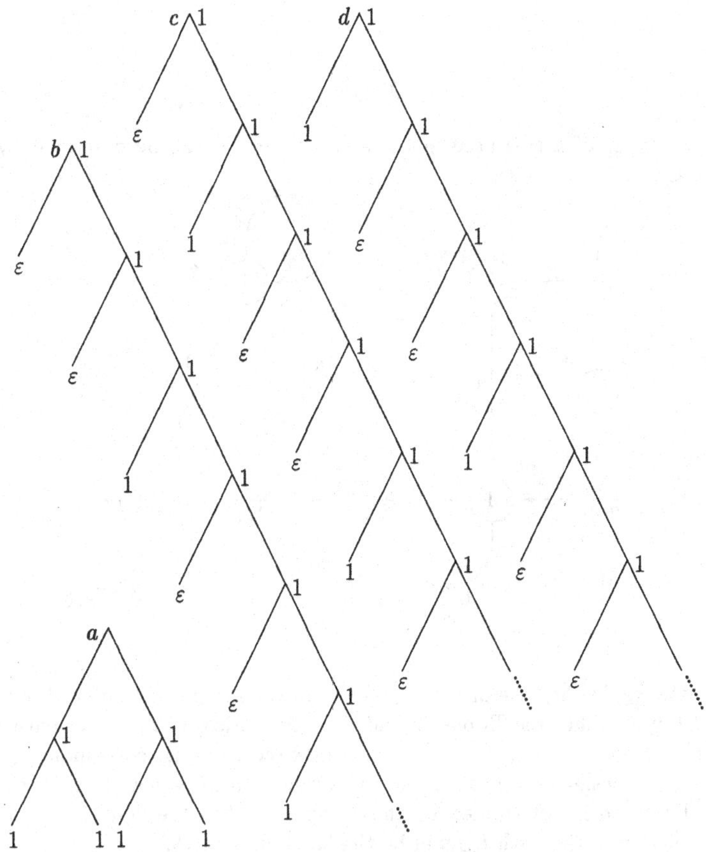

Diagram 1

On the other hand the second group from [Gr 80] is generated by two automorphisms α, β of the tree T_4 as shown on Diagram 2.

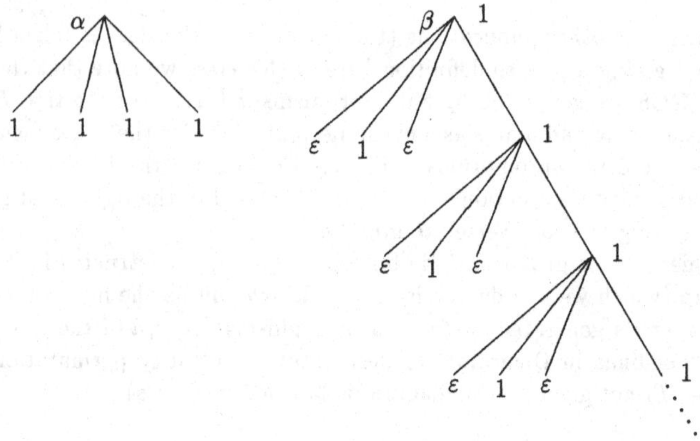

Diagram 2

The labellings of a tree used in the above examples can be described by finite automata.

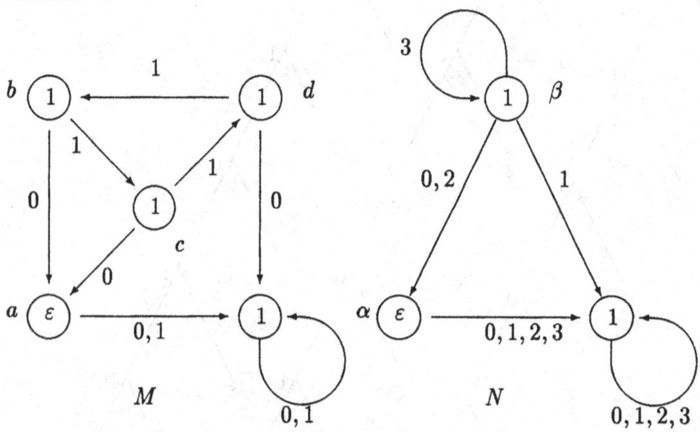

For instance, in the case of the group G and the generator b to get the label of some vertex x of the tree T_2 one should identify vertices of T_2 in the natural way with strings over the alphabet $\{0,1\}$. Then take the string w_x corresponding to the vertex x and consider the path l_{w_x} in the automaton M which starts in the state marked by b and is determined by the string w_x. Then the label (1 or ε) of the terminal vertex of the path l_{w_x} will be the label for x in T_2.

The particular labelling for the generators of G is related to the fact that G is isomorphic to a group generated by four initial automata M_a, M_b, M_c, M_d (obtained from automaton M by choosing a corresponding initial state) with respect to the operation of superposition of automata.

Similarly, the automaton N describes the labelling for generators α, β in the second example. It has only three states and this is the minimal number of states needed to produce an example of Burnside group. The method of study of the properties of the group G suggested in [Gr 80] turned out to be universal for GGS groups and was further developed in the papers of other authors.

In [Gr 80] it was first stated that GGS groups cannot be described by a finite set of defining relations. In the same paper a sketch of a proof was given that G is not a finitely presented group. Two different proofs of this statement (based on ideas different from those in [Gr 80] appeared in [Gr 84, Gr 85]. On the other hand there have been attempts to find "good" presentations of certain GGS groups. As early as the publication of it was already noted that they have a solvable word problem and hence are recursively presentable. In [Si] a presentation of a 3-group from [Gu1] is given and this is a much simpler presentation than that which lists as relators all those words which are equal to the identity in the group. However, Lysionok [Ly] found the most appropriate way to present a group from the class in question. He proved that the group G from [Gr 80] can be presented as:

$$\begin{aligned} G = \langle a,b,c,d \mid 1 = a^2 = b^2 = c^2 &= d^2 = bcd = \sigma^k((ad)^4) \\ &= \sigma^k((adacac)^4), k = 0,1,\ldots\rangle \end{aligned} \tag{1}$$

where the permutation σ is defined via

$$\sigma : \begin{cases} a \to & aca \\ b \to & d \\ c \to & b \\ d \to & c \end{cases} . \tag{2}$$

Recently this presentation was used in [Gr 97] to construct a finitely presented amenable but not elementary amenable group. This increased the interest in the question of finding presentations for other GGS groups analogous to the presentation (1).

In §3 we suggest a new approach to obtaining presentations of type (1) for GGS groups. We conjecture that any GGS group can be given by a presentation similar to (1), that is to say by a presentation with a finite set of defining words, some substitution and its iterations involved in the presentation. We propose to call such presentation an L-presentation.

In G. Baumslag's book [Ba], he draws attention to the question of the existence of good presentations for GGS groups. Also the relevance the Schur multiplier $H_2(\cdot,\mathbb{Z})$ (that is, the second integral homology group) for the question under consideration is emphasized there. It is well known that if it is infinite dimensional then there is no finite presentation for the group G. The problem of calculating $H_2(\cdot,\mathbb{Z})$ for GGS groups is also posed in [Se]. At present it is known that every GGS group is not finitely presented. In the next section we will give two simple proofs of the fact that the group G is not finitely presented. One of these proofs follows the line suggested in [Gr 80].

Let us recall that the group G has the following three fundamental properties which make it important for group theory:

(1) It is a Burnside group.

(2) It has an intermediate growth, [Gr 84].

(3) It has finite width (the factors $\gamma_k(G)/\gamma_{k+1}(G)$ are elementary abelian 2-groups and with ranks uniformly bounded by the number 3, [Ro]).

Moreover, has recently been discovered that there are only three essentially different classes of finitely generated just infinite groups: simple groups, hereditarily just infinite groups and branch groups. The group G is a typical representative of the class of branch groups and therefore the knowledge of properties of G leads to an understanding of the class of branch groups.

The main purpose of this work is to calculate the Schur multiplier of the group G and to prove that Lysionok's system of defining relations is independent. We shall now formulate the main results. To this end we will use another presentation for G:

$$\begin{aligned} G = \langle a,b,c,d \mid 1 &= a^2 = b^2 = c^2 = bcd = b^2c^2d^2(bcd)^{-2} = \sigma^k((ad)^4a^{-4}d^{-4}) \\ &= \sigma^k((adacac)^4a^{-12}c^{-8}d^{-4}),\ k = 0,1,\ldots\rangle \end{aligned} \tag{3}$$

(the fact that it is equivalent to (1) will be proved in Lemma 6). The reason for considering such a presentation will later become clear. We denote by r_n (respectively R_n), $n \geq 1$ the defining words from presentation (1) (presentation (3)) numbered in the order of their natural occurrence.

Let F be a free group with a basis a,b,c,d, $K = \langle r_n, n \geq 1\rangle_F^\sharp = \langle R_n, n \geq 1\rangle_F^\sharp$ where $\langle\cdot\rangle_F^\sharp$ denotes the operation of normal closure in F. Thus, $G \cong F/K$. We denote by $\overline{R}_n$ the image of the element R_n in the factor-group $K/[K,F]$. Let C denote an infinite cyclic group, C_2 a multiplicative group of order 2 and C_2^∞ a direct product of infinitely many copies of C_2.

Theorem 1 *There is an isomorphism*

$$K/[K,F] \simeq C^4 \times C_2^\infty. \tag{4}$$

Moreover $\overline{R}_1,\ldots,\overline{R}_4$ *are independent generators of the factor* C^4 *and* $\overline{R}_n$, $n \geq 5$ *are independent generators of the factor* C_2^∞. *Also we have isomorphisms*

$$K/K \cap F' \simeq C^4 \text{ and } K \cap F'/[K,F] \simeq C_2^\infty.$$

Theorem 2 *The Schur multiplier* $H_2(G,\mathbb{Z})$ *is isomorphic to* C_2^∞.

This statement is the corollary of the previous one and of Hopf's theorem because

$$H_2(G,\mathbb{Z}) \simeq K \cap F'/[K,F] \simeq C_2^\infty.$$

Theorem 3 *Lysionok's system of relators is independent.*

This statement is also a corollary of Theorem 1. However, we will postpone its proof until the end of §4. To conclude this section we note that it would be interesting to compute the higher homology groups $H_n(G,\mathbb{Z}), n \geq 3$ and also the cohomology groups $H^n(\widehat{G},\mathbb{Z}_2), n \geq 2$ where $\widehat{G}$ is the pro-2-group which is 2-completion of the group G.

2 On the finite presentability problem of the group G

The goal of this section is to prove that the group G is not finitely presented, and also to make necessary preparations for proving the main results. We are going to remind the reader of some of the properties of the group G (they can be found in more detail in [Gr 80, Gr 84, Ro]). The generators a, b, c, d have order 2, therefore every group element can be expressed as a positive word in the alphabet $A = \{a, b, c, d\}$. The elements b, c, d commute with each other and, together with the identity, form a subgroup isomorphic to the Klein group $C_2 \times C_2$. Therefore the following relations hold: $bc = d$, $bd = c$, $cd = b$ and any element $g \in G$ can be represented by a reduced word in the alphabet A, that is to say by a word in which a symbol a is followed by a symbol from the set $\{b, c, d\}$ and vice versa. We denote by $H < G$ a subgroup of index 2 which does not contain a. The elements in H stabilize the vertices on the first level of the tree T_2 and the embedding $\psi : H \to G \times G$ is defined by $\psi = \varphi_0 \times \varphi_1$ where the homomorphisms $\varphi_i : H \to G, i = 0, 1$ are defined as restrictions of the action of H on the subtrees in T_2 with root vertices situated at the first level in T_2. The group H is generated by the elements b, c, d, aba, aca, ada (in fact, the set c, d, aca, ada is a minimal set of generators) and $\varphi_i, i = 0, 1$ and ψ act on these generators as follows:

$$\varphi_0 : \begin{cases} b \to & a \\ c \to & a \\ d \to & 1 \\ aba \to & c \\ aca \to & d \\ ada \to & b \end{cases} \quad \text{and } \varphi_1 : \begin{cases} b \to & c \\ c \to & d \\ d \to & b \\ aba \to & a \\ aca \to & a \\ ada \to & 1 \end{cases} \tag{5}$$

$$\psi : \begin{cases} b \to & (a, c) \\ c \to & (a, d) \\ d \to & (1, b) \\ aba \to & (c, a) \\ aca \to & (d, a) \\ ada \to & (b, 1). \end{cases} \tag{6}$$

From this it follows that $\varphi_i : H \to G, i = 0, 1$ are epimorphisms and it follows from their geometric meaning that ψ is a monomorphism.

Let $B = \langle b \rangle_G^\sharp$, $D = \langle a, d \rangle_G$ be a dihedral group of order 8. It is easy to see that $B \cap D = \{1\}$ therefore $G/B \simeq D$ and $[G : B] = 8$.

Proposition 1 *$\psi(H)$ is a subgroup of index 8 in $G \times G$. A complement to it is the group $D_1 = \langle (a, 1), (d, 1) \rangle_{G \times G} \simeq D$. Additionally we have the following decomposition into a semidirect product $\psi(H) = (B \times B) \cdot D_2$, where $D_2 = \langle (a, d), (d, a) \rangle_{G \times G} \simeq D$.*

Proof Since $\psi(d) = (1, b)$ then $\psi(x^{-1}dx) = (1, y^{-1}by)$ for any element $x \in H$ with $\psi(x) = (z, y)$ for some z. Therefore the inclusion $\psi(H) \geq 1 \times B$ holds and hence $\psi(H) \geq B \times B$.

The group $\psi(H)$ also contains elements $\psi(c) = (a, d)$, $\psi(aca) = (d, a)$ and thus it contains the subgroup D_2. Since $B \cap D = \{1\}$ we have that $(B \times B) \cap D_2 = \{(1,1)\}$, that is to say $(B \times B) \cdot D_2$ is a semidirect product.

From the inclusion $\psi(H) \geq (B \times B) \cdot D_2$ and also from the fact that $|G : B| = 8$ it follows that $[(G \times G) : \psi(H)] \leq 8$. However, $\psi(H) \cap D_1 = \{(1,1)\}$. This fact is not difficult to establish by projecting all the groups under consideration onto the finite subtree in T_2 which contains the vertices from the first to the fourth level (see [Ro]). Thus all the statements of Proposition 1 hold. □

When studying the properties of the group G, an important role is played by algorithm $\mathcal{A}$ which solves the word problem. It is mentioned in [Gr 80] and [Gr 83] and described in [Gr 84]. Here we will describe it again since it is important for an understanding of the proofs given below.

To this end we denote by $P < F$ a subgroup of index 2 which consists of the elements $W \in F$ such that $\Sigma_a(W) = 0 \pmod 2$. Here W stands for a freely reduced word on the alphabet $A \cup A^{-1}$ and $\Sigma_a(W)$ denotes the exponent sum of the symbol a in the word W. It is easy to see that the group P is freely generated by the elements $a^2, b, c, d, aba^{-1}, aca^{-1}, ada^{-1}$. We denote by $\overline{\varphi}_i : P \to F$, $i = 0, 1$ the homomorphisms

$$\overline{\varphi}_0 : \begin{cases} a^2 \to & 1 \\ b \to & a \\ c \to & a \\ d \to & 1 \\ aba^{-1} \to & c \\ aca^{-1} \to & d \\ ada^{-1} \to & b \end{cases} \quad \text{and} \quad \overline{\varphi}_1 : \begin{cases} a^2 \to & 1 \\ b \to & c \\ c \to & d \\ d \to & b \\ aba^{-1} \to & a \\ aca^{-1} \to & a \\ ada^{-1} \to & 1 \end{cases} \tag{7}$$

and by $\overline{\psi} : P \to F \times F$ their direct product $\overline{\varphi}_0 \times \overline{\varphi}_1$:

$$\overline{\psi} : \begin{cases} a^2 \to & (1,1) \\ b \to & (a,c) \\ c \to & (a,d) \\ d \to & (1,b) \\ aba^{-1} \to & (c,a) \\ aca^{-1} \to & (d,a) \\ ada^{-1} \to & (b,1). \end{cases} \tag{8}$$

We draw the reader's attention to the connection between the tables (5), (6) and (7), (8).

Algorithm $\mathcal{A}$: In order to verify the equality $W \overset{G}{=} 1$, that is to say the inclusion $W \in K$, we do the following.

(1) Compute $\Sigma_a(W)$. If $\Sigma_a(W) \not\equiv 0 \pmod 2$ then $W \overset{G}{\neq} 1$.

(2) Suppose $\Sigma_a(W) \equiv 0 \pmod 2$ and that the word W does not contain the symbols a or a^{-1} so $W = W(b, c, d)$. Then we verify the relation $W = 1$ in the Klein 4 group $K = \{1, b, c, d\}$. In this case $W \overset{G}{=} 1 \Leftrightarrow W \overset{K}{=} 1$.

(3) Suppose $\Sigma_a(W) \equiv 0 \pmod 2$ and that W contains at least one symbol a or a^{-1}. Compute the words $W_i = \overline{\psi}_i(W)$, $i = 0,1$, make a free reduction of them and return to step 1 of the algorithm but this time verify two relations $W_i \overset{G}{=} 1$, $i = 0,1$ and bear in mind that $W \overset{G}{=} 1 \Leftrightarrow (W_i \overset{G}{=} 1,\ i = 0,1)$.

The algorithm described is a branching process portrayed by a binary tree.

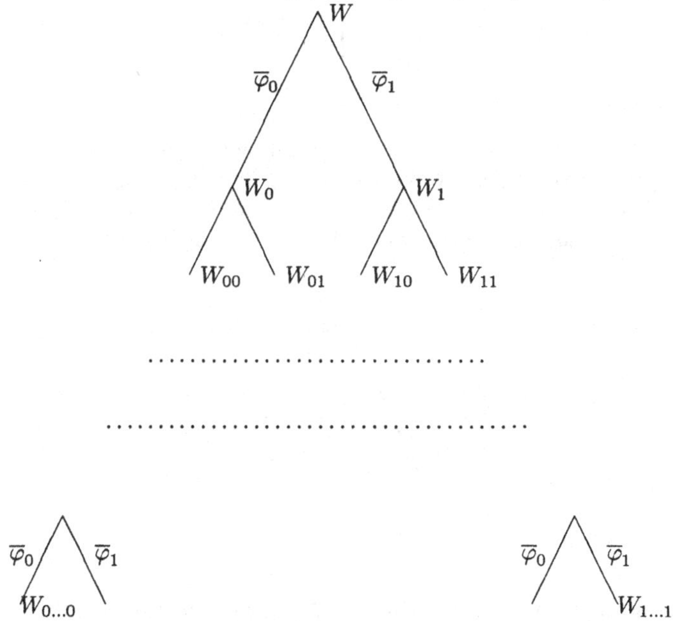

The equality $W \overset{G}{=} 1$ holds if and only if there is n such that all the words $W_{i_1\ldots i_n}$, $i_1 \ldots i_n = 0,1$ are empty. We shall call the depth of the word W which determines the identity element the smallest n with this property. Denote by $K_n < K$, $n = 0,1,\ldots$ the subgroup of F consisting of the elements of depth at most n. Obviously, $K = \cup_{n=0}^{\infty} K_n$ and $\{K_n\}_{n=0}^{\infty}$ is a non-descending sequence of subgroups and the K_n are normal in F. Indeed, if $g \in K_n$ then obviously $aga \in K_n$. If $\overline{\varphi}_i(g) = g_i$, $i = 0,1$ then

$$\overline{\varphi}_i(bgb) = \begin{cases} ag_0a \text{ if } i = 0 \\ cg_1c \text{ if } i = 1 \end{cases} \quad \overline{\varphi}_i(cgc) = \begin{cases} ag_0a \text{ if } i = 0 \\ dg_1d \text{ if } i = 1 \end{cases} \quad \overline{\varphi}_i(dgd) = \begin{cases} g_0 \text{ if } i = 0 \\ bg_1b \text{ if } i = 1 \end{cases}$$

and we can use induction on the length of g. It is easy to see that if the length of g is greater than or equal to 2 then φ_i will diminish it.

Proposition 2 *The group G is not finitely presented.*

Proof (1) It suffices to prove that $\{K_n\}_{n=0}^{\infty}$ is a strictly ascending sequence of groups. We denote by $\pi_i : F \times F \to F$, $i = 0,1$ the projection on the relevant factor. It is easy to verify that

$$\pi_0\overline{\psi}\overline{\sigma}: \begin{cases} a \to & d \\ b \to & 1 \\ c \to & a \\ d \to & a, \end{cases} \tag{9}$$

$$\pi_1\overline{\psi}\overline{\sigma}: \begin{cases} a \to & a \\ b \to & b \\ c \to & c \\ d \to & d, \end{cases} \tag{10}$$

where $\overline{\sigma} : F \to F$ is the endomorphism defined by the substitution (2).

Let $u_n = \overline{\sigma}^{(n)}((ad)^4)$, $n \geq 0$. We assert that $u_n \in K_{n+1}$ but $u_n \notin K_n$. This can be proved by induction on n. If $n = 0$, then $u_0 \in K_1$ but $u_0 \notin K_0$ which can be seen from the following diagram which describes the action of Algorithm $\mathcal{A}$ on the word u_0.

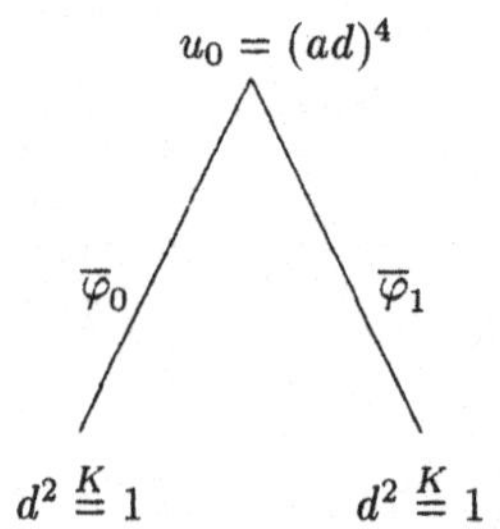

Let us assume that $u_n \in K_{n+1}$ but $u_n \notin K_n$. Then by (9) and (10) we have

$$\begin{aligned} \overline{\psi}(u_{n+1}) &= (\pi_0\overline{\psi}\overline{\sigma}\overline{\sigma}^{(n)}((ad)^4),\ \pi_1\overline{\psi}\overline{\sigma}\overline{\sigma}^{(n)}((ad)^4)) \\ &= (x^4,\ \overline{\sigma}^{(n)}((ad)^4)) \overset{G\times G}{=} (1, u_n) \end{aligned}$$

where x is some element of the dihedral group $\langle a, d\rangle$ of order 8. Thus $u_{n+1} \in K_{n+2}$ but $u_{n+1} \notin K_{n+1}$. The proposition is now proved. □

Proof (2) Suppose G is finitely presented. Then G also has a finite presentation of the following type.

$$G = \langle a, b, c, d \mid a^2 = b^2 = c^2 = d^2 = bcd = (ad)^4 = r_1 = \ldots = r_n = 1\rangle \tag{11}$$

(using Algorithm $\mathcal{A}$ it is easy to see that relation $(ad)^4 = 1$ holds in G). We shall consider only those presentations of G which are of the above type. Among them we choose one with the minimal value of the number $L = \max_{1\leq i\leq n} |r_i|$. We can assume that all the words r_i, $1 \leq i \leq n$ are reduced and are of even length. Consider the group

$$\Gamma = \langle a, b, c, d \mid a^2 = b^2 = c^2 = d^2 = bdc = (ad)^4 = 1\rangle \tag{12}$$

and let $\Delta = \langle b,c,d\rangle_\Gamma^\sharp$ be a subgroup of index 2 in Γ, $\Omega = \langle r_i,\ 1 \le i \le n\rangle_\Gamma^\sharp$.

Obviously, $G = \Gamma/\Omega$ and the inclusion $\Omega < \Delta$ holds. The group Δ is generated by the elements b, c, d, aba, aca, ada. Denote by $\tilde{\psi} : \Delta \to \Gamma \times \Gamma$ a homomorphism defined via the same rule (6) as the homomorphism ψ. Let $\tilde{\varphi}_i = \tilde{\pi}_i \circ \tilde{\psi}$, $i = 0,1$ where $\tilde{\pi}_i : \Gamma \times \Gamma \to \Gamma$, $i = 0,1$ are projections on the factors. The homomorphisms $\tilde{\varphi}_i : \Delta \to \Gamma$ act on the generators of Δ via the rules (5).

Let $r_j^{(i)} = \tilde{\varphi}_i(r_j)$, $1 \le j \le n$, $i = 0,1$. Since the words r_j are reduced we can easily see that $|r_j^{(i)}| \le (|r_j|+1)/2$, $1 \le j \le n$, $i = 0,1$. Hence, $|r_j^{(i)}| < |r_j|$ if $|r_j| \ge 2$. Since G is a 2-group [Gr 80] it is obvious that there is no presentation for G of type (11) with the property that all the defining words r_j have length 1. Therefore Proposition 2 will be proved if we prove the following statement.

Lemma 1 *If (11) is a presentation of the group G, then*

$$\begin{aligned}\langle a,b,c,d \mid a^2 &= b^2 = c^2 = d^2 = bcd = (ad)^4 \\ &= r_1^{(0)} = \ldots = r_n^{(0)} = r_1^{(1)} = \ldots = r_n^{(1)} = 1\rangle \end{aligned} \tag{13}$$

is also a presentation of G.

Proof Since the homomorphisms $\tilde{\varphi}_i$, $i = 0,1$ are defined by the same permutations as the homomorphisms φ_i, $i = 0,1$ it follows that $g \in \Omega$ if and only if $g \in \Delta$ and $g_i \in \Omega$, $i = 0,1$ where $g_i = \tilde{\varphi}_i(g)$. This implies the inclusion $\tilde{\psi}(\Omega) \le \Omega \times \Omega$. The following statement will play an important role not only here but also in the proof of Theorem 1.

Lemma 2 *The relation $\tilde{\psi}(\Omega) = \Omega \times \Omega$ holds.*

ProofLet $W \in \Omega$. The word $\sigma(W)$ defines an element of Δ. We assert that $\sigma(W) \in \Omega$ since $\tilde{\psi}(\sigma(W)) = (1, W)$. Indeed, the homomorphisms $\tilde{\pi}_i\tilde{\psi}\tilde{\sigma}$, $i = 0,1$ (where $\tilde{\sigma} : \Gamma \to \Gamma$ is the endomorphism defined by the permutation σ) act via the following rules

$$\tilde{\pi}_0\tilde{\psi}\tilde{\sigma} : \begin{cases} a \to & d \\ b \to & 1 \\ c \to & a \\ d \to & a, \end{cases} \tag{14}$$

$$\tilde{\pi}_1\tilde{\psi}\tilde{\sigma} : \begin{cases} a \to & a \\ b \to & b \\ c \to & c \\ d \to & d, \end{cases} \tag{15}$$

(these relations are analogues to (9), (10)).

Therefore $\tilde{\psi}(\tilde{\sigma}(W)) = (x, W)$ where $x \in \langle a,d\rangle_\Gamma$ is some element. It follows from this that $\psi(H)$ contains an element of the type $(\overline{x}, 1)$ where $\overline{x} \in \langle a,d\rangle_G$ is an element defined by the word x.

From Proposition 1 it follows that $\overline{x} = 1$ and therefore $x = 1$. Thus we have proved that $\tilde{\psi}(\Omega) \ge 1 \times \Omega$. Therefore $\tilde{\psi}(\Omega) \ge \Omega \times \Omega$ and hence the equality $\tilde{\psi}(\Omega) = \Omega \times \Omega$ holds. □

Any element $g \in \Omega$ can be represented as

$$g = \prod_{i=1}^{k} t_i^{-1} r_{j_i}^{\varepsilon_i} t_i,$$

where $t_i \in \Gamma$, $i = 1, \dots, k$ are some elements and $\varepsilon_i = 0, 1$, r_{j_i} are defining relations from the presentation.

If $t \in \Delta$, $\tilde{\psi}(t) = (t_0, t_1)$, $\tilde{\psi}(r) = (r_0, r_1)$ then $\tilde{\psi}(t^{-1}rt) = (t_0^{-1}r_0t_0, t_1^{-1}r_1t_1)$. If, however, $t \notin \Delta$ and $\tilde{\psi}(ta) = (t_0, t_1)$ then $\tilde{\psi}(t^{-1}rt) = (t_0^{-1}r_1t_0, t_1^{-1}r_0t_1)$.

Thus $\tilde{\psi}(g) = (g_0, g_1)$ where each of the elements g_i, $i = 0, 1$ is a product of elements of the form $t^{-1}(r_j^{(0)})^{\pm 1}t$, $t^{-1}(r_j^{(1)})^{\pm 1}t$, $t \in \Gamma$, $1 \leq j \leq n$, that is to say $g_i \in \langle r_i^{(0)}, r_i^{(1)}, 1 \leq i \leq n \rangle_\Gamma^\sharp$, $i = 0, 1$. It implies that $\Omega = \langle r_i^{(0)}, r_i^{(1)}, 1 \leq i \leq n \rangle_\Gamma^\sharp$.

Lemma 1 together with Proposition 2 are now proved. □

3 Constructing a system of defining relations for G

A canonical projection $\lambda : F \to \Gamma$, $a \to a$, $b \to b$, $c \to c$, $d \to d$ defines the sequence $\{\Omega_n\}_{n=0}^{\infty}$ of normal subgroups of Γ where $\Omega_n = \lambda(K_n)$. Since $\Omega = \lambda(K)$ then $\Omega = \cup_{n=0}^{\infty} \Omega_n$.

Let us consider a sequence $u_i = \sigma^i((ad)^4)$, $v_i = \sigma^i((adacac)^4)$, $i = 0, 1, \dots$ (note that $u_0 = 1$).

The fact that (1) is a presentation of the group G follows from the following statement.

Proposition 3 *For any $n \geq 1$ the following relation*

$$\Omega_n = \langle u_{i+1},\ v_i,\ 0 \leq i \leq n-1 \rangle_\Gamma^\sharp \tag{16}$$

holds.

Proof (outline) We will give a sketch of a proof of this statement which will be useful for understanding the proof of Theorem 1. Those statements which are given without proof are proved in [Gr 97]. They are analogues of the corresponding Lemmas from §5.

The inclusion $g \in \Omega_n$ holds if and only if the following diagram holds for g.

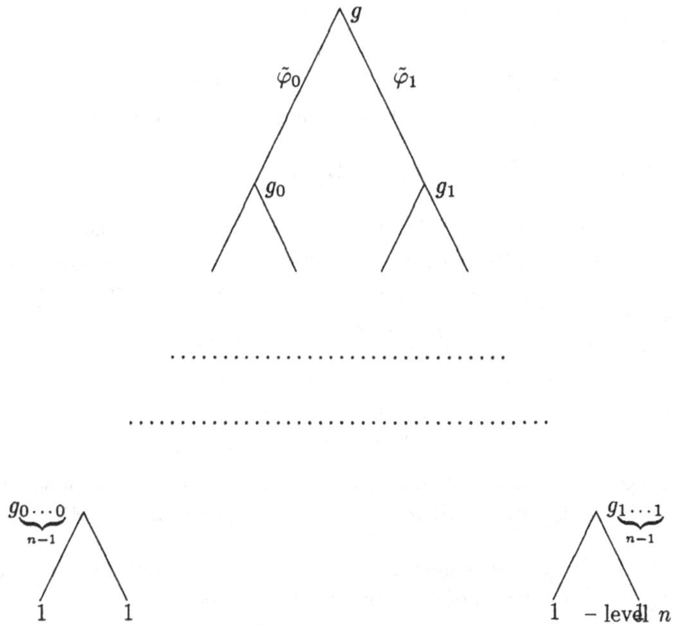

In particular $\Omega_1 = \text{Ker } \tilde{\psi}$. We will omit tilde over ψ, φ_i, π_i in the sequel.

Lemma 3 *The relation $\Omega_1 = \langle u_1, v_0 \rangle_\Gamma^\sharp$ holds.*

ProofThe fact that u_1, $v_0 \in \Omega_1$ can be verified easily. The appropriate diagrams have the form:

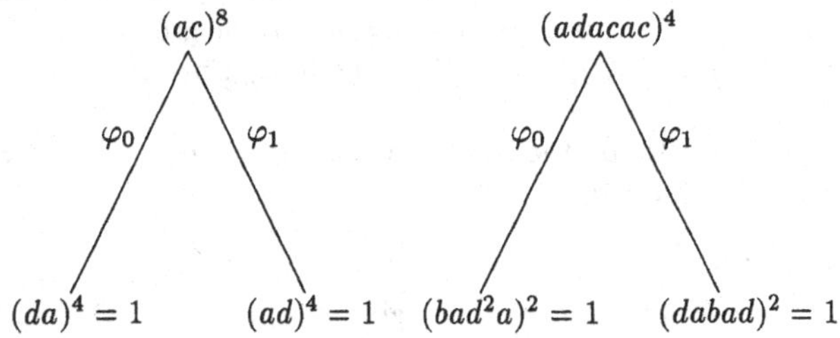

To prove the inclusion $\Omega_1 \leq \langle u_1, v_0 \rangle_\Gamma^\sharp$ we will use presentations of the groups Δ and $\psi(\Delta)$. Namely, if $\alpha = c$, $\beta = aca$, $\delta = d$, $\gamma = ada$ then

$$\Delta = \langle \alpha, \beta, \delta, \gamma \mid \alpha^2 = \beta^2 = \delta^2 = \gamma^2 = [\alpha, \delta] = [\beta, \gamma] = [\delta, \gamma] = 1 \rangle \qquad (17)$$

and if $\mu = \psi(\alpha)$, $\nu = \psi(\beta)$, $\eta = \psi(\delta)$, $\xi = \psi(\gamma)$ then

$$\begin{aligned}\psi(\Delta) = \langle \mu, \nu, \eta, \xi \mid \mu^2 &= \nu^2 = \eta^2 = \xi^2 = (\mu\nu)^4 \\ &= [\mu, \eta] = [\eta, \xi] = [\nu, \xi] = (\eta\nu\mu\xi\mu\nu)^2 = 1\rangle \end{aligned} \quad (18)$$

The rest is obvious if one notes that $(\mu\nu)^4 = \psi((ac)^8) = \psi(u_1)$, $(\eta\nu\mu\xi\mu\nu)^2 = \psi((adacac)^4) = \psi(v_0)$.

Obtaining presentation (17) is an easy exercise in using the Reidermeister-Schreier theorem. Presentation (18) is also not difficult to obtain by using the fact that the group $\psi(\Delta) = (B \times B) \cdot \Delta$ is a semidirect product where $B = \langle b \rangle_\Gamma^\sharp$, $D = \langle (a, d), (d, a) \rangle_{\Gamma \times \Gamma}$ is a dihedral group of order 8 and for B the following presentation with respect to the system of generators $\xi_1 = b$, $\xi_2 = aba$, $\xi_3 = dabad$, $\xi_4 = adabada$ is valid:

$$B = \langle \xi_1, \xi_2, \xi_3, \xi_4 \mid \xi_1^2 = \xi_2^2 = \xi_3^2 = \xi_4^2 \rangle. \quad (19)$$

Writing down a presentation for $\psi(\Delta)$ as being a semidirect product and then rewriting it in the generators μ, ν, η, ξ we obtain presentation (18). □

An alternative way of obtaining a presentation for the group $\psi(\Delta)$ is to consider it as a subgroup (of index 8) of the group $\Gamma \times \Gamma$. A presentation for $\Gamma \times \Gamma$ is obvious and again one can use the Reiedermeister-Schreier thgeorem by taking the set of elements given below as a Schreier system of representatives of $\Gamma \times \Gamma$ with respect to $\psi(\Delta)$:

$$\{(1,1),\ (a,1),\ (d,1),\ (ad,1),\ (da,1),\ (ada,1),\ (dad,1),\ (adad,1),\}.$$

Lemma 4 *For any $n \geq 0$ the equation $\psi(\Omega_{n+1}) = \Omega_n \times \Omega_n$ holds.*

Proof The inclusion $\psi(\Omega_{n+1}) \leq \Omega_n \times \Omega_n$ is obvious. By the relations (14), (15) the elements $\sigma(u_{i+1})$, $\sigma(u_i)$, $0 \leq i \leq n-1$ belong to the group Ω_{n+1} on condition that the elements u_{i+1}, v_i, $0 \leq i \leq n-1$ belong to Ω_n since we have the diagrams shown.

Thus $\psi(\Omega_{n+1}) \geq 1 \times \Omega_n$ on the condition that $\Omega_n = \langle u_{i+1},\ v_i,\ 0 \leq i \leq n-1 \rangle_\Gamma^\sharp$ and hence $\psi(\Omega_{n+1}) \geq \Omega_n \times \Omega_n$. □

Now a proof of Proposition 3 can be obtained immediately since Ω_{n+1} factorized by u_1, v_0 (which means application of (4)) gives the group $\Omega_n \times \Omega_n$ and the normal closure of the set of elements $\psi(u_{i+1}) = (1, u_i)$, $\psi(au_{i+1}a) = (u_i, 1)$, $\psi(v_i) = (1, v_{i-1})$, $\psi(av_ia) = (v_{i-1}, 1)$, $1 \leq i \leq n$ coincides with $\Omega_n \times \Omega_n$.

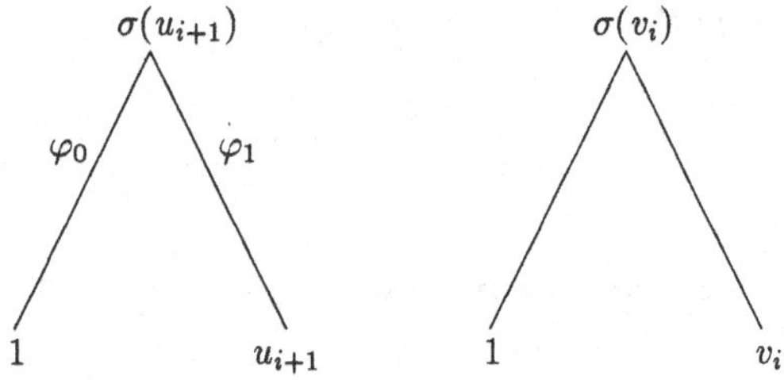

□

The scheme suggested for constructing a presentation of the group G also applies to other GGS-groups. Namely, an arbitrary GGS-group $\mathcal{T}$ acting on the tree T_{p^k} contains a subgroup $\mathcal{H}$ of index p^k which admits an embedding

$$\psi : \mathcal{H} \longrightarrow \underbrace{\mathcal{T} \times \dots \mathcal{T}}_{p^k}.$$

The image $\Psi(\mathcal{H})$ has finite index in $\mathcal{T}^{p^k}$ and has a supplementary subgroup which is normally not difficult to determine and this allows the construction of a system of Schreier representatives of $\mathcal{T}^{p^k}$ with respect to $\psi(\mathcal{H})$.

There is a finitely presented group Γ generated by the same set of symbols as $\mathcal{T}$ which covers $\mathcal{T}$ and which has a subgroup Δ of index p^k admitting the embedding

$$\tilde{\psi} : \Delta \longrightarrow \underbrace{\Gamma \times \dots \Gamma}_{p^k}.$$

It projects onto ψ under the canonical homomorphism $\Gamma \to \mathcal{T}$.

Having written down presentations of the groups Δ and $\psi(\Delta)$ one can determine a finite system of elements $r_1, \dots, r_m$ which generate Ker $\tilde{\psi}$ as a normal subgroup of Γ.

Let $\mathcal{T} = \Gamma/\Omega$. The kernel Ω can be represented as $\Omega = \cup_{n=0}^{\infty} \Omega_n$ where the normal subgroup $\Omega_n \triangleleft \Gamma$ consists of the elements for which the algorithm solving the word problem (analogous to algorithm $\mathcal{A}$ described above) finishes working at the vertices of level n of the tree T_{p^k} and which are accepted by this algorithm.

There is a substitution

$$\tau : \begin{cases} a \to a^{-1}xa \\ b \to \chi(b) \\ \vdots \\ d \to \chi(d) \end{cases}$$

where $x \in \{b, \dots, d\}$ and χ is some permutation of the set $\{b, \dots, d\}$ such that for any $g \in \Delta$ we have

$$\tilde{\psi}(\tau(g)) = (g_0, \dots, g_{p^k-2}, g)$$

and the elements g_i, $0 \le i \le p^k - 2$ belong to some finite subgroups in Γ selected in advance. Additionally the following relations hold

$$\tau(\Omega_{n+1}) = \underbrace{\Omega_n \times \ldots \times \Omega_n}_{p^k}$$

where $n = 0, 1, \ldots$.

We assert that for a suitable choice of the groups Γ, Δ and the permutation τ we have that

$$\mathcal{T} \simeq \langle a, b, \ldots, d \mid \text{ relations of } \Gamma,\ \tau^n(R_i) = 1,\ 1 \le i \le m,\ n \ge 0 \rangle$$

where $R_i = R_i(a, b, \ldots, d)$ is a word which represents an element r_i, $1 \le i \le m$. For example, consider the Gupta-Sidki p-group from [Gu] for $p \ge 5$ (where p is a prime). It is defined as the group $\mathcal{T}$ generated by the automorphisms a, b of the tree T_p given in diagram 3.

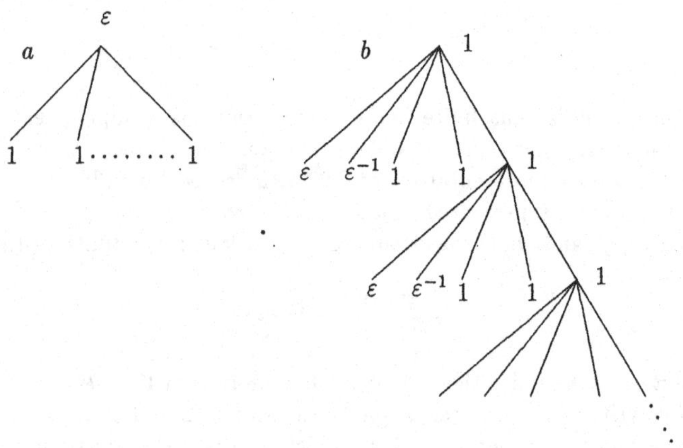

Diagram 3

Let $\Gamma = \langle a, b \mid a^p = b^p = 1 \rangle$,

$$\Delta = \langle b \rangle_\Gamma^{\#} = \langle \xi_0, \ldots, \xi_{p-1} \mid \xi_0^p = \ldots = \xi_{p-1}^p = 1 \rangle$$

where $\xi_i = a^{-i} b a^i$, $0 \le i \le p-1$.

We define the inclusion

$$\tilde{\Psi} : \Delta \longrightarrow \underbrace{\Gamma \times \ldots \times \Gamma}_{p}$$

on the generators ξ_i in the following way:

$$\bar{\xi}_0 = \tilde{\psi}(\xi_0) = (a, a^{-1}, 1, \ldots, 1, b).$$

The vector $\bar{\xi}_i = \tilde{\psi}(\xi_i)$ is obtained from vector $\bar{\xi}_0$ via the i-th power of a cyclic permutation of its co-ordinates.

In order to obtain a presentation of $\tilde{\psi}(\Delta)$ we have to compute the commutators $[\bar{\xi}_i, \bar{\xi}_j]$, $[[\bar{\xi}_i, \bar{\xi}_j], \bar{\xi}_k]$, $[[\bar{\xi}_i, \bar{\xi}_j], [\bar{\xi}_k, \bar{\xi}_l]]$ and choose those of them which are equal to the identity in the group Γ^p. Let $\{r_j(\bar{\xi}_\mu)\}_{j\in J}$ be a set of commutators of the type mentioned above which are equal to the identity in Γ^p together with the extra relator $(\bar{\xi}_0\bar{\xi}_1 \cdots \bar{\xi}_{p-1})^p$. Then $\tilde{\psi}(\Delta)$ has the presentation

$$\tilde{\psi}(\Delta) = \langle \bar{\xi}_0, \ldots, \bar{\xi}_{p-1} \mid \bar{\xi}_i^p = 1,\ r_j(\bar{\xi}_\mu) = 1,\ j \in J \rangle.$$

Express the word $r_j(\xi_\mu)$ via the generators a, b having applied the substitution $\xi_\mu \to a^{-\mu}ba^\mu$, $0 \le \mu \le p-1$. We denote the word obtained by R_j, $j \in J$. Then $\operatorname{Ker} \tilde{\psi} = \langle R_j, j \in J \rangle_\Gamma^\sharp$.

We introduce into consideration the following permutation

$$\tau : \begin{cases} a \to & a^{-1}ba \\ b \to & b. \end{cases}$$

Then

$$\tilde{\psi}(\tau(a)) = (a^{-1}, 1, \ldots, 1, b, a),\ \tilde{\psi}(\tau(b)) = (a, a^{-1}, \ldots, 1, 1, b)$$

and for an arbitrary element $g \in G$ if

$$\tilde{\psi}(\tau(g)) = (g_0, \ldots, g_{p-2}, g_{p-1}),$$

then $g_{p-1} = g$, $g_0 \in \langle a \rangle$, $g_1 \in \langle a \rangle$, $g_2 \in \{1\}, \ldots, g_{p-3} \in \{1\}$, $g_{p-2} \in \langle b \rangle$. Thus in this case the finite subgroups distinguished in Γ are cyclic groups of order p generated by the elements a and b together with the trivial subgroup. We assert that for a Gupta and Sidki p-group we have the following presentation:

$$\mathcal{T} = \langle a, b \mid a^p = b^p = \tau^n(R_j) = 1,\ j \in J,\ n \ge 0 \rangle$$

(a detailed proof of this fact will be given in a separate paper). The case $p = 3$ is special because of the difficulty of determining the substitution τ. However, the author hopes that using the approach suggested it is possible to find a good presentation in this case as well.

4 Proof of Theorem 1

We remind the reader that in §2 we used the endomorphism $\bar{\sigma} : F \to F$ defined by the permutation σ. The bar over σ will be omitted in the sequel. The proof of Theorem 1 consists of a long sequence of auxiliary statements. We will now state and prove them.

Lemma 5 *The following inclusions hold:*

$$\sigma(K) \le K \tag{20}$$

$$\sigma([K, F])) \le [K, F] \tag{21}$$

Proof The second inclusion follows from the first one. In order to prove (20) it suffices to verify that $\sigma(r_n) \in K$, $1 \le n \le 5$ (here we use the fact that $K = \langle r_n, n \ge 1\rangle_F^\sharp$ and special characteristics of the defining relations in (1)). But $\sigma(a^2) = aca^2ca \in K$, $\sigma(b^2) = c^2 \in K$ and similarly $\sigma(c^2)$, $\sigma(d^2) \in K$. □

Lemma 6 *Presentation (1) is equivalent to (3) in the sense that* $\langle r_n, n \ge 1\rangle_F^\sharp = \langle R_n, n \ge 1\rangle_F^\sharp$.

Proof Put $K_0 = \langle R_n, n \ge 1\rangle_F^\sharp$. Since obviously $R_1, \ldots, R_7$ belong to K we have $K_0 \le K$ by the previous lemma. The converse inclusion is also obvious. A direct verification shows that $r_i \in K_0$, $1 \le i \le 7$ and also that $\sigma(K_0) \le K_0$. From this the inclusion $K \le K_0$ follows. □

We will use the following notation in the sequel: $[x, y] = x^{-1}y^{-1}xy$, $x^y = x^{-1}yx$, $\approx$ is an equality sign in the group $K/[K, F]$, $\sim$ is a conjugation sign. The expressions $[x, y, z]$, $[x, y, z, w]$ denote left hand side commutators of the weight 3 and 4 respectively. It is obvious that the group $K/[K, F]$ is commutative and generated by the elements $\overline{R}_n$, $n \ge 1$.

Lemma 7 *In the group* $K/[K, F]$ *the relations* $\overline{R}_n^2 \approx 1$, $n \ge 5$ *hold.*

Proof Use the fact that x^2, $[x^\varepsilon, y^\mu]$ belong to K, where $x, y \in \{b, c, d\}$, $\varepsilon, \mu = \pm 1$.

Firstly, check that $\overline{R}_5^2 \approx 1$. To this end we express the element $\overline{R}_5$ as a product of commutators.

$$\overline{R}_5 = b^2c^2d^2d^{-1}c^{-1}b^{-1}d^{-1}c^{-1}b^{-1} \approx dcbd^{-1}c^{-1}b^{-1} = dd^{-1}cb[cb, d^{-1}]c^{-1}b^{-1}$$

$$\approx cb[c, d^{-1}]^b[b, d^{-1}]c^{-1}b^{-1} \approx [c^{-1}, b^{-1}][c, d^{-1}][b, d^{-1}] \approx [c, b][c, d][b, d].$$

Apart from that

$$1 \approx [c, b^2] = [c, b]^2[c, b, b] \approx [c, b]^2$$

and similarly $[c, d]^2 \approx 1$, $[b, d]^2 \approx 1$. Thus $\overline{R}_5^2 \approx 1$.

In order to prove the relations $\overline{R}_n^2 \approx 1$, $n \ge 6$ it suffices to establish that $\overline{R}_6^2 \approx \overline{R}_7^2 \approx 1$ since $\overline{R}_{6+2k} = \sigma_*^k(\overline{R}_6)$, $\overline{R}_{7+2k} = \sigma_*^k(\overline{R}_7)$, $k \ge 0$ where

$$\sigma_* : K/[K, F] \to K/[K, F]$$

is the endomorphism induced by σ.

We have

$$\begin{aligned} \overline{R}_6^2 &= (ad)^4a^{-4}d^{-4}(ad)^4a^{-4}d^{-4} \\ &\approx a^{-1}d^{-1}a^{-1}d^{-1}a^{-1}d^{-1}a^{-1}d^{-1}a^{-1}aadadadad \\ &= [a, (ad)^4] \approx 1. \end{aligned}$$

Finally, the element $\overline{R}_7^2$ is conjugate via $adac$ to the product

$$acadacacadacacadacaca^{-12}c^{-8}d^{-4}adacacadacacadacacadacaca^{-12}c^{-8}d^{-4}adac$$

$$\approx a^{-1}c^{-1}a^{-1}d^{-1}a^{-1}c^{-1}a^{-1}c^{-1}a^{-1}d^{-1}a^{-1}c^{-1}a^{-1}c^{-1}a^{-1}d^{-1}a^{-1}c^{-1}a^{-1}c^{-1}a^{-1}$$
$$\cdot d^{-1}a^{-1}c^{-1}a^{-1}aacadacacadacacadacacadac = [a,(acadac)^4] \approx 1.$$

□

Lemma 8 *The following isomorphism holds:*

$$\langle \overline{R}_1, \overline{R}_2, \overline{R}_3, \overline{R}_4 \rangle \simeq C^4. \tag{22}$$

Proof Denote by $\tilde{R}_1, \tilde{R}_2, \tilde{R}_3, \tilde{R}_4$ the images of the elements R_1, R_2, R_3, R_4 under the quotient map $K/[K,F] \to K/K \cap F'$. Since $KF'/F' \simeq K/K \cap F'$ and $F/F' \simeq C^4$ we have that the statement of the Lemma follows from the linear independence over $\mathbb{Z}$ of the vectors $(2,0,0,0)$, $(0,2,0,0)$, $(0,0,2,0)$, $(0,1,1,1)$ whose co-ordinates are the exponent sums of the symbols a, b, c, d in the words R_1, R_2, R_3, R_4. □

Corollary 1 *The following isomorphism holds*

$$K/[K,F] \simeq C^4 \times T, \tag{23}$$

where the factor C^4 is freely generated by the elements $\overline{R}_1, \overline{R}_2, \overline{R}_3, \overline{R}_4$ and is isomorphic to $K/K \cap F'$ and T is the torsion part generated by the elements R_n, $n \geq 5$ which is an elementary 2-group, isomorphic to the group $K \cap F'/[K,F]$.

Proof Since $A = \langle R_i \mid 1 \leq i \leq 4 \rangle$ is a torsion-free group and $B = \langle R_n,\ n \geq 5 \rangle$ is an elementary 2-group it is obvious that $K/[K,F] \simeq A \times B$. Also,

$$C^4 \simeq A \simeq KF'/F' \simeq K/K \cap F'$$

since $\overline{R}_n \in F'$ for $n \geq 6$. Finally,

$$(K/[K,F])/(K \cap F'/[K,F]) \simeq K/K \cap F' \simeq C^4$$

and hence $T = B = K \cap F'/[K,F]$ is distinguished as a direct factor as mentioned in (22). □

In the remaining part of the proof of the theorem we will show that the elements $\overline{R}_n$, $n \geq 6$ are independent in T. The first step in this direction is the following lemma.

Lemma 9 *The elements $\overline{R}_5, \overline{R}_6, \overline{R}_7, \overline{R}_8$ are independent in T.*

This statement follows from the lemma given below the proof of which is postponed until §5.

Lemma 10 *The group $Q = F'/[K,F]\cdot\gamma_5(F)\cdot F^{(2)}$ has the following presentation:* ⟨ *generators:*

a) $[a,b]$, $[a,c]$, $[a,d]$, $[b,c]$,

b) $[a,x,y]$, $x \neq y$, $x,y \in \{b,c,d\}$,

c) $[a,x,y,z]$, $x \neq y$, $y \neq z$, (x,y,z) *is not a permutation of the triple* (b,c,d) *where* $x,y \in \{b,c,d\}$, $z \in \{a,b,c,d\}$;

relations:

(i) *commutativity relations*

(ii) $[a,b]^8 = [a,c]^8 = [a,d]^4 = [b,c]^2 = 1$

(iii) $[a,b,c]^8 = [a,b,d]^8 = [a,c,b]^4 = [a,c,d]^4 = [a,d,b]^2 = [a,d,c]^2 = 1$

(iv) $[a,x,y,z]^2 = 1$ ⟩.

Apart from this the elements $\tilde{R}_5, \tilde{R}_6, \tilde{R}_7, \tilde{R}_8$ can be expressed via the generators in the following way:

$$\tilde{R}_5 = [b,c],\ \tilde{R}_6 = [a,d]^2,$$

$$\tilde{R}_7 = [a,d]^2[a,c,d]^2,\ \tilde{R}_8 = [a,c]^8,$$

where $\tilde{R}_i$, $5 \leq i \leq 8$ denotes the image of the element R_i in Q.

Let us define a homomorphism $\tilde{\psi} : P \to \Gamma \times \Gamma$ on the set of the generators of the group P by relations (8) as in the case of the homomorphism $\overline{\psi}$. When working with the homomorphism $\tilde{\psi}$ we will omit the tilde. We hope that it will not lead to any confusion.

Lemma 11 *The group Ker ψ is normal not only in P but also in F and the relation Ker $\psi = \langle R_i,\ 1 \leq i \leq 8\rangle_F^\#$ holds.*

We will also postpone proving this statement until the next section.

Lemma 12 *The following isomorphism holds:*

$$K/[K,F] \simeq C^4 \times C_2^4 \times \psi(K)/\psi([K,F]), \tag{24}$$

where the factor C^4 is generated by the elements $\overline{R}_1,\ldots,\overline{R}_4$ and the factor C_2^4 is generated by the elements $\overline{R}_5,\ldots,\overline{R}_8$.

Proof Let

$$\psi_* : K/[K,F] \to \psi(K)/\psi([K,F])$$

be the homomorphism induced by the homomorphism ψ. It follows from Lemmas 8 and 9 that

$$\mathrm{Ker}\psi_* = (\mathrm{Ker}\psi)\cdot[K,F]/[K,F] = \langle\overline{R}_1,\ldots,\overline{R}_8\rangle \simeq C^4 \times C_2^4.$$

From this and from Corollary 1 the statement of Lemma 12 follows. □

In order to describe the groups $\psi(K)$, $\psi([K,F])$ from the right hand side of (23) we introduce the notation τ_i, $i = 0,1$ for the homomorphisms $\tilde{\pi}_i\psi\sigma : F \to \Gamma$ which acts as

$$\tau_0 : \begin{cases} a \to & d \\ b \to & 1 \\ c \to & a \\ d \to & a \end{cases},$$

$$\tau_1 : \begin{cases} a \to & d \\ b \to & b \\ c \to & c \\ d \to & d \end{cases}$$

similarly to (14), (15).

Lemma 13 *The following relations hold:*

$$\psi(R_{6+2k}) = (1, R'_{6+2(k-1)}),\ \psi(R_{7+2k}) = (1, R'_{7+2(k-1)}),\ k \geq 1, \tag{25}$$

where R'_n, $n \geq 6$ is the image of R_n under the canonical homomorphism $\lambda : F \to \Gamma$.

Proof Using the fact that the image of τ_0 coincides with the group $\langle a,d\rangle_\Gamma$ we have

$$\begin{aligned} \psi(R_{6+2k}) &= \psi(\sigma^k((ad)^4a^{-4}d^{-4}) \\ &= (\tau_0(\sigma^{k-1}((ad)^4)))(\tau_0(\sigma^{k-1}(a))^{-4}\tau_0((\sigma^{k-1}(d)))^{-4}, \end{aligned}$$

$$\tau_1(\sigma^{k-1}((ad)^4a^{-4}d^{-4}))) = (1, \tau_1(R_{6+2(k-1)})) = (1, R'_{6+2(k-1)}).$$

and similarly

$$\begin{aligned} \psi(\sigma^k((adacac)^4a^{-12}c^{-8}d^{-4})) &= (1, \tau_1(\sigma^{k-1}((adacac)^4a^{-12}c^{-8}d^{-4}))) \\ &= (1, \tau_1(R_{7+2(k-1)})) = (1, R'_{7+2(k-1)}). \end{aligned}$$

□

In the sequel we will omit the prime in the right hand side of relations (25). Denote by θ the shift to the left by two units in the sequence $R_7, R_8, \ldots$ of defining words. That is $\theta R_n = R_{n-2}$, $n \geq 9$, $\theta(R_7) = \theta(R_8) = \emptyset$ (where $\emptyset$ is an empty word).

Lemma 14 *The following relations hold:*

$$\psi(K) = \Omega \times \Omega \text{ and } \psi([K,F]) = ([\Omega,\Gamma] \times [\Omega,\Gamma]) \cdot \Xi \tag{26}$$

where $\Xi \leq \Gamma \times \Gamma$ is the subgroup consisting of elements of the form (ω^{-1}, ω), $\omega \in \Omega$ and in the first relation the generators of each of the factors of Ω, as a normal subgroup in Γ, are the projections of θ-images of the generators R_n, $n \geq 9$.

Proof The first of the relations is analogous to the relation of Lemma 2 and can be proved similarly using Lemma 13. We will now prove the second relation.

Let $f \in P$, $\psi(f) = (f_0, f_1)$, $k \in K$, $\psi(k) = (k_0, k_1)$. Then

$$\psi([f,k]) = ([f_0,k_0],\ [f_1,k_1]), \tag{27}$$

in particular $\psi([f,k]) = (1,[f_1,k_1])$ if $k_0 = 1$.

For any element $f_1 \in \Gamma$ there is $f \in P$ such that $\psi(f) = (f_0, f_1)$ where $f_0 \in \Gamma$ is some element. From Lemma 13 it follows that $\psi([K,F]) \geq 1 \times [\Omega,\Gamma]$ and hence $\psi([K,F]) \geq [\Omega,\Gamma] \times [\Omega,\Gamma]$. Using the relation

$$\psi([af,k]) = \psi(f^{-1}a^{-1}k^{-1}afk) = ([f_0,f_1],\ [f_1,k_0]) \cdot (k_1^{-1}k_0, k_0^{-1}k_1) \tag{28}$$

where $f \in P, k \in K$ we obtain the inclusion

$$\psi([K,F]) \leq ([\Omega,\Gamma] \times [\Omega,\Gamma]) \cdot \Xi.$$

Substituting into (28) the identity instead of f and the element R_n, $n \geq 9$ instead of K and using Lemma 13 we obtain the inclusion $\psi([a,K]) \geq \Xi$, and therefore the second relation of (26). □

Denote by $\overline{\theta}$ a shift by two units in the sequence $\overline{R}_7$, $\overline{R}_8, \ldots$, where $\overline{R}_n$, $n \geq 7$ is the element of the group $\Omega/[\Omega,\Gamma]$ which is the image of the element R_n.

Lemma 15 *The following isomorphism holds:*

$$\psi(K)/\psi([K,F]) \simeq \Omega/[\Omega,\Gamma] \tag{29}$$

where the generator $\overline{\psi(R_n)}$, $n \geq 9$ of the group $\psi(K)/\psi([K,F])$ is mapped by this isomorphism to the generator $\theta(\overline{R}_n) = \overline{R}_{n-2}$ of the group $\Omega/[\Omega,\Gamma]$.

Proof We have already proved that $\psi(K)/\psi([K,F])$ is an elementary 2-group generated by the images $\overline{\psi(R_n)}$, $n \geq 9$ of the generators $\psi(R_n)$ of the group $\psi(K)$. If t is a generator of a group of order 2 then

$$\langle t \rangle \times \langle t \rangle / \langle (t,t) \rangle \simeq \langle t \rangle \simeq C_2.$$

By the previous lemma we obtain the relation (29) and also the rest of the statement of the lemma. □

The following proposition is a corollary of Lemmas 12, 14 and 15.

Proposition 4 *The following isomorphism holds*

$$K/[K,F] \simeq C^4 \times C_2^4 \times \Omega/[\Omega,\Gamma]$$

where the generators $\overline{R}_n$, $1 \leq n \leq 8$ are mapped by this isomorphism to the generators of the factor $C^4 \times C_2^4$ and the generators $\overline{R}_n$, $n \geq 9$ are mapped to the generators $\overline{R}_{n-2} = \overline{\theta}(\overline{R}_n)$.

A proof of the next statement can be constructed by approximately the same scheme as the proof of Proposition 4 and consists of the proofs of Lemmas 16-20.

Proposition 5 *The following isomorphism holds*

$$\Omega/[\Omega,\Gamma] \simeq C_2^2 \times \Omega/[\Omega,\Gamma] \tag{30}$$

where the factor C_2^2 is generated by the elements $\overline{R}_7$, $\overline{R}_8$ and the second factor of the right hand side of (30) is generated by $\overline{\theta}$-images of the generators $\overline{R}_n$, $n \geq 9$ of the left hand side.

Lemma 16 *The elements $\overline{R}_7$, $\overline{R}_8$ are independent in $\Omega/[\Omega,\Gamma]$.*

This statement follows from Lemma 17 given below which is similar to Lemma 10 and the proof of which is postponed until the next section.

Lemma 17 *The group $S = \Gamma'/[\Omega,\Gamma]\gamma_5(\Gamma)\Gamma^{(2)}$ has the following presentation:*
⟨ *generators:*
(a) $[a,b]$, $[a,c]$, $[a,d]$,
(b) $[a,x,y]$, $x \neq y$, $x,y \in \{b,c,d\}$,
(c) $[a,x,y,z]$, $x \neq y$, $y \neq z$, (x,y,z) *is not a permutation of the triple* (b,c,d), $x,y \in \{b,c,d\}$, $z \in \{a,b,c,d\}$;
relations:
(i) *commutativity relations*
(ii) $[a,b]^8 = [a,c]^8 = [a,d]^2 = 1$,
(iii) $[a,b,c]^8 = [a,b,d]^8 = [a,c,b]^4 = [a,c,d]^4 = [a,d,b]^2 = [a,d,c]^2 = 1$
(iv) $[a,x,y,z]^2 = 1$⟩.
Additionally the elements R_7, R_8 can be expressed via the generators in the following way:

$$R_7 = [a,c,d]^2,\ R_8 = [a,c]^4.$$

Lemma 18 *The following isomorphism holds*

$$\Omega/[\Omega,\Gamma] \simeq C_2^2 \times \psi(\Omega)/\psi([\Omega,\Gamma]) \tag{31}$$

where the elements $\overline{R}_7$, $\overline{R}_8$ are generators of the factor C_2^2 and the elements $\overline{R}_n$, $n \geq 9$ are generators of the second factor of the right hand side of (31).

Proof The proof is analogous to that of Lemma 12 and is based on an application of Lemma 17. □

Lemma 19 *The following relations hold*

$$\psi(\Omega) = \Omega \times \Omega \text{ and } \psi([\Omega,\Gamma]) = ([\Omega,\Gamma] \times [\Omega,\Gamma]) \cdot \Xi \tag{32}$$

and the generators of each factor Ω as normal group in Γ from the first relation are $\overline{\theta}$-images of the generators $\overline{R}_n$, $n \geq 9$.

The proof is analogous to that of Lemma 13 and is omitted.

Lemma 20 *The following isomorphism holds*

$$\psi(\Omega)/\psi([\Omega,\Gamma]) \simeq \Omega/[\Omega,\Gamma]$$

where the generator $\overline{\psi(R_n)}$, $n \geq 9$ *of the group* $\psi(\Omega)/\psi([\Omega,\Gamma])$ *is mapped by this isomorphism to the generator* $\overline{\theta}(\overline{R}_n) = \overline{R}_{n-2}$ *of the group* $\Omega/[\Omega,\Gamma]$.

The proof is analogous to that of Lemma 15 and is omitted.

Proposition 5 is a direct corollary of Lemmas 16-20.

Finally Theorem 1 follows from Propositions 4 and 5.

It remains to prove Theorem 3. To this end we factorize the group $M = K/[K,\Gamma]$ by introducing to it the relations $\overline{R}_i^2 = 1$, $1 \leq i \leq 4$. We denote the factor-group by $\overline{M}$ and by $m_n \in M$ the image $\overline{R}_n$, $n \geq 1$.

Then $\overline{M} \simeq C_2^\infty$ and the elements $m_n, n \geq 1$ form a basis. For the map $M \to \overline{M}$ the image of the element $\overline{r}_n \in K/[K,F]$ is the element m_n if $n \neq 4,5$ and $\overline{r}_4 \to m_2m_3m_5$, $\overline{r}_5 \to m_4$.

If we suppose that some defining word r_k belongs to the group $\langle r_n,\ n \geq 1,\ n \neq k\rangle_F^\sharp$ then it will lead to a non-trivial relation between the elements $\overline{r}_n$ in M and hence to a non-trivial relation between the elements m_n which is absurd. As G is residually finite, and therefore hopfian, Theorem 3 follows.

5 Proofs of Lemmas 10, 17 and 11

We will now embark on the proof of Lemma 10. Obviously the group Q is commutative and since $a^2, b^2, c^2, d^2 \in K$ we have that Q is generated by the elements of the form

$a_1)$ $[x,y]$, $x \neq y$, $x,y \in \{a,b,c,d\}$,

$b_1)$ $[x,y,z]$, $x \neq y$, $x,y,z \in \{a,b,c,d\}$,

$c_1)$ $[x,y,z,w]$, $x \neq y$, $x,y,z,w \in \{a,b,c,d\}$.

Using the standard relations of the commutator calculus it is easy to see that all the relations of the group Q, with respect to the system of generators given above, are the corollaries of the relations

$$[x,R_n] = 1,\ n \geq 1,\ x \in \{a,b,c,d\}, \tag{33}$$

and also of the relations $y = 1$ where $y \in \gamma_5(F')$ or $y \in F^{(2)}$.

First we will use the relations (33) when $1 \leq n \leq 7$. Let $n = 1$, $x \in \{b,c,d\}$. Then

$$1 \approx [x,a^2] = [x,a]^2[x,a,a].$$

from which it follows that $[x,a,a] \approx [a,x]^2$. Thus the generators $[x,a,a]$, $x \in \{b,c,d\}$ can be omitted from the system of generators.

Let $n = 2,3,4$, $x \in \{a,b,c,d\}$, $y \in \{b,c,d\}$. Then

$$1 \approx [x,y^2] = [x,y]^2[x,y,y]$$

Since $[x,y] \in K$ for $x,y \in \{b,c,d\}$, it follows from the previous relation that $[b,c]^2 \approx [b,d]^2 \approx [c,d]^2 \approx 1$ and also that

$$[a,b,b] \approx [a,b]^{-2}, [a,c,c] \approx [a,c]^{-2} \text{ and } [a,d,d] \approx [a,d]^{-2}. \tag{34}$$

Therefore these triple commutators can be omitted from the system of generators. Let $n = 5$. Then

$$\begin{aligned} 1 &\approx [a,bcd] = [a,cd][a,b][[a,b],cd] \\ &= [a,d][a,c][a,c,d][a,b][[a,b],d][[a,b],c][[[a,b],c],d]. \end{aligned}$$

This relation permits the omission of $[a,b,c,d]$ from the system of generators. Hence we can omit from the system of generators the elements of the form $[a,x,y,z]$ where (x,y,z) is some permutation of the triple (b,c,d).

We also have the relations

$$1 \approx [b,bcd] \approx [b,c][b,d], \; 1 \approx [c,bcd] \approx [c,d][c,b], \; 1 \approx [d,bcd] \approx [d,b][d,c]$$

from which it follows that $[c,d] \approx [b,c] \approx [b,d]^{-1}$. Thus $[c,d]$ and $[b,d]$ can be omitted from the system of generators.

Let $n = 6$. We have

$$1 \approx [a,(ad)^4] \approx [a,d]^4 \tag{35}$$

and

$$1 \approx [b,(ad)^4] \approx [b,[a,d]^2] = [b,[a,d]]^2][[[b,ad]],[a,d]] \approx [b,[a,d]]^2] = [a,d,b]^{-2}.$$

Hence $[a,d,b]^2 \approx 1$ and the relation $[a,d,c]^2 \approx 1$ is established similarly. Apart from this, a corollary of the relation (34) is the relation $[a,d,d]^2 \approx 1$.

Finally the relation $1 \approx [d,(ad)^4]$ leads to the relation $[a,d,d]^2 \approx 1$ obtained above.

Let us express the element R_7 via the generators of the group Q. We have

$$\begin{aligned} R_7 &= (adacac)^4 a^{-12} c^{-8} d^{-4} = adacacadacacadacacadacaca^{-12}c^{-8}d^{-4} \\ &= (ad)^2(ac)^2[(ac)^2,ad](ac)^2(ad)^2(ac)^2[(ac)^2,ad](ac)^2a^{-12}c^{-8}d^{-4} \\ &\approx [a,d]^2[a,c]^4[[a,c],ad]^2 \\ &\approx [a,d]^2[a,c]^4[a,c,d]^2[a,c,a]^2[a,c,a,d]^2 \approx [a,d]^2[a,c,d]^2. \end{aligned}$$

since $[a,c,a] \approx [a,c]^4$ and $[a,c]^8 \approx 1$. Therefore

$$\begin{aligned} [b,R_7] \approx [b,[a,d]^2[a,c,d]^2] &\approx [b,[a,c,d]^2][b,[a,d]^2] \approx [b,[a,d]]^2[b,[a,c,d]]^2 \\ &\approx [a,d,b]^{-2}[a,c,d,b]^{-2} \approx 1 \end{aligned}$$

using the relations obtained earlier. Thus the relation $[b,R_7] \approx 1$ is a corollary of the relations obtained before. Similarly one can check that the relations $[c,R_7] \approx 1$, $[d,R_7] \approx 1$ are corollaries of those obtained previously.

By now we have simplified the presentation of the group Q using the relations $[x, R_i] \approx 1$, $x \in \{a, b, c, d\}$, $1 \leq i \leq 7$. We will now consider the relations $[x, R_n] \approx 1$, $x \in \{a, b, c, d\}$, $n \geq 8$.

Let us establish that $R_8, R_9 \in \gamma_4(F) \bmod [U, F]F^{(2)}$ where

$$U = \langle a^2, b^2, c^2, d^2, , bcd, (ad)^4 \rangle_F^{\sharp}.$$

Indeed, the following relations hold mod $[U, F]F^{(2)}$:

$$\begin{aligned} R_8 &= (ac)^8(aca)^{-4}c^{-4} = [c, a, a, a], \\ R_9 &= (acacacabacab)^4(aca)^{-12}b^{-8}c^{-4} \\ &= ((ac)^4ab[ab, ac][[ab, ac], ab])^4(aca)^{-12}b^{-8}c^{-4} \\ &= (ac)^{16}(ab)^8[[ab, ac], ab]^4(aca)^{-12}b^{-8}c^{-4} \\ &= [c, a, a, a, a][b, a, a, a] \in \gamma_4(F). \end{aligned} \tag{36}$$

In the relations above we have used the fact that the following relations hold mod $[U, F]$:

$$(ac)^2 = a^2c^2[a, c], \ (ab)^2 = a^2b^2[a, b] \text{ and } [ab, ac] = badac \sim (ad)^2 = a^2d^2[a, d]$$

(recall that $\sim$ is a conjugacy sign). We have also used the fact that the following relation holds mod $[U, F]F^{(2)}$:

$$[ab, [ab, ac]]^2 = [ab, [ab, ac]^2][[ab, [ab, ac]], [ab, ac]]^{-1} = 1.$$

Since the inclusions $\sigma(U) \leq U$, $\sigma(\gamma_5(F)) \leq \gamma_5(F)$, $\sigma(F^{(2)}) \leq F^{(2)}$ hold we conclude that the relations $[x, R_n] = 1$, $n \geq 8$ are consequences of the relations obtained earlier. Hence the group Q has a presentation with the set of relations equivalent to the system of relations $[x, R_i] = 1$, $1 \leq i \leq 7$, that is to say it has a presentation given in the statement of Lemma 10.

The proof of Lemma 17 is analogous to the proof of Lemma 10. One should only note that $[b, c] = 1$ in Γ and so the element $[b, c]$ can be deleted from the system of generators and also that $[a, d]^2 = (ad)^4 = 1$ in Γ. Therefore R_5, R_6 become the trivial elements in S and the expressions for R_7, R_8 take the form given in the statement of Lemma 17.

Finally, let us prove Lemma 11. Ker ψ is a normal subgroup in F (not just in P) for if $g \in P, \psi(g) = 1$ then $\psi(aga) = 1$ as well. It is not difficult to check that $\langle R_i,\ 1 \leq i \leq 8 \rangle_F^{\sharp} = \langle r_i,\ 1 \leq i \leq 8 \rangle_F^{\sharp}$ and we will therefore prove that Ker $\psi = \langle r_i,\ 1 \leq i \leq 8 \rangle_F^{\sharp}$. One can pass from the system of generators $X = \{a^2, b, c, d, aba^{-1}, aca^{-1}, ada^{-1}\}$ of the free group P to the system of generators $Y = \{a^2, bcd, abcda^{-1}, c, d, aca^{-1}, ada^{-1}\}$. The images under the homomorphism ψ of the first three generators of Y are equal to 1. Denote $\alpha = c$, $\beta = aca^{-1}$, $\delta = d$, $\gamma = ada^{-1}$ and let $\overline{\psi}$ be the restriction of the homomorphism ψ to the subgroup $E = \langle \alpha, \beta, \delta, \gamma \rangle$. Now Lemma 11 follows from the next statement:

Lemma 21

$$Ker\, \overline{\psi} = \langle \alpha^2,\ \beta^2,\ \delta^2,\ \gamma^2,\ [\alpha, \beta]^2,\ [\alpha, \delta],\ [\beta, \gamma],\ [\delta, \gamma],\ (\delta\beta\alpha\gamma\alpha\beta)^2 \rangle_E^{\sharp}. \tag{37}$$

Indeed, denote by V the right hand side of the previous relation and by W – the group $\langle r_i,\ 1 \le i \le 8\rangle_F^\sharp$. A direct verification shows that $r_i \in \operatorname{Ker} \psi$, $1 \le i \le 8$, so $W \le \operatorname{Ker} \psi$. Obviously, $\operatorname{Ker} \psi = \langle a^2, bcd\rangle_F^\sharp \cdot \operatorname{Ker} \overline{\psi}$. Let us prove that $\operatorname{Ker} \overline{\psi} \le W$ on condition that lemma 21 is true. We have $\alpha^2 = c^2$, $\delta^2 = d^2$, $[\alpha, \delta] = [c, d]$, $\beta^2 = ac^2a^{-1}$, $\gamma^2 = ad^2a^{-1}$,

$$[\alpha, \beta] = [c, aca^{-1}] = (ac)^4 \bmod W$$

$$[\beta, \gamma] = a[c, d]a^{-1},$$

$$[\delta, \gamma] = [d, ada^{-1}] = (ad)^4 \bmod W$$

$$(\delta\beta\alpha\gamma\alpha\beta)^2 = (daca^{-1}cada^{-1}caca^{-1})^2 = (dacaca)^4 \bmod W.$$

That is what we had to prove to complete the proof of Lemma 11.

Let us now prove Lemma 21.

Proof We use the fact that

$$\psi(P) = \psi(E) = (B \times B) \cdot D$$

is a semi-direct product (the relation which we have already seen in §2 but in slightly different notation) where $B = \langle b\rangle_\Gamma^\sharp$, $D = \langle (a,d), (d,a)\rangle_{\Gamma\times\Gamma}$ and for B presentation (19) holds where $\xi_1 = b$, $\xi_2 = aba$, $\xi_3 = dabad$, $\xi_4 = adabada$. The last fact easily follows from the Reidermeister-Schreier theorem if we take the set $\{1, a, d, ad, da, ada, dad, adad\}$ as a system of representatives of Γ with respect to B.

In §3 we introduced the notation $\mu = \psi(c) = (a, d)$, $\nu = \psi(aca) = (d, a)$. In addition we introduce the notation $\eta_i = (1, \xi_i)$, $\zeta_i = (\xi_i, 1)$, $1 \le i \le 4$ and $\eta_1 = \eta$, $\zeta_1 = \zeta$ for η and ζ defined in §3.

Lemma 22 *The group $N = (B \times B) \cdot D$ has the following presentation:*

$$N = \langle \mu, \nu, \eta_1, \ldots, \eta_4, \zeta_1, \ldots, \zeta_4 \mid \mu^2 = \nu^2 = (\mu\nu)^4 = \eta_i^2 = \zeta_j^2 = [\eta_i, \zeta_j] = 1,$$

$$1 \le i, j \le 4,\ \eta_1^\mu = \eta_1,\ \eta_2^\mu = \eta_3\ \eta_4^\mu = \eta_4,\ \zeta_1^\mu = \zeta_2,\ \zeta_3^\mu = \zeta_4,\ \eta_1^\nu = \eta_2,\ \eta_3^\nu = \eta_4$$

$$\zeta_1^\nu = \zeta_1,\ \zeta_2^\nu = \zeta_3,\ \zeta_4^\nu = \zeta_4\rangle. \tag{38}$$

Proof Relations (38) were written down using the relations of the groups B and $D = \langle x, y \mid x^2 = y^2 = (xy)^4 = 1\rangle$ where $x = (a, d)$, $y = (d, a)$ taking into consideration the action of the generators of the group D on the generators of the group $B \times B$. We will verify this now.

From (37) it is clear that

$$B \times B \simeq \langle \eta_1, \ldots, \eta_4, \zeta_1, \ldots, \zeta_4 \mid \eta_i^2 = \zeta_j^2 = [\eta_i, \zeta_j] = 1,\ 1 \le i, j \le 4\rangle.$$

Let us calculate the action of the generators of the group D on the generators of the group $B \times B$. We have

$$\begin{aligned}
\eta_1^\mu &= (a,d)(1,b)(a,d) = (1,b) = \eta_1, \ \eta_2^\mu = (a,d)(1,aba)(a,d) = (1,dabad) = \eta_3,\\
\eta_4^\mu &= (a,d)(1,adabada)(a,d) = (1,dadabadad) = (1,adadbdada)\\
&= (1,adabada) = \eta_4,\\
\zeta_1^\mu &= (a,d)(b,1)(a,d) = (aba,1) = \zeta_2,\\
\zeta_3^\mu &= (a,d)(dabad,1)(a,d) = (adabada,1) = \zeta_4.
\end{aligned}$$

The relations for the action of ν are verified similarly. □

We rewrite presentation (38) in terms of the generators $\mu,\ \nu,\ \eta = \eta_1,\ \zeta = \zeta_1$. The rest of the generators can be expressed via these as follows:

$$\begin{aligned}
\eta_2 &= (1,aba) = (d,a)(1,b)(d,a) = \nu\eta\nu,\\
\eta_3 &= (1,dabad) = (a,d)(d,a)(a,b)(d,a)(a,d) = \mu\nu\eta\nu\mu,\\
\eta_4 &= (1,adabada) = \nu\mu\nu\eta\nu\mu\nu
\end{aligned}$$

and, similarly,

$$\zeta_2 = \mu\zeta\mu,\ \zeta_3 = \nu\mu\zeta\mu\nu,\ \zeta_4 = \mu\nu\mu\zeta\mu\nu\mu.$$

Lemma 23 *With respect to the generators $\mu,\ \nu,\ \eta,\ \zeta$ the group N has the following presentation*

$$\begin{aligned}
N = \langle \mu,\ \nu,\ \eta,\ \zeta \mid \mu^2 &= \nu^2 = \eta^2 = \zeta^2 = (\mu\nu)^4 = [\mu,\ \eta]\\
&= [\nu,\ \zeta] = [\eta,\ \zeta] = (\eta\nu\mu\zeta\mu\nu)^2 = 1.\rangle
\end{aligned} \tag{39}$$

Proof We rewrite the relations (38) via the the generators $\mu,\ \nu,\ \eta,\ \zeta$ and reduce these relations to the relations (39). The relations $\eta_i^2 = \zeta_j^2 = 1,\ 1 \le i \le 4$ can obviously be reduced to $\mu^2 = \nu^2 = 1$ which together with relations $\eta^2 = \zeta^2 = (\mu\nu)^4 = [\eta,\ \zeta] = [\mu,\ \eta] = [\nu,\ \zeta] = 1$ are part of the relations (38). For example the relation $[\mu,\eta] = 1$ is presented in (38) in the form $\eta_1^\mu = \eta_1$.

We rewrite the rest of the relations using those just obtained. This yields

$$[\eta,\ \zeta_2] = \eta\mu\zeta\mu\eta\mu\zeta\mu \sim \eta\mu^2\zeta\eta\mu^2\zeta = [\eta,\ \zeta],\ [\eta,\ \zeta_3] = (\eta\nu\mu\zeta\mu\nu)^2$$

and we have obtained the last of the relations (39). Also,

$$[\eta,\ \zeta_4] = \eta\mu\nu\mu\zeta\mu\nu\mu\eta\mu\nu\mu\zeta\mu\nu = \mu\eta\nu\mu\zeta\mu\nu\eta\nu\mu\zeta\mu\nu\mu \sim (\eta\nu\mu\zeta\mu\nu)^2$$

and reduction of the rest of the relations is conducted similarly. Ker ψ coincides with the normal closure in E of those elements which are mapped to the defining words of presentation (39). Keeping in mind that $\psi(\alpha) = \mu,\ \psi(\beta) = \nu,\ \psi(\delta) = \eta,\ \psi(\gamma) = \zeta$ and that $(\mu\nu)^4 = [\mu,\nu]^2 \bmod \langle \mu^2, \nu^2\rangle^\sharp$ we obtain the statement of Lemma 2.

Acknowledgements I express deep gratitude to Geoff Smith and Olga Tabachnikova for the warm hospitality I received during the group theory conference Groups St Andrews 97 in Bath held in during July-August 1997 where the results of this paper were presented. I also like to thank them for the help given when preparing this paper for publication; in particular - for Olga's wonderful translation of this paper into English. I woul also like to thank Said Sidki for giving me the opportunity to visit Brazil where this paper was written.

References

[Al] S. V. Aleshin. *Finite automata and the Burnside problem for periodic groups.* Mat. Zametki 11 (1972), 319-328.

[Ba] G. Baumslag. *Topics in Combinatorial Group Theory.* Birkhauser Verlag, 1993.

[Go] E. S. Golod. *On nilalgebras and finitely approximable p-groups.* Amer. Math. Soc. Transl. Ser. 2, 48 (1965), 103-106.

[Sa] E. S. Golod and I. R. Safarevich. *On class fields towers.* Amer. Math. Soc. Transl. Ser. 2, 48 (1965), 91-102.

[Gr 80] R. I. Grigorchuk. *On the Burnside problem for periodic groups.* Funktional Anal. i Prilozen. 14 (1980), 53-54.

[Gr 83] R. I. Grigorchuk. *On the Milnor problem on group growth.* Dokl. AN SSSR. V. 271 (1983), 31-33.

[Gr 84] R. I. Grigorchuk. *The growth degrees of finitely generated groups and the theory of invariant means.* Izv. AN SSSR. Ser. Matem. 48 (1984), 939-985.

[Gr 85] R. I. Grigorchuk. *The growth functions of finitely generated groups and its applications.* Doctoral dissertation. Moscow, 1985.

[Gr 97] R. I. Grigorchuk. *An example of finitely presented amenable group which does not belong to the class EG.* To appear in Mat. Sb.

[Gu1] N. Gupta and S. Sidki. *On the Burnside problem for periodic groups.* Math. Z. 182 (1983), 385-388.

[Gu2] N. Gupta *Recursively presented two generated infinite p-groups.* Math. Z. 188 (1984), 89-90.

[Ly] I. G. Lysionok. *A system of defining relations for the Grigorchuk group.* Mat. Zametki 38 (1985), 503-511.

[No] P. S. Novikov and S. I. Adyan. *On infinite periodic groups.* I, II, III. Izv. AN SSSR. Ser. matem. 32 (1968), 212-244, 251-524, 709-731.

[Ro1] A. V. Rozhkov. *Centralizers of elements in a group of tree automorphisms.* Izv. RAN, Ser. Matem. V.57 (1993), 82-105.

[Ro2] A. V. Rozhkov. *The lower central series of one group of automorphisms of a tree.* Mat. Zametki 60, No 2 (1996), 223-237.

[Se] V. Sergiescu. *A quick introduction to Burnside's problem. In: Group Theory from a Geometrical Viewpoint.* Ed. E. Ghys, A. Haefliger, A. Verjovsky. World Scientific (1991), 622-632.

[Si] S. Sidki. *On a 2-generated infinite 3-group: the presentation problem.* J. Algebra. 110 (1987), 13-23.

[Su] V. I. Suschanskii. *Periodic p-groups of permutations and the unrestricted Burnside problem.* Dokl. AN SSSR 247 (1979), 557-561.

[Ze 90] E. I. Zel'manov. *A solution of the Restricted Burnside Problem for groups of odd exponent.* Izv. AN SSSR. Ser. Matem. 54 (1990), 42-59.

[Ze 91] E. I. Zel'manov. *A solution of the Restricted Burnside Problem for 2-groups.* Mat. Sb. 182 (1991), 568-592.

ON THE DIMENSION OF GROUPS ACTING ON BUILDINGS

JENS HARLANDER

FB Mathematik, Universität Frankfurt, Robert-Mayer-Str. 8, 60054 Frankfurt/Main, Germany

1 Background

In this paper we construct upper bounds for the virtual cohomological dimension of virtually torsion-free groups that admit particularily nice actions on acyclic simplicial complexes. A G-complex X is called rigid if there is a subcomplex K such that the composition

$$K \hookrightarrow X \to X/G$$

is a simplicial isomorphism. In that case K is called a strong fundamental domain. The panels of K are the intersections of K with its G-translates.

Assume that X is a rigid acyclic G-complex that admits a strong fundamental domain with acyclic panels and let $\mathcal{H}$ be a set of subgroups of G consisting of stabilizers of vertices of such a strong fundamental domain. In that case we can define $\rho(G, \mathcal{H})$ to be the minimal dimension of such a rigid G-complex. It turns out that $\rho(G, \mathcal{H})$ can be described entirely in terms of the poset $\mathcal{H}$ (Theorem 2.8). For instance, if $\mathcal{H}$ is isomorphic to the poset of faces $\mathcal{S}(L)$ of a finite simplicial complex L, then

$$\rho(G, \mathcal{H}) = \max\{n | \exists \sigma \in \mathcal{S}(L) \text{ such that } \tilde{H}^n(lk_L(\sigma)) \neq 0\} + 1.$$

Here $lk_L(\sigma)$ denotes the link of σ in L. In case G is virtually torsion-free, the number

$$\min\{\text{vcd}(H) | H \in \mathcal{H}\} + \rho(G, \mathcal{H})$$

serves as an upper bound for the virtual cohomological dimension of G.

We apply our results to chamber-transitive type-preserving automorphism groups of buildings. Our main application is the following:

Theorem *Suppose the virtually torsion-free group G admits a chamber-transitive and type-preserving action on a contractible building Δ with associated Coxeter group W. Suppose furthermore that all special parabolic subgroups of G are finite. Then* $\text{vcd}(G) = \text{vcd}(W)$.

At this point we should also mention the following related result, which is much easier to prove: Suppose the virtually torsion-free group G admits a proper (that is, all vertex stabilizers are finite) cocompact action on a contractible building Δ of dimension n with associated Coxeter group W. Then $\text{vcd}(G) = \text{vcd}(W) = n$.

The proper cocompact action forces the building Δ to be locally finite. It follows that the appartments in Δ are n-dimensional manifolds (see Brown [4]) and that the compact support cohomology $H_c^n(\Delta)$ of Δ does not vanish in the top dimension.

The result now follows from standard arguments (see Brown [3], page 188 and page 209).

In the above Theorem the dimension of Δ can be larger than $\mathrm{vcd}(G)$. For example let W be the Coxeter group generated by the three reflections on the sides of an ideal triangle in the hyperbolic plane and let Δ be the associated Coxeter complex (Δ is the tesselated hyperbolic plane with added on cusp points). Then the hypothesis of the Theorem is satisfied (with $G = W$) and we have $1 = \mathrm{vcd}(W) < \dim(\Delta) = 2$.

The virtual cohomological dimension of Coxeter groups is well known. It was computed by M. Bestvina [2], using techniques of M. Davis [5]. In [9], H. Meinert and the author used a generalized version of the Davis-Bestvina construction in order to compute the virtual cohomological dimension of graph products with finite vertex groups. M. Davis has since observed (see [6]) that graph products admit chamber-transitive actions on buildings, so we obtain the main result on graph products in [9] as a corollary of the above Theorem. M. Davis also remarked in [6] (see Remark 10.4) on the possibility of obtaining information on the virtual cohomological dimension of chamber-transitive automorphism groups of buildings using methods as in [2] or [9].

I would like to thank Holger Meinert for helpful discussions and comments.

2 Rigid complexes and stabilizer schemes

Some of the material presented in this section is already contained in [9].

Let G be a group and X be a simplicial G-complex. A subcomplex K of X is called a *strong fundamental domain* if for every simplex σ' of X there exists a unique simplex σ of K such that $g\sigma = \sigma'$, for some $g \in G$. A G-complex that admits a strong fundamental domain is called *rigid*.

Let X be a rigid G-complex with strong fundamental domain K. A subcomplex of K of the form

$$K \cap \bigcap_{i \in I} g_i K,$$

where I is some index set and $g_i \in G$ for $i \in I$, is called a *panel* of K. If H is a subgroup of G, let P_H be the subcomplex of K containing all simplices σ of K such that $H \leq \mathrm{stab}(\sigma)$, where $\mathrm{stab}(\sigma)$ is the stabilizer of σ.

A set of subgroups $\mathcal{H}$ of G is *associated with* K if $\{\mathrm{stab}(v) | v \in V(K)\} \subseteq \mathcal{H}$ and every $H \in \mathcal{H}$ is a subgroup of some $\mathrm{stab}(v)$, $v \in V(K)$, the set of vertices of K.

Lemma 2.1 *Let X be a rigid G-complex with strong fundamental domain K. Let $\mathcal{H}$ be a set of subgroups of G associated with K.*

(i) *The subcomplexes P_H, $H \leq G$, are precisely the panels of K; in fact, $K \cap \bigcap_{i \in I} g_i K = P_H$, where H is the subgroup generated by the $g_i, i \in I$.*

(ii) *If $H \leq G$, then $P_H = P_{\bar{H}}$, where $\bar{H}$ is the intersection of all $H' \in \mathcal{H}$ that contain H.*

(iii) $P_{H_1} \cap P_{H_2} = P_{\langle H_1, H_2 \rangle}$, *where $H_1, H_2 \leq G$.*

(iv) *If the set of subgroups $\mathcal{H}$ is intersection-closed and P_H is acyclic for every $H \in \mathcal{H}$, then P_H is acyclic for any subgroup H of G.*

Proof (i) Assume σ is a simplex in $K \cap \bigcap_{i\in I} g_i K$. Then there exist simplices σ_i of K such that $\sigma = g_i\sigma_i$, $i \in I$. Since K is a strong fundamental domain we must have $\sigma_i = \sigma$ and thus $H \leq \operatorname{stab}(\sigma)$. So σ is a simplex of P_H.

Conversely assume that σ is a simplex of P_H. Then $H \leq \operatorname{stab}(\sigma)$, so $g_i\sigma = \sigma$ for all $i \in I$. Thus σ is a simplex of $K \cap \bigcap_{i\in I} g_i K$.

(ii) Since $H \leq \bar{H}$, we clearly have $P_{\bar{H}} \subseteq P_H$. In order to show the other direction let $\sigma = \{v_0, ..., v_l\}$ be a simplex of P_H. Then $H \leq \operatorname{stab}(\sigma) = \bigcap_{i=1}^{l} \operatorname{stab}(v_i)$. So $H \leq \operatorname{stab}(v_i) \in \mathcal{H}$, $i = 1, ..., l$ and thus $\bar{H} \leq \bigcap_{i=1}^{l} \operatorname{stab}(v_i) = \operatorname{stab}(\sigma)$. Hence σ is a simplex of $P_{\bar{H}}$.

(iii) Since $H_1, H_2 \leq \langle H_1, H_2\rangle$ we clearly have $P_{\langle H_1,H_2\rangle} \subseteq P_{H_1} \cap P_{H_2}$. If σ is a simplex in $P_{H_1} \cap P_{H_2}$, then $\langle H_1, H_2\rangle \in \operatorname{stab}(\sigma)$ and thus σ is a simplex of $P_{\langle H_1,H_2\rangle}$.

(iv)This is immediate from (ii). □

Proposition 2.2 *Let X_1 and X_2 be rigid G-complexes and suppose $\mathcal{H}$ is a set of subgroups associated with strong fundamental domains in X_1 and X_2 with acyclic panels. Then $H_*(X_1) \approx H_*(X_2)$.*

Proof Let $X = X_1$ and let $\mathcal{H}$ be associated with the strong fundamental domain K with acyclic panels. Let $\mathcal{U} = \{gK | g \in G\}$. Then $\mathcal{U}$ is an acyclic covering of X with acyclic intersections, since we assumed that K has acyclic panels. Thus $H_*(X) \approx H_*(N(\mathcal{U}))$, where $N(\mathcal{U})$ is the nerve of the covering (see Brown [3], page 168). We next define a complex $X(\mathcal{H})$ entirely in terms of $\mathcal{H}$ and then show that it is equal to the nerve $N(\mathcal{U})$. The vertices of $X(\mathcal{H})$ are the elements of G; a finite subset $\{g_0, ...g_n\}$ is a simplex of $X(\mathcal{H})$ if there exists $g \in G$ and $H \in \mathcal{H}$ such that $g_i \in gH$ for each i. Let us first show that $N(\mathcal{U})$ is a subcomplex of $X(\mathcal{H})$. Suppose that $\sigma = \{g_0, ..., g_l\}$ is a simplex of $N(\mathcal{U})$. Then $g_0K \cap ... \cap g_lK$ is non-empty, so $K \cap g_0^{-1}g_1K \cap ... \cap g_0^{-1}g_lK = P_H$ is non-empty, where H is the subgroup generated by the $g_0^{-1}g_i$, $i = 0, ..., l$ (see Lemma 2.1(i)). Let v be a vertex of P_H. Then $H \leq \operatorname{stab}(v) \in \mathcal{H}$ and so $g_0, ..., g_l \in g_0\operatorname{stab}(v)$. Thus σ is a simplex of $X(\mathcal{H})$.

For the other inclusion suppose that $\sigma = \{g_0, ..., g_l\}$ is a simplex of $X(\mathcal{H})$. Then, for some $g \in G$ and some $H \in \mathcal{H}$ we have $g_i \in gH$, $i = 0, ..., l$. Since $\mathcal{H}$ is associated with X, there is a vertex v of K so that $H \leq \operatorname{stab}(v)$. So $g^{-1}g_iv = v$, $i = 0, ..., l$, and hence $gv \in g_0K \cap ... \cap g_lK$. So σ is a simplex of $N(\mathcal{U})$. □

Let G be a group and let $\mathcal{H}$ be a set of subgroups. Suppose that there exists an acyclic rigid G-complex so that $\mathcal{H}$ is associated with some strong fundamental domain with acyclic panels. Let

$$\rho(G, \mathcal{H})$$

be the minimal dimension of such a G-complex. Note that in the case when G is virtually torsion-free and X is such a rigid G-complex of dimension $\rho(G, \mathcal{H})$ with fundamental domain K, we have

$$\begin{aligned} \operatorname{vcd}(G) &\leq \min\{\operatorname{vcd}(\operatorname{stab}(\sigma)) + \dim(\sigma)) | \sigma \in \mathcal{S}(K)\} \\ &\leq \min\{\operatorname{vcd}(H) | H \in \mathcal{H}\} + \rho(G, \mathcal{H}). \end{aligned} \tag{1}$$

Here $\mathcal{S}(K)$ denotes the set of simplices of K. For the first inequality see Brown [3], page 188, Exercise 4. The second inequality holds because each vertex stabilizer is contained in $\mathcal{H}$. In the remaining of this section we will describe the number $\rho(G, \mathcal{H})$ entirely in terms of the poset $\mathcal{H}$.

Let $\mathcal{P}$ be a poset. For $p \in \mathcal{P}$ we define a simplicial complex $L(p)$ as follows. The vertex set of $L(p)$ consits of elements $p' \in \mathcal{P}$ so that $p < p'$. A simplex of $L(p)$ is a finite set $\{p_0, ..., p_l\}$ such that $p < p_i$, $i = 0, ..., l$, and, for some $\bar{p} \in \mathcal{P}$, $p_i < \bar{p}$, $i = 0, ..., l$.

In case $\mathcal{P}$ is the poset of faces $\mathcal{S}(L)$ of a simplicial complex L it turns out that $L(\sigma)$, $\sigma \in \mathcal{S}(L)$, is homotopic to the link $lk_L(\sigma)$ of σ in L (recall that $lk_L(\sigma) = \{\sigma' \in \mathcal{S}(L) | \sigma \cap \sigma' = \emptyset, \sigma \cup \sigma' \in \mathcal{S}(L)\}$). This was shown in [9], Lemma 6.9.

Lemma 2.3 *If $\mathcal{P} = \mathcal{S}(L)$, the poset of faces of a simplicial complex L, then $L(\sigma)$, $\sigma \in \mathcal{S}(L)$, is homotopy equivalent to $lk_L(\sigma)$, the link of σ in L.*

Define

$$\alpha(\mathcal{P}) = \max\{n | \exists p \in \mathcal{P} \text{ such that } \tilde{H}^n(L(p)) \neq 0\} + 1.$$

In case $\mathcal{P} = \mathcal{S}(L)$, we simply write $\alpha(L)$ instead of $\alpha(\mathcal{S}(L))$. Note that, in the light of Lemma 2.3,

$$\alpha(L) = \max\{n | \exists \sigma \in \mathcal{S}(L) \text{ such that } \tilde{H}^n(lk_L(\sigma)) \neq 0\} + 1.$$

Lemma 2.4 *Let G be a group and $\mathcal{H}$ a set of subgroups that is associated with a strong fundamental domain with acyclic panels of some acyclic rigid G-complex. Then*

$$\alpha(\mathcal{H}) \leq \rho(G, \mathcal{H}).$$

Proof Let X be a rigid G-complex as in the statement with fundamental domain K. Let $H \in \mathcal{H}$ such that $\tilde{H}^n(L(H)) \neq 0$ for some n. Let

$$L = \bigcup_{H' \in V(L(H))} P_{H'}.$$

Let $\mathcal{U} = \{P_{H'} | H' \in V(L(H))\}$. It is an acyclic covering of L with acyclic intersections since we assumed acyclic panels. Thus $H^n(L) \approx H^n(N(\mathcal{U}))$. It is not difficult to check that $N(\mathcal{U})$ actually agrees with $L(H)$. Suppose $\sigma = \{H'_0, ..., H'_l\}$ is a simplex of $N(\mathcal{U})$. Then, by Lemma 2.1(iii), $P_{H'_0} \cap ... \cap P_{H'_l} = P_{\langle H'_0,...,H'_l \rangle} \neq \emptyset$. For v a vertex in the intersection we have $\langle H'_0, ..., H'_l \rangle \leq \text{stab}(v) \in \mathcal{H}$. Thus σ is a simplex of $L(H)$. For the other direction suppose that $\sigma = \{H'_0, ..., H'_l\}$ is a simplex of $L(H)$. Then there is a group $\bar{H} \in \mathcal{H}$ such that $\langle H'_0, ..., H'_l \rangle \leq \bar{H}$. Since $\mathcal{H}$ is associated with K, there exists a vertex v of K such that $\bar{H} \leq \text{stab}(v)$. Hence $v \in P_{\langle H'_0,...,H'_l \rangle} = P_{H'_0} \cap ... \cap P_{H'_l}$ and thus σ is a simplex of the nerve $N(\mathcal{U})$.

So $H^n(L) \approx H^n(N(\mathcal{U})) = \tilde{H}^n(L(H)) \neq 0$ and it follows that the dimension of L, and hence of K, is at least n. But if the dimension of both L and K is n, then non-vanishing of $H^n(L)$ implies that $H^n(K)$ does also not vanish, contradicting acyclicity of K. So $\dim(X) = \dim(K) \geq n + 1$. Thus $\dim(X) \geq \alpha(\mathcal{H})$. □

Let $\mathcal{H}$ be a finite intersection closed set of subgroups of G. In case $\rho(G,\mathcal{H})$ is defined, we will construct an acyclic rigid G-complex with acyclic panels of dimension $\alpha(\mathcal{H})$. This will show that

$$\alpha(\mathcal{H}) = \rho(G,\mathcal{H}).$$

For this it is convenient to talk about stabilizer schemes (see [9]).

Definition 2.5 A *stabilizer scheme* is a simplicial complex K together with a map G_* that assigns to a vertex v of K a subgroup G_v of G. A set of subgroups $\mathcal{H}$ is *associated with the stabilizer scheme* if $G_v \in \mathcal{H}$ for every vertex v and every $H \in \mathcal{H}$ is contained in some G_v.

Given a stabilizer scheme (K, G_*) and a subgroup H of G, we will denote the subcomplex of K spanned by vertices v such that $H \leq G_v$ by P_H and call it a panel of K. One can construct a rigid G-complex $X = X(K, G_*)$ with fundamental domain (isomorphic to) K, stabilizers stab$(v) = G_v$ for v a vertex of K, and panels agreeing with the panels defined for the stabilizer scheme. The set of vertices of X are pairs (gG_v, v), $g \in G, v \in V(K)$. A simplex of X is a finite subset $\{(gG_{v_0}, v_0), ..., (gG_{v_l}, v_l)\}$, such that $\{v_0, ..., v_l\}$ is a simplex of K. The action of G on X is simply left multiplication. The strong fundamental domain for the action consists of simplices $\{(G_{v_0}, v_0), ..., (G_{v_l}, v_l)\}$, such that $\{v_0, ..., v_l\}$ is a simplex of K, and is obviously isomorphic to K. The other statements made about X above are also clear. We call X the *realization* of the stabilizer scheme (K, G_*).

Proposition 2.6 *Let G be a group and $\mathcal{H}$ a set of subgroups which is associated to a strong fundamental domain with acyclic panels of some acyclic rigid G-complex X_1. If $\mathcal{H}$ is also associated with a stabilizer scheme (K, G_*) with acyclic panels, then the realization $X_2 = X(K, G_*)$ is acyclic.*

Proof This is immediate from Proposition 2.2. □

Suppose $\mathcal{H}$ is a finite set of subgroups of a group G. We inductively define a stabilizer scheme (K, G_*), with $G_v \in \mathcal{H}$ for $v \in V(K)$, the vertex set of K, and acyclic panels P_H, $H \in \mathcal{H}$. This presents a generalization of a construction of Bestvina [2] that is already contained in [9]. For the convenience of the reader we will recall this construction here.

First we filter $\mathcal{H}$. Let $\mathcal{H}_0$ be the set of maximal elements of the finite partially ordered set $\mathcal{H}$ (the partial ordering being inclusion). Suppose $\mathcal{H}_i$ is already constructed. Choose a maximal element H_{i+1} from $\mathcal{H} - \mathcal{H}_i$ and define $\mathcal{H}_{i+1} = \mathcal{H}_i \cup \{H_{i+1}\}$.

We inductively define stabilizer schemes $(K_i, G_{i,*})$ with $G_{i,v} \in \mathcal{H}_i$ and acyclic panels $P_{i,H}$ for $H \in \mathcal{H}_i$. Let K_0 be a set of vertices v_H in one-to-one correspondence with the elements H of $\mathcal{H}_0$. Define $G_{0,v_H} = H$. Notice that $P_{0,H} = v_H$, $H \in \mathcal{H}_0$, which is clearly acyclic. Suppose $(K_i, G_{i,*})$ is already constructed with the desired properties. Let

$$L_{i+1} = \bigcup_{H_{i+1} < H \in \mathcal{H}_i} P_{i,H},$$

$P_{i,H}$ being the panel of $H \in \mathcal{H}_i$ in K_i. If $\tilde{H}^{\dim(L_{i+1})}(L_{i+1}) = 0$, then define J_{i+1} to be a finite connected acyclic $(\dim(L_{i+1}))$-dimensional complex containing L_{i+1} as a full subcomplex (for the existence of such a complex J_{i+1} see the Lemma in Bestvina [2]). Otherwise define J_{i+1} to be the cone on L_{i+1}. Let $K_{i+1} = K_i \cup J_{i+1}$ and define $G_{i+1,v} = G_{i,v}$ for v a vertex of K_i and $G_{i+1,v} = H_{i+1}$ for every vertex v of K_{i+1} not already contained in K_i. One checks that the panel $P_{i+1,H}$ in K_{i+1} is equal to the panel $P_{i,H}$ in case $H \in \mathcal{H}_i$ and $P_{i+1,H_{i+1}} = J_{i+1}$. Let K be the union of all the K_i and let G_* be the union of all the $G_{i,*}$. A stabilizer scheme (K, G_*) obtained in this way we call a *Bestvina stabilizer scheme.*

It is important to note that the subcomplex

$$L_{i+1} = \bigcup_{H_{i+1} < H \in \mathcal{H}_i} P_{i,H}$$

of K can be written as

$$L_{i+1} = \bigcup_{H_{i+1} < H \in \mathcal{H}} P_H.$$

This is because $H_{i+1} < H \in \mathcal{H}$ already implies $H \in \mathcal{H}_i$; furthermore, by construction, if $H \in \mathcal{H}_i$, then $P_{i,H} = P_{i+1,H} = \ldots = P_H$.

Lemma 2.7 *Let $\mathcal{H}$ be a finite intersection-closed set of subgroups of a group G associated with a Bestvina stabilizer scheme (K, G_*). Then*

$$\dim(K) = \alpha(\mathcal{H}).$$

Proof Let $n+1$ be the dimension of K. Then there exists an i such that $H^n(L_i) \neq 0$, where

$$L_i = \bigcup_{H_i < H' \in \mathcal{H}} P_{H'}.$$

Then $\mathcal{U} = \{P_{H'} | H_i < H' \in \mathcal{H}\}$ is an acyclic covering of L_i with acyclic intersections (by Lemma 2.1(iv)) and so $H^n(L_i) \approx H^n(N(\mathcal{U}))$, where $N(\mathcal{U})$ is the nerve of the covering. As in Lemma 2.4 one shown that $N(\mathcal{U}) = L(H_i)$. So $H^n(L(H_i)) \neq 0$ and hence $\alpha(\mathcal{H}) \geq n+1 = \dim(K)$.

In order to show the other inequality assume that $\alpha(\mathcal{H}) = n+1$. Then, for some $H_i \in \mathcal{H}$ we have $H^n(L(H_i)) \neq 0$. Since $H^n(L(H_i)) \approx H^n(L_i)$ (see above), we see that $\dim(L_i) \geq n$. Since the cone on L_i is a subcomplex of K we see that $\dim(K) \geq n+1$. □

Theorem 2.8 *Let G be a group and $\mathcal{H}$ a finite intersection-closed set of subgroups that is associated with a strong fundamental domain with acyclic panels of some acyclic rigid G-complex. Let (K, G_*) be a Bestvina stabilizer scheme constructed from $\mathcal{H}$. Then the realization $X = X(K, G_*)$ is such a rigid G-complex of minimal dimension. In particular*

$$\alpha(\mathcal{H}) = \rho(G, \mathcal{H}).$$

Proof The realization X of the Bestvina stabilizer scheme is a rigid G-complex with acyclic panels P_H for $H \in \mathcal{H}$ by construction. Since $\mathcal{H}$ is intersection-closed, we use Lemma 2.1(iv) to conclude that all panels P_H, $H \leq G$ are acyclic. It follows from Proposition 2.2 that X is acyclic. Since $\dim(X) = \dim(K) = \alpha(\mathcal{H})$ by Lemma 2.7, we see that $\alpha(\mathcal{H}) \geq \rho(G, \mathcal{H})$, and hence, by Lemma 2.4, the two numbers are equal. □

Combining Lemma 2.3, Theorem 2.8 and inequality (1), we obtain the following

Corollary 2.9 *Let G be a virtually torsion-free group and $\mathcal{H}$ a finite set of subgroups that is associated with a strong fundamental domain with acyclic panels of some rigid acyclic G-complex. If $\mathcal{H}$ is isomorphic to the poset of faces of some finite complex L, then*

$$\begin{aligned} \mathrm{vcd}(G) \;\leq\; & \max\{\mathrm{vcd}(H) | H \in \mathcal{H}\} \\ & + \max\{n | \exists \sigma \in \mathcal{S}(L) \text{ such that } \tilde{H}^n(lk_L(\sigma)) \neq 0\} + 1. \end{aligned}$$

3 Coxeter complexes

Suppose that Σ is a contractible Coxeter complex, that is a thin chamber complex which admits sufficiently many foldings (see Tits [11] or Brown [4]).

Fix a chamber c. Let W be the subgroup of $Aut(\Sigma)$ generated by the reflections on the sides of c. The group W acts on Σ with strong fundamental domain c. Let $\{v_0, ..., v_n\}$ be the set of vertices of c. For J a proper subset of $I = \{0, ...n\}$ define W_J to be the stabilizer of the face $c_{I-J} = \{v_i | i \in I - J\}$ of c (note that $W_\emptyset = 1$). Such subgroups are called *parabolic with respect to c*. Let $\mathcal{H}_\Sigma$ be the set of all parabolic subgroups of W.

Let $L = L(W, I)$ be the simplicial complex on vertices I and simplices the spherical subsets of I. A subset $J \subsetneq I$ is called *spherical* if W_J is finite. Let $\mathcal{S}^f$ be the set of spherical subsets of I. Let $\mathcal{H}_\Sigma^f$ be the set of *special parabolic subgroups* W_J, $J \in \mathcal{S}^f$. Note that the assignement $J \rightarrow W_J$ defines a poset isomorphism between $\mathcal{S}^f$ and $\mathcal{H}_\Sigma^f$ (see [4], the Lemma on page 58). In particular we have $\alpha(\mathcal{H}_\Sigma^f) = \alpha(\mathcal{S}^f) = \alpha(L)$.

The complex Σ is a contractible rigid W-complex with contractible panels (the panels are full subcomplexes of a simplex). Since c is a fundamental domain and $\mathrm{stab}(v_i) = W_{I-\{i\}}$, the complex Σ can be described as the complex on vertices $wW_{I-\{i\}}$, $w \in W, i \in I$, and simplices $\{wW_{I-\{i\}} | i \in J\}$, $w \in W, J \subsetneq I$.

In [5] Davis considers the complex $\Sigma(W, I)$ on vertices wW_J, $w \in W, J \in \mathcal{S}^f$ and simplices of the form $\{wW_{J_0}, ..., wW_{J_n}\}$ with $w \in W$ and $J_0 \subsetneq ... \subsetneq J_n \in \mathcal{S}^f$. This complex can easily be shown to be a subcomplex of the baricentric subdivision of Σ.

Theorem 3.1 (i) (Davis [5]) *The complex $\Sigma(W, I)$ is a contractible rigid W-complex and the subgroup set $\mathcal{H}_\Sigma^f$ is associated with a strong fundamental domain with contractible panels.*

(ii) $\mathrm{vcd}(W) \leq \alpha(L) = \max\{n | \exists\sigma \in \mathcal{S}(L) \textit{ such that } \tilde{H}^n(lk_L(\sigma)) \neq 0\} + 1.$

Proof (i) The only difficult statement here is contractibility of $\Sigma(W, I)$. For that the reader is refered to [5].

The subcomplex on vertices W_J, $J \in \mathcal{S}^f$ and simplices $\{W_{J_0}, ..., W_{J_n}\}$ with $J_0 \subsetneq ... \subsetneq J_n \in \mathcal{S}^f$ is a strong fundamental domain. Stabilizers of he vertices of this fundamental domain are exactly the elements of $\mathcal{H}^f_\Sigma$.

In order to show that panels are contractible note that, in the light of Lemma 2.1(iv), it suffices to show that P_H is contractible for $H \in \mathcal{H}^f_\Sigma$. So suppose $H = W_J \in \mathcal{H}^f_\Sigma$. Now P_{W_J} is the full subcomplex of the fundamental domain on vertices $W_{J'}$ such that W_J is a subgroup of the stabilizer of the vertex $W_{J'}$, which is the subgroup $W_{J'}$. Thus P_{W_J} is the cone over some subcomplex of the fundamental domain with cone point W_J, and thus is contractible.

(ii) Follows from (i) and Corollary 2.9. □

It turns out that equality holds in *(ii)* of the above Proposition. In [2] Bestvina shows that the realization X of a Bestvina stabilizer scheme is a proper W-complex of minimal dimension (see also the proof below). Using our computation of the dimension of X, we obtain the following :

Theorem 3.2 $\mathrm{vcd}(W) = \alpha(L) = \max\{n | \exists\sigma \in \mathcal{S}(L) \textit{ such that } \tilde{H}^n(lk_L(\sigma)) \neq 0\} + 1.$

Proof Again let X be the realization of a Bestvina stabilizer scheme and let $n = \dim(X) = \alpha(L)$. In [2] Bestvina shows that the compact support cohomology $H^n_c(X)$ is non-trivial (see also [9]). Since

$$H^n(W, \mathbb{Z}W) \approx H^n_c(X)$$

(see Brown [3] page 209) it follows that $\mathrm{vcd}(W) \geq n = \alpha(L)$. □

Remarks (1) A description of the virtual cohomological dimension for right-angled Coxeter groups (more generally for graph products with finite non-trivial vertex groups) as in Theorem 3.2 is already given in [9]. .

(2) In [7] Davis completely determines the abelian structure of the cohomology of Coxeter groups with group ring coefficients. See also Dicks and Leary [8] where the W-module structure of the cohomology is computed for Coxeter groups for which the associated complex L is a manifold.

4 Buildings

Let Δ be a contractible building and G be a group acting on Δ with a single chamber c as strong fundamental domain (equivalently we could say the action is chamber-transitive and type-preserving). Let $\{v_0, ..., v_n\}$ be the set of vertices of c and $I = \{0, ..., n\}$. For $J \subseteq I$ let c_J be the face $\{v_i | i \in J\}$ of c. Let Σ be an appartment of Δ containing c and let W be the Coxeter group generated by the reflections on the sides of c. We can define a set map

$$\phi : W \to G$$

as follows: for $w \in W$ choose an element $\phi(w) \in G$ so that $wc = \phi(w)c$. We will need this map in the proof of Theorem 4.2 below.

As before, $\mathcal{S}^f$ denotes the set of spherical subsets of I (with respect to W). For J a proper subset of I define $G_J = \text{stab}(c_{I-J})$. Such groups are called *parabolic subgroups* of G (with respect to the choosen chamber c). Let $\mathcal{H}_\Delta$ be the set of parabolic subgroups and $\mathcal{H}^f_\Delta = \{G_J | J \in \mathcal{S}^f\}$ be the set of *special parabolic subgroups* . One can again check that the assignement $G_J \to J$ defines an isomorphism between the poset $\mathcal{H}^f_\Delta$ and the poset of spherical subsets $\mathcal{S}^f$, the poset of faces of L. In particular we have

$$\alpha(\mathcal{H}^f_\Delta) = \alpha(L) = \text{vcd}(W).$$

The last equality is Theorem 3.2.

Since c is a strong fundamental domain and $\text{stab}(v_i) = G_{I-\{i\}}$, the G-complex Δ can the described as the simplicial complex on vertices $gG_{I-\{i\}}, g \in G, i \in I$ and simplices the finite subsets $\{gG_{I-\{i_0\}}, ..., gG_{I-\{i_k\}}\}$, $g \in G, i_j \in I, j \in \{0, ..., k\}$. In [6] Davis considers the complex $\Delta(G, I)$ on vertices gG_J, $g \in G, J \in \mathcal{S}^f$ and simplices the finite subsets $\{gG_{J_0}, ..., gG_{J_k}\}$, $g \in G, J_0 \subsetneq ... \subsetneq J_k \in \mathcal{S}^f$. Again it is not difficult to check that $\Delta(G, I)$ is isomorphic to a subcomplex of the baricentric subdivision of Δ.

Theorem 4.1 (i) (Davis [6]) *The complex $\Delta(G, I)$ is a contractible rigid G-complex and $\mathcal{H}^f_\Delta$ is associated with some strong fundamental domain with contractible panels.*

(ii) *If G is virtually torsion-free, then*

$$\text{vcd}(G) \leq \text{vcd}(W) + \max\{\text{vcd}(G_J) | J \in \mathcal{S}^f\}.$$

Proof (i) Contractibility of $\Delta(G, I)$ is shown in [6], using metric arguments. The remaining statements are clear (see also the proof of Theorem 3.1).

(ii) Follows from (i), Corollary 2.9 together with Theorem 3.2. □

Theorem 4.2 *If G is virtually torsion-free and all special parabolic subgroups are finite, then* $\text{vcd}(G) = \text{vcd}(W)$.

Proof By Theorem 4.1(ii) it suffices to show that $\text{vcd}(W) \leq \text{vcd}(G)$. Let (K_1, W_*), (K_2, G_*) be Bestvina stabilizer schemes constructed from $\mathcal{H}^f_\Sigma, \mathcal{H}^f_\Delta$ respectively. Since these posets are isomorphic (they are both isomorphic to $\mathcal{S}^f$) and the construction of a Bestvina stabilizer scheme depends only on the isomorphism type of the poset, we may assume that $K_1 = K_2 = K$. Let X_Σ, X_Δ be the realization of $(K, W_*), (K, G_*)$ respectively. Now X_Σ is the simplicial complex with vertices (wW_v, v), $v \in V(K), w \in W$, and simplices $\{(wW_{v_0}, v_0), ..., (wW_{v_k}, v_k)\}$,

$w \in W, \{v_0, ..., v_k\} \in \mathcal{S}(K)$. Similarily, the vertices of X_Δ are (gG_v, v), $g \in G, v \in V(K)$ and it's simplices are $\{(gG_{v_0}, v_0), ..., (gG_{v_k}, v_k)\}$, $g \in G, \{v_0, ..., v_k\} \in \mathcal{S}(K)$. One can check that the map $\phi : W \to G$, defined at the beginning of this section, defines a simplicial embedding $\phi : X_\Sigma \to X_\Delta$ sending wW_v to $\phi(w)G_v$, which, in turn, induces a homomorphism

$$\phi_* : H_c^*(X_\Delta) \to H_c^*(X_\Sigma).$$

This homomorphism is onto in the top dimension $n = \dim(K)$. Since $H_c^n(X_\Sigma) \neq 0$ (see the proof of Theorem 3.2), we conclude that $H_c^n(X_\Delta) \neq 0$. Hence $H^n(G, \mathbb{Z}G) \neq 0$ by [3], page 209, and so $\mathrm{vcd}(G) \geq n$. Since $n = \alpha(\mathcal{H}_\Sigma^f) = \alpha(L) = \mathrm{vcd}(W)$, the proof is complete. □

5 Graph products

Let Γ be a finite simplicial graph with vertex set $I = \{0, ..., n\}$ and let $(G(i))_{i \in I}$ be a collection of non-trivial groups. The graph product $G = G(\Gamma)$ is the group generated by the vertex groups $G(i)$, $i \in I$, with the added relations that elements of distinct adjacent vertex groups commute. Graph products have turned out to be important for constructing groups with unusual properties. See [1], [2], [5], [10].

Taking $G(i)$ to be cyclic of order two for each i we obtain the right angled Coxeter group $W = W(\Gamma)$.

For $J \subseteq I$ denote by $G(J)$ the subgroup of G generated by the $G(i)$, $i \in J$.

Theorem 5.1 (Davis [6]) *The graph product G admits a chamber-transitive and type-preserving action on a building.*

Davis considers a certain chamber system $\mathbf{C} = (G, \{1\}, (G(i))_{i \in I})$ for which he defines a W-valued distance function $\delta : G \times G \to W$ which makes $\mathbf{C}$ into a combinatorial building. One obtains a building Δ in the classical sense in the following well known way. Let $V(\Delta)$ be the set of $I - \{i\}$ residue classes of chambers, $i \in I$. A set of elements from $V(\Delta)$ forms a simplex of Δ if and only if the corresponding equivalence classes have non-empty intersection. In our case this leads to a building Δ on vertices $gG(I - \{i\}), i \in I, g \in G$ and simplices of the form $\{gG(I - \{i\}) | i \in J \subseteq I\}$. It is clear that the action is chamber-transitive and type-preserving on Δ by multiplication.

Let us determine the parabolic subgroups with respect to the chamber $c = \{G(I - \{i\}) | i \in I\}$. For $J \subseteq I$ the group G_J is the stabilizer of the face $c_{I-J} = \{G(I - \{i\}) | i \in I - J\}$, which is $G(J)$. The special parabolic subgroups are of the form $G(J)$, $J \in \mathcal{S}^f$ (with respect to $W = W(\Gamma)$).

Note that $J \in \mathcal{S}^f$ if and only if any two elements of J (vertices of Γ) are connected by an edge in Γ. Thus the simplicial complex $L = L(W, I)$ with faces $\mathcal{S}^f$ is the full simplicial complex over the graph Γ.

Theorem 5.2 (Harlander, Meinert [9]) *Let $G = G(\Gamma)$ be a graph product with finite non-trivial vertex groups. Let $W = W(\Gamma)$ be the associated right angled*

Coxeter group and let L be the full simplicial complex over Γ. *Then*

$$\mathrm{vcd}(G) = \mathrm{vcd}(W) = \alpha(L) = \max\{n | \exists \sigma \in \mathcal{S}(L) \text{ such that } \tilde{H}^n(lk_L(\sigma)) \neq 0\} + 1.$$

Proof If G is finite the statement is clear (in this case L is a simplex and hence $\alpha(L) = 0$). So suppose G is infinite. The W is infinite as well and the building Δ for G is contractible (since it is non-spherical). If $J \subseteq S^f$ then $G(J) = \prod_{i \in J} G(i)$ is finite. So G admits a chamber-transitive and type-preserving action on a building with finite special parabolic subgroups. The result now follows from Theorem 4.2. □

References

[1] M. Bestvina, N. Brady, *Morse theory and finiteness properties of groups*, Invent. Math. 129 (1997), 445-470.

[2] M. Bestvina, *The virtual cohomological dimension of Coxeter groups*, in Geometric Group Theory vol. 1 (ed. G.A. Niblo, M.A. Roller), London Math. Soc. Lecture Note Series 181, Cambridge University Press 1993.

[3] K.S. Brown, Cohomology of Groups, Springer-Verlag 1982.

[4] K. S. Brown, Buildings, Springer-Verlag 1989.

[5] M.W. Davis, *Groups generated by reflections and aspherical manifolds not covered by Euclidian space*, Ann. of Math. 117 (1983), 293-324.

[6] M.W. Davis, *Buildings are CAT(0)*, to appear in the Proceedings of the LMS Durham Symposium on Geometric and Cohomological Group Theory 1994 (ed. G. Niblo, P. Kropholler, R. Stöhr), Cambridge University Press.

[7] M.W. Davis, *The cohomology of Coxeter groups with group ring coefficients*, preprint, Ohio State University, 1996.

[8] W. Dicks, I.J. Leary, *On subgroups of Coxeter groups*, to appear in the Proceedings of the LMS Durham Symposium on Geometric and Cohomological Group Theory 1994 (ed. G. Niblo, P. Kropholler, R. Stöhr), Cambridge University Press.

[9] J. Harlander, H. Meinert, *Higher generation subgroup sets and the virtual cohomological dimension of graph products of finite groups*, J. London Math. Soc. (2) 53 (1996), 99-117.

[10] G. Mess. *Examples of Poincare duality groups*, Proc. Am. Math. Soc. 110 (1990), 1145-1146.

[11] J. Tits, Buildings of Spherical Type and Finite BN-Pairs, Lecture Notes in Mathematics 386, Springer 1974.

DADE'S CONJECTURE FOR THE SIMPLE HIGMAN-SIMS GROUP

NABILA MOHAMED HASSAN* and ERZSÉBET HORVÁTH†

*Eötvös Loránd University, Department of Algebra and Number Theory, H-1088 Budapest, Múzeum krt. 6-8, Hungary
†Technical University of Budapest, Department of Algebra, H-1521 Budapest, Műegyetem rkp. 3-9, Hungary

Abstract

The topic of this paper is belonging to the program to study Dade's conjecture on counting the characters in blocks of finite groups. The radical $\mathcal{U}$-chains of the sporadic simple Higman-Sims group and its double cover are classified up to conjugacy. With the help of them the invariant projective form of Dade's conjecture is confirmed for the Higman-Sims group. For calculations the GAP system [21] is used.
AMS subject classification: 20C15, (20C20, 20C25).
Keywords: Dade's conjecture, characters, blocks, defects, p-chains, radical p-subgroups, covering.

1 Introduction

According to [6], the final form of Dade's conjecture, the so-called "inductive conjecture" holds for arbitrary finite groups, if it holds for all non-abelian finite simple groups.

It has been verified for a number of simple groups so far, including the Mathieu groups, the first three Janko groups, $Sz(q)$, $PSL_2(q)$, McL, the Tits group and the Held group, see [2], [3], [4], [6], [11], [12], [14], [15], [17] and [18]. It has been confirmed partially for some other classes of groups, as well, see e.g. [1], [20].

The Higman-Sims simple group can be obtained as a rank 3 primitive permutation group of degree 100 in which the stabilizer of a point has orbits of length $1, 22, 77$ and is isomorphic to the Mathieu group M_{22}. For further information see e.g. [5], [13] and [22]. The Higman-Sims group has a Schur multiplier of order 2 and outer automorphism group of order 2. According to [6], the inductive conjecture is equivalent to the invariant projective one, for all finite groups G, such that the outer automorphism group, $\mathrm{Out}(G)$, has cyclic Sylow q-subgroups for each prime q. Hence, in our case it is sufficient to check the invariant projective conjecture.

There are several forms of Dade's conjecture, see [8]. The invariant projective conjecture implies both the invariant and the projective conjecture. These conjectures all imply the weakest one, the ordinary conjecture. Since positive results of each weaker form of the conjecture are partial proofs of the stronger forms, we first checked the ordinary and the invariant forms, and then the projective and the invariant projective one. The main steps of this proof is summarized in Theorems 4.9 and 5.3.

The order of HS is $2^9 \cdot 3^2 \cdot 5^3 \cdot 7 \cdot 11$. Since Dade has proved in [9] that the invariant projective form of Dade's conjecture holds for all blocks with cyclic defect groups, we only need to verify the conjecture for the primes 2, 3 and 5.

Now we mention a few notations used in this paper: Let G be a finite group, p a prime number. R denotes a complete local discrete valuation ring with quotient field F of characteristic zero and residue class field $F' = R/J(R)$ of characteristic p. We assume that F, F' are both splitting fields for every subgroup of G. The set of all irreducible F-characters of G is denoted by $Irr(G)$. If B is a p-block of G, then $Irr(B)$ stands for the set of all irreducible F-characters of G, belonging to the block B. If $N \triangleleft G$ and $\theta \in Irr(N)$, then $Bl(G|\theta)$ is the set of blocks of G, lying over θ. ν_p is the p-adic exponential valuation of R. $O_p(G)$ is the largest normal subgroup of G of p-power order. For further notations and the basic definitions of modular representation theory the reader is referred to [19].

2 Definitions and known results

In this section we give the basic definitions and formulate those forms of Dade's conjecture that we will need later.

Definition 2.1 A *radical p-subgroup* P of G is p-subgroup of G satisfying

$$O_p(N_G(P)) = P.$$

Definition 2.2 A *p-chain* of G is a non-empty, strictly increasing chain

$$C : P_0 < P_1 < \ldots < P_n \tag{1}$$

of p-subgroups P_i of G, its *length* $|C| = n$. Denote by C_i the *initial subchain*

$$C_i : P_0 < P_1 < \ldots < P_i$$

of C of length i and by C^i the *final subchain*

$$C^i : P_i < \ldots < P_n$$

of C of length $n - i$.

The group G acts by conjugation on the family $\mathcal{P}(G)$ of all p-chains of G. We denote by C^g the image of C under the action of $g \in G$.

$$C^g : P_0^g < P_1^g < \ldots < P_n^g. \tag{2}$$

The stabilizer of C under the conjugation action of G is

$$N_G(C) = N_G(P_1) \cap N_G(P_2) \cap \ldots \cap N_G(P_n).$$

If P is normal in G, then G acts also on the set $\mathcal{P}(G|P)$ of all p-chains beginning with P.

Definition 2.3 A *radical p-chain* of G is a p-chain $C : P_0 < P_1 < \ldots < P_n$ of G satisfying

(a) $P_0 = O_p(G)$

(b) $P_i = O_p(N_G(C_i))$, $i = 0, \ldots, n$.

We denote the set of all radical p-chains of G by $\mathcal{R}(G)$.

Definition 2.4 An *$\mathcal{U}$-chain* of G is a p-chain $C : P_0 < P_1 < \ldots < P_n$ of G satisfying the condition that every P_i is a radical p-subgroup of G, $i = 0, \ldots, n$ and $P_0 = O_p(G)$. We denote the set of all $\mathcal{U}$-chains of G by $\mathcal{U}(G)$.

Definition 2.5 An *elementary p-chain* of G is a p-chain $C : P_0 < P_1 < \ldots < P_n$ of G such that $P_0 \triangleleft G$, P_i/P_0 is an elementary abelian p-subgroup of G for $i = 0, \ldots, n$. We will denote by $\mathcal{E}(G|P)$ the set of all elementary p-chains of G beginning with P.

Definition 2.6 An *$\mathcal{N}$-chain* of G is a p-chain $C : P_0 < P_1 < \ldots < P_n$ of G with P_i normal in P_n for $i = 0, \ldots, n$. We will denote by $\mathcal{N}(G|P)$ the set of all $\mathcal{N}$-chains of G beginning with P.

In the various forms of Dade's conjecture, some alternating sums are stated to be zero. The general properties of the functions that occur in these sums and the general form of these sums are described in the following:

Properties and notations 2.7 Let E be a finite group, let G be a normal subgroup of E, let $P \leq G$ be a p-subgroup and let H be a subgroup of $N_E(P)$. Then H acts by conjugation (2) on the family $\mathcal{P}(G|P)$ of all p-chains of G beginning with P.

If $\mathcal{F}$ is a H-invariant subfamily of $\mathcal{P}(G|P)$, then $\mathcal{F}/H$ will denote an arbitrary set of representatives of H-orbits in $\mathcal{F}$.

Let $\mathcal{Q}$ be the field of rational numbers and let f be a function $f : \mathcal{P}(G|P) \to \mathcal{Q}$. Assume that $f(C) = f(C')$ whenever C and C' are conjugate in H or, when the normalizers $N_E(C)$ and $N_E(C')$ are equal. We will denote by $\mathcal{S}(f, \mathcal{F}/H)$ the alternating sum

$$\mathcal{S}(f, \mathcal{F}/H) = \sum_{C \in \mathcal{F}/H} (-1)^{|C|} f(C).$$

In the special case when $E = H = G$ and $O_p(G) = 1$, Knörr and Robinson, [16], showed that

$$\mathcal{S}(f, \mathcal{P}(G|1)/H) = \mathcal{S}(f, \mathcal{U}(G)/H) = \mathcal{S}(f, \mathcal{E}(G|1)/H) = \mathcal{S}(f, \mathcal{N}(G|1)/H).$$

Dade [6] showed that in this case also $\mathcal{S}(f, \mathcal{N}(G|1)/H) = \mathcal{S}(f, \mathcal{R}(G)/H)$.

These equations are also true for arbitrary E and H satisfying the properties of 2.7.

Now we formulate the conjectures of Dade.

Dade's ordinary conjecture 2.8 Let G be a finite group with $O_p(G) = 1$, and let B be a p-block of G with defect $d(B) > 0$. Let d be a non-negative integer. Then

$$\sum_{C \in \mathcal{F}/G} (-1)^{|C|} k(N_G(C), B, d) = 0,$$

where $k(N_G(C), B, d)$ is the number of characters in the set

$$\{\chi \in Irr(N_G(C)) | b(\chi)^G = B, d(\chi) = d\}.$$

Here $b(\chi)$ denotes the block of $N_G(C)$ containing χ, $d(\chi)$ stands for the defect of χ and $\mathcal{F}$ is any of the families $\mathcal{R}(G), \mathcal{U}(G), \mathcal{E}(G|1), \mathcal{N}(G|1)$ or $\mathcal{P}(G|1)$.

For the invariant conjecture, one has to embed G as a normal subgroup into an extension group E. If the centre of G is trivial, then we can identify G with its inner automorphism group, and so $G \triangleleft \text{Aut}(G)$. In this case, we may assume that $E = \text{Aut}(G)$. Then E acts on the chains in $\mathcal{E}(G), \mathcal{U}(G), \mathcal{N}(G)$ or $\mathcal{R}(G)$. For each chain C, $N_G(C) \triangleleft N_E(C)$ and so for each $\psi \in Irr(N_G(C))$ there exists an inertia subgroup $T(\psi)$ in $N_E(C)$. Of course $N_G(C) \leq T(\psi)$. For each chain C and for each subgroup H of E containing G, let $k(C, B, d, H)$ denote the number of irreducible characters ψ of $N_G(C)$, with $d(\psi) = d$ and $GT(\psi) = H$, which belong to the p-block $b = b(\psi)$ of $N_G(C)$ with $b^G = B$.

Dade's invariant conjecture 2.9 Let $O_p(G) = 1$ and $d(B) > 0$. Then

$$\sum_{C \in \mathcal{F}/G} (-1)^{|C|} k(C, B, d, H) = 0,$$

where $\mathcal{F}$ is any of $\mathcal{R}(G), \mathcal{U}(G), \mathcal{E}(G|1), \mathcal{N}(G|1), \mathcal{P}(G|1)$, d is any nonnegative integer, H is any subgroup of E containing G.

It is well-known, see e.g. [10], that projective representations with factor set α are equivalent to ones that can be lifted to ordinary representations of the covering group G^* of G got as a central extension of G by $Z^* = \langle \alpha \rangle$, where α can be chosen up to equivalence so that its order is equal to the order of its image in the Schur multiplier of G. In this correspondence, irreducible projective representations of G lift to irreducible representations of G^* and if we fix a faithful irreducible character ζ of Z^*, then those irreducible characters of G^*, that come from lifting projective characters of G, are just those, which lie above ζ, while the factor set belonging to the central extension G^* of G, is just $\zeta^{-1}(\alpha)$.

Remark 2.10 According to Proposition 2.2 of [7], if $K = K_p \times Z$, where $K \triangleleft G$, K_p is a p-subgroup and Z is a central p'-subgroup of G, then if $\overline{G} = G/K$ then there is an inclusion preserving bijection between p-subgroups of $\overline{G}$ and p-subgroups of G containing K_p, which induces a length preserving bijection on p-chains C of $\overline{G}$ and those p-chains C^* of G that $K_p \leq P_0$. In this correspondence normalizers of p-subgroups and p-chains correspond to similar normalizers in the other group, radical p-subgroups correspond to radical p-subgroups etc. Applying this to $K :=$

Z^*, $G := G^*$ and $\overline{G} := G$, we get that $\mathcal{R}(G^*/Z^*)$ and $\mathcal{R}(G^*)$, $\mathcal{U}(G^*/Z^*)$ and $\mathcal{U}(G^*)$, $\mathcal{P}(G^*/Z^*|PZ^*/Z^*)$ and $\mathcal{P}(G^*|P)$, etc. correspond to each other.

As a corollary one gets like in Proposition 2.10 [7], for $K := O_p(G)$, $\overline{G} := G/O_p(G)$, that chains of $\mathcal{P}(G/O_p(G)|\overline{1})$ and $\mathcal{P}(G|O_p(G))$ correspond to each other in a similar way as above, and if $f : \mathcal{P}(G|O_p(G)) \to \mathcal{Q}$ like above, then

$$\sum_{C \in \mathcal{P}(G|O_p(G))/H} (-1)^{|C|} f(C) = \sum_{C \in \mathcal{R}(G)/H} (-1)^{|C|} f(C) = \sum_{C \in \mathcal{U}(G)/H} (-1)^{|C|} f(C).$$

So the functions of 2.7 can be calculated on any of these chains.

Write $k(N_{G^*}(C^*), B^*, d^*, \zeta)$ for

$$|\{\psi \in Irr(N_{G^*}(C^*)) | d(\psi) = d^*, b(\psi)^{G^*} = B^*, (\psi_{Z^*}, \zeta) \neq 0\},$$

where ζ is a faithful irreducible character of Z^* and $d^* \geq \nu_p(|Z^*|)$. Now we state Dade's projective conjecture using ordinary characters of the covering group G^* above.

Dade's projective conjecture 2.11 Let G be a finite group with $O_p(G) = 1$. Let Z^* be a cyclic subgroup of the Schur multiplier of the group G and let G^* be a central extension of G with Z^*. Let ζ be a faithful irreducible character of Z^*. Then

$$\sum_{C^* \in \mathcal{F}(G^*|O_p(G^*))/G^*} (-1)^{|C^*|} k(N_{G^*}(C^*), B^*, d^*, \zeta) = 0$$

for each block $B^* \in Bl(G^*|\zeta)$ with $d(B^*) > \nu_p(|Z^*|)$, where $\mathcal{F}$ can be any of $\mathcal{R}, \mathcal{U}, \mathcal{E}, \mathcal{N}, \mathcal{P}$.

We formulate the invariant projective conjecture:

Dade's invariant projective conjecture 2.12 Let G be a finite with $O_p(G) = 1$. Let Z^* be a cyclic subgroup of the Schur multiplier of the group G and let G^* be a central extension of G with Z^*. Let ζ be a faithful irreducible character of Z^*. Let E^* be a group containing G^* as a normal subgroup, such that E^* is acting on Z^* trivially. Then for every subgroup H^* of E^* containing G^*,

$$\sum_{C^* \in \mathcal{F}(G^*|O_p(G^*))/G^*} (-1)^{|C^*|} k(C^*, B^*, d^*, H^*, \zeta) = 0,$$

where $k(C^*, B^*, d^*, H^*, \zeta)$ denotes the number of irreducible characters ψ of the group $N_{G^*}(C^*)$ of defect d such that $b(\psi)^{G^*} = B^*$, $(\psi_{Z^*}, \zeta) \neq 0$, and for the inertial group $T^*(\psi)$ of ψ in $N_{E^*}(C^*)$, $T^*(\psi)G^*$ is equal to H^*. The block $B^* \in Bl(G^*|\zeta)$ satisfies $d(B^*) > \nu_p(|Z^*|)$, and $\mathcal{F}$ is any of $\mathcal{R}, \mathcal{U}, \mathcal{E}, \mathcal{N}, \mathcal{P}$.

3 Preliminary lemmas

In the following we shall check Dade's conjecture with the help of $\mathcal{U}$-chains. By experience there are not so many of them as of elementary chains or of $\mathcal{N}$-chains, and they can be constructed easier than radical p-chains. So we need an algorithm, with the help of that one can construct a full set of representatives of conjugacy classes of $\mathcal{U}$-chains in a finite group. Our algorithm uses the following:

Lemma 3.1 *Let G be a finite group, p a prime. Let $R(G)$ denote a set of representatives of radical p-subgroups of G that properly contain $O_p(G)$, under conjugation with G. Let $R(G, P_1, \ldots, P_n)$ denote a set of representatives of radical p-subgroups of G that properly contain $O_p(G)$ and that are proper subgroups of P_1, under conjugation with $\bigcap_{i=1}^n N_G(P_i)$.*

Set

$$\mathcal{U}_0(G) = \{O_p(G)\} \quad \text{and} \quad \mathcal{U}_1(G) = \{O_p(G) < P | P \in R(G)\},$$

and define inductively

$$\mathcal{U}_{n+1}(G) = \bigcup \{O_p(G) < Q < P_1 < \cdots < P_n |$$

$$O_p(G) < P_1 < \cdots < P_n \in \mathcal{U}_n(G) \text{and} Q \in R(G, P_1, \ldots, P_n)\}.$$

Then $\mathcal{U}_n(G)$ is a set of representatives of $\mathcal{U}$-chains of length n under G-conjugation.

Proof The statement holds clearly for $n = 0$ and $n = 1$. Assume that it holds for n. Then we show that

(1) the chains in $\mathcal{U}_{n+1}(G)$ are pairwise non-conjugate in G, and

(2) each $\mathcal{U}$-chain of length $n+1$ is G-conjugate to a chain in $\mathcal{U}_{n+1}(G)$.

(1) Take two G-conjugate chains

$$O_p(G) < Q < P_1 < \cdots < P_n \text{ and } O_p(G) < Q' < P_1' < \cdots < P_n'$$

in $\mathcal{U}_{n+1}(G)$. By the induction hypothesis, the chains

$$O_p(G) < P_1 < \cdots < P_n \text{ and } O_p(G) < P_1' < \cdots < P_n'$$

are G-conjugate if and only if they are equal, so each conjugating element lies in $\bigcap_{i=0}^n N_G(P_i)$. Hence $Q = Q'$ by the construction of $\mathcal{U}_{n+1}(G)$.

(2) Let

$$C: O_p(G) < Q < P_1 < \cdots < P_n$$

be a $\mathcal{U}$-chain of length $n+1$. By the induction hypothesis, the $\mathcal{U}$-chain

$$O_p(G) < P_1 < \cdots < P_n$$

is G-conjugate to a chain in $\mathcal{U}_n(G)$, so C is G-conjugate to a chain in $\mathcal{U}_{n+1}(G)$ by construction. □

The G-classes of $\mathcal{U}$-chains can be enumerated if the sets $R(G, P_1, \ldots, P_n)$ and $R(G)$ can be computed. We can get the first set for example by calculating the conjugacy classes of subgroups of P_1 above $O_p(G)$, selecting those classes that contain radical p-subgroups in G, and then selecting representatives under conjugation with $\bigcap_{i=1}^n N_G(P_i)$. $R(G)$ can be computed similarly, by calculating the conjugacy classes of subgroups of a fixed Sylow P above $O_p(G)$, and selecting those classes that contain radical p-subgroups in G.

Remark 3.2 (a) If C is a p-chain $C : P_0 < P_1 < \ldots < P_n$ with final subchains C^i, $i = 0, \ldots, n$ and normalizers of the final subchains, $G_i = N_G(C^i)$ $i = 0, \ldots, n$, together with $G_{n+1} = G$. Then

$$P_i C_{G_{i+1}}(P_i) \leq N_{G_{i+1}}(P_i) = N_G(C^i) \quad (i = 0, \ldots, n).$$

So by Theorem 3.6 and Corollary 3.7 in Chapter 5 of [19], if $b_i \in Bl(N_G(C^i))$ then the induced block $b_i^{G_{i+1}}$ is defined for $i = 0, \ldots, n$ and so by induction, b^G is defined for each block b of $N_G(C)$. This result was first proved in Lemma 3.2 of [16], a version of it on twisted group algebras can be found in Proposition 10.14 of [7], from which this proof can be deduced.

(b) This is well-known, for example see Lemma 3.3 of Chapter 5 in [19], that the defect group of the induced block, b^G contains a conjugate of the defect group of b, and so $d(b^G) \geq d(b)$. As the defect of a block is the maximum of defects of irreducible ordinary characters of the block, $k(N_G(C), B, d) = 0$ if $d > d(B)$, especially if $d(B) = 0$ then B can be induced only from blocks of defect zero, and $k(N_G(C), B, d) = 0$ for each positive integer d.

Now we formulate some other results that make it easier to compute the numbers $k(N_G(C), B, d)$. First we prove a generalization of Brauer's Third Main Theorem to chain normalizers:

Lemma 3.3 *Let C be a p-chain in G, $C : P_0 < P_1 < \ldots < P_n$. Then a p-block b of $N_G(C)$ is the principal block if and only if $b^G = B_0$, where B_0 is the principal p-block of G.*

Proof Let $G_0, G_1, \ldots, G_{n+1}$ be the final subchain normalizers $G_i = N_G(C^i)$ for $i = 0, \ldots, n$ together with $G_{n+1} = G$. Then we have that $G_0 = N_G(C) = N_G(C^0) = N_G(P_0) \cap N_G(C^1) = N_{G_1}(P_0) \leq G_1 = N_G(P_1) \cap N_G(C^2) = N_{G_2}(P_1) \leq G_2 = N_{G_3}(P_2) \leq G_3 \ldots \leq G_n = N_G(P_n) \leq G$. By Brauer's Third Main Theorem, see e.g. Chapter 5, Theorem 6.1 in [19], the principal block b_i of $N_{G_{i+1}}(P_i)$ induces the principal block b_{i+1} of $G_{i+1} = N_{G_{i+2}}(P_{i+1})$ for $i = 0, \ldots, n-1$, and $b_n \in Bl(N_G(P_n))$ induces the principal block B_0 of G. Thus the principal block of $N_G(C)$ induces to the principal block of G. Conversely by the other part of the Third Main Theorem, for each $i = 0, \ldots, n$ the principal block of $G_i = N_{G_i+1}(P_i)$ is induced only from the principal block b_{i-1} of $N_{G_i}(P_{i-1})$ for $i = 1, \ldots, n$ and $B_0 \in Bl(G)$ is induced only from the principal block of $G_n = N_G(P_n)$. Hence if a block b of $N_G(C)$ induces to the principal block B_0 of G then b is the principal block of $N_G(C)$. □

Corollary 3.4 *Let B_0 be the principal p-block of G and let b_o be the principal p-block of the chain normalizer $N_G(C)$. Let k be the function from 2.8. Then $k(N_G(C), B_0, d) = |\{\psi \in Irr(b_0)|d(\psi) = d\}|$.*

If G has only two p-blocks B_0 and B then

$$k(N_G(C), B, d) = |\{\psi \in Irr(N_G(C))\backslash Irr(b_0)|d(\psi) = d\}|.$$

If additionally we know that $N_G(C)$ has at most two blocks, then if the other block is b, then $k(N_G(C), B, d) = |\{\psi \in Irr(b)|d(\psi) = d\}|$. On the other hand, if $N_G(C)$ has just one block, then $k(N_G(C), B, d) = 0$ for each nonnegative integer d. Similar statements hold for the functions k of 2.9, 2.11 and 2.12, as well.

The next lemma generalizes Lemma 6.9 of [6] for arbitrary p-chains:

Lemma 3.5 *Let b be a p-block of $N_G(C)$, where $C : P_0 < P_1 < \ldots < P_n$ is a p-chain in G. Then a defect group of b^G contains a conjugate of P_n. Especially if for the function k of 2.8 and $k(N_G(C), B, d) \neq 0$, then P_n is contained in some defect group of B. Similar statements hold for functions k of 2.9, 2.11 and 2.12, as well.*

Proof With the notation of the above lemma, if we take a block $b = b_0$ of $N_G(C) = N_G(C^0) = N_{G_1}(P_0)$ and $b_i = (b_0)^{N_{G_{i+1}}(P_i)}$, then by Theorem 2.8 in Chapter 5 of [19], we have that P_i is contained in every defect group of b_i, for $i = 0, .., n-1$. If $i = n$, then b_n is a block of $N_{G_{n+1}}(P_n) = N_G(P_n)$. So we get here, that any defect group of b_n contains P_n and hence by Remark 3.2 (b), a defect group of $b^G = {b_n}^G$ contains a conjugate of P_n. If $k(N_G(C), B, d) \neq 0$, then $B = b^G$ for some block b of $N_G(C)$, and so a defect group of B contains a conjugate of P_n, in other words, P_n is contained in some defect group of B. □

Corollary 3.6 *If C is a p-chain of G, $C : P_0 < P_1 < \ldots < P_n$ with $|P_n| > p^{d(B)}$ for each nonprincipal p-block B of G, then $N_G(C)$ has just one block.*

Proof If $b \in Bl(N_G(C))$, then the defect group of b^G contains a conjugate of P_n, thus $|P_n| \leq p^{d(b^G)}$. By our assumption, then b^G is the principal block B_0 of G. Hence Lemma 3.3 gives us, that b is the principal block of $N_G(C)$. □

4 Invariant conjecture for HS

We applied the algorithm of Lemma 3.1 in GAP to construct a full set of representatives of HS-conjugacy classes of $\mathcal{U}$-chains in HS. For this a faithful degree 100 permutation representation of $HS2$ can be used, of which HS is its derived subgroup. The faithful degree 100 representation of $HS2$ can be obtained from a faithful degree 1408 permutation representation of $2HS2$, as action on the cosets of a subgroup of index 100, as the centre of $2HS2$ is contained in every subgroup of index 100. The degree 1408 permutation representation is the representation of $2HS2$ on the cosets of a subgroup of type $U_3(5)$, whose index is 1408. This representation can be obtained from the matrix representation of $2HS2$ of degree 56 over $GF(5)$, which can be found in [22].

$C_0 : 1$	$C_{17} : 1 < R2_3 < R256_2$	$C_{34} : 1 < R2_1^k < R4 < R512$
$C_1 : 1 < R2_1$	$C_{18} : 1 < R4 < R256_2$	$C_{35} : 1 < R2_1^g < R128 < R512$
$C_2 : 1 < R4$	$C_{19} : 1 < R64_1 < R256_2$	$C_{36} : 1 < R16 < R128 < R512$
$C_3 : 1 < R16$	$C_{20} : 1 < R64_2 < R256_2$	$C_{37} : 1 < R64_1 < R128 < R512$
$C_4 : 1 < R64_1$	$C_{21} : 1 < R2_1 < R512$	$C_{38} : 1 < R2_1 < R256_1 < R512$
$C_5 : 1 < R64_2$	$C_{22} : 1 < R2_2 < R512$	$C_{39} : 1 < R2_1^w < R256_1 < R512$
$C_6 : 1 < R128$	$C_{23} : 1 < R2_3 < R512$	$C_{40} : 1 < R16 < R256_1 < R512$
$C_7 : 1 < R256_1$	$C_{24} : 1 < R4 < R512$	$C_{41} : 1 < R64_2 < R256_1 < R512$
$C_8 : 1 < R256_2$	$C_{25} : 1 < R16 < R512$	$C_{42} : 1 < R2_3 < R256_2 < R512$
$C_9 : 1 < R512$	$C_{26} : 1 < R64_1 < R512$	$C_{43} : 1 < R2_1 < R256_2 < R512$
$C_{10} : 1 < R2_1 < R4$	$C_{27} : 1 < R64_2 < R512$	$C_{44} : 1 < R4 < R256_2 < R512$
$C_{11} : 1 < R2_1^g < R128$	$C_{28} : 1 < R128 < R512$	$C_{45} : 1 < R64_1 < R256_2 < R512$
$C_{12} : 1 < R16 < R128$	$C_{29} : 1 < R256_1 < R512$	$C_{46} : 1 < R64_2 < R256_2 < R512$
$C_{13} : 1 < R64_1 < R128$	$C_{30} : 1 < R256_2 < R512$	$C_{47} : 1 < R2_1 < R4 < R256_2 < R512$
$C_{14} : 1 < R2_1 < R256_1$	$C_{31} : 1 < R2_1 < R4 < R256_2$	$C_{48} : 1 < R2_1^h < R4 < R256_2 < R512$
$C_{15} : 1 < R16 < R256_1$	$C_{32} : 1 < R2_1 < R4 < R512$	$C_{49} : 1 < R2_1^k < R4 < R256_2 < R512$
$C_{16} : 1 < R64_2 < R256_1$	$C_{33} : 1 < R2_1^h < R4 < R512$	

Table 1

4.1 $p = 2$ ordinary conjecture

We calculated a fixed Sylow 2-subgroup S of HS. We got the representatives of S-conjugacy classes of nontrivial radical 2-subgroups of HS in S: $\{R2_1, R2_2, R2_3, R4, R16, R64_1, R64_2, R128, R256_1, R256_2, R512\}$. Here $R512 = S$ and in general Rn_i denotes a representative of the i-th S-conjugacy class of radical 2-subgroups of order n.

Remark 4.1 We do not add the full structural information about these groups. With this we would like to emphasize that in the GAP programs we actually did not use any special properties of them, just those which the programs could deduce from the normalizers of these groups and from their character tables.

A maximal set of in HS non-conjugate elements of the above set is $\{R2_1, R4, R16, R64_1, R64_2, R128, R256_1, R256_2, R512\}$. A full set of representatives of HS-conjugacy classes of $\mathcal{U}$ 2-chains in HS is sown in Table 1. Here $R2_1^g$ etc. denote suitable conjugates of $R2_1$.

In HS there are two 2-blocks, B_0 and B_1. $d(B_0) = 9$ and B_0 has defect group $S \in \mathrm{Syl}_2(HS)$, $d(B_1) = 2$ and so B_1 has defect group $R4$, which is elementary abelian of order 4. We were putting in the following table the numbers $k(N_{HS}(C_i), B_j, d)$, where $i \in \{0, \ldots, 49\}$, $j \in \{0, 1\}$ and $d \in \{0, \ldots, 9\}$. We omitted the zero columns and rows of zero contribution. Chains are belonging to one row if and only if their normalizers are equal in HS. In fact in the table below chains in different rows are not even conjugate in HS. The values were calculated with GAP, and are shown in Table 2. The last row contains the alternating sum of the ordinary Dade's conjecture. The first column contains the sign with that the values of the respective row have to be taken after taking into consideration the chain lengths.

Remark 4.2 Applying Corollary 3.6 and the fact that HS has two 2-blocks with defects 9 and 2 respectively, we get that except for the chains C_0, C_1, C_2 and C_{10}, the chain normalizers $N_{HS}(C_i)$ have only one block. By Corollary 3.4, if b_0^i is the

Sign		$d=2$	$d=4$	$d=5$	$d=6$	$d=7$	$d=8$	$d=9$
		B_1	B_0	B_0	B_0	B_0	B_0	B_0
+	C_0 :	4	1	0	1	0	10	8
−	C_1 :	4	2	4	16	0	0	0
−	C_2 :	4	16	0	0	0	0	0
−	C_3 :	0	1	0	0	4	16	0
−	C_4 :	0	1	0	5	4	10	8
−	C_5 :	0	0	0	1	0	10	8
+	C_6, C_{12}, C_{13} :	0	1	2	0	12	16	0
+	C_8, C_{19}, C_{20} :	0	0	0	5	4	10	8
+	C_{10}:	4	16	0	0	0	0	0
+	C_{11}:	0	4	16	0	0	0	0
+	C_{15}:	0	0	0	4	4	16	0
+	C_{17}:	0	2	4	16	0	0	0
+	C_{18}:	0	16	0	0	0	0	0
−	$C_{24}, C_{31}, C_{32}, C_{33}, C_{34}$, $C_{44}, C_{47}, C_{48}, C_{49}$:	0	16	0	0	0	0	0
−	$C_{25}, C_{28}, C_{36}, C_{37}, C_{40}$,:	0	0	2	4	12	16	0
−	C_{35}:	0	4	16	0	0	0	0
SUM		0	0	0	0	0	0	0

Table 2

principal block of $N_{HS}(C_i)$ then $k(N_{HS}(C_i), B_0, d) = |\{\psi \in Irr(b_0^i) | d(\psi) = d\}|$, where according to Remark 3.2 (b), $d \leq 9$ gives only nonzero values. By the same Corollary 3.4, for $i \notin \{0,1,2,10\}$, $k(N_{HS}(C_i), B_1, d) = 0$ for each positive integer d. On the other hand, for $i \in \{0,1,2,10\}$, $N_{HS}(C_i)$ has two blocks, b_0^i and b_1^i, hence

$$k(N_{HS}(C_i), B_1, d) = |\{\psi \in Irr(b_1^i) | d(\psi) = d\}|,$$

for each positive integer $d \leq 2$, for other values of d these numbers are zero.

We have:

Theorem 4.3 *In HS the ordinary Dade's conjecture holds for the prime* 2.

4.2 $p = 2$ invariant conjecture

For considering Dade's invariant conjecture to HS, it is necessary to extend HS with its outer automorphism group. This extension is $HS2$, and it is acting by conjugation on the $\mathcal{U}$-chains of HS. Taking a chain C, its normalizer in HS, $N_{HS}(C)$ has index at most 2 in $N_{HS2}(C)$. Thus each irreducible character of $N_{HS}(C)$ has inertia group either $N_{HS}(C)$, or $N_{HS2}(C)$. Then for the inertia subgroup $T(\psi)$ of a character $\psi \in Irr(N_{HS}(C))$, we get that $T(\psi)HS$ is either HS or $HS2$, thus we have to check the invariant conjecture for $H = HS$ and $H = HS2$. As the sums belonging to H and $HS2$ add up to the alternating sum of the ordinary conjecture, using Theorem 4.3, we have to calculate only one of these, say the case $H = HS2$.

We put the numbers $k(C, B, d, HS2)$ in Table 3. It is zero in the event that $N_{HS2}(C) \leq HS$, otherwise it is the number of irreducible characters ψ of $N_{HS}(C)$ of defect d, such that the block of ψ, $b \in Bl(N_{HS}(C))$, induces B and ψ is invariant

Sign		d=2	d=4	d=5	d=6	d=7	d=8	d=9
		B_1	B_0	B_0	B_0	B_0	B_0	B_0
+	C_0 :	2	1	0	1	0	6	8
−	C_1 :	2	2	4	8	0	0	0
−	C_2 :	2	8	0	0	0	0	0
−	C_3 :	0	1	0	0	4	16	0
−	C_4 :	0	1	0	3	4	6	8
−	C_5 :	0	0	0	1	0	6	8
+	C_6, C_{12}, C_{13} :	0	1	2	0	12	16	0
+	C_8, C_{19}, C_{20} :	0	0	0	3	4	6	8
+	C_{10}:	2	8	0	0	0	0	0
+	C_{11}:	0	4	16	0	0	0	0
+	C_{15}:	0	0	0	4	4	16	0
+	C_{17}:	0	2	4	8	0	0	0
+	C_{18}:	0	8	0	0	0	0	0
−	$C_{24}, C_{31}, C_{32}, C_{44}, C_{47}$:	0	8	0	0	0	0	0
−	$C_{25}, C_{28}, C_{36}, C_{37}, C_{40}$:	0	0	2	4	12	16	0
−	C_{35}:	0	4	16	0	0	0	0
SUM		0	0	0	0	0	0	0

Table 3

in $N_{HS2}(C)$. We omitted the zero columns and rows with zero contribution. Chains are belonging to the same row if and only if their have equal normalizers in $HS2$.

We have:

Theorem 4.4 *In HS the invariant Dade's conjecture holds for the prime* 2.

4.3 $p = 3$ ordinary and invariant conjecture

We got that a full set of representatives of HS-conjugacy classes of nontrivial radical 3-subgroups of HS is $\{R3,\ R9\}$, where Rn is of order n. A full set of representatives of HS-conjugacy classes $\mathcal{U}$ 3-chains is:

$$C_0 : 1 \qquad C_2 : 1 < R9$$
$$C_1 : 1 < R3 \quad C_3 : 1 < R3 < R9.$$

In HS there are six 3-blocks B_0, B_1 and B_2 with defects $d(B_0) = 2$, $d(B_1) = 2$, $d(B_2) = 1$, respectively, the other 3 blocks are of defect zero. Here B_0 and B_1 have defect groups $R9 \in \mathrm{Syl}_3(HS)$ and B_2 has defect group $R3$.

We put the numbers $k(N_{HS}(C_i), B_j, d)$ in the first part of Table 4, and the numbers $k(C_i, B_j, d, HS2)$ in the second part of Table 4, where $i \in \{0, \ldots, 3\}$, $j \in \{0, \ldots, 2\}$ and $d \in \{0, \ldots, 2\}$. We omit the zero columns.

Remark 4.5 HS has six 3-blocks, B_0, B_1, B_2, B_3, B_4, B_5 with defects 2, 2, 1, 0, 0, 0, respectively. In this case, $N_{HS}(C_0)$ has six, $N_{HS}(C_1)$ has three, $N_{HS}(C_2)$ and $N_{HS}(C_3)$ have two 3-blocks. By Corollary 3.4, if b_0^i is the principal block of $N_{HS}(C_i)$, then $k(N_{HS}(C_i), B_0, d) = |\{\psi \in Irr(b_0^i) | d(\psi) = d\}|$, where according to Remark 3.2 (b), $d \le 2$ gives only nonzero values. For $i = 1$, it is also true that ${b_1^1}^{HS} = B_1$ and ${b_2^1}^{HS} = B_2$, so $k(N_{HS}(C_1), B_1, d) = |\{\psi \in Irr(b_1^1) | d(\psi) = d\}|$, for

Sign		d=1	d=2	d=2
		B_2	B_0	B_1
+	C_0 :	3	9	9
−	C_1 :	3	9	9
−	C_2 :	0	9	9
+	C_3 :	0	9	9
SUM		0	0	0

Sign		d=1	d=2	d=2
		B_2	B_0	B_1
+	C_0 :	3	9	3
−	C_1 :	3	9	9
−	C_2 :	0	9	3
+	C_3 :	0	9	9
SUM		0	0	0

Table 4

Sign		$d=1$	$d=2$	$d=3$
		B_1	B_0	B_0
+	C_0 :	5	4	13
−	C_1 :	5	20	0
−	C_2 :	0	4	13
+	C_3 :	0	20	0
SUM		0	0	0

Sign		$d=1$	$d=2$	$d=3$
		B_1	B_0	B_0
+	C_0 :	5	2	9
−	C_1 :	5	10	0
−	C_2 :	0	2	9
+	C_3 :	0	10	0
SUM		0	0	0

Table 5

each positive integer $d \leq 2$, $k(N_{HS}(C_1), B_2, d) = |\{\psi \in Irr(b_2^1)|d(\psi) = d\}|$, for each positive integer $d \leq 1$, for other values of d, these numbers are zero. For $i \in \{2,3\}$, as $d(b_1^i) = 2$, we get that ${b_1^i}^{HS} = B_1$ and

$$k(N_{HS}(C_i), B_1, d) = |\{\psi \in Irr(b_1^i)|d(\psi) = d\}|,$$

for each positive integer $d \leq 2$, for other values of d these numbers are zero. On the other hand, $k(N_{HS}(C_i), B_2, d) = 0$, for every d, as 3.5 tells us.

We have:

Theorem 4.6 *In HS the invariant Dade's conjecture holds for the prime* 3.

4.4 $p = 5$ ordinary and invariant conjecture

We obtained that a full set of representatives of nontrivial radical 5-subgroups is $\{R5,\ R125\}$, where Rn is a subgroup of order n. We got that a full set of representatives of $\mathcal{U}$ 5-chains in HS is the following:

$$C_0 : 1 \qquad C_2 : 1 < R125$$
$$C_1 : 1 < R5 \quad C_3 : 1 < R5 < R125$$

In HS there are four 5-blocks, two of them, B_0 and B_1 with defects $d(B_0) = 3$, $d(B_1) = 1$ and defect groups $R125$ and $R5$, respectively, the other two blocks are of defect zero.

We put the numbers $k(N_{HS}(C_i), B_j, d)$ in the first part of Table 5, and the numbers $k(C_i, B_j, d, HS2)$ in the second part of Table 5, where $i \in \{0,\ldots,3\}$, $j \in \{0,1\}$ and $d \in \{0,\ldots,3\}$. We omit the zero columns.

Remark 4.7 Applying Corollary 3.6 and the fact that HS has four 5-blocks B_0, B_1, B_2 and B_3 with defects 3, 1, 0, 0, respectively, we get that except for the chains C_0 and C_1, the chain normalizers $N_{HS}(C_i)$, have only one block. By Corollary 3.4, if b_0^i is the principal block of $N_{HS}(C_i)$, then $k(N_{HS}(C_i), B_0, d) = |\{\psi \in Irr(b_0^i) | d(\psi) = d\}|$, where according to Remark 3.2 (b), $d \leq 3$ gives only nonzero values. By the same Corollary 3.4, for $i \notin \{0,1\}$ $k(N_{HS}(C_i), B_1, d) = 0$, for each positive integer d. On the other hand, $N_{HS}(C_1)$ has two blocks, b_0^1 and b_1^1, hence $k(N_{HS}(C_1), B_1, d) = |\{\psi \in Irr(b_1^1) | d(\psi) = d\}|$, for $d = 1$, for other values d, these numbers are zero.

We have:

Theorem 4.8 *In HS the invariant Dade's conjecture holds for the prime* 5.

Taking into consideration the results of Theorem 4.4, 4.6, 4.8 and that for prime divisors of $|HS|$ with cyclic defect groups Dade's invariant conjecture holds, we get the following:

Theorem 4.9 *In HS the invariant Dade's conjecture holds for each prime divisoı of the order of HS. Especially the ordinary Dade's conjecture is true, as well.*

5 The invariant projective conjecture for *HS*

As the Schur multiplier of HS is of order 2, we have to check the invariant projective conjecture for two central extensions. The first is the central extension with the trivial group, in which case we get back HS, the other is the central extension:

$$1 \to Z(2HS) \to 2HS \to HS \to 1.$$

Here $Z(2HS) = O_2(2HS)$, which we shall denote by O_2. And these we have to extend by automorphisms acting on the centre trivially. If we extend by the trivial subgroup, we get back the situation of the invariant conjecture, which we considered already. The central extension given by $2HS$, extended by the above mentioned automorphisms, gives the group $2HS2$. This group is acting on $\mathcal{U}$-chains of $2HS$. Let C^* be such a chain. It can be obtained from the corresponding chain C of HS, as by Remark 2.10 radical p-subgroups of $2HS$ are in one-to-one correspondence with radical p-subgroups of HS under the map taking the Sylow p-subgroup of the its inverse image in $2HS$. This gives a correspondence of chains, too.

5.1 $p = 2$ projective conjecture

The conjugacy class representatives of $\mathcal{U}$ 2-chains, being inverse images of those in the case of the ordinary conjecture, look similarly, just each subgroup is twice as big, and each chain starts with $O_2(G^*)$, which is of order 2.

In $2HS$ there are two 2-blocks B_0 and B_1 with defects $d(B_0) = 10$ and $d(B_1) = 3$. We put the numbers $k(N_{2HS}(C_i^*), B_j, d, \zeta)$, for $i \in \{0, \ldots, 49\}$, $j \in \{0,1\}$, $d \in \{1, \ldots, 10\}$ and ζ, the faithful irreducible character of $O_2(2HS)$, in Table 6. We omit the zero columns and those rows, whose contribution is zero. Chains are belonging to one row if and only if their normalizers in $2HS$ are equal. The values were calculated with GAP.

Sign		d=2	d=4	d=5	d=6	d=7	d=8
		B_1	B_0	B_0	B_0	B_0	B_0
+	C_0^* :	3	0	0	4	7	4
−	C_1^* :	1	0	2	8	0	0
−	C_2^* :	3	12	0	0	0	0
−	C_3^* :	0	0	0	6	2	8
−	C_4^* :	0	0	0	4	5	12
−	C_5^* :	0	0	0	0	7	4
+	$C_6^*, C_{12}^*, C_{13}^*$:	0	0	0	4	10	8
+	$C_8^*, C_{19}^*, C_{20}^*$:	0	0	0	0	5	12
+	C_{10}^*:	1	4	0	0	0	0
+	C_{11}^*:	0	0	4	16	0	0
+	C_{15}^*:	0	0	0	2	6	8
+	C_{17}^*:	0	0	2	8	0	0
+	C_{18}^*:	0	12	0	0	0	0
−	$C_{24}^*, C_{31}^*, C_{32}^*, C_{33}^*, C_{34}^* C_{44}^*, C_{47}^*, C_{48}^*, C_{49}^*$:	0	4	0	0	0	0
−	$C_{25}^*, C_{28}^*, C_{36}^*, C_{37}^*, C_{40}^*$:	0	0	0	0	14	8
−	C_{35}^*:	0	0	4	16	0	0
SUM		0	0	0	0	0	0

Table 6

5.2 $p = 2$ invariant projective conjecture

The normalizer $N_{2HS}(C^*)$ has index at most 2 in its normalizer in $N_{2HS2}(C^*)$. So we have, that the subgroup H of the invariant projective conjecture is either $2HS$ of $2HS2$. As the sums for these groups add up to the sum in the projective conjecture, it is enough to calculate only one of these, say the case, when $H = 2HS2$. We put in Table 7 the numbers of $k(C^*, B, d, 2HS2, \zeta)$. This is zero if $N_{2HS2}(C^*) \leq 2HS$. Otherwise it denotes the number of those irreducible characters $\psi \in Irr(N_{2HS}(C^*))$ of defect d, such that the block b of ψ induces B, ψ is invariant in $N_{2HS2}(C^*)$ and the restriction of ψ to $Z^* = O_2(2HS)$ contains ζ as a constituent. Here ζ is as before the unique faithful irreducible character of $O_2(2HS)$. According to the conjecture, the alternating sums for each column has to be zero. We omitted the zero columns, zero rows and rows with zero sign. Chains are belonging to one row if and only if they have equal normalizers in $2HS2$. The values were calculated with GAP.

5.3 $p = 3$ projective and invariant projective conjecture

The representatives of conjugacy classes of $\mathcal{U}$-chains of $2HS$ are in one-one correspondence with those of HS. The groups in these chains C_i^* are of the same order as those in HS, being Sylow 3-subgroups in their inverse images.

In $2HS$ there are five 3-blocks of positive defect B_0, B_1, B_2, B_3 and B_4 with defects $d(B_0) = 2$, $d(B_1) = 2$, $d(B_2) = 1$, $d(B_3) = 2$ and $d(B_4) = 1$, the other nine blocks are of defect zero.

We put the numbers $k(N_{2HS}(C_i^*), B_j, d, \zeta)$ in the first part of Table 8, and the numbers $k(C_i^*, B_j, d, 2HS2, \zeta)$ in the second part of Table 8, where $i \in \{0, \ldots, 3\}$, $j \in \{0, \ldots, 4\}$ and $d \in \{1, 2\}$. We omit the zero columns.

Sign		d=2	d=5	d=6	d=7
		B_1	B_0	B_0	B_0
+	C_0^* :	1	0	0	5
−	C_1^* :	1	2	0	0
−	C_2^* :	1	0	0	0
−	C_3^* :	0	0	2	2
−	C_4^* :	0	0	0	3
−	C_5^* :	0	0	0	5
+	$C_6^*, C_{12}^* C_{13}^*$:	0	0	0	2
+	$C_8^*, C_{19}^*, C_{20}^*$:	0	0	0	3
+	C_{10}^*:	1	0	0	0
+	C_{15}^*:	0	0	2	6
+	C_{17}^*:	0	2	0	0
−	$C_{25}^*, C_{28}^*, C_{36}^*, C_{37}^*, C_{40}^*$:	0	0	0	6
SUM		0	0	0	0

Table 7

Sign		d=1	d=2
		B_4	B_3
+	C_0^* :	3	9
−	C_1^* :	3	18
−	C_2^* :	0	9
+	C_3^* :	0	18
SUM		0	0

Sign		d=1	d=2
		B_4	B_3
+	C_0^* :	1	3
−	C_1^* :	1	0
−	C_2^* :	0	3
+	C_3^* :	0	0
SUM		0	0

Table 8

Sign		$d=2$	d=3
		B_2	B_2
+	C_0^* :	4	13
−	C_1^* :	20	0
−	C_2^* :	4	13
+	C_3^* :	20	0
SUM		0	0

Sign		$d=2$	d=3
		B_2	B_2
+	C_0^* :	2	3
−	C_1^* :	2	0
−	C_2^* :	2	3
+	C_3^* :	2	0
SUM		0	0

Table 9

Remark 5.1 In $2HS$ the 3-blocks B_0, B_1 and B_2 dominate the blocks of HS with the same indices. The columns of these blocks in the above tables are zero. $N_{2HS}(C_1^*)$ has six 3-blocks, b_0^1, b_1^1, b_2^1, b_3^1, b_4^1, b_5^1, with defects 2, 2, 2, 2, 1, 1, respectively. ${b_0^1}^{2HS} = B_0$, ${b_1^1}^{2HS} = B_1$, ${b_2^1}^{2HS} = B_3$, ${b_3^1}^{2HS} = B_3$, ${b_4^1}^{2HS} = B_2$ and ${b_5^1}^{2HS} = B_4$. $N_{2HS}(C_2^*)$ has three 3-blocks b_0^2, b_1^2 and b_2^2, with defects 2, 2, 2, respectively. ${b_0^2}^{2HS} = B_0$, ${b_1^2}^{2HS} = B_1$, ${b_2^2}^{2HS} = B_3$. $N_{2HS}(C_3^*)$ has four 3-blocks, b_0^3, b_1^3, b_2^3, b_3^3, with defects 2, 2, 2, 2, respectively. ${b_0^3}^{2HS} = B_0$, ${b_1^3}^{2HS} = B_1$, ${b_2^3}^{2HS} = B_3$ and ${b_3^3}^{2HS} = B_3$.

5.4 $p = 5$ projective and invariant projective conjecture

The representatives of conjugacy classes of $\mathcal{U}$-chains of $2HS$ are in one-one correspondence with those of HS. The groups in these chains C_i^* are of the same order as those in HS, being Sylow 5-subgroups in their inverse images.

In $2HS$ there are three 5-blocks with positive defects, B_0, B_1 and B_2 with defects $d(B_0) = 3$, $d(B_1) = 1$ and $d(B_2) = 3$, the other three blocks are of defect zero. So the Sylow is the defect group of B_0 and B_2, and the radical 5-subgroup of order 5 is the defect group of B_1.

We put the numbers $k(N_{2HS}(C_i^*), B_j, d, \zeta)$ in the first part of Table 9, and the numbers $k(C_i^*, B_j, d, 2HS2, \zeta)$ in the second part of Table 9, where $i \in \{0, \ldots, 3\}$, $j \in \{0, \ldots, 2\}$ and $d \in \{1, \ldots, 3\}$. We omit the zero columns.

Remark 5.2 In $2HS$ the 5-blocks, B_0 and B_1, dominate the blocks of HS with the same indices. The columns of these blocks in the above tables are zero. $N_{2HS}(C_1^*)$ has three 5-blocks b_0^1, b_1^1, b_2^1 with defects 2, 2, 1 respectively. ${b_0^1}^{2HS} = B_0$, ${b_1^1}^{2HS} = B_2$ and ${b_2^1}^{2HS} = B_1$. $N_{2HS}(C_2^*)$ has two 5-blocks, b_0^2 and b_1^2, with defects 3, 3, respectively. ${b_0^2}^{2HS} = B_0$, ${b_1^2}^{2HS} = B_2$. $N_{2HS}(C_3^*)$ has two 5-blocks, b_0^3 and b_1^3, with defects 2, 2 respectively. ${b_0^3}^{2HS} = B_0$, ${b_1^3}^{2HS} = B_2$.

Taking into consideration the results on blocks with cyclic defect groups and Theorem 4.9, we get:

Theorem 5.3 *In HS the invariant projective Dade's conjecture holds for each prime.*

Acknowledgements The authors would like to express their gratitude to Professor Herbert Pahlings and Priv.Doz.Dr. Meinolf Geck for suggesting the topic and for their advices. We are very grateful to Professor E.C. Dade for pointing out an error in one of the proofs in an earlier version of this paper and for his comments and suggestions. Thanks are due to Dr. Péter Hermann, the PhD supervisor of the first author, for his support in this work. We would like to thank Thomas Breuer for his help concerning the GAP system and for his suggestions.

References

[1] J. An, Dade's conjecture for Chevalley groups $G_2(q)$ in non-defining characteristics, *Canadian J. Math.* **48** (1996), 673-691.

[2] J. An, Dade's conjecture for the Tits group, *New Zealand J. Math.* **25** (1996), 107-131.

[3] J. An, The Alperin and Dade conjectures for the simple Held group, *Journal of Algebra* **189** (1997), 34-57.

[4] J. An and M. Conder, The Alperin and Dade conjectures for the simple Mathieu groups, *Communications in Algebra* **23** (1995), 2797-2823.

[5] J. Conway, R. Curtis, S.Norton, R. Parker and R. Wilson, Atlas of finite groups, Oxford University Press, New York, London, Oxford, 1985.

[6] E. Dade, Counting characters in blocks I., *Invent. Math.* **109** (1992), 187-210.

[7] E. Dade, Counting characters in blocks II., *J. Reine Angew. Math.* **448** (1994), 97-190.

[8] E. Dade, Counting characters in blocks. II.9, preprint.

[9] E. Dade, Counting characters in blocks with cyclic defect groups I., *Journal of Algebra* **186** (1996), 934-969.

[10] L. Dornhoff, Group representation theory, Part A, Marcel Dekker Inc., New York, 1971.

[11] G. Entz, Dades Vermutung für gewöhnliche und projektive Charaktere, Diplomarbeit, Aachen, 1997.

[12] G. Entz and H. Pahlings, The Dade conjecture for the McLaughlin group, preprint.

[13] G. Higman and C.C. Sims, A simple group of order 44, 352, 000, *Math. Z.* **105** (1968), 110-113.

[14] J. F. Huang, Verification of the McKay-Alperin-Dade conjecture for the covering groups of the Mathieu group M_{22}, Thesis, Urbana, 1992.

[15] J. F. Huang, Counting characters in blocks of M_{22}, *Journal of Algebra* **191** (1997), 1-75.

[16] R. Knörr and G. Robinson, Some remarks on a conjecture of Alperin, *J. London Math. Soc.* **(2) 39** (1989), 48-60.

[17] S. Kotlica, Verification of Dade's conjecture for Janko group J_3, *Journal of Algebra* **187**, (1997), 579-619.

[18] J. Murray, Dade's conjecture for the McLaughlin simple group. Thesis. University of Illinois at Urbana-Champaign, 1997.

[19] H. Nagao and Y. Tsushima, Representations of finite groups, Academic Press, New York 1989.

[20] J.B. Olsson and K. Uno, Dade's conjecture for general linear groups in the defining characteristic, *Proc. London Math. Soc.* **(3) 72** (1996), 359-384.

[21] M. Schönert et. al., GAP-Groups, Algorithms and Programming, *Lehrstuhl D für Mathematik, RWTH-Aachen*, fifth edition, 1995.

[22] R.A. Wilson, An Atlas of Group Representations, `http://www.mat.bham.ac.uk/R.A.Wilson/mtx-atlas, 1996.`

ON THE F^*-THEOREM

MARTIN HERTWECK and WOLFGANG KIMMERLE

Mathematisches Institut B, Universität Stuttgart, Pfaffenwaldring 57, D–70550 Stuttgart, Germany

1 Introduction

With respect to the structure of torsion subgroups of integral group rings the following conjecture due to Zassenhaus has been over the last twenty years in the middle of the research. The conjecture may be stated as follows.

(ZC) Let $\mathbb{Z}G$ be the integral group ring of the finite group G. Denote the units of augmentation 1 by $V(\mathbb{Z}G)$ and let H be a subgroup of $V(\mathbb{Z}G)$ of the same order as G. Then there exists a central automorphism σ of $\mathbb{Z}G$ with $\sigma(G) = H$.

The conjecture is also of interest for more general coefficient rings than $\mathbb{Z}$. We say that (ZC) holds for a group ring RG, if the content of the conjecture holds in RG. It has been shown by Roggenkamp and Scott that (ZC) does not hold for any finite group [9], [13]. But, if (ZC) is true, it gives a strong answer to the isomorphism problem of integral group rings. For a recent survey about the Zassenhaus conjecture and related questions we refer to [6]. For the newest developments with respect to the isomorphism problem see [2], [3] rsp. The most far reaching result obtained in the positive direction is the following one.

Theorem 1.1 ([15], [12]) *Denote by $\pi(G)$ the set of primes dividing the order of G. Let S be the semilocal ring $\mathbb{Z}_{\pi(G)}$. Then (ZC) holds for SG provided the generalized Fitting subgroup $F^*(G)$ is a p-group.*

Theorem 1.1 is a consequence of the following result which we call the F^* - Theorem.

Theorem 1.2 ([15] [12]) *Let G be a finite group with a normal p -subgroup N such that $C_G(N) \subset N$. Let $\alpha \in \operatorname{Aut}_n(\hat{\mathbb{Z}}_pG)$. Assume that α stabilizes $I_{\hat{\mathbb{Z}}_p}(N)\cdot\hat{\mathbb{Z}}_pG$. Then α is the composition of an automorphism induced from a group automorphism of G followed by an inner automorphism.*

The proof of Theorem 1.2 has not been completely published as well as the proof of Theorem 1.1. Some ideas and related topics are mentioned in [15] and the main ingredients are given in [11]. However one of the main ingredients [11, Theorem 4.10] is at least in the generality as stated false in the case when $p = 2$. Note that for p odd the statement is correct. For the convenience of the reader we restate this ingredient [11, Theorem 4.10].

Let p be a prime and let P and Q be p-groups. Let R be an unramified extension of $\hat{\mathbb{Z}}_p$ with residue field k. Let M be an $R(P \times Q)$-module which is as RQ-module free of finite rank. Assume that $\overline{M/(I_R(Q)\cdot M)}$ is a kP - permutation module. Then $\overline{M}$ is a $k(P\times Q)$-permutation module (- denotes the reduction of R-modules to k-modules).

Example 1.3 Let $P =< x >$ and $Q =< y >$ be cyclic groups of order 2. Denote by R the 2 - adic integers and let $M = R^4$. Define an action of $P \times Q$ on M by letting x and y acting from the right via

$$x := \begin{pmatrix} 1 & 0 & 1 & 1 \\ 0 & 1 & 1 & 1 \\ 0 & 0 & -1 & 0 \\ 0 & 0 & 0 & -1 \end{pmatrix}$$

and

$$y := \begin{pmatrix} 0 & 1 & 0 & 0 \\ 1 & 0 & 0 & 0 \\ 0 & 0 & 0 & 1 \\ 0 & 0 & 1 & 0 \end{pmatrix}.$$

Then M is as an RQ-module free of rank 2 and $\overline{M/M \cdot I_R(Q)}$ is as a trivial kP-module of course a permutation module. But as one easily sees by calculating the orbits $\overline{M}$ is not a $k(P \times Q)$-permutation module.

The original proof of Theorem 1.2 is affected in that the proof is complete only in the case when $O_p(G)$ is a Sylow p-subgroup or in the case when p is odd. For $p = 2$ a gap remains.

The goal of this note is to establish results which close this gap at least in the case when the automorphism α permutes the class sums of G in RG. More precisely our repair permits, together with the parts of the original proof which are not affected by [11, Theorem 4.10], a complete proof of the following result.

Theorem 1.4 *Let $R = \hat{\mathbb{Z}}_p$ and let G be a finite group whose generalized Fitting subgroup is a p-group. Let α be a normalized automorphism of RG which permutes the class sums of p-elements of G. Then $\alpha(G)$ is conjugate to G within the units of RG.*

The assumption that $R = \hat{\mathbb{Z}}_p$ may be weakened, cf. 3.3. We shall restrict attention however mainly to the case studied by A.Weiss because this case suffices for the applications to integral group rings.

If S is an integral domain of characteristic zero and no prime divisor of $|G|$ is invertible in S, then each augmentation preserving S-algebra automorphism of SG permutes the class sums of G. Therefore Theorem 1.1 follows from Theorem 1.4.

Theorem 1.1 plays in many results on integral group rings of various classes of finite groups an important role. In order to underline the importance we mention the following ones and we would like to point out that their proof is complete by using Theorem 1.4.

Theorem 1.5 ([5]) *The isomorphism problem of integral group rings has a positive solution for nilpotent by abelian groups.*

Theorem 1.6 ([8]) *Let G be a finite soluble group and let H be group basis of $\mathbb{Z}G$. Then Sylow subgroups of H and G are conjugate within $\mathbb{Q}G$.*

Moreover in [7] finite groups are studied as projective limits of certain quotients. The resulting Čech cohomology style obstruction theory for the isomorphism problem of such groups is clearly only powerful if a result like Theorem 1.4 holds. Eventually the relationship of Theorem 1.1 to conjugacy problems of defect groups in blocks is described in [16].

This note is organized as follows. Proposition 3.1, which closes the gap in the proof of Theorem 1.1, is established in section 3. Also we prove there related results about automorphisms of group rings which are of independent interest. Section 2 is devoted to collecting the necessary ingredients and to discussing the affected part in the proof of Theorem 1.2.

2 The critical point in the proof

The following two results of A.Weiss are essential in the original proof of Theorem 1.2 and they are again of importance for our arguments.

Theorem 2.1 ([18, Theorem 2]) *Let $R = \hat{\mathbb{Z}}_p$ and let G be a finite p-group. Let M be an RG-lattice. Let N be a normal subgroup of G. Assume that*

(i) *M regarded by restriction as RN - module is free and*

(ii) *that $M/I_R(N)M$ is a permutation RG/N-module.*

Then M is a permutation RG-module.

Theorem 2.1 is based on the following result.

Theorem 2.2 ([18, Theorem 3]) *Let G be a finite p-group and let $R = \hat{\mathbb{Z}}_p[\zeta]$, where ζ denotes a primitive p-th root of unity. Put $\pi = 1 - \zeta$. Let M be as R-module free of finite rank. Assume that $\overline{M} = M/\pi \cdot M$ is an F_pG-permutation module. Then M is a generalized permutation RG-module.*

For the discussion of the proof of the F^*-Theorem we fix the following notation: $P \in \mathrm{Syl}_p(G)$, $\alpha \in \mathrm{Aut}_n(RG)$, $R = \hat{\mathbb{Z}}_p$, $K = \mathrm{Quot}(R)$, $k = F_p$.

The proof of the F^*-Theorem is divided into three steps. The first step is to show that the automorphism α may be modified by an inner automorphism such that the modified α fixes P. This is done in three parts. The crucial one is Step 1b).

For an augmentation preserving R-algebra homomorphism $\beta : RP \longrightarrow RG$ consider RG as a $P \times P$-module via the action

$$(x, y) \bullet m = x \cdot m \cdot \beta(y^{-1}).$$

We denote this module by ${}_1RG_\beta$ and we use this notation also in the case when G acts from the left via multiplication and the action from P on the right comes from the restriction of an R-algebra automorphism of RG to RP.

Step 1a) Show that α may be modified by an inner automorphism to an automorphism again denoted by α which fixes N.

Step 1b) Show that ${}_1RG_\alpha$ is a permutation $P \times P$ - module.

Step 1c) Vertex arguments and the indecomposability of RG show then that ${}_1RG_\alpha$ is a permutation $G \times P$-module. Here G acts from the left by ordinary multiplication and P as before from the right via $\alpha(P)$. It follows then immediately that P and $\alpha(P)$ are conjugate in RG.

The second step of the proof of the F^*-Theorem proceeds by showing that α may be modified to an automorphism which is restricted to the identity on P. In the final step one shows that such an automorphism is indeed given by conjugation with a unit in RG.

For the discussion of Step 1b) consider the following diagram.

$$\begin{array}{ccc} {}_1RG_\alpha & \longrightarrow & {}_1RG/N_\alpha \\ \downarrow & & \downarrow \\ {}_1kG_\alpha & \longrightarrow & {}_1kG/N_\alpha \end{array}$$

Note that α induces automorphisms on $kG, kG/N$ and RG/N because it fixes the kernels $I_R(N) \cdot RG, I_k(N) \cdot kG$ and pRG. We denote the induced automorphisms for simplicity also by α.

By Step 1a) one may assume that α stabilizes N. This enables the use of the following lemma ([4], [11, Lemma 23]) of Coleman.

Lemma 2.3 *Let G be a finite group, let P be a p-subgroup of G and let S be an integral domain, in which the prime p is not a unit. Denote the normalized unit group of SG by $V = V(SG)$. Then*

$$N_V(P) = N_G(P) \cdot C_V(P).$$

The Coleman Lemma shows that for all $g \in G$ there are $h_g \in G$ and $x_g \in C_{V(RG)}(N)$ such that $\alpha(g) = h_g \cdot x_g$. By assumption $C_G(N) \subset N$. Hence α induces on kG/N a group automorphism of G/N. In particular ${}_1kG/N_\alpha$ is a permutation module. The aim of [11, Theorem 4.10] was then to conclude that ${}_1kG_\alpha$ is also a permutation module. From this it would follow that ${}_1RG_\alpha$ is a permutation $P \times P$-module.

We shall go in the diagram above the other way round. Under the additional assumption that α permutes the class sums of 2-elements it follows from Theorem 2.2 and Proposition 3.1 that ${}_1RG/N_\alpha$ is a permutation module. Hence we may apply Theorem 2.1 and get that ${}_1RG_\alpha$ is a permutation $P \times P$-module. Hence Step 1b) is established in this case.

3 Modularly trivial automorphisms

Let A be a ring and let G be a group. A generalized AG-permutation module is an AG-module which is induced from a 1-dimensional AU - module for some subgroup U of G. Since such a 1-dimensional module is determined by a homomorphism $\phi : U \longrightarrow A^*$ we use for generalized permutation modules the notation $A_\phi G/U$. In the permutation module case, i.e. the case when ϕ is trivial, we simply write AG/U.

Proposition 3.1 *Let G be a p-group. Let A be an integral domain of characteristic zero. Let m be a maximal ideal which contains p and put $k = A/m$. Denote by K the quotient field of A. Suppose that M is an AG-permutation module and N a generalized AG - permutation module such that $kM \cong kN$ and $KM \cong KN$. Then $M \cong N$ and N is therefore an AG- permutation module.*

Proof Let $N = A_{\phi_1}G/U_1 \oplus ... \oplus A_{\phi_h}G/U_h$ and $M = AG/V_1 \oplus ... \oplus AG/V_l$. Note that k has characteristic p. Thus $k_{\phi_i}G/U_i \cong kG/U_i$. Since G is a p-group all the permutation modules kG/U_i and kG/V_j are indecomposable. By the Krull - Schmidt Theorem it follows from $kM \cong kN$ that $h = l$ and after renumbering we may assume that U_i is conjugate to V_i for $1 \leq i \leq l$. Note that $K_{\phi_j}G/U_j$ permits a non-trivial homomorphism into K regarded as a trivial module only if ϕ_j is trivial. Thus $\mathrm{Hom}_{KG}(KN, K) \cong K^w$, where w denotes the number of homomorphisms ϕ_j which are trivial. On the other hand $\mathrm{Hom}_{KG}(KM, K) \cong K^l$. Now it follows from $KM \cong KN$ that $w = l$. By the previous $h = l$ and therefore we get that N is a permutation module. Since U_j and V_j are conjugate we get finally that $M \cong N$. □

The final part of this section is not necessary for the proof of the F^*-Theorem but is related to the previous discussion.

Proposition 3.2 *Let G be a finite group. Let $A = \hat{\mathbb{Z}}_p$. Put $k = A/pA$ and denote the quotient field of A by K.*

(i) *Suppose that α is a normalized A-algebra automorphism of AG which induces on kG the identity. Assume that p is odd or that α permutes the class sums of 2-elements of G. Then α is conjugation with a unit in AG.*

(ii) *Let Q be a p-subgroup of G. Suppose that $\alpha : AQ \longrightarrow AG$ is an A-algebra homomorphism which induces mod p the inclusion of kQ into kG. Assume that p is odd or that α fixes for the irreducible characters of G the values of the elements of Q. Then α is conjugation with a unit in AG of the form $1 + pu$.*

Proof We prove first part (ii). By assumption the A-algebra homomorphism α induces the inclusion of kQ into kG. Then $k \otimes {}_1AG_\alpha$ and $k \otimes {}_1AG_1$ are isomorphic. Thus $k \otimes {}_1AG_\alpha$ is a permutation kG-module. It follows from Theorem 2.2 that ${}_1A(\zeta)G_\alpha$ is a generalized permutation $A(\zeta)(Q \times Q)$-lattice. If p is odd then A contains no non-trivial p-th roots of unity. Note that $pA(\zeta) \subset (1 - \zeta)A(\zeta)$. Hence ${}_1AG_\alpha$ is a permutation module.

Assume that $p = 2$. By assumption Q and $\alpha(Q)$ have the same ordinary K - characters on the simple KG-modules. Thus Q and $\alpha(Q)$ are conjugate in KG. Consequently $K \otimes {}_1AG_\alpha$ and $K \otimes {}_1AG_1$ are isomorphic. Now by Proposition 3.1 it follows that ${}_1AG_\alpha$ is a permutation $A(Q \times Q)$-module.

From now on p is an arbitrary prime. Since ${}_1AG_\alpha$ is a permutation module for $Q \times Q$ it is a permutation module for the diagonal subgroup ΔQ. Hence

$$H^1(\Delta Q, {}_1AG_\alpha) = 0.$$

Consider the inner derivation $\delta_1 : G \longrightarrow {}_1AG_\alpha$ defined by $g \mapsto (g-1) \bullet 1 = g \cdot \alpha(g)^{-1} - 1$. Because α is on kQ the inclusion $\delta_1(g) \in p \cdot AG$. Hence δ_1 divided by p is again a derivation and because of $H^1(\Delta Q, {}_1AG_\alpha) = 0$ this derivation is of the form $\delta_u(g) = (g-1) \bullet u$ for some $u \in AG$. It follows that

$$g \cdot (1 - p \cdot u) \cdot \alpha(g)^{-1} = 1 - pu.$$

But $1 - pu$ is a unit and hence α is conjugation with this unit.

In order to show (i) we start with a normalized A-algebra automorphism α. Let P be a Sylow p-subgroup of G. Note that in the case $p = 2$ by assumption α permutes the class sums of the elements of P and because α is modulo 2 the inclusion we get that α fixes the G-class sum of each element of Q. Hence on the simple KG-modules P and $\alpha(P)$ have the same ordinary K -characters. Thus we may apply part (ii) to the restriction of α to P. As in the proof of part (ii) it follows that

$$H^1(P\ ,\ {}_1AG_\alpha) = 0.$$

For each $n \in \mathbf{N}$ we have the following commutative diagram

$$\begin{array}{ccc} H^1(G, {}_1AG_\alpha) & \longrightarrow & H^1(P, {}_1AG_\alpha) \\ \lambda_n \downarrow & & \downarrow \\ H^1(G, {}_1(A/p^nA)G_\alpha) & \xrightarrow{\kappa_n} & H^1(P, {}_1(A/p^nA)G_\alpha) \end{array}$$

Note κ_n is injective. Thus it follows that the image of λ_n is zero. Analogously as in the proof of (ii) we get now for each $n \in \mathbf{N}$ that

$${}_1(A/p^nA)G_\alpha \cong {}_1(A/p^nA)G_1.$$

By [10], see also [1, (30.14)], it follows that

$${}_1AG_\alpha \cong {}_1AG_1$$

and therefore G and $\alpha(G)$ are conjugate in AG. □

Remarks 3.3 (i) Proposition 3.2 holds in a wider context. The ring $\hat{\mathbf{Z}}_p$ may be replaced by a complete Dedekind domain A of finite rank over $\hat{\mathbf{Z}}_p$ with certain properties. The behaviour of generalized permutation lattices in this more general context has been proved in [13, VI] and in order to prove the Proposition 3.2 for A as above one replaces the use of Theorem 2.2 by the use of [13, VI,Theorem 1.3].

(ii) In [17] part (i) of Proposition 3.2 has been proved under the stronger hypothesis that α induces already the identity on $\hat{\mathbf{Z}}_p/p^{m+1}\hat{\mathbf{Z}}_p$, where p^m is the order of a Sylow p - subgroup of G.

(iii) The condition on the class sums of 2-elements in Proposition 3.2 is necessary. This is shown by the following example. Let $G = < x >$ be the cyclic group of order 6. Put $a = x^3$ and $b = x^2$. Then

$$y = (-\frac{1}{3} + \frac{2}{3}b + \frac{2}{3}b^2)a$$

is a normalized unit of $V(\hat{\mathbf{Z}}_2G)$ of order 2. $H = < y, b >$ is a group basis of $\mathbf{Z}_2G$. But $H = G$ modulo 2.

References

[1] C. W. Curtis and I. Reiner, Methods of Representation Theory I, Wiley Interscience, New York 1981.

[2] M. Hertweck, Two non-isomorphic finite groups with isomorphic integral group rings, to appear in Comptes Rendus, Paris.

[3] M. Hertweck, A solution of the isomorphism problem for integral group rings, Preprint.

[4] S. Jackowski and Z. Marciniak, Group automorphisms inducing the identity map on cohomology, J. Pure Appl. Algebra **44** (1987), 241–250.

[5] W. Kimmerle, Class Sums of p-elements, in: [14], 117–124.

[6] W. Kimmerle, On Automorphisms of $\mathbb{Z}G$ and the Zassenhaus Conjectures, CMS Conference Proceedings Vol. **18** (1996), 383–397.

[7] W. Kimmerle and K. W. Roggenkamp, Projective limits of group rings, J. Pure Appl. Algebra **88** (1993) 119–142.

[8] W. Kimmerle and K. W. Roggenkamp, A Sylowlike theorem for integral group rings of finite solvable groups, Arch.Math. **60** (1993), 1–6.

[9] L. Klingler, Construction of a counterexample to a conjecture of Zassenhaus, Comm. Algebra, **19** (1991), 2303–2330.

[10] J. M. Maranda, On p-adic integral representations of finite groups, Canad. J. Math. **5** (1953), 344–355.

[11] K. W. Roggenkamp, The Isomorphism Problem for Integral Group Rings of Finite Groups. Proceedings of the Int. Congres of Mathematicians, Kyoto 1990, Springer 1991, 369–380.

[12] K. W. Roggenkamp and L. L. Scott, A strong answer to the isomorphism problem for finite p-solvable groups with a normal p-subgroup containing its centralizer, manuscript 1987.

[13] K. W. Roggenkamp, Units and the isomorphism problem, Part I of [14].

[14] K. W. Roggenkamp and M. Taylor, Group rings and class groups, DMV-Seminar 18, Birkhäuser, Basel Boston, Berlin 1992.

[15] L. L. Scott, Recent progress on the isomorphism problem, Proc. Symposia in Pure Math. **47** (1987) 259–274.

[16] L. L. Scott, Defect groups and the isomorphism problem, Représentations linéaires des groupes finis, Proc. Colloq. Luminy, France 1988, Astèrisque **181-182** (1990), 257–262.

[17] S. K. Sehgal, Isomorphisms of p-adic Group Rings, J. Number Theory **2** (1970), 500–508.

[18] A. Weiss, p-adic rigidity of p-torsion, Ann. Math. **127** (1987), 317–332.

COVERING NUMBERS FOR GROUPS

MARCEL HERZOG

School of Mathematical Sciences, Raymond and Beverly Sackler Faculty of Exact Sciences, Tel-Aviv University, Tel-Aviv, Israel

Let G be a group and suppose that there exists an integer r such that

$$C^r = \{c_1 c_2 \cdots c_r \mid c_i \in C\} = G$$

for all non-trivial conjugacy classes $C \neq 1$ of G. The least such integer we call *the covering number* of G, and we denote it by $cn(G)$. This notion was first introduced in 1985 in [AH], a monograph edited by Zvi Arad and the author. We emphasize that $cn(G)$ doesn't always exist and our first open problem is: *determine all groups G for which $cn(G)$ exists.*

It is clear that for $cn(G)$ to exist, G needs to be a non-abelian simple group or $G = 1$. For finite groups the answer is known.

Theorem 1 ([AHS]) *Let $G \neq 1$ be a finite group. Then $cn(G)$ exists if and only if G is a non-abelian simple group.*

This follows from the fact that if A is any non-empty subset of a finite group G, then there exists an integer $n = n(A)$ such that A^n is a subgroup of G. In particular, if C is a conjugacy class of G, then there exists $n = n(C)$ such that C^n is a subgroup of G. If $C \neq 1$ is a conjugacy class of a simple group, then $1 < C^n \trianglelefteq G$ and hence $C^n = G$. Now

$$cn(G) = max\{n(C) \mid C \neq 1 \text{ conjugacy classes of } G\}.$$

This maximum exists, since G is a finite group.

So $cn(G)$ exists for all finite non-abelian simple groups. But it exists also for some infinite groups. Nick Gordeev has shown in [G] that $cn(G)$ exists for every simple algebraic group G over an algebraically closed field of characteristic 0 and moreover for these groups

$$cn(G) \leq 4 \cdot rank(G)\ .$$

If we require the classes C in the definition of $cn(G)$ to be non-central, rather than non-trivial, then $cn(G)$ may exist for quasi-simple groups G (i.e. G is perfect and $G/Z(G)$ is simple). In a paper under preparation, Ellers, Gordeev and myself prove the following theorem:

Theorem 2 ([EGH]) *There is a positive integer d such that*

$$cn(G) \leq d \cdot rank(G)$$

for every quasi-simple non-twisted Chevalley group or quasi-simple finite twisted Chevalley group G.

Here d is a fixed constant which is independent of the type, rank or field of G. All twisted and non-twisted finite simple groups of Lie-type belong to the set considered.

The constant d emerging from our calculations is very large. However, we believe that d should be small.

On the other hand, $cn(G)$ does not always exist for simple groups G. For example, if $G = A_\infty$ is the alternating group on $\{1, 2, 3, \ldots\}$ consisting of even permutations of finite support, then clearly $C^n \neq A_\infty$ for all conjugacy classes C of A_∞ because of support considerations.

If $cn(G)$ exists, it is quite difficult to determine its value. It was shown in [AHS] that if G is a finite simple group and $k = k(G)$ is the number of conjugacy classes of G, then:

$$cn(G) \leq \frac{4}{9}k^2 \ .$$

This bound is too large; the proper bound should be linear in k. Indeed, if each element of G is a commutator, then we proved in [AHS] that:

$$cn(G) \leq 2(k-1) \ .$$

Already in 1951 Ore [O] conjectured that every element of a finite non-abelian simple group is a commutator. An even stronger conjecture was stated by John Thompson: If G is a finite non-abelian simple group, then there exists a conjugacy class C in G such that $C^2 = G$. An important step toward proving the Thompson conjecture was recently achieved by Eric Ellers and Nick Gordeev, who proved in [EG] that this conjecture holds for all Chevalley groups over fields K satisfying $|K| > 8$. The conjecture is known to hold for the sporadic and the alternating finite simple groups.

The exact values of the $cn(G)$ were determined for certain families of finite simple groups and for the sporadic simple groups. In [Z], Ilan Zisser determined the covering numbers of all the sporadic simple groups. They range between 2 and 6. In [D], Yoav Dvir proved that

$$cn(A_5) = 5 \quad \text{and for } n \geq 6 \quad cn(A_n) = [\frac{n}{2}] \ .$$

The brackets denote the integral part of $\frac{n}{2}$. In [ACM] it was shown that

$$cn(Sz(2^{2n+1})) = 3 \quad \text{for all } n \ .$$

It had been known for some time that $cn(PSL(2,q)) = 3$. Recently, Arie Lev proved [L2] that

$$cn(PSL(n,q)) = n \quad \text{for} \quad n \geq 3 \ .$$

A.Lev proved more. Let K be any field, finite or infinite. Then

$$cn(PSL(n,K)) = n \quad \text{provided that} \quad |K| \geq 4 \ \text{ and } \ n \geq 4 \ .$$

The minimal value of $cn(G)$ for a simple group G is 2. In a surprising result, Arad, Chillag and Moran [ACM] showed that if G is a finite simple group then

$$cn(G) = 2 \quad \text{if and only if} \quad G = J_1$$

where J_1 is the first Janko simple group. For infinite simple groups G, $cn(G) = 2$ is not that rare. It follows from Theorem 5 in Lev's paper [L1] that for all algebraically closed fields K we have

$$cn(PSL(2, K)) = 2 .$$

We conclude this note with a second open problem: *determine all simple groups G satisfying $cn(G) = 2$.*

References

[ACM] Z. Arad, D. Chillag and G. Moran, Chapter 4 in [AH].

[AH] Z. Arad and M. Herzog (Eds.), *Products of Conjugacy Classes in Groups*, Lecture Notes in Math. 1112, Springer-Verlag, New York, 1985.

[AHS] Z. Arad, M. Herzog and J. Stavi, Chapter 1 in [AH].

[D] Y. Dvir, Chapter 3 in [AH].

[EG] E. Ellers and N. Gordeev, *On the Conjectures of J. Thompson and O. Ore*, Trans. Amer. Math. Soc., to appear.

[EGH] E. Ellers, N. Gordeev and M. Herzog, *Covering Numbers for Chevalley groups*, in preparation.

[G] N. Gordeev, *Products of Conjugacy Classes in Algebraic Groups I*, J. Algebra **173** (1995), 715–744.

[L1] A. Lev, *Products of Cyclic Conjugacy Classes in the Groups PSL(n,F)*, Linear Algebra Appl. **179** (1993), 59–83.

[L2] A. Lev, *The Covering Number of the Groups $PSL_n(F)$*, J. Algebra **182** (1996), 60-84.

[O] O. Ore, *Some Results on Commutators*, Proc. Amer. Math. Soc. **2** (1951), 307-314.

[Z] I. Zisser, *The Covering Numbers of the Sporadic Simple Groups*, Israel J. Math. **67** (1989), 217–224.

CHARACTERIZING SUBNORMALLY CLOSED FORMATIONS

MARK C. HOFMANN

Department of Mathematics and Computer Science, Skidmore College, Saratoga Springs, NY 12866, U.S.A.

1 Introduction

All groups considered in this paper are finite.

In *Zur Theorie der endlichen aüflosbaren Gruppen* [10] Gaschütz defined a formation as a class of groups $\mathcal{F}$ with the following properties:

1) every homomorphic image of an $\mathcal{F}$−group is an $\mathcal{F}$−group.

2) If G/M and G/N are $\mathcal{F}$−groups then $G/M \cap N$ is an $\mathcal{F}$−group.

In [10] a formation $\mathcal{F}$ is said to be saturated if the group G belongs to $\mathcal{F}$ whenever $G/\Phi(G)$ is in $\mathcal{F}$. In the universe of finite solvable groups the definition of a saturated formation provided a context in which to study extensions of the properties of Carter subgroups to classes other than the nilpotent groups. The importance of saturated formations and their relationship to $\mathcal{F}$ -covering subgroups is well known. *Zur Theorie der endlichen aüflosbaren Gruppen* also provided the spur for other developments in finite group theory. In [10] Gaschütz proved that if $\mathcal{F}$ is a saturated formation that contains the nilpotent groups, then $\mathcal{F}$ has the following property:

∗ For $H \trianglelefteq G$, if $H/H \cap \Phi(G)$ belongs to $\mathcal{F}$ then H belongs to $\mathcal{F}$.

Note that any formation $\mathcal{F}$ that has the property (*) must necessarily be saturated and contain all groups of prime order. Thus $\mathcal{F}$ must contain the nilpotent groups. Consequently. the property (*) characterizes the saturated formations of finite solvable groups that contain the nilpotent groups.

In *Praefrattinigruppen* [9] Gaschütz constructed an analogue to the Frattini subgroup. The prefrattini subgroups are a conjugate class of subgroups which are preserved under homomorphism, cover Frattini chief factors, avoid complemented chief factors and are analogous in ways to system normalizers and Carter subgroups of a group. In [12] T. Hawkes defined $\mathcal{F}$-prefrattini subgroups for a saturated formation $\mathcal{F}$. For a given Sylow system $\mathcal{S}$ of a group G and a saturated formation $\mathcal{F}$, the $\mathcal{F}$−prefrattini subgroup of G associated with $\mathcal{S}$ is $W^{\mathcal{F}}(\mathcal{S}) = \cap\{M \mid M$ an $\mathcal{F}$−abnormal maximal subgroup of G, $S^p \leq M, S^p \in \mathcal{S}\}$. This conjugacy class of subgroups avoid complemented $\mathcal{F}$- eccentric chief factors and cover all other chief factors. Moreover an $\mathcal{F}$-prefrattini subgroup is the product of an $\mathcal{F}$-system normalizer and a Gaschütz prefrattini subgroup.

Further studies of the relationship between Frattini like subgroups and classes of groups have included two directions. The first being generalizations of Gaschütz's and Hawkes' results on prefrattini subgroups to larger domains. Notably P. Förster extended Hawkes' results to Schunck classes [8] and more recently Ballester-Bolinches and Ezquerro have defined similar prefrattini subgroups in the general class $\mathcal{E}$ of finite groups [2]. A second direction of development has been to characterize formations through the behavior of Frattini type normal subgroups, in the manner that the behavior of the Frattini subgroup characterizes the saturated formations containing the nilpotent groups. For example for a formation $\mathcal{H}$, M. Hale defines $\Phi_{\mathcal{H}}(G)$ as the intersection of the $\mathcal{H}$-abnormal maximal subgroups of G [11]. In the finite solvable universe, Hale proves that the following property characterizes $\mathcal{H}$ as a subnormal closed saturated formation which contains the nilpotent groups:

If $H \lhd G$ and $H/H \cap \Phi_{\mathcal{H}}(G) \in \mathcal{H}$, then $H \in \mathcal{H}$ [11].

In [5] this result is extended to the universe of all finite groups $\mathcal{E}$. The focus of this work is to define a subgroup $X_{\mathcal{H}}(G)$ which characterizes the subnormally closed tn-formations in the same manner as $\Phi_{\mathcal{H}}(G)$ characterizes subnormally closed saturated formations. This is accomplished by using the prefrattini subgroup of Ballester-Bolinches and Ezquerro to define yet another analogue to Gaschütz's prefrattini subgroup, the $P_{\mathcal{H}}$−subgroup of a group. For a formation $\mathcal{H}$ in the universe $\mathcal{S}$ of finite solvable groups, the $P_{\mathcal{H}}$ -groups are a class of cover-avoid subgroups whose relationship to a prefrattini subgroup and $\mathcal{H}$- residual defines certain classes of formations (e.g the saturated, totally nonsaturated, and locally constructed formations; see [14], [15]). In this article the definition of a $P_{\mathcal{H}}$−subgroup is extended to the class of all finite groups $\mathcal{E}$.

Most notation and terminology is standard and can be found in [7] . A finite group is *primitive* if it has a core free maximal subgroup. The class of all primitive groups is denoted by $\mathcal{P}$ and $\mathcal{P} = \mathcal{P}_1 \cup \mathcal{P}_2 \cup \mathcal{P}_3$ for which $\mathcal{P}_1$ is the class with an abelian socle, $\mathcal{P}_2$ is the class with precisely one nonabelian minimal normal subgroup, and $\mathcal{P}_3$ is the class with two nonabelian minimal normal subgroups. The definition of a $P_{\mathcal{H}}$ -group in the class $\mathcal{E}$ depends on the results of Ballester-Bolinches and Ezquerro [2] on $\mathcal{G}$-prefrattini subgroups for a Schunck class $\mathcal{G}$.

As in [2], let $M(G)$ denote the set of maximal subgroups of a group G and define an equivalence relation $\sim$ on $M(G)$ by $A \sim B$ if and only if $\mathrm{core}A = \mathrm{core}B$ for $A, B \in M(G)$. Denote by $M_1(G)$ the possibly empty set $\{A < \cdot G \mid G/\mathrm{core}A \in \mathcal{P}_1\}$ and analogously set $M_2(G) = \{A < \cdot G \mid G/\mathrm{core}A \in \mathcal{P}_2\}$ for the classes of primitive groups $\mathcal{P}_1, \mathcal{P}_2$ of type 1 and type 2 respectively. A maximal subgroup A is said to be *monolithic*, provided that $A \in M_1(G) \cup M_2(G)$. Consider a complete set T of representatives for the relation $\sim$. The subgroup U of G is *admissible* for T if U is the intersection of some elements from T. A *system* of maximal subgroups of *type 1* of a group G is a set $\{G\} \cup T$ for $T \subseteq M_1$ such that if $A = UN \in M_1(G)$ for $N \lhd G$ and U admissible for T, then $A \in T$. A *system* of maximal subgroups of *type 2* is a set $\{G\} \cup T$ for $T \subseteq M_2(G)$. A *system* Γ of maximal subgroups is the union of the two systems of type 1 and type 2.

For a Schunck class $\mathcal{G}$, a maximal subgroup M is $\mathcal{G}$-*normal* if $G/\mathrm{core}M \in \mathcal{G}$ and $\mathcal{G}$-ab*normal* otherwise. For a system Γ of maximal subgroups and a Schunck class $\mathcal{G}$, the $\mathcal{G}$-*prefrattini* subgroup of G associated with Γ is $W(G, \mathcal{G}, \Gamma) = \cap\{A \in \Gamma \mid A$ is $\mathcal{G}$-abnormal in $G\}$ [2]. Denote by $W(\Gamma)$ the $\{1\}$-prefrattini subgroup of G associated with Γ or simply the prefrattini subgroup of G associated with Γ. (This is consistent in usage with soluble groups since the Gaschütz prefrattini subgroups are contained in the intersection of the normal maximal subgroups.) Since the normal maximal subgroups of G are one-element equivalence classes, then each is in Γ . So $\Phi(G) \subseteq \cap\{M \mid M < \cdot G \text{ and } M \triangleleft G\} \subseteq W(\Gamma)$.

A group G is quasinilpotent if each automorphism induced on each chief factor H/K by an element of G is an inner automorphism of H/K (see Section X.13 [6]). Let Σ denote the class of quasinilpotent groups. For a nonempty formation $\mathcal{H}$, let $\Sigma\circ\mathcal{H}$ represent the class of groups with quasinilpotent $\mathcal{H}$-residual. By Lemma X.13.3 [7], Σ is a formation. Consequently $\Sigma\circ\mathcal{H}$ is a formation (Theorem IV.1.8 [7]) and $\mathcal{G} = E_\Phi(\Sigma \circ \mathcal{H})$ is a saturated homomorph, *i.e.* a Schunck class. Note that $\mathcal{H} \subseteq \Sigma \circ \mathcal{H} \subseteq \mathcal{G}$. Whenever $\mathcal{H} = \{1\}$, then $\mathcal{G} = E_\Phi(\Sigma)$.

2 $P_{\mathcal{H}}$-subgroups

Definition 2.1 For a nonempty formation $\mathcal{H}$, $\mathcal{G} = E_\Phi(\Sigma \circ \mathcal{H})$ and a system Γ of maximal subgroups, the $P_{\mathcal{H}}$ *-subgroup* of G associated with Γ is defined as $P_{\mathcal{H}}(G,\Gamma) = W(G,\mathcal{G},\Gamma) \cap G^{\mathcal{H}}$. In particular, let $P(\Gamma)$ denote the case for $\mathcal{H} = \{1\}$, i.e. $\mathcal{G} = E_\Phi(\Sigma)$.

Since the $\mathcal{G}$-prefrattini subgroup does not necessarily have the cover and avoidance property in general, then neither should this be expected for $P_{\mathcal{H}}(G,\Gamma)$. However note that for a subgroup N normal in G, $P_{\mathcal{H}}(G/N, \Gamma N/N) = [W(G,\mathcal{G},\Gamma)N/N] \cap [NG^{\mathcal{H}}/N] = [W(G,\mathcal{G},\Gamma)N \cap NG^{\mathcal{H}}]/N$ by Proposition 4.2 [2]. If either $N \subseteq \text{core} W(G,\mathcal{G},\Gamma)$ or $N \subseteq G^{\mathcal{H}}$, then $P_{\mathcal{H}}(G/N,\Gamma N/N) = NP_{\mathcal{H}}(G,\Gamma)/N$.

Theorem 2.2 *Consider a formation $\mathcal{H}$ and a system Γ of maximal subgroups in a group G.*

(a) *$P_{\mathcal{H}}(G,\Gamma)$ covers the chief factor H/K if and only if H/K is covered by $G^{\mathcal{H}}$ and H/K is either Frattini or $\mathcal{G}$-central.*

(b) *$G \notin \mathcal{H}$ if and only if $P_{\mathcal{H}}(G,\Gamma)$ covers at least one chief factor $G^{\mathcal{H}}/K$.*

Proof Part (a) is the result of $W(G,\mathcal{G},\Gamma)$ covering the Frattini and $\mathcal{G}$-central supplemented chief factors. For (b), consider a chief factor $G^{\mathcal{H}}/K$ for $G \notin \mathcal{H}$. If $G^{\mathcal{H}}/K$ is supplemented in G by a subgroup $M < \cdot G$, then either $G^{\mathcal{H}}\text{core } M/\text{core} M$ is semi-simple or abelian. In either case, $G^{\mathcal{H}}\text{core} M/\text{core} M \in \Sigma$. Hence $G/\text{core} M \in \mathcal{G}$. Then $G^{\mathcal{H}}/K$ is either Frattini or supplemented and $\mathcal{G}$-central. Therefore $G^{\mathcal{H}}/K$ is covered by $P_{\mathcal{H}}(G/K,\Gamma K/K) = KP_{\mathcal{H}}(G,\Gamma)/K$. If $G \in \mathcal{H}$, then $P_{\mathcal{H}}(G,\Gamma) = \{1\}$. □

Corollary 2.3 *A group G belongs to a formation $\mathcal{H}$ if and only if $P_{\mathcal{H}}(G,\Gamma) = \{1\}$.*

Of the chief factors covered by $G^{\mathcal{H}}$, $P_{\mathcal{H}}(G,\Gamma)$ covers the Frattini and $\Sigma\circ\mathcal{H}$ -central chief factors and avoids the complemented $\Sigma \circ \mathcal{H}$-eccentric chief factors. If a factor H/K is supplemented by a maximal subgroup $A \in M_2(G)$, it may not necessarily be avoided by $P_{\mathcal{H}}(G,\Gamma)$. Moreover, $\text{core}P_{\mathcal{H}}(G,\Gamma) = \text{core}W(G,\mathcal{G},\Gamma) \cap G^{\mathcal{H}}$.

Theorem 2.4 *Let M be a quasinilpotent normal subgroup in a group G such that $M \subseteq G^{\mathcal{H}}$ and $M \cap \text{core}P_{\mathcal{H}}(G,\Gamma) = \{1\}$ for a formation $\mathcal{H}$ and system Γ of maximal subgroups. Then $P_{\mathcal{H}}(G,\Gamma)$ covers no G-chief factor H/K contained in M.*

Proof Assume H/K is a G-chief factor in M that is covered by $P = P_{\mathcal{H}}(G,\Gamma)$. Then $H \cap \Phi(G) \subseteq M \cap \Phi(G) \subseteq M \cap \mathrm{core}P_{\mathcal{H}}(G,\Gamma) = \{1\}$. Moreover $H = K \times N$ for $N \cdot\!\lhd G$. By part (b) of the Lemma in [16], H/K is supplemented in G. So H/K is not Frattini, nor is N. By Theorem 2.2, H/K is $\mathcal{G}$-central. From G-isomorphism, N is $\mathcal{G}$-central. Consequently $N \subseteq M \cap \mathrm{core}P = \{1\}$. The conclusion follows from the contradiction. □

The generalized Fitting subgroup $F^*(G)$ of G is the set of all elements $x \in G$ which induce an inner automorphism on each chief factor of G [6].

Corollary 2.5 *If* $\mathrm{core}P_{\mathcal{H}}(G,\Gamma) = \{1\}$, *then* $P_{\mathcal{H}}(G,\Gamma)$ *covers no chief factors in* $F^*(G)$.

Proof Set $P = P_{\mathcal{H}}(G,\Gamma)$. Note that $P \cap F^*(G) = P \cap G^{\mathcal{H}} \cap F^*(G) = P \cap F^*(G^{\mathcal{H}})$. So P avoids all chief factors between $F^*(G)$ and $F^*(G^{\mathcal{H}})$. By Theorem 2.4, the G-chief factors in $F^*(G^{\mathcal{H}})$ are not covered by P. Hence P does not cover any G-chief factor in a G-chief series from $F^*(G)$ through $F^*(G^{\mathcal{H}})$ to $\{1\}$. □

Definition 2.6 Consider a formation $\mathcal{H}$ and a group G.

(a) $X^*_{\mathcal{H}}(G)$ is the terminal member of the series $\{1\} = X_0 \subseteq X_1 \dots$ defined by $X_i/X_{i-1} = F^*(G/X_{i-1})$ if $\mathrm{core}P_{\mathcal{H}}(G/X_{i-1}, \Gamma X_{i-1}/X_{i-1}) = X_{i-1}$ for $i = 1, \dots$ and $X_i = X_{i-1}$ otherwise.

(b) $X_{\mathcal{H}}(G)$ is the terminal member of the series $\{1\} = X_0 \subseteq X_1 \dots$ defined by $X_i/X_{i-1} = F^*(G/X_{i-1})$ if $\mathrm{core}\Phi(G/X_{i-1}) \cap (G/X_{i-1})^{\mathcal{H}} = X_{i-1}$ for $i = 1, \dots$ and $X_i = X_{i-1}$ otherwise.

Both $X^*_{\mathcal{H}}(G)$ and $X_{\mathcal{H}}(G)$ are well-defined characteristic subgroups of G such that $X^*_{\mathcal{H}}(G) \subseteq X_{\mathcal{H}}(G)$. Theorem 3.5 of [14] is generalized to the next result by Corollary 2.5.

Theorem 2.7 $G \in \mathcal{H}$ *if and only if* $X^*_{\mathcal{H}}(G) = G$.

3 Totally nonsaturated formations

A *totally nonsaturated formation* (*tn-formation*) is a nonempty formation defined as follows: If $G/N \in \mathcal{H}$ but $G \notin \mathcal{H}$ for a minimal normal subgroup N in a group G, then $N \subseteq \Phi(G)$. This formation, introduced for solvable groups in [14] and investigated further in [2], is 'dual' to the concept of a saturated formation. The concept of a tn-formation was extended to the class $\mathcal{E}$ in [2].

For a formation $\mathcal{H}$, let $\sigma\mathcal{H} = \{G \mid$each G-chief factor in $G^{\mathcal{H}}$ is supplemented in $G\}$. From Lemma 2.2 in [13], $\sigma\mathcal{H}$ is a formation. From Theorem 5.3 in [2], $\mathcal{H}$ is a tn-formation if and only if $\mathcal{H} = \sigma\mathcal{H}$. Hence the formation of groups which split over each normal subgroup, the formation of nC-groups, is contained in $\mathcal{H}$. Consequently for a tn-formation $\mathcal{H}$ and the nC-groups $\mathcal{F}$, both $\Phi(G)$ and $G^{\mathcal{H}}$ are contained in $G^{\mathcal{F}}$, possibly properly.

Theorem 5.3 of [2] partially extends the results of Theorem 4.2 in [14] to a more general setting. With the introduction of $P_{\mathcal{H}}(G,\Gamma)$ for nonsolvable groups, the parallel can be completed.

Theorem 3.1 *For a formation $\mathcal{H}$, a group G, a system Γ of maximal subgroups of G, and $W(\Gamma) = W(G,1,\Gamma)$, the following statements are equivalent:*

(a) *$\mathcal{H}$ is a tn-formation.*

(b) *$P_{\mathcal{H}}(G,\Gamma) = W(\Gamma) \cap G^{\mathcal{H}}$ for each group G.*

(c) *$P(G,\Gamma) = W(\Gamma) \cap G^{\sigma\mathcal{H}}$ for each group G.*

Proof Assume (b) is valid. For $P = P_{\mathcal{H}}(G,\Gamma) = W(\Gamma) \cap G^{\mathcal{H}}$, consider $G^{\mathcal{H}}$ to be minimal normal in G. Since $G \notin \mathcal{H}$, by Theorem 2.2(b), P covers $G^{\mathcal{H}}$. So $G^{\mathcal{H}} \subseteq \Phi(G)$. Hence $\mathcal{H}$ is a tn-formation. Therefore (b) implies (a).

Assume that $\mathcal{H}$ is a tn-formation. Let N be minimal normal in G, $N \subseteq G^{\mathcal{H}}$ and inductively assume that $P_{\mathcal{H}}(G/N, \Gamma N/N) = W(G/N, 1, \Gamma N/N) \cap (G/N)^{\mathcal{H}} = W(\Gamma)N/N \cap G^{\mathcal{H}}N/N = [W(\Gamma)N \cap G^{\mathcal{H}}]/N$. If $N \subseteq \Phi(G)$, then $N \subseteq P$ by Theorem 2.2(a). Therefore $P = W(\Gamma) \cap G^{\mathcal{H}}$. So (a) implies (b) whenever $N \subseteq P$.

Next consider N not contained in P and N nonfrattini. A maximal subgroup M from Γ that defines P either supplements N or contains N. Consequently the intersection of $G^{\mathcal{H}}/N$ with those maximal subgroups from Γ that contain N coincides with PN/N. It suffices to prove that N is supplemented by a monolithic $\mathcal{G}$-abnormal maximal subgroup for $\mathcal{G} = E_{\Phi}(\Sigma \circ \mathcal{H})$.

Let M be a maximal subgroup from the above that supplements N and set $C = \mathrm{core}M$. Then $G/C \in \mathcal{P}_1 \cup \mathcal{P}_2$. Since $G^{\mathcal{H}}$ is not contained in C, $G/C \notin \mathcal{H}$. If $G/C \in \mathcal{G}$, then $G/C \in \Sigma \circ \mathcal{H}$. So $CG^{\mathcal{H}}/C \in \Sigma$. Since $\Phi(G/C) = \{C\}$, then part (a) of the Lemma in [16] implies $CG^{\mathcal{H}}/C$ is a direct product of minimal normal subgroups each of which is supplemented in G/C. Therefore $CG^{\mathcal{H}}/C$ is a supplemented G-chief factor. This contradicts $\mathcal{H}$ being a tn-formation. So $G/C \notin \mathcal{G}$. Consequently (a) and (b) are equivalent in both cases.

If $\mathcal{H}$ is a tn-formation, then $\mathcal{H} = \sigma\mathcal{H}$. By the above, $P_{\mathcal{H}}(\Gamma) = W(\Gamma) \cap G^{\sigma\mathcal{H}}$. On the other hand if $P_{\mathcal{H}}(\Gamma) = W(\Gamma) \cap G^{\sigma\mathcal{H}}$, then $G \in \sigma\mathcal{H}$ implies $P_{\mathcal{H}}(\Gamma) = \{1\}$. By Corollary 2.3, $G \in \mathcal{H}$. Hence $\mathcal{H} = \sigma\mathcal{H}$. So ($b$) and (c) are equivalent. □

From Definition 2.6, one concludes that $\mathcal{H}$ is a tn-formation if and only if $X^*_{\mathcal{H}}(G) = X_{\mathcal{H}}(G)$ for each group G.

Theorem 3.2 *For a group G and subnormally closed tn-formation $\mathcal{H} = s_n\mathcal{H}$, $X_{\mathcal{H}}(G) \in \mathcal{H}$.*

Proof Consider the series $\{1\} \subseteq F_1^*(G) \subseteq \ldots \subseteq F_r^*(G) = G$ and the case $K = X_{\mathcal{H}}(G) \neq \{1\}$. By induction, $X_{\mathcal{H}}(G/F_1^*(G)) = K/F_1^*(G) \in \mathcal{H}$. Since $\mathcal{H} = s_n\mathcal{H}$, $K^{\mathcal{H}} \subseteq G^{\mathcal{H}}$. Hence $K^{\mathcal{H}}$ is quasinilpotent and $K^{\mathcal{H}} \cap \Phi(G) \subseteq G^{\mathcal{H}} \cap \Phi(G) = \{1\}$. Consider a G-chief factor of the form $K^{\mathcal{H}}/L$. By the Lemma in [16], $K^{\mathcal{H}}/L$ is supplemented in G. Since $\mathcal{H}$ is a tn-formation, $K^{\mathcal{H}}/L \subseteq \Phi(K/L) \subseteq \Phi(G/L)$. A contradiction arises unless $K^{\mathcal{H}} = \{1\}$. Therefore $X_{\mathcal{H}}(G) \in \mathcal{H}$. □

Theorem 3.3 *The following are equivalent for a nonempty formation:*
(a) $\mathcal{H} = s_n\mathcal{H}$ *is a tn-formation.*
(b) *If for a normal subgroup* N *in a group* G, $N/N \cap X_{\mathcal{H}}(G) \in \mathcal{H}$, *then* $N \in \mathcal{H}$.

Proof Assume that (*b*) is valid and that $G \in \mathcal{H}$. Then $G \supseteq X_{\mathcal{H}}(G) \supseteq X^*_{\mathcal{H}}(G) = G$. So $N/N \cap X_{\mathcal{H}}(G) = N \in \mathcal{H}$ for each subgroup N normal in G. Therefore $\mathcal{H}$ is sn-closed. Suppose $G \notin \mathcal{H}$ but G contains a minimal normal subgroup N such that $G/N \in \mathcal{H}$. If N is not Frattini, then $N \subseteq F^*(G)$ and $G/F^*(G) \in \mathcal{H}$. Therefore $G = X_{\mathcal{H}}(G)$. So $G \in \mathcal{H}$. Hence $N \subseteq \Phi(G)$. Therefore $\mathcal{H}$ is an sn-closed tn-formation. So (*b*) implies (a). Consider (a) to be valid and that the Frattini chief factor $N^{\mathcal{H}}/K$ is in $X_{\mathcal{H}}(G)$ for a subgroup N normal in G. Since $\mathcal{H}$ is sn-closed, $N^{\mathcal{H}} \subseteq G^{\mathcal{H}} \cap X_{\mathcal{H}}$. Hence for some j, $N^{\mathcal{H}} \subseteq X_{j+1}$ but $N^{\mathcal{H}}$ is not contained in $X_j = X$. Then $N^{\mathcal{H}}/N^{\mathcal{H}} \cap X \cong XN^{\mathcal{H}}/X \subseteq X_{j+1}/X$. Since $\mathcal{H}$ is a tn-formation, X_{j+1}/X contains a Frattini chief factor. By Theorem 2.2(a), that chief factor is covered by $P(G/X, \Gamma X/X)$. But this contradicts Corollary 2.5 with respect to $F^*(G/X) = X_{j+1}/X$. So (a) implies (*b*). □

As a generalization of the Zacher {1}-series [4], define $\delta(G)$ to be the terminal member of the series $\{1\} = F_0 \subseteq F_1 \subseteq \dots$ such that $F_i/F_{i-1} = F^*(G/F_{i-1})$ if $\Phi(G/F_{i-1}) = F_{i-1}$ and $F_i = F_{i-1}$ otherwise. For some $r \geq 0$, $F_r = \delta(G)$ but $F_{r-1} \neq \delta(G)$. By the Lemma in [14] and the strengthened form of the Jordan-Holder theorem [7], $\delta(G)$ contains no Frattini chief factors. One notes that $\delta(G) \subseteq X_{\mathcal{H}}(G)$. From these observations and the fact that $\mathcal{H} = s_n\mathcal{H}$, the next theorem is valid in a manner of proof similar to Theorem 3.3.

Theorem 3.4 *The following are equivalent for a nonempty formation:*
(a) $\mathcal{H} = s_n\mathcal{H}$ *is a tn-formation.*
(b) *If for a normal subgroup* N *in a group* G, $N/N \cap \delta(G) \in \mathcal{H}$, *then* $N \in \mathcal{H}$.

References

[1] A. Ballester-Bolinches, Maximal subgroups and formations, J. Pure Appl. Algebra 61 (1989), 223-232.
[2] A. Ballester-Bolinches and L. M. Ezquerro, On maximal subgroups of finite groups, Comm. Alg. 19(8) (1991), 2373-2394.
[3] A. Ballester-Bolinches and M. D. Perez-Ramos, On $\mathcal{F}$-Subnormal subgroups and Frattini-like subgroups of a finite group, Glasgow Math. J. 36 (1994), 241-247.
[4] H. Bechtell, Locally complemented formations, J. Algebra 106 (1987), 413-429.
[5] H. Bechtell and M. C. Hofmann, Kernels of formations, Problems in Algebra, Gomel: University Press 9, (1996) - to appear.
[6] N. Blackburn and B. Huppert, Finite Groups III, Springer-Verlag, Berlin, 1982.
[7] K. Doerk and T. Hawkes, Finite Soluble Groups, De Gruyter, Berlin, 1992.
[8] P. Förster, Prefrattini subgroups, J. Austral. Math. Soc. (Series A) 34 (1983), 234-247.
[9] W. Gaschütz, Praefrattinigruppen, Arch. Math. 13 (1962), 418-426.
[10] W. Gaschütz, Zur Theorie der endlichen aüflosbaren Gruppen, Math. Z. 80 (1963), 300-305.

[11] M. Hale, Normally closed saturated formations, Proc. Amer. Math. Soc. 33 (1972), 337-342.
[12] T. O. Hawkes, Analogues of prefrattini subgroups, Proc. Internat. Conf. Theory of Groups, Austral. Nat. Univ. Canberra (1965), 145-150.
[13] U. C. Herzfeld, Frattini classes of formations of finite groups, Boll. U.M.I. (7) 2-B(1988), 601-611.
[14] M. C. Hofmann, A residual generating prefrattini subgroup, Arch. Math. 48 (1987), 199-207.
[15] M. C. Hofmann, Locally constructed formations, Arch. Math. 53 (1989), 528-537.
[16] M. C. Hofmann, The normal complemented formation, Comm. Alg. 23(14) (1995), 5499-5501

SYMMETRIC WORDS IN A FREE NILPOTENT GROUP OF CLASS 5

WALDEMAR HOŁUBOWSKI

Institute of Mathematics, Silesian Technical University, ul.Kaszubska 23, 44-101 Gliwice, Poland

Abstract

A binary word $w(x,y)$ is called a 2−symmetric word for a group G if $w(g,h) = w(h,g)$ for all g,h in G. In this note we describe 2−symmetric words in a free nilpotent group of class 5.

1 Introduction

Symmetric words in a group G are closely connected with fixed points for automorphisms permuting generators, with symmetric operations in universal algebras and symmetric identities.

For a given G, n−symmetric words form a group $S^{(n)}(G)$ introduced in [P1], where $S^{(n)}(G)$ was described for nilpotent groups of class ≤ 3 (all $S^{(n)}(G)$ are abelian in this case) and $S^{(2)}(D_p)$ where D_p is the dihedral group of order $2p$ (p−prime) [P1],[P2]. $S^{(2)}(G)$ for a free metabelian group and a free soluble group of class 3 were found in [MS]. In [H] there is a description of $S^{(2)}(G)$ and $S^{(3)}(G)$ for G a free metabelian group and a free metabelian of nilpotency class c (for arbitrary c). Since any two-generator nilpotent group of class 4 is metabelian this also covers the case of $S^{(2)}(G)$ for G free nilpotent of class 4.

In this note we give a description of $S^{(2)}(G)$ for G free nilpotent group of class 5 by means of basic commutators. The proof is rather technical, based on identities in nilpotent groups of class 5.

We denote by $[x,y] = x^{-1}y^{-1}xy$ the commutator of elements x,y. Left-normed commutators are defined by $[x_1,\ldots,x_n,x_{n+1}] = [[x_1,\ldots,x_n],x_{n+1}]$. Also $[x,_{(r+1)}y] = [[x,_r y],y]$, $r \geq 0$, for $[x,_0 y] = x$. We use notation $\binom{n}{i} = \frac{1}{i!}\cdot n(n-1)\cdots(n-i+1)$ and $\phi(n) = \sum_{i=1}^{n-1}\binom{i}{2}$, $\psi(n) = \sum_{i=1}^{n-1}\phi(i)$, $\chi(n) = \sum_{i=1}^{n-1} i^2$.

We need some well known identities:

(1) $[x^{-1},y] = [x,y]^{-1}[y,x,x^{-1}]$,

(2) $[x,y^{-1}] = [x,y]^{-1}[y,x,y^{-1}]$,

(3) $[xy,z] = [x,z][x,z,y][y,z]$,

(4) $[x,yz] = [x,z][x,y][x,y,z]$, and

(5) $[[x,y],z^x][[z,x],y^z][[y,z],x^y] = 1$ (Jacobi identity),

valid in arbitrary group.

2 The main result

Let F_2 be a free group generated by x, y.

Definition 1 A word $w(x,y) \in F_2$ is called a 2-symmetric word for a group G if $w(g,h) = w(h,g)$ for all g, h in G. If $w(g,h) = 1$ for all g, h we call $w(x,y)$ a trivial symmetric word.

By another words: $w(x,y)$ is a non-trivial 2-symmetric word for G if G satisfies the identity $w(x,y) = w(y,x)$ but does not satisfy the identity $w(x,y) = 1$.

So we can consider $G = F_2/\gamma_6(F_2)$ and perform calculations in F_2 modulo $\gamma_6(F_2)$.

Definition 2 A word $w(x,y) \in F_2$ is called positive if modulo F_2' it has the form $x^\alpha y^\alpha$, $\alpha \geq 0$.

Our main result here is:

Theorem *A positive binary word w is a non-trivial 2-symmetric word in a free nilpotent group of class 5 if and only if w has the form*

$$\begin{aligned} w(x,y) \;=\;& x^\alpha y^\alpha [y,x]^\beta [y,x,x]^{\gamma_1} [y,x,y]^{\gamma_2} [y,x,x,x]^{\delta_1} [y,x,x,y]^{\delta_2} [y,x,y,y]^{\delta_3} \\ & [y,x,x,x,x]^{\epsilon_1} [y,x,x,x,y]^{\epsilon_2} [y,x,x,y,y]^{\epsilon_3} [y,x,y,y,y]^{\epsilon_4} \\ & [[y,x,x],[y,x]]^{\kappa_1} [[y,x,y],[y,x]]^{\kappa_2} \end{aligned}$$

where

$$2\beta = \alpha^2,\ \gamma_1 + \gamma_2 = \alpha\binom{\alpha}{2},\ \delta_1 + \delta_3 = \alpha\phi(\alpha),\ 2\delta_2 = \binom{\alpha}{2}^2,\ \epsilon_1 + \epsilon_4 = \alpha\psi(\alpha),$$

$$\epsilon_2 + \epsilon_3 = \binom{\alpha}{2}\phi(\alpha),\ \phi_1 - \phi_2 - \gamma_2 + \delta_2 + \epsilon_3 = 2\binom{\alpha}{2}^2 + \binom{\alpha}{2}\phi(\alpha) - \beta\alpha\binom{\alpha}{2},$$

$$\phi_2 - \phi_1 - \gamma_1 + \delta_2 + \epsilon_2 = \binom{\alpha}{2}^2 + \binom{\alpha}{2}\phi(\alpha) + \alpha\chi(\alpha) - \beta\alpha\binom{\alpha}{2}.$$

Proof Every word in F_2 modulo F_2' is of the form $x^\alpha y^\beta$. If $w(x,y)$ is symmetric in F_2 modulo $\gamma_6(F_2)$, then also modulo F_2' and hence $\alpha = \beta$. We shall assume that $\alpha \geq 0$.

We fix a standard order for the basic commutators. Every word modulo $\gamma_6(F_2)$ is a product of basic commutators of weight ≤ 5 and such a product is unique [N]. So any binary word for G has a form $w(x,y) = x^{\alpha_1} y^{\alpha_2} \cdot \prod d_i$, where d_i are commutators of the weight ≥ 2.

We write it implicitly

$$\begin{aligned} w(x,y) \;=\;& x^\alpha y^\alpha [y,x]^\beta [y,x,x]^{\gamma_1} [y,x,y]^{\gamma_2} [y,x,x,x]^{\delta_1} [y,x,x,y]^{\delta_2} [y,x,y,y]^{\delta_3} \\ & [y,x,x,x,x]^{\epsilon_1} [y,x,x,x,y]^{\epsilon_2} [y,x,x,y,y]^{\epsilon_3} [y,x,y,y,y]^{\epsilon_4} \\ & [[y,x,x],[y,x]]^{\phi_1} [[y,x,y],[y,x]]^{\phi_2}. \end{aligned}$$

We now get $w(y,x)$ by permuting generators. We have

$$\begin{aligned} w(y,x) &= y^{\alpha}x^{\alpha}[x,y]^{\beta}[x,y,y]^{\gamma_1}[x,y,x]^{\gamma_2}[x,y,y,y]^{\delta_1}[x,y,y,x]^{\delta_2}[x,y,x,x]^{\delta_3} \\ &\quad [x,y,y,y,y]^{\epsilon_1}[x,y,y,y,x]^{\epsilon_2}[x,y,y,x,x]^{\epsilon_3}[x,y,x,x,x]^{\epsilon_4} \\ &\quad [[x,y,y],[x,y]]^{\phi_1}[[x,y,x],[x,y]]^{\phi_2} \\ &= x^{\alpha}y^{\alpha}[y^{\alpha},x^{\alpha}][x,y]^{\beta}[x,y,y]^{\gamma_1}[x,y,x]^{\gamma_2}[x,y,y,y]^{\delta_1}[x,y,y,x]^{\delta_2} \\ &\quad [x,y,x,x]^{\delta_3}[x,y,y,y,y]^{\epsilon_1}[x,y,y,y,x]^{\epsilon_2}[x,y,y,x,x]^{\epsilon_3}[x,y,x,x,x]^{\epsilon_4} \\ &\quad [[x,y,y],[x,y]]^{\phi_1}[[x,y,x],[x,y]]^{\phi_2}. \end{aligned}$$

The idea of the proof is rewriting this as a product of basic commutators and comparing exponents.

As a first step we use $y^{\alpha}x^{\alpha} = x^{\alpha}y^{\alpha}[y^{\alpha},x^{\alpha}]$. For the next step we need the following lemmas.

Lemma 1 *The following identities hold in the nilpotent group of class* 5

(6) $[x,y,x] = [y,x,x]^{-1}[[y,x,x],[y,x]]$

(7) $[x,y,y] = [y,x,y]^{-1}[[y,x,y],[y,x]]$

(8) $[x,y,y,y] = [y,x,y,y]^{-1}$

(9) $[x,y,x,x] = [y,x,x,x]^{-1}$

(10) $[x,y,y,x] = [y,x,x,y]^{-1}[[y,x,x],[y,x]]^{-1}[[y,x.y],[y,x]]^{-1}$

(11) $[x,y,x,x,x] = [y,x,x,x,x]^{-1}$

(12) $[x,y,y,x,x] = [y,x,x,x,y]^{-1}[[y,x,x],[y,x]]^{-1}$

(13) $[x,y,y,y,x] = [y,x,x,y,y]^{-1}[[y,x,y],[y,x]]^{-1}$

(14) $[x,y,y,y,y] = [y,x,y,y,y]^{-1}$.

Proof Identities (6) – (9) follow easily from (1) – (4). Expanding commutator $[xy,yx]$ in two ways, first applying (3) and next (4), a second time using first (4) and next (3) we obtain

$$[xy,yx] = [xy,x][xy,y][xy,y,x] = [x,y,y][x,y,x][x,y,y,x]$$

and

$$[xy,yx] = [x,yx][x,yx,y][y,yx] = [x,y][x,y,x][x,y,y][x,y,x,y][y,x].$$

So we have

$$[x,y,y,x] = [y,x,x][y,x,y][x,y][x,y,x][x,y,y][x,y,x,y][y,x]$$

and finally

$$\begin{aligned} [x,y,y,x] &= [x,y,x,y][[y,x,x],[x,y]][[y,x,y],[x,y]] \\ &= [y,x,x,y]^{-1}[[y,x,x],[y,x]]^{-1}[[y,x,y],[y,x]]^{-1}. \end{aligned}$$

Identities (11) – (14) follow immediately from (1) – (5) and (10). □

Lemma 2 *The following identity holds in the nilpotent group of class* 5 *for all natural* k, n

$$\begin{aligned}[y^k, x^n] = \; & [y,x]^{kn}[y,x,x]^{k\binom{n}{2}}[y,x,y]^{\binom{k}{2}n}[y,x,x,x]^{k\phi(n)}[y,x,x,y]^{\binom{k}{2}\binom{n}{2}} \\ & \cdot[y,x,y,y]^{\phi(k)n}[y,x,x,x,x]^{k\psi(n)}[y,x,x,x,y]^{\binom{k}{2}\phi(n)} \\ & \cdot[y,x,x,y,y]^{\phi(k)\binom{n}{2}}[y,x,y,y,y]^{\psi(k)n}[[y,x,x],[y,x]]^{2\binom{k}{2}\binom{n}{2}+\binom{k}{2}\phi(n)} \\ & \cdot[[y,x,y],[y,x]]^{\binom{k}{2}\binom{n}{2}+\phi(k)\binom{n}{2}+\chi(k)n}.\end{aligned}$$

Proof First, we prove by induction that

$$[y^k, x] = [y,x]^k[y,x,y]^{\binom{k}{2}}[y,x,y,y]^{\phi(k)}[y,x,y,y,y]^{\psi(k)}[[y,x,y],[y,x]]^{\chi(k)}.$$

This follows easily from the identities

$$\begin{aligned}[y,x,y,y^k] &= [y,x,y,y]^k[y,x,y,y,y]^{\binom{k}{2}} \\ [y,x,y^k] &= [y,x,y]^k[y,x,y,y]^{\binom{k}{2}}[y,x,y,y,y]^{\phi(k)}\end{aligned}$$

and from the inductive step

$$[y^{k+1}, x] = [y,x][y,x,y^k][y^k,x] = [y,x][y^k,x][y,x,y^k][[y,x,y],[y,x]]^{k^2}.$$

Since

$$[y^k, x^{n+1}] = [y^k, x^n][y^k, x][y^k, x, x^n]$$

we get by induction the required conclusion using the identities

$$\begin{aligned}[y,x,x^n] &= [y,x,x]^k[y,x,y,y]^{\binom{k}{2}}[y,x,y,y,y]^{\phi(k)} \\ [y^k,x,x,x] &= [y,x,x,x]^k[y,x,x,x,x]^{\binom{k}{2}}[[y,x,x],[y,x]]^{\binom{k}{2}} \\ [y^k,x,x] &= [y,x,x]^k[y,x,x,y]^{\binom{k}{2}}[y,x,x,y,y]^{\phi(k)}[[y,x,x],[y,x]]^{2\binom{k}{2}} \\ & \quad [[y,x,y],[y,x]]^{\phi(k)+\binom{k}{2}} \\ [y^k,x,x^n] &= [y,x,x]^{kn}[y,x,x,x]^{k\binom{n}{2}}[y,x,x,y]^{\binom{k}{2}n}[y,x,x,x,x]^{k\phi(n)} \\ & \quad [y,x,x,x,y]^{\binom{k}{2}\binom{n}{2}}[y,x,x,y,y]^{\phi(k)n}[[y,x,x],[y,x]]^{2\binom{k}{2}n+\binom{k}{2}\binom{n}{2}} \\ & \quad [[y,x,y],[y,x]]^{(\phi(k)+\binom{k}{2})n},\end{aligned}$$

which finishes the proof of Lemma 2. □

To finish the proof of the Theorem we rewrite all commutators in $w(y,x)$ (given above) as the products of basic commutators using Lemma 1 and 2 and comparing exponents of each basic commutator we obtain the equalities needed. □

Remark The group $S^{(2)}(F_2/\gamma_6(F_2))$ is not abelian. This follows since a 2−symmetric word

$$w_1 = [y,x,x]^{\gamma}[y,x,y]^{-\gamma}[[y,x,x],[y,x]]^{\phi}[[y,x,y],[y,x]]^{\phi+\gamma}$$

does not commute with any 2−symmetric positive word for $\alpha > 0$. It is interesting to find the nilpotency class of $S^{(2)}(F_2/\gamma_6(F_2))$ however it seems to be very tiresome.

References

[Ha] M. Hall, *Theory of groups*, MacMillan Comp., New York, 1959.

[H] W. Hołubowski, *Symmetric words in metabelian groups*, Comm. Algebra, 23 (14) (1995), 5161-5167.

[MS] O. Macedońska and D. Solitar, *On binary σ−invariants words in a group*, Contemporary Math., 169 (1994), 431-449.

[N] H. Neumann, *Varieties of groups*, Springer V., Berlin-Heidelberg-New York, 1967.

[P1] E. Plonka, *Symmetric operations in groups*, Colloq. Math., 21 (1970), 179-186.

[P2] E. Płonka, *Symmetric words in nilpotent groups of class* ≤ 3, Fund. Math., 97 (1977), 95-103.

A NON-RESIDUALLY FINITE SQUARE OF FINITE GROUPS

TIM HSU* and DANI WISE†[1]

*Department of Mathematics, University of Michigan, Ann Arbor, MI 48109, U.S.A.
† Department of Mathematics, Cornell University, White Hall, Ithaca, NY 14853, U.S.A.

Abstract

We construct a non-positively curved non-residually finite square of finite groups whose vertex groups are of order 288, 288, 576, and 576. (In contrast, an earlier such example constructed by the authors had vertex groups of order between 2^{60} and 2^{150}.) In doing so, we demonstrate a new, more geometric method for embedding the fundamental group of a complete squared complex in the fundamental group of a square of finite groups.

1 Introduction

A *triangle* of groups is a diagram of group inclusions like the diagram T shown in Figure 1. The *fundamental group* of T, or $\pi_1(T)$, is the colimit of the diagram T. In other words, $\pi_1(T)$ is the group given by the presentation whose generators are the elements of X, Y, and Z, and whose relators are the multiplication tables for X, Y, and Z and the relations induced by the inclusions in the diagram. Note that in general, X, Y, and Z will not be subgroups of $\pi_1(T)$, since the other relations may make these groups collapse.

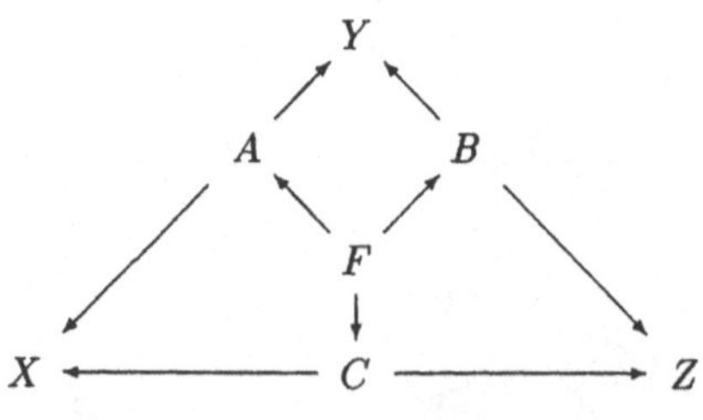

Figure 1. A triangle of groups T

Triangles of groups, especially triangles of finite groups, arise naturally in many places. For instance, many buildings can be formed from triangles of groups (see Brown [4], Ronan [15], and Tits [18]). In another context, triangles of groups play a key role in the proof of "uniqueness theorems" for sporadic finite simple groups (see Aschbacher and Segev [2] and Ivanov [10]).

Geometrically, if T does not collapse (i.e., if X, Y, and Z are subgroups of $\pi_1(T)$), we see that $\pi_1(T)$ naturally acts on a triangulated complex, with quotient

[1]The second author is supported by NSF grant no. DMS-9627506.

a triangle. This complex is called the *universal cover* of T, or $\widetilde{T}$. In that case, for any triangle Δ in the universal cover of T, the stabilizers of the vertices of Δ will be (isomorphic to) X, Y, and Z. Likewise, the stabilizers of the edges of Δ will be A, B, and C, and the stabilizer of all of Δ will be F.

One important situation where we are guaranteed to avoid collapse is the class of *non-positively curved triangles of groups*, as defined by Stallings [17]. More generally, there is a theory of non-positively curved *orbihedra* of groups, due to Haefliger [7], which deals with a diagram of inclusions corresponding to the cells of an arbitrary complex. In particular, one can consider non-positively curved squares, pentagons, etc., of groups. Note that for $n \geq 4$, it is very easy to construct lots of non-positively curved n-gons of groups; for instance, any n-gon of groups ($n \geq 4$) whose vertex groups are semidirect products of its edge groups is non-positively curved, since the Gersten-Stallings angle at each vertex is $\pi/2$. (See Stallings [17] for the definition of the Gersten-Stallings angle.)

Since the fundamental group of a (nondegenerate) non-positively curved polygon of finite groups is always infinite, it is natural to ask:

> *Which finiteness properties does the fundamental group of a non-positively curved polygon of finite groups have? For instance, is such a group residually finite, or virtually torsion-free?*

In particular, until recently, it was not known whether every non-positively curved polygon of finite groups has a residually finite fundamental group. Some positive results concerning the residual finiteness of certain classes of polygons of groups are due to Allenby and Tang [1] and Kim [11]. More recently, the authors [9] obtained non-positively curved squares and triangles of finite groups with non-residually finite fundamental groups.

Now, philosophically, the main idea behind [9] is that, just as the fundamental groups of finite 1-complexes are almost the same as amalgamated free products of finite groups, the fundamental groups of finite non-positively curved squared complexes are almost the same as those of non-positively curved squares of finite groups. In this paper, we illustrate this idea by showing, in a very natural way:

Theorem 1.1 *There exists a non-positively curved square of groups whose vertex groups have order* 288, 288, 576, *and* 576, *and whose fundamental group is non-residually finite.*

In contrast, our previous example of a non-positively curved non-residually square of finite groups [9] had vertex groups of order between 2^{60} and 2^{150}.

After introducing a certain complete squared complex X whose fundamental group has useful subgroup inseparability properties (Section 2), we describe a natural way of embedding $\pi_1(X)$ in a square of groups whose edge groups are free, and whose vertex groups are semidirect products (Section 3). An algebraic trick then allows us to make the edge and vertex groups finite (Section 4). Note that compared to our earlier embedding trick (the First Main Theorem in [9]), the embedding here is more natural, more geometric, and gives slightly nicer looking results in terms of the induced map on the universal cover.

In any case, our embedding gives us a non-positively curved square of (small) finite groups R with the same subgroup inseparability property that X has, so by "doubling" R, we obtain a nonresidually finite example (Section 5). In conclusion, we list some related open problems of interest (Section 6).

2 The complete squared complex X

In this section, we briefly describe some material that we will need from [19]. The interested reader can find a more elaborate discussion there.

Definition 2.1 A combinatorial 2-complex Y whose 2-cells are squares is said to be a *complete squared complex* (CSC) if the link of each vertex of Y is a complete bipartite graph.

The simplest example of a CSC is the direct product of two graphs. An equivalent definition to the one given above is that a squared 2-complex is a CSC if it is locally isomorphic to the direct product of two trees. In fact, it is easy to prove that:

Theorem 2.2 *Let Y be a squared 2-complex, and let $\widetilde{Y}$ denote the universal cover of Y. Then Y is a CSC if and only if $\widetilde{Y}$ is isomorphic to the direct product of two trees.*

We now give an example of a CSC X which is not the direct product of two graphs. X may be described as the result of gluing 6 squares to a bouquet of 5 edges in the manner defined by Figure 2. We think of the 1-skeleton of X as the union of a bouquet H of 2 horizontal circles, labelled $\{x, y\}$, and a bouquet V of 3 vertical circles, labelled $\{a, b, c\}$.

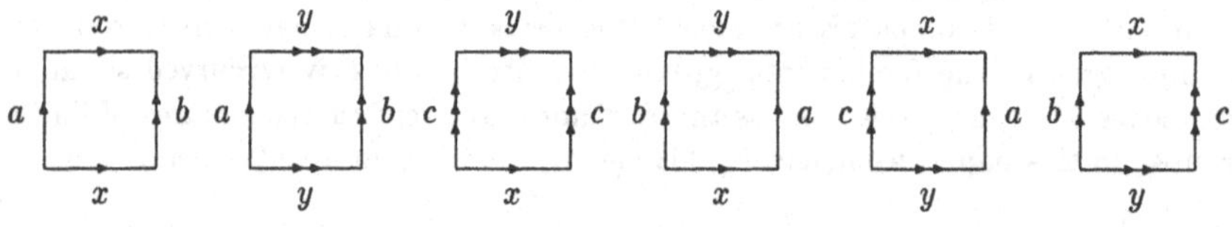

Figure 2. The six squares of X

Since X has only one 0-cell, $\pi_1(X)$ is generated by $\{a, b, c, x, y\}$, and we may read the following relators for $\pi_1(X)$ off of Figure 2:

$$\begin{array}{lll} xb = ax & yb = ay & xc = cy \\ xa = by & ya = cx & yc = bx. \end{array} \tag{2.1}$$

Because of Theorem 2.2, $\widetilde{X}$ is isomorphic to a direct product of trees $\widetilde{V} \times \widetilde{H}$. Accordingly, we obtain the following corollary for use in Section 3.

Corollary 2.3 *Each element of $\pi_1(X)$ can be expressed as a product $v \cdot h$, where v and h are reduced combinatorial paths in V and H, respectively (i.e., $v \in \pi_1(V)$ and $h \in \pi_1(H)$).*

To describe the most useful property of the complex X (Theorem 2.5), we need the following definition.

Definition 2.4 A group G is said to be *subgroup separable* with respect to a subgroup C provided that for each element $g \in G - C$, there is a finite index subgroup $K \subset G$ containing C such that $g \notin K$. A group which is subgroup separable with respect to the trivial subgroup is said to be *residually finite.*

Equivalently, G is subgroup separable with respect to C if, for every $g \in G - C$, there is some finite quotient of G in which the image of g is not contained in the image of C.

Theorem 2.5 *The group $\pi_1(X)$ is not subgroup separable with respect to $\langle a, b, c \rangle$. Specifically, the element $xy^{-1} \notin \langle a, b, c \rangle$ cannot be separated from $\langle a, b, c \rangle$ in any finite quotient of $\pi_1(X)$.*

The proof of Theorem 2.5 is based on the following contradiction. On the one hand, careful analysis of covers of the complex X shows that if $\langle a, b, c \rangle$ could be separated from xy^{-1} in a finite quotient, then X would have a finite cover which is the direct product of two graphs. On the other hand, one can deduce from certain aperiodic tilings called *anti-tori* which occur in $\widetilde{X}$, that X has *no* such finite cover which is a product. More specifically, there is a particular anti-torus in $\widetilde{X}$ which implies that that a and y do not have commuting non-trivial powers a^n, y^m. Since a and y are vertical and horizontal, this clearly precludes the existence of a finite product cover.

3 A "free" embedding of $\pi_1(X)$

We now wish to transfer the subgroup inseparability property of $\pi_1(X)$ to the fundamental group of some non-positively curved square of groups R by embedding $\pi_1(X)$ in $\pi_1(R)$. The idea behind this embedding may be thought of in terms of an equivariant map from $\widetilde{X}$ to $\widetilde{R}$. We send each 1-cell in $\widetilde{X}$ to a pair of opposite "wall-crossing paths" in $\widetilde{R}$. In other words, each generator of $\pi_1(X)$ is sent to the product of a pair of elements stabilizing opposing edges of R. This idea is depicted in Figure 3, where the heavy solid lines represent a typical square of $\widetilde{X}$, the light solid lines represent our subdivision of this square, and the dashed lines represent a portion of $\widetilde{R}$. (The annotations in Figure 3 will be explained below.)

We can implement this idea algebraically in the following way. Let

$$\begin{aligned} E &= \langle x_e, y_e \rangle & N &= \langle a_n, b_n, c_n \rangle \\ W &= \langle x_w, y_w \rangle & S &= \langle a_s, b_s, c_s \rangle \end{aligned} \qquad (3.1)$$

be free groups on the indicated generating sets, and let R be the square of groups

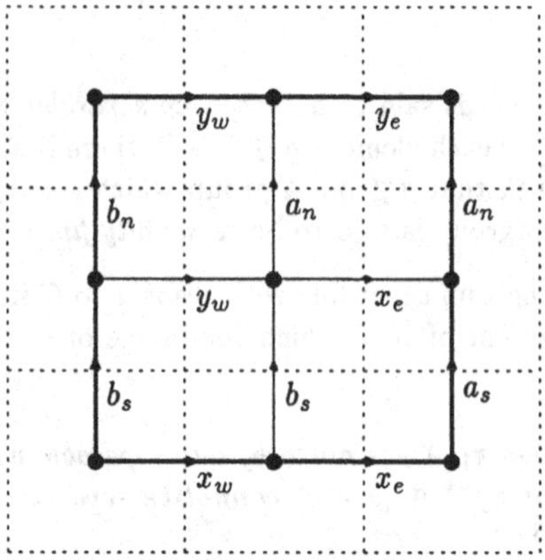

Figure 3. Embedding picture/van Kampen diagram for the fourth square of X

given by:

$$\begin{array}{ccccc} N \rtimes W & \leftarrow & N & \rightarrow & E \rtimes N \\ \uparrow & & \uparrow & & \uparrow \\ W & \leftarrow & 1 & \rightarrow & E \\ \downarrow & & \downarrow & & \downarrow \\ W \rtimes S & \leftarrow & S & \rightarrow & S \rtimes E. \end{array} \tag{3.2}$$

Note that, as mentioned in the introduction, R is non-positively curved.

The semidirect products which form the vertex groups of R are defined by relations described by the following mnemonic, with one relation per vertex group per 2-cell of X.

- The relation of $S \rtimes E$ arising from each 2-cell of X imitates the way the west edge of the 2-cell is changed when it is pushed through the South edge to the East edge.
- The relation of $E \rtimes N$ arising from each 2-cell of X imitates the way the south edge of the 2-cell is changed when it is pushed through the East edge to the North edge.
- The relation of $N \rtimes W$ arising from each 2-cell of X imitates the way the West edge of the 2-cell is changed when it is pushed through the North edge to the east edge.
- The relation of $W \rtimes S$ arising from each 2-cell of X imitates the way the South edge of the 2-cell is changed when it is pushed through the West edge to the north edge.

More precisely, these relations are given by the following array, where the rows correspond with the six 2-cells of X, in the order given in Figure 2, and the columns correspond to the vertex groups $S \rtimes E$, $E \rtimes N$, $N \rtimes W$, and $W \rtimes S$, in that order.

$$x_e^{-1}a_s x_e = b_s \qquad b_n^{-1}x_e b_n = x_e \qquad x_w^{-1}a_n x_w = b_n \qquad a_s^{-1}x_w a_s = x_w \tag{3.3}$$
$$y_e^{-1}a_s y_e = b_s \qquad b_n^{-1}y_e b_n = y_e \qquad y_w^{-1}a_n y_w = b_n \qquad a_s^{-1}y_w a_s = y_w \tag{3.4}$$
$$x_e^{-1}c_s x_e = c_s \qquad c_n^{-1}x_e c_n = y_e \qquad y_w^{-1}c_n y_w = c_n \qquad c_s^{-1}x_w c_s = y_w \tag{3.5}$$
$$x_e^{-1}b_s x_e = a_s \qquad a_n^{-1}x_e a_n = y_e \qquad y_w^{-1}b_n y_w = a_n \qquad b_s^{-1}x_w b_s = y_w \tag{3.6}$$
$$y_e^{-1}c_s y_e = a_s \qquad a_n^{-1}y_e a_n = x_e \qquad x_w^{-1}c_n x_w = a_n \qquad c_s^{-1}y_w c_s = x_w \tag{3.7}$$
$$y_e^{-1}b_s y_e = c_s \qquad c_n^{-1}y_e c_n = x_e \qquad x_w^{-1}b_n x_w = c_n \qquad b_s^{-1}y_w b_s = x_w \tag{3.8}$$

Now let ϕ be the map from $\pi_1(X)$ to $\pi_1(R)$ defined by

$$\begin{array}{ccc} \phi(a) = a_s a_n & \phi(b) = b_s b_n & \phi(c) = c_s c_n \\ \phi(x) = x_w x_e & & \phi(y) = y_w y_e. \end{array} \tag{3.9}$$

We first claim that ϕ is a homomorphism. To verify this claim, it is enough to verify that the defining relations (2.1) for $\pi_1(X)$ are respected by ϕ. It is not too hard to do this directly, as each row of (3.3)–(3.8) yields the fact that ϕ respects the corresponding defining relation in (2.1). More instructively, consider Figure 3 again. The van Kampen diagram contained in Figure 3 shows that the relation from square number 4 is respected by ϕ, and similar van Kampen diagrams show that the other relations are respected. (Note that the southeast corner of the van Kampen diagram in Figure 3 is obtained from relation number 4 for $S \rtimes E$, the northeast corner from $E \rtimes N$, and so on.)

Perhaps more importantly, Figure 3 shows that ϕ induces a map from $\widetilde{X}$ to $\widetilde{R}$ which is an embedding on each square of $\widetilde{X}$, so at this point, one might optimistically hope that ϕ actually induces an embedding of all of $\widetilde{X}$ in $\widetilde{R}$. However, the induced map on universal covers is not an embedding, or even a local embedding, for the following reason.

Let Γ be the figure eight graph, and let Γ' be the edge of groups with trivial edge group and both vertex groups isomorphic to Z_2^2 (where Z_2 denotes the group of order 2). Let one vertex group of Γ' be denoted by $\langle a,b\rangle$, and let the other be $\langle c,d\rangle$. Recall that $\pi_1(\Gamma) = F_2 = \langle x,y\rangle$, the free group on two generators, and $\pi_1(\Gamma') = Z_2^2 * Z_2^2$. Consider the homomorphism $\rho : \pi_1(\Gamma) \to \pi_1(\Gamma')$ defined by $\rho(x) = ac$, $\rho(y) = bd$. Now, as shown in Figure 4, ρ naturally induces a map from Γ (dotted curves on the left side of Figure 4) to Γ' (solid lines on the right side of Figure 4). However, the map on universal covers is not an embedding, since it is not a local embedding in the neighborhood of a 0-cell of $\widetilde{\Gamma}$, as can be seen in Figure 4. The analogous local problem occurs with the induced map on universal covers $\widetilde{X} \to \widetilde{R}$, and so this induced map is also not an embedding.

Nevertheless, we can see that ϕ itself is still an embedding as follows: First, consider the free products $N * S$ and $W * E$ which correspond to obvious suborbihedra of R. We claim that the intersection of the corresponding subgroups of $\pi_1(R)$ is trivial. To see this, observe that an element of $N * S$ must stabilize the associated vertical tree subspace of $\widetilde{R}$, and similarly, an element of $W * E$ must stabilize the

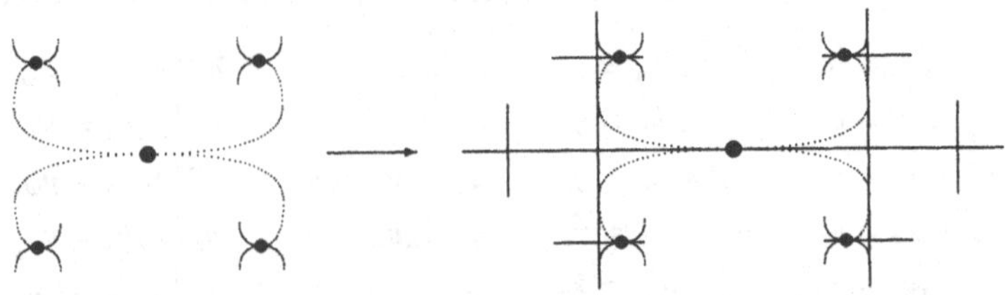

Figure 4. The induced map on universal covers

associated horizontal tree subspace of $\widetilde{R}$. An element in both subgroups must stabilize both trees, and therefore must stabilize their intersection, which is the center of a square, since $\widetilde{R}$ is a product. Therefore, since the center of a square has trivial stabilizer, an element in both subgroups must be trivial.

Next, observe that ϕ maps $\pi_1(V)$ to $N * S$ and ϕ maps $\pi_1(H)$ to $W * E$. Furthermore, by the normal form theorem for free products, both of these (restricted) maps are injections, just as in the case of $\pi_1(\Gamma) \to \pi_1(\Gamma')$ above.

So now suppose that $\phi(g) = 1$ for some $g \in \pi_1(X)$. By Corollary 2.3, $g = vh^{-1}$ for some elements $v \in \pi_1(V)$ and $h \in \pi_1(H)$, which means that $\phi(v) = \phi(h)$. However, since $\phi(v) \in N * S$ and $\phi(h) \in W * E$, from the above argument, $\phi(v) = 1 = \phi(h)$. Therefore, since ϕ is injective on $\pi_1(V)$ and $\pi_1(H)$, we conclude that $v = 1$ and $h = 1$, and so $g = vh^{-1} = 1$.

We therefore obtain:

Theorem 3.1 *ϕ embeds $\pi_1(X)$ as a subgroup of $\pi_1(R)$.*

Remark 3.2 Although we did not obtain an equivariant embedding of $\widetilde{X}$ in $\widetilde{R}$, we note that because the vertex groups are semidirect products of the edge groups, it follows that $\widetilde{R}$ is the product of two trees. Again, this shows how the geometry of $\widetilde{X}$ is duplicated in $\widetilde{R}$. Furthermore, we note that ϕ actually induces a quasi-isometry.

Finally, since Theorem 2.5 implies that xy^{-1} cannot be separated from $\langle a, b, c\rangle$ in any finite quotient of $\pi_1(X)$, $\phi(xy^{-1})$ cannot be separated from $\langle \phi(a), \phi(b), \phi(c)\rangle$ in any finite quotient of $\pi_1(R)$. Let C denote the amalgamated free product $(E \rtimes N) *_E (S \rtimes E)$ corresponding to the east edge of R. Since $\phi(xy^{-1}) \notin C$ and $\langle \phi(a), \phi(b), \phi(c)\rangle \subseteq C$, we have:

Corollary 3.3 *Let C be the amalgamated free product corresponding to the east edge of R. Then $\pi_1(R)$ is not C-separable.*

Remark 3.4 The construction in this section may be generalized in the following way. Note that (3.3)–(3.8) give complete definitions for $S \rtimes E$, $E \rtimes N$, $N \rtimes W$, and $W \rtimes S$ precisely because X is a CSC with just one 0-cell, which means that any "north" edge of X meets any "east" edge of X in precisely 1 square, and so on. In the general case of a CSC with more than one 0-cell, if we perform the same procedure, with the addition of commuting relations corresponding to edges which do not meet in a square, we get an injective homomorphism from the fundamental groupoid $\pi(X)$ into the fundamental group of a square of groups R, with the geometry shown above.

Remark 3.5 J. McCammond (personal communication) has observed that the group with defining relations (3.3)–(3.8) is a LOG group (as defined by Howie [8]). More generally, Remark 3.4 implies that the fundamental groupoid of any CSC may be embedded in a LOG group.

4 Making the embedding finite

Having embedded $\pi_1(X)$ in the fundamental group of a non-positively curved square of groups R, we now use an algebraic trick to construct a quotient $\overline{R}$ of R which is a non-positively curved square of *finite* groups.

The idea is that the semidirect products described by (3.3)–(3.8) really just describe how the action of E permutes the generators of S, the action of N permutes the generators of E, and so on. If we write these permutations down, for E and W, we have, in cycle form:

$$x_e = (a_s, b_s) \qquad y_e = (a_s, b_s, c_s) \tag{4.1}$$

$$x_w = (a_n, b_n, c_n) \qquad y_w = (a_n, b_n), \tag{4.2}$$

and for N and S, we have:

$$a_n = (x_e, y_e) \qquad b_n = () \qquad c_n = (x_e, y_e) \tag{4.3}$$

$$a_s = () \qquad b_s = (x_w, y_w) \qquad c_s = (x_w, y_w). \tag{4.4}$$

This is our first approximation for the edge groups of $\overline{R}$.

The problem with our first approximation is that the permutations indicated by x_e and y_e are no longer automorphisms of the group $\langle a_s, b_s, c_s \rangle$, and so on. However, we may fix this problem by "symmetrizing" the generators of the edge groups with respect to all possible permutations of the generators. That is, let

$$\begin{aligned} x_e &= ((a_s, b_s)\qquad, (a_s, b_s, c_s)) \\ y_e &= ((a_s, b_s, c_s), (a_s, b_s)\qquad) \end{aligned} \tag{4.5}$$

be the indicated elements of $S_3 \times S_3$, where S_n denotes the symmetric group of degree n, and similarly, let

$$\begin{aligned} x_w &= ((a_n, b_n, c_n), (a_n, b_n)\qquad) \\ y_w &= ((a_n, b_n)\qquad, (a_n, b_n, c_n)), \end{aligned} \tag{4.6}$$

$$\begin{aligned} a_n &= ((x_e, y_e), (x_e, y_e), ()\qquad, ()\qquad, (x_e, y_e), (x_e, y_e)) \\ b_n &= (()\qquad, (x_e, y_e), (x_e, y_e), (x_e, y_e), (x_e, y_e), ()\qquad) \\ c_n &= ((x_e, y_e), ()\qquad, (x_e, y_e), (x_e, y_e), ()\qquad, (x_e, y_e)), \end{aligned} \tag{4.7}$$

and

$$\begin{aligned} a_s &= ((), (), (x_w, y_w), (x_w, y_w), (x_w, y_w), (x_w, y_w)) \\ b_s &= ((x_w, y_w), (x_w, y_w), (), (x_w, y_w), (), (x_w, y_w)) \\ c_s &= ((x_w, y_w), (x_w, y_w), (x_w, y_w), (), (x_w, y_w), ()) \end{aligned} \tag{4.8}$$

be the indicated elements of either S_3^2 (in (4.6)) or S_2^6 (in (4.7) and (4.8)).

Let E, N, W, and S again be defined by (3.1), replacing the free generators with the permutations in (4.5)–(4.8), and let the semidirect products $S \rtimes E$, $E \rtimes N$, $N \rtimes W$, and $W \rtimes S$ again be given by (3.3)–(3.8). (Note that E acts on S, N acts on E, and so on, by factoring through the "erase all coordinates but the first" homomorphism.) A brief calculation shows that $E \cong W \cong S_3^2$, and $N \cong S \cong Z_2^3 \subseteq S_2^6$.

Finally, let $\overline{R}$ be the square of finite groups shown in (3.2). Repeating the arguments of the previous section, we see that (3.9) defines an injective homomorphism ϕ from $\pi_1(X)$ to $\pi_1(\overline{R})$. Therefore, summarizing the results of Sections 3 and 4, we have:

Theorem 4.1 *Let $\overline{R}$ be the non-positively curved square of finite groups whose edge and vertex groups are described by* (4.5)–(4.8) *and* (3.2)–(3.8), *and let C be the amalgamated free product corresponding to the east edge of $\overline{R}$. Then $\pi_1(\overline{R})$ is not C-separable.*

Remark 4.2 Continuing the observation in Remark 3.4, we note that the "symmetrizing" idea in this section also works if we start with an arbitrary CSC. In particular, if the initial CSC has h horizontal and v vertical 1-cells, then E and W will be subgroups of $S_v^{h!}$, N and S will be subgroups of $S_h^{v!}$, and the vertex groups will be subgroups of either $S_v \wr S_h$ or $S_h \wr S_v$. These groups will generally be very large as h and v increase; however, since $h = 2$ and $v = 3$ for the complex X, our edge and vertex groups are relatively small.

5 Getting a non-residually finite example

To get our non-residually finite example, let $G = \pi_1(\overline{R})$, and let C be the amalgamated free product corresponding to the east edge of $\overline{R}$. Then, since G is not C-separable, it follows from Long and Niblo [13] (see also [19]) that $G *_C G$ (with both embeddings of C in G the same), the *double* of G along its subgroup C, is not residually finite. It therefore remains only to embed the double of G along C in an appropriate non-positively curved square of finite groups.

Now in general, the double $(A *_B C) *_C (A *_B C)$ of the group $A *_B C$ along C is an index 2 subgroup of $A *_B (C \times Z_2)$. In our case, G splits as $A *_B C$ where $A = (N \rtimes W) *_W (W \rtimes S)$, $B = N * S$, and $C = (E \rtimes N) *_E (S \rtimes E)$. Therefore, the double $G *_C G$ is an index 2 subgroup of the group $A *_B (C \times Z_2)$ which is the fundamental group of the non-positively curved square of groups D indicated in the diagram below.

$$\begin{array}{ccccc} N \rtimes W & \leftarrow & N & \rightarrow & (E \rtimes N) \times Z_2 \\ \uparrow & & \uparrow & & \uparrow \\ W & \leftarrow & 1 & \rightarrow & E \quad \times Z_2 \\ \downarrow & & \downarrow & & \downarrow \\ W \rtimes S & \leftarrow & S & \rightarrow & (S \rtimes E) \times Z_2 \end{array} \tag{5.1}$$

Since the orders of the vertex groups of D are $288 = 8 \cdot 36$, 288, $576 = 288 \cdot 2$, and 576, Theorem 1.1 follows.

6 Some related open problems

In closing, we mention several open problems about polygons of finite groups.

- *Are the fundamental groups of negatively curved polygons of finite groups residually finite?* Note that this is a special case of the question: *Are word hyperbolic groups residually finite?* (See Gromov [6].)
- *Are the fundamental groups of non-positively curved polygons of groups Hopfian?* Note that Meier [14] has proved this in the building sub-case, and Sela [16] proved this for word-hyperbolic groups. An example of a compact non-positively curved 2-complex with non-Hopfian fundamental group is given in [21].
- *Are the fundamental groups of non-positively curved polygons of finite groups virtually torsion-free?* That is, do such groups contain torsion-free subgroups of finite index? Note that examples of non-positively curved orbihedra whose fundamental groups are not virtually torsion-free were given in [19], [20].

References

[1] R. B. J. T. Allenby and C. Y. Tang, *On the residual finiteness of certain polygonal products*, Canad. Math. Bull. **32** (1989), 11–17.

[2] M. Aschbacher and Y. Segev, *Uniqueness of sporadic groups*, Groups, Combinatorics and Geometry [12], pp. 1–11.

[3] M. R. Bridson, A. Haefliger, *Metric spaces of non-positive curvature*, to appear.

[4] K. S. Brown, *Buildings*, Springer-Verlag, 1989.

[5] E. Ghys, A. Haefliger, and A. Verjofsky (eds.), *Group Theory from a Geometric Viewpoint*, World Sci. Pub., 1991.

[6] M. Gromov, *Hyperbolic groups*, Essays in group theory (S. M. Gersten, ed.), Springer-Verlag, 1987.

[7] A. Haefliger, *Complexes of groups and orbihedra*, Group Theory from a Geometric Viewpoint [5], pp. 504–540.

[8] J. Howie, *On the asphericity of ribbon disc complements*, Trans. Amer. Math. Soc. **289** (1985), no. 1, 281–302.

[9] T. Hsu and D. T. Wise, *Embedding theorems for non-positively curved polygons of finite groups*, J. Pure Appl. Alg. **123** (1998), 201–221.

[10] A. A. Ivanov, *A geometric characterization of the Monster*, Groups, Combinatorics and Geometry [12], pp. 1–11.

[11] G. Kim, *On polygonal products of finitely generated abelian groups*, Bull. Austral. Math. Soc. **45** (1992), 453–462.

[12] M. W. Liebeck and J. Saxl, *Groups, Combinatorics, and Geometry*, Cambridge Univ. Press, 1992.

[13] D. D. Long and G. Niblo, *Subgroup separability and 3-manifold groups*, Math. Z. **207** (1991), 209–215.

[14] J. Meier, *Endomorphisms of discrete groups acting chamber transitively on affine buildings*, Int. J. Alg. Comp. **3** (1993), 357–364.

[15] M. A. Ronan, *Lectures on buildings*, Perspectives in Math. **7**, Academic Press, 1989.

[16] Z. Sela, *Endomorphisms of hyperbolic groups I: The Hopf property*, preprint.

[17] J. Stallings, *Non-positively curved triangles of groups*, Group Theory from a Geometric Viewpoint [5], pp. 91–103.

[18] J. Tits, *Buildings and group amalgamations*, Proceedings of Groups - St. Andrews 1985 (E. F. Robertson and C. M. Campbell, eds.), Cambridge Univ. Press (1986), 110–127.

[19] D. T. Wise, *Non-positively curved squared complexes, aperiodic tilings, and non-residually finite groups*, Ph.D. thesis, Princeton Univ., 1996.

[20] D. T. Wise, *A $CAT(0)$ 2-complex with no finite covers*, preprint.

[21] D. T. Wise, *A non-Hopfian automatic group*, J. Algebra **180** (1996), 845–847.

Printed in the United States
By Bookmasters